MW00835294

"*Bassett's Environmental Health Procedures* provides practical guidance on implementing the wide and diverse range of legislation in the field of Environmental Health. I would thoroughly recommend it to all Environmental Health Practitioners."

Alan Higgins, independent Environmental and Waste Consultant

BASSETT'S ENVIRONMENTAL HEALTH PROCEDURES

Environmental health law is a wide-ranging, detailed and complex body of law within the UK. *Bassett's Environmental Health Procedures* is an established and essential reference source which provides an accessible entry into enforcement and administrative procedures for environmental health. The main legal procedures used in the environmental health field are presented as flow charts supported by explanatory text.

This ninth edition refines the structure introduced in the eighth edition, with each chapter addressing a single topic. It has introduced the titles of the corresponding legislation in Scotland and Northern Ireland where there is such legislation. The book has been updated throughout to reflect new practices, legislation and statutory guidance. Specifically, the ninth edition contains new content on antisocial behaviour and significant updates to sections on:

- Enforcement and administration
- Environmental protection
- Food safety
- Housing
- Public health.

Environmental health officers/practitioners and students will find this book invaluable. It will also be an essential reference for all those whose responsibilities demand they keep abreast of current environmental health practices.

Bill Bassett authored the first seven editions of *Environmental Health Procedures*, the first of which appeared in 1983. His dedication to navigating and clarifying the relevant regulations and laws in this book has aided countless environmental health practitioners in the UK and beyond. Bill passed the baton to Tim Deveaux in 2013, but remained on hand to help both the new author and the publisher throughout. The ninth edition continues in the spirit of Bill's classic book, and is the second to be named after him. Bill sadly died in 2017 and his legacy remains a testament to a leading, visionary and dedicated environmental health practitioner.

Tim Deveaux is an independent environmental health practitioner with 36 years of public sector experience in a wide range of areas in environmental health and sustainability. His experience includes housing, drainage, food safety, health and safety, air pollution and noise, infectious disease control, developing and implementing strategies to change behaviour, and protecting the environment and human health, particularly in planning for, and implementing, climate change and sustainability solutions.

CX 10 23 2010 184

BASSETT'S ENVIRONMENTAL HEALTH PROCEDURES

Ninth edition

TIM DEVEAUX

Routledge
Taylor & Francis Group

LONDON AND NEW YORK

Ninth edition published 2020
by Routledge
2 Park Square, Milton Park, Abingdon, Oxon, OX14 4RN

and by Routledge
52 Vanderbilt Avenue, New York, NY 10017

Routledge is an imprint of the Taylor & Francis Group, an informa business

© 2020 Tim Deveaux

The right of Tim Deveaux to be identified as author of this work has been
asserted in accordance with sections 77 and 78 of the Copyright,
Designs and Patents Act 1988.

All rights reserved. No part of this book may be reprinted or reproduced or
utilised in any form or by any electronic, mechanical, or other means,
now known or hereafter invented, including photocopying and recording,
or in any information storage or retrieval system, without permission
in writing from the publishers.

Trademark notice: Product or corporate names may be trademarks or
registered trademarks, and are used only for identification and
explanation without intent to infringe.

First edition published by Chapman & Hall 1983
Eighth edition published by Routledge 2014

British Library Cataloguing-in-Publication Data
A catalogue record for this book is available from the British Library

Library of Congress Cataloging-in-Publication Data
Names: Bassett, W. H., author. | Deveaux, Tim, author.
Title: Bassett's environmental health procedures / W.H. Bassett &
Tim Deveaux.
Description: Ninth edition. | Abingdon, Oxon; New York, NY :
Routledge, [2020]
Identifiers: LCCN 2019018440| ISBN 9780367183288 (hbk) |
ISBN 9780429060847 (ebk)
Subjects: LCSH: Public health laws—Great Britain. |
Environmental law—GreatBritain. |
Environmental health—Great Britain—Administration. | MESH:
Environmental Health—legislation & jurisprudence—Great Britain. |
Public Health—legislation & jurisprudence—Great Britain.
Classification: LCC KD3359 .B37 2020 | DDC 344.4104—dc23
LC record available at https://lccn.loc.gov/2019018440

ISBN: 978-0-367-18328-8 (hbk)
ISBN: 978-0-429-06084-7 (ebk)

Typeset in Times New Roman
by Swales & Willis Ltd, Exeter, Devon, UK

Contents

Preface

Environmental health law is a wide-ranging, detailed and complex body of law within the UK. *Bassett's Environmental Health Procedures* is an established and essential reference source which provides an accessible entry into enforcement and administrative procedures for environmental health. The main legal procedures used in the environmental health field are presented as flow charts supported by explanatory text.

Each edition of this work records the ever-increasing breadth of environmental health law and complexity of its content. This edition continues to reflect these trends and also indicates just how substantial legislative changes in environmental health law have been since the publication of the eighth edition in 2013. These changes have not been peripheral but have, again, affected in a major way the main activities of environmental health departments, as outlined below.

I have tried to incorporate the very helpful comments of three reviewers of the proposal to produce the ninth edition. Most suggestions have been adopted and I hope that their constructive comments have been satisfied.

The structure of this ninth edition has been refined for ease of use. Main topics are set into parts. Each part is divided into chapters addressing a single topic. The procedures are based on legislation which applies to England and Wales as much of the legislation of these two countries is common. Scotland and Northern Ireland prepare and enact their own legislation and much of it is similar to that of England and Wales. The relevant legislation for England and Wales is quoted in the extent section of each procedure. Where similar legislation is enacted in Scotland and Northern Ireland, the main titles of the relevant legislation are stated in the References section. The differences between the four countries have not been highlighted. I would submit that to include this would make the book far too complex and bulky to be useful.

The publication of a sister book – *Procedures for Licensing Authority Officers* – is a reflection of the broadening influence and activity of environmental health practitioners. This has resulted in alcohol licensing and gambling licensing being withdrawn from *Bassett's*. Many environmental health practitioners are heavily involved with licensing matters. One book of procedures to cover every issue environmental health practitioners get involved with is not enough! The book also sees the introduction of a part on antisocial behaviour. There are issues throughout the book that could be construed as antisocial behaviour issues, but the organisation of the procedures is one person's interpretation of how issues fit together.

The book has been divided into the following parts:

- Legal structure and process
- Environmental protection
- Food safety
- Health and safety
- Housing and health
- Antisocial behaviour

- Public health
- Public safety.

The book has been updated throughout to reflect emerging issues, new practices, legislation and statutory guidance, including:

- Regulators' Code
- Public health in local government
- Mental health and environmental health
- Interventions
- Partnership working
- Recording decisions
- Legislation in Scotland and Northern Ireland
- Antisocial behaviour
- Registration and approval of animal by-product plants
- Food standards, food information and food additives
- Rogue landlords
- Compulsory purchase
- Disabled facilities grants
- Energy efficiency
- Fixed penalty notices
- Protection from eviction
- Renewal areas
- Restoration of electricity and gas
- Smoke and carbon monoxide alarms
- Control of horses
- Public health funerals
- Hoarding.

Much legislation uses 'he' when referring to a person and for the purpose of this book that convention is generally followed, so when 'he' is mentioned it means 'he or she'.

The text has also been updated to remove typos and to give greater clarity and accuracy.

Environmental health officers/practitioners and students will find this book essential reading. It will also be a valuable reference for all those whose responsibilities demand they keep abreast of current environmental health practices.

It should be stressed that this work is an interpretation of the law and is not the law itself. The recent legislation on which the procedures are based involves both a change of policy approach to the legal requirements and a new enforcement regime, and together this appears to be more complicated and less easy to interpret than the legislation replaced. The editor would appreciate any comments on the work, and suggestions for alternative interpretations would be very helpful.

Tim Deveaux
September 2019

Extract from the preface to the first edition (1983)

The idea for this work came after many exasperating years of having to wade through Acts of Parliament or legal encyclopediae in order to find the legal answer to a fairly simple practical problem which faced me as an environmental health officer. The author has also been conscious of a need for a reference text for both practising and pupil environmental health officers, community physicians and others which gives basic information about environmental health procedures in a simple form. It has also been suggested to the author that other people outside the environmental health profession, including lawyers, law centre administrators, surveyors, estate agents, elected local authority members and members of the general public who become involved in environmental health problems, would welcome a procedural analysis of this type.

This book attempts to meet all those needs and provide a means of identifying procedures relating to the many facets of environmental health practice.

Each of the main procedures has been reduced to a diagrammatic form and included with it is text on the basic issues concerned with that procedure. By this means, the steps involved in the different procedures can be quickly followed and individual stages identified. Related procedures have been cross-referenced. At the beginning of each chapter, the general procedural provisions and definitions relating to the procedures contained in that chapter are indicated. Where a procedure differs either procedurally or in definition, this is indicated either in the diagram or in the text.

W.H. Bassett
Exeter
1983

Acknowledgements

I would like to thank the following environmental health practitioners for their advice, assistance and invaluable input into this edition: Alistair Bullock and Simon Griffiths for access to legal information; Tim Everett for works in default, statutory nuisances and general issues; Craig Rudman on licensing; Elaine Rudman on enforcement frameworks; Carol Deveaux and Trevor Greener on food safety; Alan Higgins on environmental protection; Julie Barratt on legal issues; and Zena Lynch on housing.

How this book is structured

The procedures interpreted in this book are based on primary and secondary legislation as it applies in England. In some cases, they will also apply in the rest of the UK but this varies because of the separate legislative powers of the National Assemblies in Northern Ireland, Scotland and Wales. In Northern Ireland and Scotland, the primary legislation as well as the secondary legislation may be different. The Welsh Assembly is able to pass Acts on any matters that are not reserved to the UK Parliament by the Government of Wales Act 2006 (as amended by the Wales Act 2017).

The first part of the book outlines the framework in which environmental health operates by describing current enforcement policies, the relationship with key agencies and new responsibilities for local authorities in the form of public health.

The procedures are divided into parts based on traditional environmental health 'functions' such as environmental protection, food safety, health and safety, housing, public health, public safety and a new part on antisocial behaviour.

In the case of some procedures relating to food safety, the environmental health service will be directly enforcing European Commission (EC) regulations without there being implementing regulations in the UK. However, this may change when/if the UK leaves the European Union (EU).

In each procedure, the legislation involved is given under a section titled 'References'. This is followed by a section headed 'Extent' which gives the applicability of the procedure to the countries of the UK. Then a section titled 'Scope' gives a brief introduction to and outline description of the procedure.

Most of the legislation also involves a flow chart which outlines the key stages in the procedure of using the legislation to obtain compliance.

PART 1

CURRENT ENVIRONMENTAL HEALTH LEGAL STRUCTURE, PROCESS AND POLICIES

Chapter 1

ENFORCEMENT AND COMPLIANCE FRAMEWORK

THE FRAMEWORK FOR LOCAL AUTHORITY ENFORCEMENT, OPERATIONAL AND HEALTH IMPROVEMENT POLICIES

In undertaking the enforcement of environmental health legislation and forming their individual enforcement policies, local authorities (LAs) need to take account of government policy, government agencies and initiatives taken at a national level which seek to guide their enforcement policies. The most significant are outlined below and it will be seen that, in total, they form a very substantial influence over local authority enforcement activity.

General guidance about the role of enforcement, regulation and compliance in local government

The Legislative and Regulatory Reform Act 2006 imposes a duty on any person exercising a specified regulatory function to have regard to the five principles of good regulation. The principles provide that regulatory activities should be carried out in a way that is:

- transparent;
- accountable;
- proportionate;
- consistent; and
- targeted only at cases in which action is needed (section 21).

Section 22 of the Act enables a Minister of the Crown to issue a Code of Practice relating to the exercise of regulatory functions (the 'Regulators' Compliance Code'). This section imposes a duty on any person exercising a specified regulatory function to have regard to the Regulators' Compliance Code when determining general policies or principles and in setting standards or giving general guidance about the exercise of other regulatory functions.

The Legislative and Regulatory Reform (Regulatory Functions) Order 2007 specifies the regulatory functions to which these duties apply.

The Office for Product Safety and Standards was formed in January 2018 to enhance protections for consumers and the environment and drive increased productivity, growth and business confidence. They do this by advising on local regulation: practice and strategy, local regulation: primary authority, local and national regulation: practice and context and national regulation: enforcement services. The Regulators' Compliance Code is issued through these enforcement services.

Regulators' Code – April 2014

Regulators must have regard to this Code when developing policies and operational procedures that guide their regulatory activities. They must also have regard to the Code when
setting standards or giving guidance which will guide the regulatory activities of other regulators. If a regulator concludes, on the basis of material evidence, that a specific provision
of the Code is either not applicable or is outweighed by another relevant consideration,
the regulator is not bound to follow that provision, but should record that decision and the
reasons for it.

1. Regulators should carry out their activities in a way that supports those they regulate
   to comply and grow. This includes avoiding imposing unnecessary regulatory burdens;
   choosing proportionate approaches; and considering how they might support or enable
   economic growth for compliant businesses.
     Regulators should ensure that their officers have the necessary knowledge and skills to
   support those they regulate.
2. Regulators should provide simple and straightforward ways to engage with those they
   regulate and hear their views. They should consider the impact on business and engage
   with business representatives including two-way dialogue. They should clearly explain
   what the non-compliant item or activity is, the advice being given, actions required or
   decisions taken and the reasons for these. This does not preclude immediate enforcement
   action. The regulator must provide an impartial and clearly explained route to appeal
   against a regulatory decision or a failure to act in accordance with this Code, and also
   have a clearly explained complaints procedure.
3. Regulators should base their regulatory activities on risk. They should take an
   evidence-based approach to determining the priority risks and allocate resources
   where they would be most effective in addressing those priority risks. Risk and review
   must be considered at every stage of their decision-making processes. Compliance
   records, earned recognition and relevant external verification should be considered
   in determining risk.
4. Regulators should share information (lawfully) about compliance and risk, including the
   principle of 'collect once, use many times' when requesting information.
5. Regulators should ensure clear information, guidance and advice are available to help
   those they regulate meet their responsibilities to comply. When providing advice and
   guidance, legal requirements should be distinguished from suggested good practice. They
   should consult those they regulate about the guidance they produce to ensure that it meets
   their needs. Regulators should work collaboratively to assist those regulated by more
   than one regulator. They should consider advice provided by other regulators and discuss
   conflicting advice with the other regulator to reach agreement.
6. Regulators should ensure that their approach to their regulatory activities is transparent, publishing a set of clear service standards, and setting out what those they regulate
   should expect from them. It should be easily accessible, including being available at a
   single point on the regulator's website that is clearly signposted, and it should be kept
   up to date. Regulators should ensure that their officers act in accordance with their
   published service standards, including their enforcement policy. They should regularly
   publish details of their performance against their service standards, including feedback
   received from those they regulate and complaints about them and appeals against their
   decisions.

Performance management and competence

The Local Authority Regulatory Services Excellence Framework 2010 guides local authorities on how to measure service delivery quality based on the achievement of set targets, identified through consultation with their customers. Many have adopted quality management systems, some obtaining certificated systems through British Standards and ISO.

'An Introductory Guide to Performance Management in Local Authority Trading Standards and Environmental Health Enforcement Services' is a best-practice guide published in 1999. It aims to assist local authorities with best value targeted performance management and includes matters relating to enforcement. The guide identifies the following themes as being central to good performance:

- Protection of the wider agenda (i.e. councils leading and energising local communities)
- Transparency and consistency
- Quality and value
- Delivery and review.

The 'Common Approach to Competence for Regulators' in 2011, which incorporates the Regulators' Development Needs Analysis (RDNA) and Guide for Regulators Information Point (GRIP) tools, are available to ensure professional competency in regulatory activities.

Local authority enforcement policies

Working within the directions and advice contained in the framework described above, each enforcing authority should adopt its own enforcement policy to guide departments and individual officers as to the way in which enforcement is to be carried out. This should include how decisions are to be reached regarding which enforcement option to adopt in individual cases.

Such policies will also include provisions for dealing with disputes and complaints arising from enforcement activity.

Councils can have one generic enforcement policy which covers the work of every area of the service, but if they do that, it is likely they will need further policies that specify things such as their graded response to dealing with breaches of legislation and prosecutions, or when and how a fixed penalty notice will be issued. It might also be necessary to have to produce separate 'mini policies' for each service if those services apply their enforcement procedures slightly differently. There are also a number of statutory policies required, e.g. licensing policy and gambling statement of principles.

In terms of the content of our enforcement policies, we will have to include information on:

- The areas regulated
- How services are delivered
- How to work with businesses and the public
- What help can be offered to businesses and the public
- Inspections and compliance visits
- How to deal with non-compliance and the different methods available for dealing with non-compliance
- How they choose to publicise our enforcement actions

- Partnership working
- Information about enforcement officers
- How to make a complaint.

The Commissioner for Local Administration (the Ombudsman)

Each procedural flow chart indicates the stages at which the courts may become involved in dealing with appeals, prosecutions etc. In addition to this, any local authority action, or lack of it, may be subject to investigation by the Local Government Ombudsman. Whilst the Ombudsman cannot challenge the decision itself, e.g. not to serve a statutory notice, he may do so in respect of the way in which the decision was reached. Whilst he does not have the power to change the local authority decision, the Ombudsman can issue a public report criticising the authority and is also able to indicate a level of compensation which he feels ought to be made.

AGENCIES THAT WORK WITH ENVIRONMENTAL HEALTH PROFESSIONALS

Environment Agency

Some local authorities work with the Environment Agency in issues such as fly tipping by shared enforcement under a strategic enforcement policy in a geographical area. Other areas of similar work could be the focus of co-ordinated enforcement. The Environment Agency Enforcement and Sanctions policy, April 2018 states their policies and practices in enforcement.

The Food Standards Agency

The Agency set and monitor standards of performance for food authorities in enforcing food safety laws. They may make reports on that performance, including guidance on improving standards, and direct that the food authority should publish, within a specified time, details of actions they will take to comply. They may also require each food authority to provide information and to make its records available for inspection.

In 2000, the Agency published a Framework Agreement on Official Feed and Food Controls by Local Authorities, which applies to the whole of the UK. It has been amended several times, the last in April 2010. This has four main elements:

- The standard – which sets out the requirements for the planning, organisation, management and delivery of local authority food law enforcement services.
- Service planning guidance – on matters that include:

  1. Service aims and objectives
  2. Background
  3. Service delivery
  4. Resources
  5. Quality assessment
  6. Review.

- A monitoring scheme – which sets out the arrangements for the Agency to obtain enforcement information from the local authorities. The Agency collects key data on how each local authority is delivering feed and food law enforcement, on an annual basis. This is achieved through the Local Authority Enforcement Monitoring System (LAEMS), which was introduced in April 2008.
- An audit scheme – which provides for a rolling programme of audits by the Agency. Authorities are audited against the Feed and Food Law Enforcement Standard in the Framework Agreement, which sets out the minimum standards of performance expected from local authorities across the full range of their feed and food law enforcement activities. The scheme is implemented on a UK basis, with the Agency in England, Scotland, Wales and Northern Ireland each coordinating their own audit programme.

In addition, statutory Codes of Practice issued under section 40 of the Food Safety Act 1990 contain mandatory provisions as to how and when certain actions are to be taken. These are identified in the appropriate procedures.

Enforcement policy is contained in National Enforcement Priorities 2018–2019 (England and Wales) and the FSA National Targeted Monitoring Strategy (England) – July 2018.

Earned recognition

The Food Standards Agency is exploring the concept of earned recognition.

Earned recognition aims to reduce the burden for compliant businesses which allows enforcement activity to concentrate on less compliant businesses. Those who qualify for earned recognition will benefit by receiving less frequent visits by the enforcement authority. The FSA review how they approach assurance through the Regulating Our Future (ROF) programme which aims to formalise private assurance schemes operating in food safety.

Guidance on Food Standards Agency Approved Assurance Schemes in the Animal Feed and Food Hygiene at the Level of Primary Production Sectors – April 2018.

The guidance on approved assurance schemes covers the following points:

- criteria details
- process for scheme approval
- governance arrangements
- exchange of information with enforcement authorities (FSA website).

Enforcement authority verification

To ensure continued compliance and confidence after approval of an assurance scheme for earned recognition, the FSA will continue to verify compliance with the criteria for earned recognition.

This includes a check carried out by enforcement authorities which is based on a percentage of assurance scheme members to verify business compliance. Positive verification will enable the FSA to have continued confidence in the assurance scheme and the ability to justify approved scheme status (FSA website).

Food hygiene rating scheme

The scheme gives businesses a rating from 5 to 0 which is displayed at their premises and online so consumers can make more informed choices about where to buy and eat food.

Ratings are a snapshot of the standards of food hygiene found at the time of inspection. It is the responsibility of the business to comply with food hygiene law at all times. The scheme is set out in law in Wales and Northern Ireland but display of the rating sticker is voluntary in England (FSA website).

The Health and Safety Executive (HSE)

Section 18(4) of the Health and Safety at Work Act 1974 requires local authorities to act in accordance with guidance issued under that section by the HSE. This includes guidance on enforcement policies.

The latest HSE Statement on Enforcement Policy was published in October 2015, which local authorities are required to reflect in their policies. The main principles they follow are:

(a) proportionality about how the law is applied and compliance secured
(b) targeting enforcement action
(c) consistency of approach
(d) transparency about operation and what is expected
(e) accountability for actions and helping duty holders understand what is expected of them.

This Enforcement Policy Statement is in accordance with the Legislative and Regulatory Reform Act 2006 and the Regulators Code 2014. The HSE have also produced an Enforcement Management Model (operational version 3.2) which is useful for other enforcement agencies.

The HSE have advised local authorities on the following elements in their enforcement policy, which are essential for a local authority to discharge its duty as an enforcing authority under the Health and Safety at Work Act 1974:

- Risk-based intervention plans focused on tackling specific risks;
- Reserving unannounced proactive inspection only for the activities and sectors published by HSE or where intelligence (national and local) suggests risks are not being effectively managed;
- Having trained and competent officers who can exercise professional judgement to:

    o Differentiate between different levels of risk or harm;
    o Decide how far short a business has fallen from managing the risks it creates effectively; and
    o Apply proportionate decision making in accordance with the LA's Enforcement Policy, HSE's Enforcement Policy Statement and Enforcement Management Model.

- Ensuring nationally published guidance is applied appropriately to address both local and national priorities;
- Taking full account of primary authority;
- Setting clear expectations for delivery;
- Allowing appropriate comparison and transparency, via publication annually of health and safety inspection data;
- Having a clear and easily accessible enforcement policy;
- Providing easily accessible information on the services and advice available to businesses including pointing to nationally available material;

- Having arrangements for keeping employees, their representatives and victims or their families informed;
- Having easily accessible complaints procedures;
- Benchmarking their performance against other LAs via data returns to HSE and peer review.

Both the Food Standards Agency and the HSE are working to target inspections of high-risk activities and premises, with dwindling resources.

A joint statement between the Food Standards Agency and the HSE, produced in 2011, aims to reduce inspections of business and introduce more local authority efficiencies to help concentrate resources. It states that combining interventions (health and safety, and food safety inspections) would only be appropriate where the local authority's health and safety and their food intervention programmes coincide.

Guidance produced in May 2011 ('Joint Guidance for Reduced Proactive Inspections') is intended to assist local authorities in determining the use of proactive interventions to achieve both improved health and safety outcomes for each locality and the government's regulatory policy.

Local authorities must also consider the **National Local Authority Enforcement Code, Health and Safety at Work, England, Scotland & Wales** issued by the HSE in May 2013.

The Code sets out what is meant by 'adequate arrangements for enforcement'. It focuses on four objectives:

(a) Clarifying the roles and responsibilities of business, regulators and professional bodies to ensure a shared understanding on the management of risk;
(b) Outlining the risk-based regulatory approach that LAs should adopt with reference to the Regulators' Code, HSE's Enforcement Policy Statement and the need to target relevant and effective interventions that focus on influencing behaviours and improving the management of risk;
(c) Setting out the need for the training and competence of LA H&S regulators linked to the authorisation and use of HSWA powers; and
(d) Explaining the arrangements for collection and publication of LA data and peer review to give assurance on meeting the requirements of the Code.

Supplementary guidance has been produced to help LAs understand and meet the requirements of the Code. It includes guidance on the following issues:

(a) LA regulatory interventions and the role of inspection
(b) Dealing with matters of evident concern
(c) Delivery of local and national priorities
(d) The role of LAs in providing advice and support to business
(e) Identification of poor performers.

The list of activities/sectors suitable for targeting proactive inspection by LAs referred to in the Code is also available.

Local authority regulators are competent professionals granted powers and duties to deliver proportionate and targeted enforcement. It is vital that local authority regulatory resource is used consistently and to best effect by targeting specific risks or focusing on specific outcomes. Local authorities should use the full range of regulatory interventions available to influence behaviours and the management of risk with proactive inspection utilised only for premises with higher risks or where intelligence suggests that risks are not being effectively managed.

The Code provides direction to LAs on meeting these requirements, and reporting on compliance. The Code is given legal effect as HSE guidance to LAs under section 18(4)(b) of Health and Safety at Work etc. Act 1974 (HSWA) and applies to England, Wales and Scotland.

Public Health England

Public Health England (PHE) is an executive agency of the Department of Health and Social Care, and a distinct organisation with operational autonomy. It provides government, local government, the NHS, Parliament, industry and the public with evidence-based professional, scientific expertise and support.

It has eight local centres, plus an integrated region and centre for London, and four regions (North of England, South of England, Midlands and East of England, and London).

It works closely with public health professionals in Wales, Scotland and Northern Ireland, and internationally.

Public Health England is responsible for:

- making the public healthier and reducing differences between the health of different groups by promoting healthier lifestyles, advising government and supporting action by local government, the NHS and the public
- protecting the nation from public health hazards
- preparing for and responding to public health emergencies
- improving the health of the whole population by sharing our information and expertise, and identifying and preparing for future public health challenges
- supporting local authorities and the NHS to plan and provide health and social care services such as immunisation and screening programmes, and to develop the public health system and its specialist workforce
- researching, collecting and analysing data to improve our understanding of public health challenges, and come up with answers to public health problems.

Its priorities are set out in its Strategic Plan which sets out how it intends to achieve its aims by a certain date (currently 2020).

Its annual business plan outlines the actions it will take to protect and improve the public's health and reduce inequalities (Gov.uk).

PUBLIC HEALTH IN LOCAL GOVERNMENT

Each upper-tier and unitary local authority in England has a duty to take such steps as it considers appropriate for improving the health of the people in its area.

Local authorities fulfil this duty by commissioning a range of services from a range of providers from different sectors, working with clinical commissioning groups and representatives of the NHS Commissioning Board to create as integrated a set of services as possible.

However, they can also influence public health through the way they operate the planning system, policies on leisure, key partnerships with other agencies, for example on children's and young people's services, and through developing a diverse provider market for public health improvement activities.

Local authorities also ensure the health needs of disadvantaged areas and vulnerable groups are addressed, as well as giving consideration to equality issues. The goal should be to improve the health of all people, but to improve the health of the poorest, fastest.

Local political leadership is critical in ensuring that public health receives the focus it needs. The role of the Cabinet lead for health within the council is critical and there should be a much broader engagement in this agenda among all local political leaders.

Commissioning

Responsibility and resources for commissioning public health services are devolved to local government. The aim is to create a set of responsibilities which clearly demonstrate local authorities' leadership role in:

- tackling the causes of ill-health, and reducing health inequalities
- promoting and protecting health
- promoting social justice and safer communities.

Local authorities work with clinical commissioning groups to provide as much integration across clinical pathways as possible, maximising the scope for upstream interventions. The health and wellbeing board is critical in driving this agenda (see p. 15).

Local authorities commission, rather than directly provide, the majority of services, giving opportunities to engage local communities and the third sector more widely in the provision of public health, and to deliver the best outcomes.

This right to provide enables staff to consider a wide range of options, including social enterprise, staff-led mutuals, joint ventures and partnerships.

The aim is to drive up quality, empower individuals and enable innovation. Local authorities also aim to provide a vehicle to improve access, address gaps and inequalities and improve quality of services where users have identified variable quality in the past.

Local authorities should decide which services to prioritise for choice on a diverse provider model based on local needs and priorities. This is informed by the joint strategic needs assessment and early and continuing engagement with health and wellbeing boards.

Health protection plans

Local authorities and the Director of Public Health acting on its behalf must take steps to ensure that plans are in place to protect the local population from threats ranging from relatively minor outbreaks to full-scale emergencies, and to prevent, as far as possible, those threats arising in the first place. The scope of this duty includes local plans for immunisation and screening, as well as plans for the prevention and control of infection, including those that are healthcare-associated.

Emergency response plans

A lead Director of Public Health coordinates the public health input to planning, testing and responding to emergencies across the local authorities in the Local Resilience Forum (LRF) area. Public Health England provides the health protection services, expertise and advice.

The NHS Commissioning Board appoints a lead director for NHS emergency preparedness and response at the LRF level, and provides necessary support to enable planning and response to emergencies that require NHS resources.

Local Health Resilience Partnerships (LHRPs) bring together the health sector organisations involved in emergency preparedness and response at the LRF level. LHRPs consist of emergency planning leads from health organisations in the LRF area and will ensure effective planning, testing and response for emergencies.

LHRPs enable all health partners to input to the LRF and in turn provide the LRF with a clear, robust view of the health economy and the best way to support LRFs to plan for and respond to health threats.

A lead director appointed by the NHS Commissioning Board and the lead Director of Public Health act as co-chairs at the LHRP during emergency planning.

Plans for screening and immunising the local population

The NHS Commissioning Board is responsible for delivery of the national screening and immunisation programmes. Public Health England provides public health advice on the specification of national programmes, and also a quality assurance function with regard to screening.

Directors of Public Health advise, for example, on whether screening or immunisation programmes in their area are meeting the needs of the population, and whether there is equitable access. They provide challenge and advice to the NHS Commissioning Board on its performance, for example through the joint strategic needs assessment and discussions at the health and wellbeing board on issues such as raising uptake of immunisations and screening, and how outcomes might be improved by addressing local factors. They also have a role in championing screening and immunisation, using their relationships with local clinicians and clinical commissioning groups, and in contributing to the management of serious incidents.

This local authority role in health protection planning is not a managerial, but a local leadership function. It rests on the personal capability and skills of the local authority Director of Public Health and his or her team to identify any issues and advise appropriately.

Population healthcare advice to the NHS

Local authorities provide population healthcare advice to the NHS.

Clinical commissioning groups require a range of information and intelligence support via both the population healthcare advice service based in local authorities and other commissioning support services such as from Public Health England, where appropriate.

Local authority responsibilities and operational framework

Local authorities' statutory public health responsibilities are set out in the Health and Social Care Act 2012 (the '2012 Act') and Standard Note: SN06844 13 March 2014.

There is a duty for all upper-tier and unitary local authorities in England to take appropriate steps to improve the health of the people who live in their areas (section 12). The Secretary of State continues to have overall responsibility for improving health – with national public health functions delegated to PHE.

Section 12 lists some of the steps to improve public health that local authorities and the Secretary of State are able to take, including:

- providing information and advice;
- providing services or facilities designed to promote healthy living;
- providing services or facilities for the prevention, diagnosis or treatment of illness;
- providing financial incentives to encourage individuals to adopt healthier lifestyles;
- providing or participating in the provision of training; and
- providing assistance to help individuals minimise risks to health arising from their accommodation or environment.

Section 12(4) gives local authorities powers to make grants or lend money to organisations or individuals in order to improve public health and it is for the local authority to determine the appropriate terms of such grants or loans.

Health is a devolved matter in Scotland, Wales and Northern Ireland, although the devolved administrations currently retain substantially the same legislative framework.

Local authorities have key responsibilities across the three domains of public health – health improvement, health protection and healthcare.

Local authorities are responsible for:

- tobacco control and smoking cessation services
- alcohol and drug misuse services
- public health services for children and young people aged 5–19 (including Healthy Child Programme 5–19) (and in the longer term all public health services for children and young people)
- the National Child Measurement Programme
- interventions to tackle obesity such as community lifestyle and weight management services
- locally led nutrition initiatives
- increasing levels of physical activity in the local population
- NHS Health Check assessments
- public mental health services
- dental public health services
- accidental injury prevention
- population-level interventions to reduce and prevent birth defects
- behavioural and lifestyle campaigns to prevent cancer and long-term conditions
- local initiatives on workplace health
- supporting, reviewing and challenging delivery of key public health funded and NHS-delivered services such as immunisation and screening programmes
- comprehensive sexual health services (including testing and treatment for sexually transmitted infections, contraception outside of the GP contract and sexual health promotion and disease prevention)
- local initiatives to reduce excess deaths as a result of seasonal mortality
- the local authority role in dealing with health protection incidents, outbreaks and emergencies
- public health aspects of promotion of community safety, violence prevention and response
- public health aspects of local initiatives to tackle social exclusion
- local initiatives that reduce public health impacts of environmental risks.

The Local Authorities (Public Health Functions and Entry to Premises by Local Healthwatch Representatives) Regulations 2013 make provision for the steps to be taken by local authorities in exercising their public health functions.

The list of commissioning responsibilities above is not exclusive. Local authorities may choose to commission a wide variety of services under their health improvement duty. This is intended to encourage locally driven solutions, underpinned by a robust analysis of the needs and assets of the local population.

Local authorities are also responsible for commissioning comprehensive open-access accessible and confidential contraception and sexually transmitted infection (STI) testing and treatment services, for the benefit of all persons of all ages present in the area.

Public Health England (PHE) provides evidence, advice and support to local authorities about fulfilling their new public health responsibilities.

A Public Health Toolkit for local authorities in England has been produced by the Department of Health, the Local Government Association and PHE.

Charges for local authority public health functions

These regulations also cover the making and recovery of charges in respect of the exercise of local authorities' public health functions. Regulation 9 provides for a local authority to charge for certain actions in its health improvement duty. Charges may be made or recovered only where the provision of the information, advice, services, facilities or training in question have been requested by, or agreed with, the person to whom they are provided. However, the charging regulations mean that when local authorities provide services as part of the comprehensive health service, these services must be free at the point of use just as they were when provided by the NHS, except in some limited circumstances.

Duties of Directors of Public Health

Section 30 requires each upper-tier local authority, acting jointly with the Secretary of State, to appoint a Director of Public Health whose role is integral to the new duties for health improvement and health protection. The responsibilities of Directors of Public Health include:

(a) the health improvement duties placed on local authorities;
(b) the exercise of any public health functions of the Secretary of State which the Secretary of State requires the local authority to exercise by regulations;
(c) any public health activity undertaken by the local authority under arrangements with the Secretary of State;
(d) local authority functions in relation to planning for, and responding to, emergencies that present a risk to public health;
(e) the local authority role in co-operating with police, probation and prison services in relation to assessing risks of violent or sexual offenders; and
(f) other public health functions that the Secretary of State may specify in regulations.

Section 31 requires local authorities to have regard to guidance from the Secretary of State when exercising their public health functions; in particular this power requires local authorities to have regard to the Department of Health's **Public Health Outcomes Framework** (PHOF). This framework sets out the Government's overarching vision for public health in

England, the desired outcomes and the indicators that will be used to measure improvements to and protection of health. Improving outcomes and supporting transparency provide summary technical specifications of public health indicators.

Section 237 also requires local authorities to comply with **National Institute for Health and Care Excellence (NICE)** recommendations to fund treatments under their public health functions.

Section 29 assigns oral public health services to local authorities, such as water fluoridation, and gives them a duty to help deliver and sustain good health among the prison population.

However, NHS England is responsible for commissioning all NHS dental services including those carried out in hospitals and high street dental practices, and is required to commission services to meet the needs of the local population, for both urgent and routine dental care.

Clinical commissioning groups (CCGs) are responsible for commissioning the promotion of opportunistic testing and treatment of sexually transmitted infections, while local authorities commission testing of sexually transmitted infections, including HIV. Local authorities also commission sexual health advice, prevention and promotion.

One of the aims of transferring public health responsibilities to local authorities was to better integrate health and social care services and other activities that affect health such as housing and maintenance of public spaces. For example, health and wellbeing boards (HWBs), hosted by local authorities, have a duty to encourage integrated working. Organisations should work together to improve outcomes for people and local authorities are encouraged to innovate and make decisions according to the needs the people in their area.

Health and Wellbeing Boards (HWBs)

Health and wellbeing boards were established under the Health and Social Care Act 2012 to act as a forum in which key leaders from the local health and care system could work together to improve the health and wellbeing of their local population, reduce health inequalities and promote the integration of services. They became fully operational on 1 April 2013 in all 152 local authorities with adult social care and public heath responsibilities.

Each top-tier and unitary authority has an HWB which has strategic influence over commissioning decisions across health, social care and public health.

Statutory board members include a locally elected councillor, a Healthwatch representative, a representative of a clinical commissioning group, a director of adult social care, a director of children's services and a director of public health.

HWB members from across local government and the health and care system work together to identify local needs, improve the health and wellbeing of their local population and reduce health inequalities.

The HWB is a key forum for encouraging commissioners from the NHS, councils and wider partners to work in a more joined-up way. Central to achieving this is the HWB's responsibility for producing a Joint Strategic Needs Assessment (JSNA) and a Joint Health and Wellbeing Strategy (JHWS).

Local authorities will also have a statutory function to provide public health advice to clinical commissioning groups, while HWBs will have to monitor performance.

The boards have very limited formal powers. They are constituted as a partnership forum rather than an executive decision-making body.

However, boards do not have powers to take on a full range of commissioning functions, though some could be achieved by, for example, designating the board as a joint committee of the local authority cabinet and CCG governing body and/or by considering the delegation of specific functions to the board. The Cities and Local Government Devolution Act 2016 allows for some NHS functions to be transferred to local authorities.

In most cases, health and wellbeing boards are chaired by a senior local authority elected member. The board must include a representative of each relevant CCG and local Healthwatch, as well as local authority representatives. The local authority has considerable discretion in appointing additional board members. Most have chosen not to invite providers to become formal members, though many engage with providers in other ways.

MENTAL HEALTH AND ENVIRONMENTAL HEALTH

Health and wellbeing is influenced by the wider physical environment. By addressing the wider determinants of health, including food safety, housing standards, health and safety, air quality, noise and environment issues generally, environmental health makes a fundamental contribution to the maintenance and improvement of public health.

The natural and 'manmade' environment is fundamentally important to both our physical and psychological wellbeing, so actions that promote and protect our environment help to increase our ability to flourish in life. In turn, people and communities that are flourishing, i.e. have high levels of wellbeing, tend to be environmentally responsible in their behaviour and can, therefore, contribute to environmental sustainability.

Housing type and quality, neighbourhood quality, noise, overcrowding, indoor air quality and light have all been linked to personal mental health.

The most important factors that operated independently were neighbour noise, sense of overcrowding in the home, escape facilities, such as green spaces and community facilities, and fear of crime. There is a need to intervene on both design and social features of residential areas to promote mental wellbeing (van Kamp and Davies, 2008).[1]

Noise

Currently there is no conclusive evidence that there is a direct association between environmental noise and mental health. However, lack of evidence should not preclude the notion that noise can affect mental health. I would submit that noise nuisance affects quality of life and wellbeing and logically, in turn, mental health. I cite the World Health Organization – excessive noise seriously harms human health and interferes with people's daily activities at school, at work, at home and during leisure time. It can disturb sleep, cause cardiovascular and psycho-physiological effects, reduce performance and provoke annoyance responses and changes in social behaviour (WHO).

Housing

Housing is a fundamentally key factor in people's mental health. People with housing problems are at greater risk of mental health problems. Good-quality, affordable and safe housing is a vital component in good mental health, as well as supporting those with existing mental health conditions.

Research shows that those who are homeless, or at risk of homelessness, are much more likely to experience mental distress and a significant number do not access the support they need. (The impact of housing problems on mental health, April 2017 – shelter.org.uk.)

Natural environments

There is growing evidence to suggest that exposure to natural environments can be associated with mental health benefits. Proximity to greenspace has been associated with lower levels of stress (Thompson et al., 2012[2]) and reduced symptoms of depression and anxiety (Beyer et al., 2014[3]), while interacting with nature can improve cognition for children with attention deficits (Taylor and Kuo, 2009[4]) and individuals with depression (Berman et al., 2012[5]).

Air quality

Air pollution has been associated with adverse neurological and behavioural health effects in children and adults. Studies show even at low levels of pollution relatively small increases in air pollution are associated with a significant increase in treated psychiatric problems. There is a growing body of evidence that air pollution can affect mental and cognitive health and that children are particularly vulnerable to poor air quality (Oudin et al., 2016[6]).

Artificial light

Artificial lighting is associated with a range of ill-health effects, both physical and mental, such as eye strain, headaches, fatigue and also stress and anxiety in more high-pressured work environments. As we spend much of the day in artificial lighting, there is evidence that the lack of natural sunlight has an adverse effect on the body and the mind, and can result in conditions such as seasonal affective disorder (SAD).

Health and safety

Back in the second century it was recognised that work assists mental health. 'Work is nature's physician and is essential to human happiness' (Galen, 129–199 AD).

Work is good for our mental wellbeing, provided we are not exposed to excessive stress. It gives us a sense of purpose and connection. However, the number of people experiencing mental ill-health at work is increasing – one in six employees are now likely to experience some form of mental distress at any one time. This suggests that the causes and effects of mental ill-health at work are often not well managed. Mental health at work is about recognising that mental health is just as important as physical health when it comes to keeping people safe and well at work.

When someone is experiencing difficulties with their mental health at work, early intervention and support are essential and help maximise the likelihood of a positive outcome. However, because of the stigma, fear and misunderstanding that still surround mental health, concerns are often not addressed in a timely or effective manner. One third of people with enduring mental health conditions say they have been dismissed or forced to resign from employment. The impact of this is huge, both for individuals and the organisations that employ them (Mental Health and Safety).

The Stevenson Farmer 'Thriving at Work' review 2017

The 'Thriving at Work' report sets out a framework of actions – called 'Core Standards' – that the reviewers recommend employers of all sizes can and should put in place. The core standards have been designed to help employers improve the mental health of their workplace and enable individuals with mental health conditions to thrive.

By taking action on work-related stress, either through using the HSE Management Standards or an equivalent approach, employers will meet parts of the core standards framework. Work–life balance is also a concept that will have a positive effect on mental health. A key way to protect your mental health against the potential detrimental effects of work-related stress is to ensure you have a healthy work–life balance (Mental Health Foundation).

Alcohol and mental health

The reason we drink and the consequences of excessive drinking are linked with our mental health. Mental health problems not only result from drinking too much alcohol, they can also cause people to drink too much. A major reason for drinking alcohol is to change our mood – or our mental state. Alcohol can temporarily alleviate feelings of anxiety and depression and people often use it a form of 'self-medication' in an attempt to cheer themselves up or sometimes help with sleep. Evidence shows that people who consume high amounts of alcohol are vulnerable to increased risk of developing mental health problems and alcohol consumption can be a contributing factor to some mental health problems, such as depression (Mental Health Foundation).

Gambling

Pathological gambling has been associated with serious mental illnesses, sometimes as the cause and other times as the result of an untreated mental illness. **Depression and anxiety** are two of the most common mental illnesses associated with gambling addiction (Mental Health Foundation).

Smoking and mental health

Most adults in the UK are aware of the physical health risks of smoking tobacco, but research shows that smoking also affects people's mental health.

Social and psychological factors play a part in keeping smokers smoking. Although many young people experiment with cigarettes, other factors influence whether someone will go on to become a regular smoker. These include having friends or relatives who smoke and their parents' attitude to smoking. As young people become adults, they are more likely to smoke if they misuse alcohol or drugs or live in poverty. These factors make it more likely that someone will encounter stress. Most adults say that they smoke because of habit or routine and/or because it helps them relax and cope with stress ('self-medication').

Research into smoking and stress has shown that instead of helping people to relax, smoking actually increases anxiety and tension. Nicotine creates an immediate sense of relaxation so people smoke in the belief that it reduces stress and anxiety. This feeling of relaxation is temporary and soon gives way to withdrawal symptoms and increased cravings.

In the UK, smoking rates among adults with depression are about twice as high as among adults without depression. People with depression have particular difficulty when they try to stop smoking and have more severe withdrawal symptoms during attempts to give up.

Most people start to smoke before they show signs of depression so it is unclear whether smoking leads to depression or depression encourages people to start smoking. The most likely explanation is that there is a complex relationship between the two (Mental Health Foundation).

OPTIONS FOR OBTAINING COMPLIANCE AND ENVIRONMENTAL HEALTH INTERVENTIONS

The procedures and flow charts in this book indicate the statutory remedies/provisions prescribed in the legislation. There are also a number of other remedies/provisions available depending upon particular circumstances and the local authorities' policies, and the options are given below. In addition, there are a number of statutory remedies that are available generally and these are also included here.

No action

In some procedures, once a local authority is satisfied that a particular set of circumstances exists, it is required by law to take the next step of statutory action, usually the service of a statutory notice – for example, in the case of the existence of a statutory nuisance under section 80(1) of the Environmental Protection Act 1990. These situations are usually identified in the legislation by the use of the word 'shall' in respect of action required of the local authority. In other cases, the action to be taken, or a decision to take no action, is at the discretion of the local authority.

Oral warning

It may be felt that, in the circumstances of a particular contravention, it can be adequately remedied by speaking to the person responsible and asking for action to be taken. Such actions should be recorded.

Informal letters/notices

These may be used to confirm the existence of contraventions and to ask for them to be remedied. Informal notices are in effect a letter given the form of a notice but nevertheless are not part of a statutory procedure and no offences are committed by not complying with them.

Service of statutory notices

These are notices served under the provisions of a particular procedure. They must be in accordance with the requirements of that procedure, and with the general provisions covering such notices, and are legally enforceable. Non-compliance will lead to the possibility of prosecution and/or the taking of further steps specified in the legislation. They are the most usual form of statutory remedy used in environmental health law.

Formal cautions

These are applicable to the enforcement of all criminal law and may sometimes be appropriate in the enforcement of environmental health legislation. Their purpose is to deal quickly and simply with less serious offences, diverting them from unnecessary court action and reducing the chances of re-offending. The detailed provisions are contained in Simple Cautions for Adult Offenders, 13 April 2015 and in HSE LAC 22/19. The essential elements are as follows:

1. Cautioning decisions are at the discretion of the enforcing authority.
2. Cautions should be used only where they are likely to be effective and their use is appropriate for the offence.
3. They should not be used for indictable-only offences.
4. Consideration should be given to:

- the nature and extent of the harm caused by the offence;
- whether the offence was racially motivated;
- any involvement of a breach of trust;
- the existence of a systematic and organised background;
- the views of the persons offended against.

5. Cautions should be recorded.
6. Multiple cautioning of the same offender should not generally be used.
7. Before a caution is administered:

- there must be evidence of the offender's guilt;
- the offender must admit the offence;
- the offender must understand the significance of the caution and give informed consent to being cautioned;
- consideration must be given to the public interest.

8. Officers giving the caution, which is of course done in person, should hold a position of seniority.

Whilst continued offending after the receipt of a formal warning is not an offence in itself, the fact will both guide subsequent decisions regarding that individual person, organisation or company and be taken into account by a court in any later proceedings.

Regulatory penalties and sanctions

Part 3 of the Regulatory Enforcement and Sanctions Act 2008 gives regulators sanctioning powers as follows:

- fixed monetary penalties – such fines will be imposed by a regulator in respect of low-level instances of non-compliance;
- discretionary requirements, which include:
  - variable monetary penalties – requiring a person to pay a monetary penalty, the value of which is determined by the regulator;
  - compliance notices – requiring a non-compliant business to undertake certain actions to bring themselves back into compliance; and
  - restoration notices – requiring a person to undertake certain actions to restore the position, as far as possible, to the way it would have been had regulatory non-compliance not occurred.
- stop notices – requiring a person to cease an activity that is causing serious harm or presents a significant risk of causing serious harm and has given rise, or is likely to give rise, to regulatory non-compliance; and
- enforcement undertakings – an agreement offered by a person to a regulator to take specific actions related to what the regulator suspects to be an offence.

Fixed penalty offences

More specifically the Environmental Protection Act 1990, the Clean Neighbourhood and Environment Act 2005 and the Anti-Social Behaviour, Crime and Policing Act 2014 make fixed penalty notices available as a discretionary power for authorised officers in some enforcement procedures as an alternative to prosecution and fine. The local authority may specify the amount of the fine to be collected detailed in the Environmental Offences (Fixed Penalties) (England) Regulations 2017. Where the local authority does not take this option, a default amount is specified in the regulations. The receipts from such notices may be used to defray the cost of designated environmental functions and are detailed in the Environmental Offences (Use of Fixed Penalty Receipts) Regulations 2007.

Each local authority needs to decide in their enforcement policy how many fixed penalty tickets can be issued to the same person before further action should be taken. The discretion of the individual officer can determine the most appropriate course of action in every case, but local authorities should consider the practicalities of enforcing any policy they decide to adopt in this area.

Mortgagees

Mortgagees usually regard any action that attracts a statutory notice as being action that devalues a property. The Housing Act 2004 and the Town and Country Planning Act 1990 actually require that notices are not only served on owners and occupiers but also on the mortgagee, so that they can take such steps as they consider appropriate to protect their investment.

Where legislation demands that notices are served on mortgagees, the local authority must do that, but where the legislation is silent on the point, there is nothing to preclude the local authority from sending the mortgagee a copy of any notice or letter for information. Any mortgagee who considers that their investment is being jeopardised will normally write to the mortgagor, requiring them to take such steps as are necessary to comply with the notice or be regarded as defaulting on the mortgage.

Clearly this only works where a property is subject to mortgage, which can be determined by a simple Land Registry check.

Works in default

Carrying out works in default (WID) is often a cost-effective option when statutory notices are not complied with and, in some cases, when urgent steps need to be taken. The local authority has these powers under a variety of different legislation, with Environmental Health, Building Control, Planning and Highways being the main service areas for such rechargeable works. In some situations, the costs incurred become an automatic charge on the property. In most cases, the local authority can sue for debt in the civil courts.

The process followed needs to be clear and well documented, as any costs being recovered potentially have to be justified to the courts. Where the local authority decides to delay reclaiming the costs and relies on a charge on the premises, the case needs to be monitored at least yearly to ensure that time limits are not breached and that the debt plus interest is still covered by the value of the property. When there is no charge on the premises, there are still several options to collect the debt. Local authorities are entitled to include in the costs all its own reasonable costs in carrying out the WID, so that potentially some of the normal service budget can be reclaimed.

As with other means of enforcement, how the local authority will use these powers should be included in its relevant policies. Carrying out WID is always discretionary,[7] even where the statutory notice is a mandatory one, so the local authority needs to show it has exercised its discretion properly.

The main environmental health powers to do WID are covered by:

- Public Health Acts 1936 (particularly Part XII) and 1961;
- Prevention of Damage by Pests Act 1949;
- Building Act 1984 (various, section 107 for the creation of a charge);
- Environmental Protection Act 1990 (section 81 as amended);
- Housing Act 2004 (schedule 3, part 3).

This list is not exhaustive, and various Local Government Acts as well as local legislation create other opportunities.

Time limits

All civil law cases are subject to the appropriate time limits for bringing the action. Under section 5 of the Limitation Act 1980, all 'simple' debt/contract cases have to be brought within six years of the cause of action. For WID, the courts have said this means six years from the date the local authority completes the relevant works.[8] Where a charge is created, this is covered by section 20 of the Act, which gives 12 years for these additional powers to be used. Again for WID, the courts have confirmed that the 12 years runs from when the demand becomes operative[9] – which is immediately upon service for the Public Health Act and Building Act (notices served on owners), and after 21 days (if no appeal) for the Environmental Protection Act (owners again) and Housing Act. In any event, interest is generally only recoverable once the demand has been served.

Why use WID?

It may be a more effective sanction. It gets the job done – with likely more satisfied tenants/neighbours etc. A prosecution and conviction for an offence may still not lead to the works being done. It may cost the recipient more and therefore be a more effective deterrent than if they are simply fined.

It may be quicker than prosecuting – in some areas, getting a defended prosecution heard can take many months. It may be easier than prosecuting due to the lower standard of proof and rules of evidence. It may be cheaper than prosecuting, and there is a better chance of getting your costs back.

In most cases where there is an offence as well as WID powers, it is possible to pursue both. The local authority should cover the interplay between these sanctions in its relevant policies. As a professional, it is submitted that getting the job done and seeing the benefits the client experiences is wonderful job satisfaction.

What to consider when using WID

- Was the original notice correctly served and signed by the right person? If not, then as there is never an obligation to appeal against an invalid notice, it is unlikely any costs can be recovered if the local authority does the works[10] (so remember to check, including delegated powers for both the notice and the power to authorise WID);

- The likelihood the works will otherwise get done, including the track record of the recipient;
- The scale of the works;
- The ability to specify the right works – if the property has deteriorated since the service of the notice and/or more extensive repairs are now found, it may be necessary to re-serve the notice;
- The impact on the owners, occupiers and neighbours – this should now include any implications under the Human Rights Act 1998;
- The contribution to any corporate policies and objectives – this could include dealing with empty and eyesore properties.

It may be sensible to warn those affected that the local authority is minded to take such action and give them a short time to make representations. There is often a formal right of appeal against the demand after the works are done so the local authority needs to be able to show it has acted reasonably. Someone who has such a right of appeal cannot use judicial review to challenge such decisions,[11] but those without such a right (e.g. any tenants) may be able to use this procedure.

Records

It is necessary to correctly record all enforcement decisions where these count as executive decisions, even when taken under delegated powers (in England). Some types of legislation (e.g. housing) are always executive decisions; others (e.g. nuisance) may be under the local choice arrangements. In practice, it is sensible to record all WID decisions in case of challenge, particularly in England where other legal requirements require a decision that affects an individual to also be recorded in the same way, whether or not it is an executive function. The courts have determined that failure to do so may make the decision invalid.

The arrangements to carry out the works will also need to be recorded, including details of any tendering or use of approved lists of contractors. These must conform to the local authority's own standing orders on letting contracts.

Records of all the time spent on arranging the works are needed, including letting the contracts, supervising the works and liaising with owners and tenants etc. The costs can also include preliminary inspections, highways management, security, consulting fees, scaffolding, roof removal works and road closures.[12]

Local authorities have often underestimated the true costs, including overheads and support staff. They are all potentially recoverable. Local authorities often do not use time recording for their environmental health staff, so a system for WID can be adopted based on arrangements used for other professional staff, such as lawyers.

Remember – the local authority does not usually have the ability to add costs in after the demand has been served!

Once all the details have been collated and the total charge calculated, the demand should be served on the right person (which may not be the person who received the original notice), and any others who are entitled to receive a copy.

The original statutory notice should have been included on the Local Land Charges register maintained by the local authority. Details of the WID expenses should also be lodged with the Local Land Charges to warn anyone carrying out a search before buying the property. Failure to do this may make the costs irrecoverable against a new owner. Following the implementation of Section 34 of the Infrastructure Act 2015, Local Land Charges are due to be digitised and transferred to the Land Registry, but kept as a separate register. The phased programme has been delayed and as of April 2019 only five English authorities have been so affected.

Remember – in the appropriate cases, an automatic charge on the property comes into existence as soon as the demand is served/becomes operative. It is not necessary to register these at the Land Registry. Such a charge binds all estates and interests in the property and takes precedence over any existing mortgage etc.[13]

Getting the money back where there is no automatic charge on the property

An early decision is needed on pursuing debt through civil courts if the demand is not successful in getting payment or an agreed payment plan.

Action to collect a debt is taken in court (fast track <£25,000, small claims <£10,000). Both of these are likely to be dealt with by the County Court, and small claims can be dealt with without lawyers (and the legal costs cannot be reclaimed). Complex cases and those above £25,000 in value may be dealt with in the High Court.

Once a County Court Judgement is obtained, the debt can be collected in several ways:

- through bailiffs;
- attachment of earnings;
- third-party debt order;
- applying for a Charging Order – if the debtor has assets such as property or shares (interim and final orders have to be registered at Land Registry!). This has the same effect as a charge on a property but it is at the discretion of the court and ranks behind existing mortgages etc. Such an Order can be sought against any asset owned by the debtor, not just the property which is the subject of the WID action.

If all this fails, the local authority can factor the debts (i.e. sell them on to someone else). If the debt has to be written off, this is likely to impact negatively on the service accounts. While environmental health practitioners (EHPs) may think that debt collection is not their problem, failure to run the process properly may discourage the local authority from using it in future.

Getting the money back where there is an automatic charge on the premises

Apart from using the normal debt collection process, the local authority can also enforce such a charge by:

- taking possession and selling the property (this is what banks etc. tend to do);
- putting in a receiver and collecting any rents;
- seeking a vesting order, which transfers the property to the local authority without further payment (this is foreclosure, and it wipes out any other equity in the property).

County Court procedure for recovering a debt

The local authority should attempt to recover the outstanding debt using the County Court. If the identity of the debtor is known, action can be commenced against him by summons alleging that the debt is owed to the local authority. The summons, with particulars of the claim attached, is issued by the court and the procedure then effectively runs itself. Should the debtor fail to respond to the summons within the prescribed time, the local authority can gain judgement in default. The judgement can then be enforced by the local authority using the court bailiffs.

The process is not without difficulties. The debtor may seek to defend the action, in which case a hearing of the matter will follow. The debtor may take no part in the process at all, and after judgement is awarded for the local authority and bailiffs engaged, it may become clear that the debtor has no goods against which the debt can be levied. The local authority may end up with nothing at the end of the process; however, the advantage of a County Court Judgement over a charge on property is that the judgement is personal, and restrictive on the debtor. He will be prevented from obtaining or extending a mortgage and from obtaining credit, which will often focus the mind of the debtor rather more than a charge on a property.

Prosecutions

Environmental health law is a branch of criminal law, and prosecution of offenders in the courts is the usual last stage for most of the procedures dealt with here. The decision to prosecute or not is a matter of discretion for the local authority, guided by its own enforcement policy, which will have taken into account statutory and other guidance from the Health and Safety Executive and the Food Standards Agency. The Code for Crown Prosecutors issued by the Director of Public Prosecutions, whilst mainly aimed at police enforcement, provides useful guidance on the process of deciding whether prosecution should be taken.

Cases taken for offences against Housing Acts procedures are normally a matter initially for the first tier and upper tribunal, whereas most other environmental health procedures are dealt with at the initial stages by the magistrates' court. Higher courts will become involved in the event of appeals against decisions made by these courts, although there are situations (e.g. with the abatement of statutory nuisances) where a local authority may proceed directly to the High Court in order to seek a remedy.

Closure/withdrawal of licence etc.

It is sometimes the case that, in addition to prosecution to exact a penalty, the local authority may give consideration to the cancellation of the licence/approval it may have issued authorising the activity where the offence has taken place, e.g. approval of meat product establishments and pet shop licences. Forfeiture, detention and banning orders etc. may also be available.

PRIMARY AUTHORITIES

References

Regulatory Enforcement and Sanctions Act 2008
Co-ordination of Regulatory Enforcement Functions Regulations 2017
Draft Statutory Guidance for Primary Authority from 1 October 2017, Department for Business, Energy and Industrial Strategy

Extent

These provisions apply in England, Wales, Scotland and Northern Ireland.

Scope

Where a person or company carries on an activity in the area of two or more local authorities, and each of those authorities has the same relevant function in relation to that activity, the two parties can agree to set up a primary authority agreement.

Primary authorities for regulated persons and regulated groups

The Secretary of State can nominate a qualifying regulator (local authority, or a specified regulator) to be the 'primary authority' for the exercise of the specified partnership functions for a person or group (a 'direct primary authority' or 'co-ordinated primary authority') (section 23A(1) and (2)).

A function can be specified only if:

- it is a relevant function of the primary authority, and is, or would be, exercisable by the primary authority in relation to the regulated person or a member of the regulated group; or
- it is a relevant function of a qualifying regulator other than the primary authority, and is, or would be, exercisable by that other regulator in relation to the regulated person or a member of the regulated group and is equivalent to a relevant function of the primary authority (section 23A(3)–(5)).

Nomination of primary authorities

A qualifying regulator can only be nominated as a direct primary authority if the regulator and the regulated person have agreed in writing to the nomination (section 23B(1)).

The Secretary of State can only nominate a qualifying regulator as a co-ordinated primary authority if:

(a) there is a co-ordinator of the regulated group, and
(b) the regulator and the co-ordinator have agreed in writing to the nomination (section 23B(2)).

The Secretary of State must maintain a register of nominations and make the register available for inspection free of charge and can revoke the nomination at any time (section 23B(3) and (4)).

Membership of a regulated group

Where a qualifying regulator is nominated as a co-ordinated primary authority, the co-ordinator of the regulated group concerned must maintain a list of members of the group (section 23D(2)).

The list must include, in relation to each member, the member's name and address, when the person became a member and, if applicable, when the person ceased to be a member (section 23D(3)).

The co-ordinator of the regulated group must make a copy of the list available free of charge, on request, to the Secretary of State, the primary authority, and a qualifying regulator who has a function which is both a relevant function of the regulator and a partnership function (section 23D(4)).

The copy must be made available within three days of the request being received by the co-ordinator (section 23D(5)).

Primary authority advice and guidance

The direct primary authority has the function of giving advice and guidance to the regulated person in relation to each partnership function and to other qualifying regulators as to how they should exercise it in relation to the regulated person (section 24A(1)).

The co-ordinated primary authority has the function of giving advice and guidance to the co-ordinator of the regulated group and to other qualifying regulators in relation to each partnership function, as to how they should exercise it in relation to a member of the group (section 24A(2))

The primary authority can make arrangements with the regulated person or the co-ordinator of the regulated group as to how the authority will discharge its functions (section 24A(3)).

The co-ordinator of the regulated group must notify any advice or guidance given to the co-ordinator to those members of the group to whom the co-ordinator considers it can be relevant (section 24A(4)).

Advice or guidance can be given to qualified regulators only with the consent of the Secretary of State (section 24A(5)).

Enforcement action by primary authority

If the primary authority proposes to take enforcement action against the regulated person or a member of the regulated group, it must notify either in writing before taking the proposed enforcement action. It cannot take the action during the referral period (period in which the regulated person or the member can refer the action to the Secretary of State) unless notified in writing by the regulated person or the member (section 25B(1) and (2)).

Enforcement action other than by primary authority

Where a qualifying regulator other than the primary authority proposes to take enforcement action against the regulated person or a member of the regulated group, the enforcing authority must notify the primary authority in writing before taking the proposed enforcement action, and cannot take the action during the relevant period (section 25C(1) and (2)).

If the enforcing authority fails to notify the primary authority of the proposed enforcement action, but the primary authority is notified of it by the regulated person or the member or the co-ordinator of the regulated group, the primary authority must notify the enforcing authority in writing that the enforcing authority is prohibited from taking the action during the relevant period (section 25C(3)).

If the primary authority determines, within the relevant period, that the proposed enforcement action is inconsistent with advice or guidance previously given by it (generally or specifically), it can direct the enforcing authority in writing not to take the action (section 25C(4)).

Any such direction must be given as soon as is reasonably practicable, and in any event within the relevant period (section 25C(5)).

If the enforcing authority is directed not to take the proposed enforcement action, and continues to propose to take the action, it must inform the regulated person or the member, and it cannot refer the action to the Secretary of State unless notified in writing by the regulated person or the member that no such reference is to be made (section 25C(6)).

Relevant case – R (on application of Hull City Council) v Secretary of State for Business, Innovation and Skills, Queens Bench Division, April 2016.

Inspection plans

Where a partnership function consists of an '*inspection function*', the primary authority can make an inspection plan (section 26A(1)).

An '*inspection plan*' is a plan containing recommendations as to how the inspection function should be exercised by an inspecting regulator in relation to the regulated person or a member of the regulated group (section 26A(2)).

A person is an '*inspecting regulator*' if the person is a qualifying regulator, and the inspection function is a relevant function of the person (section 26A(3)).

An inspection plan can in particular:

(a) set out what an inspection should consist of;
(b) set out the frequency with which inspections should be carried out;
(c) set out the circumstances in which they should be carried out;
(d) require the inspecting regulator to provide the primary authority with a report on the inspecting regulator's exercise of the inspection function (section 26A(4)).

Before making an inspection plan the primary authority must consult the regulated person or the co-ordinator of the regulated group (section 26A(5)).

When making an inspection plan the primary authority must take into account any relevant recommendations relating to inspections which are published about a regulatory function by a person other than an inspecting regulator (section 26A(6)).

When it has made an inspection plan, the primary authority can apply to the Secretary of State for consent to the plan (section 26A(7)).

If the Secretary of State consents to a plan, the primary authority must notify the plan to the regulated person or the co-ordinator of the regulated group, and inspecting regulators (section 26A(8)).

If, in the case of a regulated group, an inspection plan is notified to the co-ordinator, the co-ordinator must:

(a) notify the plan to those members of the group to whom the co-ordinator considers it can be relevant,
(b) prepare a list of the names and addresses of those members,
(c) ensure, as far as is reasonably practicable, that the list is accurate and kept up to date,
(d) provide the primary authority with the list, including any updates to it, and
(e) notify any member whose name is included in the list or removed from it of the inclusion or removal (section 26A(9)).

If a list or update is provided to the primary authority, it must notify the list or update to inspecting regulators (section 26A(10)).

Effect of inspection plans

If the Secretary of State consents to an inspection plan, the primary authority must have regard to the plan when it exercises the inspection function in relation to the regulated person, or a member of the regulated group whose name is included in the list provided to the primary authority (section 26B(1)).

If an inspection plan of the primary authority is notified to an inspecting regulator, the inspecting regulator must follow the plan unless:

(a) the inspecting regulator has notified the primary authority in writing of the way in which it proposes to exercise the function, and the primary authority has notified the regulator in writing that the primary authority consents to that proposed exercise, or

(b) in the case of a regulated group, the member's name is not included in the list notified to the inspecting regulator (section 26B(2)).

A notification by an inspecting regulator must include reasons for not following the plan (section 26B(3)).

A primary authority is to be treated as having given the notification of consent if it is notified by the inspecting regulator, and it fails to notify the inspecting regulator in writing within five days of receiving the notification, whether it consents (section 26B(4)).

Revocation and revision of inspection plans

A primary authority can, with the consent of the Secretary of State, revoke an inspection plan (section 26C(1)).

If a primary authority revokes an inspection plan, it must notify the regulated person or the co-ordinator of the regulated group and inspecting regulators that the plan is no longer in effect (section 26C(2)).

Where the revocation of an inspection plan is notified to the co-ordinator of a regulated group, the co-ordinator must notify the revocation to those members of the group to whom the co-ordinator considers it can be relevant (section 26C(3)).

A primary authority can from time to time revise an inspection plan (section 26C(4)).

Power to charge

The primary authority can charge the regulated person or regulated group fees as the authority considers to represent the costs reasonably incurred by it in the exercise of these functions (section 27A(1)).

Support of primary authority by other regulators – supporting regulators

A supporting regulator can do anything which it considers appropriate for the purpose of supporting the primary authority in the preparation of:

(a) advice or guidance in relation to the partnership function, or

(b) an inspection plan in relation to the partnership function (section 28A(2)).

The supporting regulator must act consistently with any advice or guidance, or any inspection plan:

(a) which is subsequently given or made in relation to the partnership function, and

(b) to which the supporting regulator has consented (section 28A(3)).

But, in the case of a regulated group, this duty applies to the exercise of the designated function in relation to a member of the group only (section 28A(4) and (5)).

If the supporting regulator provides support and the regulated person or the co-ordinator of the regulated group has agreed in writing to the provision of that support, it can charge fees (section 28A(6)).

In the case of a regulated group, the co-ordinator of the group must make a copy of the group membership list or a copy of a list of group members to whom an inspection plan can be relevant available free of charge (section 28A(7)).

The copy must be made available within three days of the request (section 28A(8)).

Other regulators to act consistently with primary authority advice etc. – complementary regulators

A complementary regulator must act consistently with primary authority advice and guidance in the exercise of the designated function in relation to the regulated person or a member of the regulated group (section 28B(2)).

But, in the case of a regulated group, the duty applies to the exercise of the designated function in relation to a member of the group only if the complementary regulator is aware that the member belongs to the group (section 28B(3)).

In the case of a regulated group, the co-ordinator of the group must make a copy of the group membership list maintained or a copy of a list of group members to whom an inspection plan can be relevant available free of charge (section 28B(6)).

The copy must be made available within three days of the request (section 28B(7)).

Complementary regulators are not defined.

Primary authority enforcement action inconsistent with another authority's advice etc.

Where enforcement action is proposed and it is inconsistent with advice or guidance previously given (generally or specifically) by another qualifying regulator, Section 25C applies in relation to the proposed enforcement action as if the primary authority which gave the notification were an enforcing authority and section 25B no longer applies (section 29A(2)).

Concurrent duties to notify primary authorities of enforcement action

Where a qualifying regulator is nominated as a co-ordinated primary authority and an enforcing authority proposes to take enforcement action and it is required to notify the co-ordinated primary authority of the proposed enforcement action, that requirement to notify the co-ordinated primary authority does not apply if condition A or B is met (section 29B(1) and (2)).

Condition A – the enforcing authority is required to notify another qualifying regulator of the proposed enforcement action because of that other qualifying regulator's nomination as a direct primary authority (section 29B(3)).

Condition B – condition A is not met; the enforcing authority is required to notify at least one other qualifying regulator of the proposed enforcement action because of that other regulator's nomination as a co-ordinated primary authority and the enforcing authority has notified that other regulator or (if there is more than one) it has notified at least one of them (section 29B(4)).

Enforcement action notified to a primary authority inconsistent with another authority's advice etc.

Where a qualifying regulator is nominated as a direct primary authority or a co-ordinated primary authority that primary authority ('PA1') is notified of enforcement action that an enforcing authority proposes to take against the person pursuant to the function, and PA1 decides not to direct the enforcing authority not to take the enforcement action, and does not refer the action to the Secretary of State PA1 must, within the relevant period, take reasonable steps to find out if:

(a) another qualifying regulator nominated as the primary authority ('PA2') for the exercise of the function in relation to the person has previously given advice or guidance (generally or specifically), and

(b) the person considers the proposed enforcement action to be inconsistent with that advice or guidance (section 29C(1) and (2)).

If PA1 is of the view that such advice or guidance has previously been given and that the person considers the proposed enforcement action to be inconsistent with it, PA1 must refer the action to PA2, and notify the enforcing authority and the person that it has done so (section 29C(3)).

If this applies, the reference of the proposed enforcement action by PA1 to PA2 is to be treated as a notification given by the enforcing authority to PA2 and applies to PA2 as the primary authority and ceases to apply in relation to PA1 as the primary authority (section 29C(4)).

The Co-ordination of Regulatory Enforcement Regulations 2017 create the framework under which primary authority operates. It sets out which national regulators can support primary authorities when they are developing advice or guidance or inspection plans for businesses. It also defines the scope of primary authority in Scotland and Northern Ireland, sets out which types of enforcement action fall within the scope of the scheme and outlines the process for dealing with disputes relating to proposed enforcement action by enforcing authorities.

Definitions

Designated function, in relation to a supporting regulator, means a regulatory function exercised by that regulator and specified by the Secretary of State by regulations (section 28A(10)).

Enforcement action means:

(a) action which relates to securing compliance with a restriction, requirement or condition in the event of breach (or putative breach) of a restriction, requirement or condition;

(b) action taken with a view to, or in connection with, the imposition of a sanction (criminal or otherwise) in respect of an act or omission;

(c) action taken in connection with the pursuit of a remedy conferred by an enactment in respect of an act or omission (section 25A(1)).

Regulated person and regulated group, if the Secretary of State is satisfied, is a person who carries on, or proposes to carry on, an activity (section 22A(1)).

Legal structure, process and policies

FC1 Primary authorities – enforcement action against regulated person – sections 25A–25C Regulatory Enforcement and Sanctions Act 2008

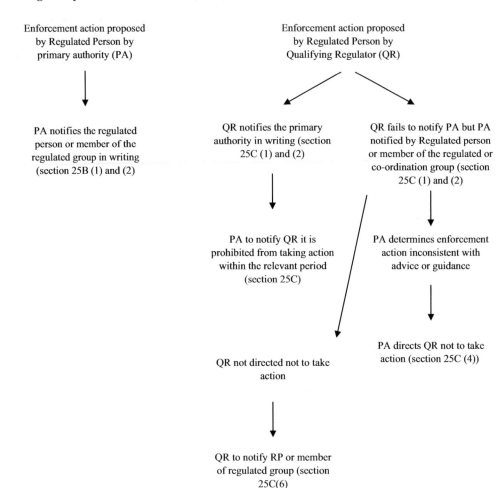

FC2 Inspection plans – section 26A–26C Regulatory Enforcement and Sanctions Act 2008

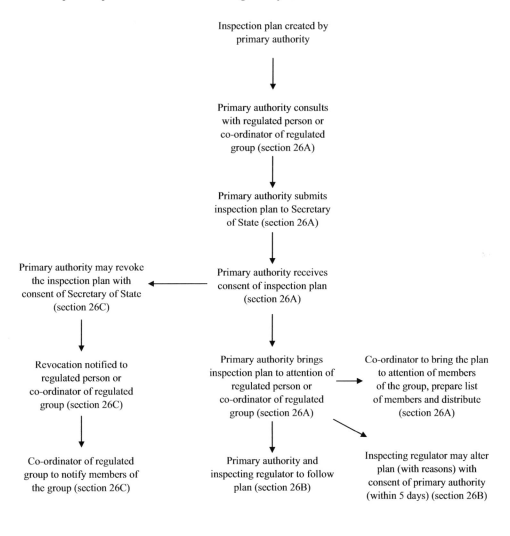

A group of persons is a 'regulated group' if the Secretary of State is satisfied that a member of the group carries on, or proposes to carry on, an activity (section 22A(2)).
Relevant period means the period which:

(a) begins when the primary authority is notified of the proposed enforcement action or the enforcing authority is notified that it is prohibited from taking the action during the relevant period

(b) ends:

  (i) at the end of the fifth working day after the day on which the period begins, or at such later time as the Secretary of State can direct, or

  (ii) if earlier, when the enforcing authority is notified in writing by the primary authority that no direction is to be given (section 25C(9)).

Qualifying regulator means a local authority, or a specified regulator (section 22B(1)).

Specified regulator means a person (other than a local authority) who has regulatory functions, and is specified by regulations made by the Secretary of State (section 22B(4)).

Supporting regulators – The Competition and Markets Authority, The Food Standards Agency, The Gambling Commission, The Health and Safety Executive, The Secretary of State.

ALTERNATIVE INTERVENTIONS

References

Effective Strategies and Interventions: environmental health and the private housing sector – Jill Stewart

Nudge: Improving Decision about Health, Wealth and Happiness – Richard Thaler and Cass Sunstein

Changing behaviours in public health – To nudge or to shove? – Local Government Association

An intervention can mean several things in different contexts. It is the act or fact of intervening or the interposition or interference of one state or person in the affairs of another.

In the medical sense, intervention is the interposition or interference in the affairs of another to accomplish a goal or end.

Environmental health interventions include the provision of safe drinking water, improved sanitation and hygiene practices, vector control, reduced exposure to air pollution, food safety control, health and safety at work controls, solid waste management, and controlling water hazards (e.g. drowning), and can also include traffic safety and environmental noise reduction.

Changes in the environment can have dramatic effects on the health of entire populations. For example, access to clean water, sanitation, education and food regulation are the most important factors in explaining the increases in longevity and decreases in infant mortality that took place in the world during the 20th century. Also, it is often easier to impact human health through environmental changes than it is to affect health through other types of interventions, such as changes in lifestyle or the provision of medical treatment and care.

Environmental interventions make changes in structures, processes or mechanisms in a person's environment in order to achieve health in a local sense. Environmental interventions that affect large groups of people can be defined as public health interventions. Environmental health does not stand still. It evolves as it tackles current problems in environmental and public health within the local or national population. Research drives this evolution forward. Evidence-based research identifies problems which need to be addressed.

Environmental health research can be defined as the systematic study of how the environment affects human health. Traditionally, the fields of epidemiology, occupational health, toxicology, ecology and public health have focused on the relationship between the environment and human health. Because almost all diseases have an environmental component, environmental health research now encompasses such fields as cardiology, oncology, gastroenterology, neurology, endocrinology, embryology and pulmonology. For example, in understanding cardiovascular and respiratory diseases as well as lung cancer, it is important to investigate all of the different factors that may increase one's risk of contracting such diseases, including genetic predispositions, diet, exercise, stress and environmental exposure to carcinogens, such as radiation or chemicals that cause cancer. If one understands how the environment affects the cardiovascular and respiratory systems, one may be able to make recommendations for changes in the environment designed to reduce the incidence of this disease, such as improving air quality and reducing exposure to carcinogens.

Behavioural change

In attempts to change behaviour of the general public, the traditional approach is that in cases where something causes serious harm, such as drug use, restricting choice or even a ban is appropriate; however, some say the state should leave it to individual choice.

But there are a variety of ways in which behaviour can be influenced, from encouraging and incentivising people, through to subtly guiding choice in a certain direction.

This can include enticing people to take up activities or using subliminal marketing. Alternatively, it can involve making an environment less conducive to someone making an unhealthy choice.

This is known as behavioural change and there has been growing interest in the issue among policy-makers. Policies should develop where authorities should act as a guiding hand, 'nudging' citizens in the right direction.

However, there are a full range of interventions that can influence behaviour.

The spectrum has been set by the Nuffield Council on Bioethics in its 'ladder of intervention'. The ladder works from bottom to top.

- Eliminate choice
- Restrict choice
- Guide choice by disincentives
- Guide choice by incentives
- Guide choice by changing default policy
- Enable choice
- Provide information
- Do nothing.

Techniques like direct incentives, such as vouchers in return for healthy behaviour, are being labelled hugs, while the tougher measures that restrict choice, like restricting takeaways from schools, are shoves. Bans, such as the restriction on smoking in public places, are simply known as smacks.

Public health teams in local authorities take a strategic lead for their area, such as setting policy and evaluating schemes, as well as playing a part in organising the interventions along with other partners from the private, public and voluntary sectors.

Nudging works in the retail world, but its effectiveness in encouraging good behaviour is still emerging with several intervention policies being introduced which seem to be working.

Local authorities could adopt the following policies for using behavioural change to improve the health of their population:

- Ensure their strategies and interventions meet local needs, identified through the JSNAs and other local data
- Appoint a strategic local authority lead to address behavioural change
- Ensure the content, scale and intensity of each intervention is proportionate and that behaviour-change interventions aim to both initiate and maintain any change
- Base interventions on a proper assessment of the target group, where they are located and the behaviour which is to be changed
- Allocate time and funds for independent evaluation of the outcomes of any behaviour-change service
- Take account of – and resolve – problems that prevent people from changing their behaviour
- Train staff to help people change their behaviour
- Consider how interventions should be complemented by other measures, including regulation
- Harness the power of the community – some areas have appointed champions among their local population.

Planning and environmental health

The six tests for the efficacy of planning conditions are that they must be:

- necessary – would the planning permission be refused if the condition was not attached? If yes, the condition is valid; if no, it is not;
- necessary to planning;
- relevant to the development to be permitted – its purpose is to ensure that conditions imposed are directly related to controlling the development, and cannot be used to achieve something that is covered by other legislation. You may not, for example, use a planning condition to prevent emission of dark smoke from a premise, since the Environmental Protection Act covers such emissions and to try to control them by planning condition is therefore unenforceable;
- enforceable;
- precise;
- reasonable in all other aspects.

How can EHPs get maximum benefit from planning conditions?

EHPs must identify where problems may arise during the development, or once the development is operational, and design a condition that is enforceable and will ameliorate the problems identified. For example, they may consider problems caused by noise or dust during the course of a development, and design a condition to deal with this – it may, for instance, cover hours of working, or suppression techniques for dust or lights on site, or number of vehicle movements permitted where residential streets are to be used. This is lawful since statutory nuisance law will not provide a remedy and, subject to the wording being precise and reasonable, will be enforceable.

1. Ensure consultation with planners at a sufficiently early stage.
2. Consider any environmental health problems that may arise either during the construction or during the operation of the development.
3. Check that there are other powers that will sort out the identified problems.
4. Draft an appropriate and lawful condition.
5. Engage with planners to ensure that the proposed condition and the reasons for it are understood and the condition is incorporated into the planning permission.

Section 106 Town and Country Planning Act 1990

Section 106 allows for a planning obligation or planning gain. Planning gain can only be sought if:

- it is relevant to planning and is directly related to the proposed development;
- it would make the proposed development acceptable in planning terms; and
- planning conditions or other procedures cannot achieve this.

Therefore, a planning obligation should represent a benefit for the land and/or the locality. Developers cannot be asked to solve existing problems, e.g. to provide a new facility, although they may be asked to contribute towards resolving an existing problem if the proposed development would make things worse.

A Section 106 Agreement cannot require the developer to do something that can be required by other legislation, but what it can do is achieve something by way of public health gain.

PARTNERSHIP WORKING

Effective partnerships can help improve the health and wellbeing of individuals and communities. They may work horizontally (across the community between government departments and other services) or vertically with opportunities to collaborate across public, private and community sectors. Partnerships can more fully tackle determinants of health and improve equity, as well as engaging hard-to-reach groups and underserved populations because they can operate and deliver at organisational or service user levels. As well as streamlining processes, effective partnerships can create shared learning and training opportunities, encourage innovation and creativity, build sustainable relationships for the future, and remove barriers and enhance ethics. They can be both proactive and design better living environments in the first place, or reactive, and respond better to stressors faced by individuals and communities.

Multi-agency working brings results. For example, private sector housing can be used to support criminal activity such as money laundering, people trafficking, sexual exploitation and modern slavery. Organisations such as local authorities, police, immigration service and other enforcement agencies gather intelligence and take action which is co-ordinated and robust. Punishment of offenders and the finances gained from these activities can be used to expand enforcement teams.

It is very important that colleagues in other services such as health and social care are aware of the powers of environmental health professionals and how they can support vulnerable residents by considering housing needs (e.g. heating improvements). The ability to engage and build relationships between professions is critical. Finding a connecting with professionals who are willing to listen and think differently is very important (Stewart and Lynch, 2018[14]).

Partnerships can be formed between individuals, agencies or organisations with a shared interest. There is usually an overarching purpose for partners to work together and a range of specific objectives.

Some partners will help you generate ideas, or develop content; others will help you to design your engagement activity; some will be able to share their skills and knowledge to ensure your activity is a success and others may be prepared to put resources into the activity. Partners can also help you develop relationships with different audiences.

You won't have to look too far to find relevant people to work with – colleagues in your own department, those in other areas of your institution, e.g. your widening participation, outreach, engagement, marketing or volunteering teams, community organisations, schools, museums, libraries, science centres, local councils and arts venues – there are lots of people who might want to work with you. However, before you get started it is important to think through why you want to work in partnership – and why your partners might want to work with you.

Partnerships can:

- **Inspire great ideas** – involving partners can help with the creative process. It can also be enjoyable working as part of a team.
- **Add another dimension** – colleagues from a different field might complement your activity, thereby adding more interest for prospective audiences.
- **Share their experiences** – working with people who have previously taken part in engagement activities can help guide you through unfamiliar processes and also assist with your own professional development.
- **Provide essential insights** – partners may have expertise in a host of relevant things such as understanding the needs of your target audience, timings and logistics, and suggestions of others you could work with.
- **Increase capacity** – partners can help you to reach a wider audience.
- **Provide an opportunity** – many partners will be looking for people to contribute to their own activities. It may be that you could support what they are doing, rather than invent something of your own.
- **Help you reach new audiences** – partners may already have access to members of your target audience, such as schools, museums, libraries or charitable organisations.
- **Provide expertise** – partners can be crucial in providing expertise you do not possess.
- **Strengthen relationships** – working with partners can deepen and strengthen your relationships as well as introducing you to new people and new ideas. This may lead to other projects in the future.

There are some guiding principles and issues to consider when planning partnerships:

- Be open to exploring new partnerships, particularly with community and voluntary sector organisations, as well as statutory/public sector partners.
- Invest time in building these partnerships and testing joint working opportunities before launching into larger-scale projects.
- Ensure appropriate due diligence is carried out in line with the scale of the project and the risk.
- Use community action investment to stimulate new partnerships and collaboration within the market – this is likely to lead to longer-term and more sustainable solutions.
- Ensure that partners are actively engaged in the governance, planning and monitoring arrangements and are able to take accountability for delivery.

Chapter 2

COMMON PROVISIONS IN ENVIRONMENTAL HEALTH LEGISLATION

The following sections include provisions which are common to most of the legislation covered in this book. They are presented in a way which shows a common interpretation of the provision across many pieces of legislation, but in some procedures some of them may be highlighted in a little more detail.

Each piece of legislation has very similar provision for the following aspects of enforcement:

- The **authority** given the power to enforce the legislation.
- How **powers** can be **delegated** to officers in the authority.
- The **authorisation** of officers to discharge the duty provided by the legislation.
- The **powers of entry** for those officers in exercising their duty under the legislation.
- The **authentication of documents** produced by the authority or the officer.
- **Service of notices** by officers on behalf of themselves or the authority
- **Information requests** regarding ownership of property or businesses.

ENFORCING AUTHORITIES

There is no single, legal definition of an environmental health authority and responsibility for the enforcement of each procedure is identified either in the particular piece of primary legislation or in the regulations made under it. However, there are only minor differences in the definitions of local authorities in the many pieces of legislation. References to licensing authorities, food authorities, housing authority etc. usually mean a local authority.

Generally, the lowest tier of local authority for the area, excluding parish councils, enforces the environmental health law dealt with in this book. The authorities charged with the implementation of most of the procedures in the book are:

(a) In England:

    (i) Unitary authorities
    (ii) District Councils in areas where there is a County Council
    (iii) County Borough Councils
    (iv) The Common Council of the City of London
    (v) The London Boroughs

(b) In Wales and Scotland, the unitary councils

(c) In Northern Ireland, the City, Borough and District councils and the Community Planning Statutory partners

(d) The Council of the Isles of Scilly for:

    (i) Caravan Sites and Control of Development Act 1960

    (ii) Health and Safety at Work etc. Act 1974

    (iii) Local Government (Miscellaneous Provisions) Act 1982

    (iv) Public Health (Control of Diseases) Act 1984

    (v) Environmental Protection Act 1990

    (vi) Food Safety Act 1990

    (vii) Noise Act 1993

    (viii) Criminal Justice and Public Order Act 1994

    (ix) Sunday Trading Act 1994

    (x) Licensing Act 2003

    (xi) Environmental Permitting Regulations 2010

(e) In London, the Sub-Treasurer of the Inner Temple or Under-Treasurer of the Middle Temple for:

    (i) Control of Pollution Act 1974

    (ii) Health and Safety at Work etc. Act 1974

    (iii) Public Health (Control of Diseases) Act 1984

    (iv) Clean Air Act 1993

    (v) Noise Act 1993

    (vi) Licensing Act 2003.

Where there is a substantial difference in definition, it is given in the relevant chapter.

A list of the definition sources in legislation can be found in Appendix 1.

DELEGATION OF AUTHORITY

The Local Government Act 1972 allows a local authority to delegate its powers and responsibilities to committees, sub-committees and officers. In respect of environmental health law enforcement, a clear scheme of delegation to officers is vital if investigation and remedial action is to be swift and effective. Such a scheme will require the formal approval of the individual local authority. It will include matters like the exercise of powers of entry and inspection, the service of statutory notices and the institution of legal proceedings. There is also a need for the scheme to include the designation of officers to whom the Act or regulations concerned give authority to act, e.g. the designation of proper officers, authorised officers and officers who will be inspectors under the Health and Safety at Work etc. Act 1974.

AUTHORISATION OF OFFICERS

Local authorities must have a system of authorisations and delegations. Environmental health practitioners have the power to carry out their functions because that power is given to them by the local authority by whom they are employed.

They are authorised by their employing local authority to carry out a number of functions, and whilst they are carrying out that function they have the right to do whatever the statute says they can and no more. This authorisation should be a council minute signed by the leader or chair, naming the authorised person by full name, stating that they are authorised by the council, and stating for what purpose the person is authorised.

If an environmental health practitioner is not authorised, anything done by way of enforcement, whether serving a notice or bringing proceedings, is void – it has no lawful standing.

Such actions taken by unauthorised officers open the council and the officer up to claims for trespass, suing for the cost of doing any work done required by the officer and any losses to the business or even misrepresentation.

Practitioners should *always* have a certified copy of the authorisation with them while at work and acting on behalf of the local authority.

The keeping of accurate records of authorisation is vital, not least because the Office of Surveillance Commissioners (OSC) will be expecting to see records of authorisations. Environmental Health departments may wish to design their own forms, but as a guide there are excellent Authorisation, Renewal of Authorisation and Cancellation of Authorisation forms on the Home Office website (www.gov.uk/government/organisations/home-office/series/ripa-forms~2), which can be tailored to suit the needs of an environmental health service.

It must be noted that the specific authorisation under the Health and Safety Work Act 1974 gives the inspector the powers rather than the local authority (p. 240).

POWERS OF ENTRY

The basis of environmental health practitioners' powers of entry is sections 287–289 Public Health Act 1936. Each piece of legislation has a specific power of entry and most are a version or extended version of this particular section – see Appendix 3.

Below is a description of that power of entry. The elements within it are provided in some form or another in the relevant legislation.

An authorised officer of a council, producing if required a duly authenticated document showing his authority, has a right to enter any premises at all reasonable hours for the following purposes:

(a) ascertaining if there has been a contravention of this Act;
(b) seeing if circumstances exist which would authorise or require the council to take any action or execute work;
(c) to take any action or execute any work authorised or required;
(d) generally for the council's performance of its functions.

For premises other than a factory, workshop or workplace, at least 24 hours' notice must be given before entry may be demanded.

Where:

(a) admission has been refused; or
(b) refusal is apprehended; or
(c) the premises are unoccupied; or
(d) the occupier is temporarily absent; or

(e) the case is one of urgency; or

(f) application for admission would defeat the object of entry,

the local authority, following notice of intention to any occupier, may make application to a Justice of the Peace by way of sworn information in writing, and the Justice of the Peace may authorise entry by an authorised officer, if necessary by force. The authorised officer may take with him any other persons as may be necessary but must leave any unoccupied premises as secure as he found them (Public Health Act 1936 section 287).

The penalty for obstructing any person acting in the execution of these Acts, regulations and orders is a maximum level 1 on the standard scale (section 288).

Where an owner is prevented from executing any work which he has been required to undertake, he may apply to a magistrates' court for an order to the occupier to permit execution of the works (section 289).

AUTHENTICATION OF DOCUMENTS

The following provisions of the Local Government Act 1972 will apply unless a particular piece of legislation has provisions which otherwise cover these issues.

1. Any notice, order or other document which a local authority is authorised or required by or under any enactment (including any enactment in this Act) to give, make or issue may be signed on behalf of the authority by the proper officer of the authority.

2. Any document purporting to bear the signature of the proper officer of the authority shall be deemed, until the contrary is proved, to have been duly given, made or issued by the authority of the local authority. In this subsection the word 'signature' includes a facsimile of a signature by whatever process reproduced.

3. Where any enactment or instrument made under an enactment makes, in relation to any document or class of documents, provision with respect to the matters dealt with by one of the two foregoing subsections, that subsection shall not apply in relation to that document or class of documents.

4. In this section 'local authority' includes a joint authority, an economic prosperity board, a combined authority, a joint waste authority, a police and crime commissioner and the Mayor's Office for Policing and Crime. (Section 234 Local Government Act 1972.)

SERVICE OF NOTICES BY LOCAL AUTHORITIES

Generally, notices must be in writing, dated and signed by an authorised officer. A facsimile signature can be used. A notice must be served on the person responsible, which is usually defined in the legal provision in each Act or regulation. Usually the notice will specify the contravention of the law or a defect which needs to be remedied. It may also specify the works to be done to rectify the contravention or defect.

Addressing a notice on individuals

The notice is addressed to the individual concerned, using the full name, not initials, and the envelope addressed likewise. Any failure to comply with the notice can be dealt with against the person named on the notice and the person to whom it is addressed can be prosecuted.

Bodies corporate

Notices can be served on and summonses issued against a number of bodies corporate, including limited companies, unincorporated bodies and local authorities. The name on the notice should be that of the company, with its registered office being given as its address. Naming the company ensures that it is the company that is required to comply with the requirements of the notice.

The name and address on the envelope is a different matter. A company secretary is the legal persona of a company upon whom notices can be served. They can be sent the notice in an envelope and will accept delivery on behalf of the company. Envelopes containing notices should therefore be addressed to the company secretary with the company name and address. This ensures that somebody with appropriate authority accepts the notice. In the case of local authorities, it is usually the chief legal officer who is authorised to accept documents on behalf of the authority, and documents addressed to that person by name or by job title are deemed as having been properly served on the authority concerned.

In the cases of unincorporated bodies, clubs and school governing bodies, it is wise to ask on whom service of the notice may be served. Where the body has a formal constitution, this person should be identifiable; where it does not, naming all of the managing committee on the face of the notice and serving each individual ensures that service is properly made.

Some Acts will specify that service shall be served in a particular way and on a particular party, and failing to comply with that laid-down method will mean that effective service will not be achieved.

If service of the document cannot be proved, no prosecution, based on allegation of non-compliance with the requirements of the notice, can proceed.

A notice that has been addressed on its face to the wrong party is a ground of appeal against it. Addressing a notice to the company secretary rather than to the company itself gives rise to that ground of appeal.

Service

This may be effected by one of the following:

(i) delivery to the person;

(ii) in the case of service on an officer of the council, leaving it or sending it in a pre-paid letter addressed to him at his office;

(iii) in the case of any other person, leaving it or sending it in a pre-paid letter addressed to him at his usual or last known residence;

(iv) for an incorporated company or body, by delivering it to its secretary or clerk at its registered or principal office, or sending it in a pre-paid letter addressed to him at that office;

(v) in the case of an owner by virtue of the fact he receives the rack-rent as agent for another person, by leaving it or sending it in a pre-paid letter addressed to him at his place of business;

(vi) in the case of the owner or occupier of a premises, where it is not practicable after reasonable enquiries to ascertain his name and address or where the premises is unoccupied, by addressing it to the 'owner' or 'occupier' of the premises (naming them) to which the notice relates and delivering it to some person on that premises, or, if there is no-one to whom the notice can be delivered, by affixing it or a copy of it to a conspicuous part of the premises (Public Health Act 1936, section 285).

Proof of receipt and/or service of notices should be obtained.

The Public Health Act 1936 (sections 283 to 285) is the basis of service of notice in environmental health law and notices served in any of these ways will be held in later proceedings to have been properly served, but it is possible to serve in some other way and prove service in subsequent proceedings.

The main provision for service of notice is found in the Local Government Act 1972 and, unless otherwise excluded by a particular provision of another Act, is available in addition to methods of service available under other legislation.

1. Subject to subsection (8) below, subsections (2) to (5) below shall have effect in relation to any notice, order or other document required or authorised by or under any enactment to be given to or served on any person by or on behalf of a local authority or by an officer of a local authority.
2. Any such document may be given to or served on the person in question either by delivering it to him, or by leaving it at his proper address, or by sending it by post to him at that address.
3. Any such document may:
   (a) in the case of a body corporate, be given to or served on the secretary or clerk of that body;
   (b) in the case of a partnership, be given to or served on a partner or a person having the control or management of the partnership business.
4. For the purposes of this section and of section 26 of the Interpretation Act 1889 (service of documents by post) in its application to this section, the proper address of any person to or on whom a document is to be given or served shall be his last known address, except that:
   (a) in the case of a body corporate or their secretary or clerk, it shall be the address of the registered or principal office of that body;
   (b) in the case of a partnership or a person having the control or management of the partnership business, it shall be that of the principal officer of the partnership;
      and for the purposes of this subsection the principal office of a company registered outside the United Kingdom or of a partnership carrying on business outside the United Kingdom shall be their principal office within the United Kingdom.
5. If the person to be given or served with any document mentioned in subsection (1) above has specified an address within the United Kingdom other than his proper address within the meaning of subsection (4) above as the one at which he or someone on his behalf will accept documents of the same description as that document, that address shall also be treated for the purposes of this section and section 26 of the Interpretation Act 1889 as his proper address.
6. ... (Repealed)
7. If the name or address of any owner, lessee or occupier of land to or on whom any document mentioned in subsection (1) above is to be given or served cannot after reasonable enquiry be ascertained, the document may be given or served either by leaving it in the hands of a person who is or appears to be resident or employed on the land or by leaving it conspicuously affixed to some building or object on the land.
8. This section shall apply to a document required or authorised by or under any enactment to be given to or served on any person by or on behalf of the chairman of a parish

meeting as it applies to a document so required or authorised to be given to or served on any person by or on behalf of a local authority.

9. The foregoing provisions of this section do not apply to a document which is to be given or served in any proceedings in court.
10. Except as aforesaid and subject to any provision of any enactment or instrument excluding the foregoing provisions of this section, the methods of giving or serving documents which are available under those provisions are in addition to the methods which are available under any other enactment or any instrument made under any enactment.
11. In this section 'local authority' includes a joint authority, an economic prosperity board, a combined authority, a joint waste authority, a police and crime commissioner and the Mayor's Office for Policing and Crime. (Section 233 Local Government Act 1972.)

Where notices are served by email, EHP's should obtain agreement from the recipient that they are prepared to accept notices in that way. They also could serve a section 9 witness statement (p. 62) to confirm that agreement.

The statement should also say that was achieved by sending the recipient and email, specifying the time and to which email address it was sent. A screenshot of the EHP's 'sent' mailbox, showing the email, the addresses and the time of sending, should be exhibited as should any attachments that went with the email so the court can see what was served as well as when service was achieved.

The provisions for the service of notice are included in some other Acts and Regulations. A list of the sources of these provisions is included in Appendix 2.

Further guidance on the service of notices can be obtained from www.justice.gov.uk – Courts – Procedure rules – Civil – Rules & Practice Directions – Part 6 – Service of Documents.

INFORMATION REGARDING OWNERSHIPS ETC.

The Local Government (Miscellaneous Provisions) Act 1976 makes provision for a local authority to require information to be provided about the ownership of property etc. Where a local authority, in performing functions under any legislation, requires information regarding interests in any land, it may serve a notice, specifying the land and the legal provision containing the particular function under which it is acting, requiring the person concerned to declare, within not less than 14 days, the nature of his interest and the names and addresses of all persons who he believes have interests.

The notice may be served on any of the following:

(a) the occupier;
(b) the freeholder, mortgagee or lessee;
(c) the person directly or indirectly receiving the rent;
(d) the person authorised to manage the land or arrange for its letting.

Failure to comply with the notice or the making of false statements carries a maximum penalty of level 5 on the standard scale (section 16).

These powers are given to county councils, district councils, London borough councils, the Common Council, the Council of the Isles of Scilly, police authorities, joint authorities, police and crime commissioners, the Mayor's Office for Policing and Crime, an economic prosperity board, a combined authority, joint waste authorities, the London Fire and Emergency Planning Authority and parish and community councils and others (section 44(1)).

THE ENVIRONMENTAL INFORMATION REGULATIONS 2004

These provide a general right of access to environmental information, subject only to withholding in the public interest where that outweighs the public interest in disclosure, and applies to many of the procedures dealt with in this work.

RECORDING DECISIONS

Reference

Local Authorities (Executive Arrangements) (Meetings and Access to Information) (England) Regulations 2012. (Similar provisions apply in Wales, Scotland and Northern Ireland.)

These regulations apply to **enforcement** actions being chosen, not to decisions to take no action. Since 2012 all such decisions taken by officers under delegated authority are subject to the same rules. That means for certain enforcement actions decisions made by individuals or committees etc. need to be recorded and made available to the public.

All decisions under Housing Acts are executive decisions. So these rules apply to decisions to serve a notice, prosecute or carry out works in default. The responsibility for other Environmental Health enforcement actions are often decided locally, some being executive decisions, some not.

The Openness of Local Government Bodies Regulations 2014 requires decision-making officers to produce a written record of any of the following decisions delegated to an officer:

(i) **granting a permission or licence**;
(ii) affect the rights of an individual; or
(iii) award a contract or incur expenditure which, in either case, materially affects that relevant local government body's financial position.

The written record must contain the following information:

(a) the date the decision was taken;
(b) a record of the decision taken along with reasons for the decision;
(c) details of alternative options, if any, considered and rejected; and
(d) the names of any member of the relevant local government body who has declared a conflict of interest in relation to the decision.

It is submitted that authorities should use the same pro forma for both executive and non-executive decisions so there is a clear record of what was done and why. The courts may decide in due course that failure to record such enforcement decisions may invalidate the decision and makes works in default costs irrecoverable.

The general principle of the Regulations is for the public to have access to meetings and documents where a local authority executive, committee or individual is taking an executive decision.

The regulations set out:

- the circumstances during which the public must be excluded from meetings
- the formalities to be complied with before a private meeting is held
- formalities to be complied with before a public meeting takes place
- rules relating to access to the agenda and reports for executive meetings.

They provide for specific requirements relating to executive decisions which are key decisions. They set out the meaning of key decision and the publicity requirements in relation to key decisions. There are exceptions to these requirements.

They require decisions to be recorded in a written statement and set out the documents which must be made available for inspection by the public.

Members of the local authority and of overview and scrutiny committees are given additional rights to access documents and there are additional rights of members of overview and scrutiny committees in relation to decisions that the committee is scrutinising and in certain circumstances the committee can access exempt or confidential information.

An overview and scrutiny committee can require the executive to make a report to the local authority on matters which have not been dealt with as a key decision and which an overview and scrutiny committee consider should have been treated. There is a reporting requirement that the executive provides reports to the local authority on all matters which have been treated as urgent.

The regulations state the general principles applicable relating to the non-disclosure of confidential, exempt information or the advice of a political adviser or assistant. They establish the manner in which documents are required to be available for inspection held at the offices of the local authority. There are also offences where documents have not been made available for inspection as required under the regulations.

Chapter 3

PROFESSIONAL PRACTICE

PROFESSIONAL PRACTICE

Legal process

Legal process is to make sure, before something is done by 'authority', whether a local council or the police, that the views, concerns and objections of individuals and groups who are affected by the decision are heard and the rights of those whose freedoms, whether personal or over property or goods, may be curtailed by the action are protected. EHPs need to understand their responsibilities and duties in order to comply with the requirements of the legal process.

Enforcement policy

Local authorities must have an enforcement policy; however, it must be noted that an enforcement policy guides – it does not direct. It is among other issues to be taken into account when considering how to decide on action. A local authority should have regard to its enforcement policy, as a guide promoting consistency and equity of treatment. Decisions about prosecution are for the prosecutor to make. Enforcement policies are drafted and agreed, taking into account the various government and agency policies and guidelines (see p. 6).

Local authorities should agree, through its cabinet, a policy to demonstrate that their procedures comply with the right to a fair trial (Article 6 of the European Convention on Human Rights).

The policy should describe the manner in which defendants will be dealt with, and the considerations that will be taken into account when determining disposal of matters.

Examples of matters to be taken into account are:

- previous Home Office cautions;
- the severity of the offence alleged;
- previous convictions for like offences;
- previous convictions by other departments of the council which may lead to the view that the individual will not cooperate with the council;
- the degree of criminality involved or likelihood of re-offending.

A policy will ensure that the local authority itself is satisfied that all those against whom it is considering taking action are fairly treated.

When the policy has been approved and adopted, there needs to be a good and compelling reason for deviation from its policies.

The local authority can clearly demonstrate that its reasoning is Article 6 compliant if challenged by a defending solicitor, or by an aggrieved 'victim' who considers that an alternative course to that adopted should have been taken.

This policy is for guidance only, and in every case officers must use discretion. In most cases, the discretion will determine that the course identified in the policy is appropriate, but where it is not, the reasoned justification for deviation should be recorded and can then be cited where allegations of rigid adherence to policy without consideration of other factors are made.

Consider the risk to complainants in the event of them being identified by those they have complained of, and the steps they have taken to prevent this happening. The policy should clearly outline how you will protect identities. A simple example is, in the event of noisy parties, it is not wise to park outside the house of the complainant, go into the house and then immediately go to the house of the party-giver; instead, the policy should advise parking some distance away, and not attending one premises directly from the other.

The policy should consider the wording to be used in letters, indicating that the complainant's name and address should not be revealed and that all letters should be worded in such a way as to prevent identification of complainants – geographical references to the location of complainants should be as oblique as possible or, preferably, no mention should be made of the complainant at all and reliance should be placed on the evidence of environmental health officers.

Competence

Officers who are authorised by the local authority must hold a relevant qualification and be sufficiently competent in the particular field of operation. Local authorities must ensure that the officers operating on their behalf are competent to do whatever is required of them. Government agencies, such as the Food Standards Agency and the Health and Safety Executive, set criteria and guidance for local authorities to guide them in ensuring officers are competent to carry out the function assigned to them. It will be necessary to show that knowledge and skills are held and practised, refresher training has been done and practical skills are up to date, as well as showing how knowledge is maintained.

Professionalism and competence may be demonstrated in many ways; however, it is submitted that holding and maintaining the highest level of membership of your professional body offers one of them, and because it is only available to a limited group, it offers one of the most valuable ones available.

Inspections

General inspections are a matter of professional training and application of the law. Such detail is not relevant to this book; however, sometimes EHPs may see or seriously suspect child abuse in the course of a visit or inspection. In preparation for such events you should make sure you have had some training in recognising abuse as defined in guidance and also that you know what to do if you see it. Make sure you know what the protocol is in your place of work about reporting your suspicions.

Inspections which may lead to potential prosecution

When carrying out an inspection which is likely to lead to a prosecution to establish matters of fact, speak to no-one about offences, but note all of the potential offences. Questions as to identity of persons and possibly ownership of the premises can be asked, depending on the charges being contemplated, but nothing about the offences observed.

The potential defendant can then be asked to attend for an interview under caution where questions about each of the offences observed can be put to him. There is no guarantee of results from this process as defendants may not accept the invitation, or may give no comment during interviews. The alternative is to lay informations citing all of the offences observed and run the risk of defences being laid out in court.

Police and Criminal Evidence Act (PACE): PACE notebooks and contemporaneous notes

The PACE (Police and Criminal Evidence) page-numbered notebook helps to demonstrate that contemporaneous notes were made when the writer says that they were.

The role of the PACE notebook is to tie the inspection sheet to the date of the inspection. The inspecting officer should record in his or her PACE notebook the date and time and address of the inspection and the fact that an inspection sheet (whether electronic or paper) was used for the details of the inspection. The inspection sheet can be given a unique code or reference, which should appear both on the sheet and in the PACE notebook, and which will tie the two together. The PACE notebook should record time in and time out and the inspection sheet should show the detail of the inspection and its findings. The content of the PACE notebook can either be exhibited or alternatively listed as unused material in a court case. The defence should know of its existence and they can see it if they want to.

In court, a well-kept PACE notebook shows that the officer has credibility, knows what they are doing and knows why it is important. Well-presented and completed PACE notebooks also protect officers from interrogation about when notes were made. They protect not just the evidence but also the officer giving it.

Good contemporaneous notes allow officers to refresh their memory before giving evidence about what they saw, did and heard, and are a valuable resource when writing witness statements.

Witness statements are disclosed; contemporaneous notes are disclosable. Descriptions recorded in the note should appear in the same way in the statement.

Note: Dating and timing of input of information into a data collector is not usually a problem; data collectors usually have a date and time-recording mechanism which, subject to the usual rules on the admissibility of computer-generated information, will be sufficient to establish when the information was input.

Caution

A person who is suspected of an offence must be cautioned before any questions about it (or further questions, if it is his answers to previous questions that provide grounds for suspicion) are put to him for the purpose of obtaining evidence which may be given to a court in a prosecution.

It is only necessary to caution before asking questions that may give rise to answers of evidential value, whether directly or indirectly. It is not necessary to caution before asking questions for other purposes – for example, to establish identity, or ownership or responsibility for a vehicle.

Wording of the caution

'You do not have to say anything, but it may harm your defence if you do not mention, when questioned, something you later rely on in court. Anything you do say may be given in evidence.'

The Code of Practice on PACE notes that minor deviations from its exact wording do not constitute breaches of the requirements of the Code of Practice, provided the general sense of the caution remains.

It remains at the absolute discretion of the person being questioned as to whether they say anything. What do you do if the defendant clearly does not understand? Do not, under any circumstances, try to explain the caution to the defendant yourself. If he, or she, has a solicitor present, let the solicitor explain it. If the defendant does not have a solicitor, terminate the interview and let him or her get one.

The purpose of the caution is to protect the defendant. The duty of the judge in court is to ensure that the prosecutor has obtained the evidence with which he intends to convict the defendant in a fair and lawful manner. A court that feels that a lack of understanding on the part of the defendant was exploited, whether intentionally or not, is likely to rule that the evidence has been unfairly obtained and is inadmissible.

Correspondence

Emails

Unlike a telephone conversation, which is generally not recorded, an email has permanent form; once generated, it is discoverable, whether as part of a case file or under the Freedom of Information Act.

The email may contain information that is defamatory or damaging to the local authority case. It may contain personal comments about the defendant or, more importantly, it may hold information that could be damaging to the substance of the council case.

Emails should be treated with the same care as a letter or memo. Language should be formal and polite, and the normal rules of professional courtesy should apply. Generally, they should relate to a single issue, so that they can be printed off, or retained in the relevant file. Multi-issue emails should be avoided, since they will have to be filed in a number of files, and will introduce unrelated material into those files. Emails need to be treated as what they are – documents of potential evidential value – and given the respect they merit in consequence.

Without prejudice

A letter written and headed 'without prejudice' will contain something which, were it to come into the public domain, could potentially be damaging for the writer, such as an admission against his interest, or a proposal for settlement of a matter. If the letter is accepted, this is the end of the matter; if the letter is not accepted, the fact that it is written 'without prejudice' means that the recipient cannot later use the letter in evidence against the writer to demonstrate the offer made, or the weakness of the writer's position.

The use of 'without prejudice' correspondence or discussions means that lawyers can deal with each other on a pragmatic basis, and can discuss matters candidly without damaging their client's position whilst doing so.

Covert surveillance

In some situations, it may be necessary for local authority enforcement officers to act in a covert manner in the investigation of possible breaches of environmental health legislation, e.g. the illegal sale of food that is unfit for human consumption.

Regulation of Investigatory Powers Act 2000 (RIPA)

The law governing the use of covert surveillance techniques by public authorities is contained in the Regulation of Investigatory Powers Act 2000 (RIPA). When public authorities, such as local authorities, need to use covert surveillance techniques to obtain private information about someone, they do it in a way that is necessary, proportionate and compatible with human rights.

RIPA applies to a wide range of investigations in which private information might be obtained, including terrorism, crime, public safety and emergency services.

RIPA's guidelines and codes apply to actions such as:

- intercepting communications, such as the content of telephone calls, emails or letters;
- acquiring communications data – the 'who, when and where' of communications, such as a telephone billing or subscriber details;
- conducting covert surveillance, either in private premises or vehicles (intrusive surveillance) or in public places (directed surveillance);
- the use of covert human intelligence sources, such as informants or undercover officers;
- access to electronic data protected by encryption or passwords.

The Regulation of Investigatory Powers (Directed Surveillance and Covert Human Intelligence Sources) Order 2010 stated that such surveillance carried out by local authorities must be by officers of at least director, head of service, service manager or equivalent status, or by an officer designated as being responsible for the management of an investigation. Guidance to local authorities is given in the Home Office guidance to local authorities in England and Wales on the judicial approval process for RIPA and the crime threshold for directed surveillance.

Local authorities are required to obtain judicial approval prior to using covert surveillance techniques. Local authority authorisations and notices under RIPA will only be given effect once an order has been granted by a Justice of the Peace in England and Wales, a sheriff in Scotland and a district judge (magistrates' court) in Northern Ireland.

Additionally, local authority use of directed surveillance under RIPA is limited to the investigation of crimes which attract a six-month or more custodial sentence, with the exception of offences relating to the underage sale of alcohol and tobacco.

Generally, local authorities are only concerned with carrying out directed surveillance 'for the purpose of preventing and detecting crime or preventing disorder' – for example, making observations in respect of fly tipping or the sale or distribution of counterfeit goods.

When considering an allegation of statutory nuisance, evidence is collected to support the serving of a notice, not to prevent or detect crime, or prevent public disorder. That being the case, it is not appropriate to use the powers in RIPA to authorise directed surveillance since the activity falls outside the terms of the specific grounds available.

It is not until the notice is served and, subject to withstanding any appeal against it, has taken force that RIPA becomes relevant. Once evidence is being gathered, this shows that the terms of the notice are being breached and the evidence is being gathered *'for the purposes of detecting crime'*, i.e. the proving that the notice is being breached and therefore an offence is being committed.

If an officer wants to carry out an investigation, he/she must first ask him/ herself if what he/she wants to do constitutes covert directed surveillance. If it does, RIPA applies and he/ she will need to have that activity authorised. He/she then needs to consider whether the purpose of the surveillance is to prevent or detect crime, or to prevent disorder. If it is neither of these, then the proposed surveillance cannot be authorised under RIPA and therefore cannot lawfully be carried out. The simplest test is to ask whether you could commence proceedings based on the evidence you would obtain. If you can, RIPA authorisation is required. If you cannot, you cannot legitimise the proposed covert directed surveillance and should not therefore carry it out.

Directed surveillance must be:

(i)   covert;
(ii) not intrusive;
(iii) for the purpose of a specific investigation; and
(iv) carried out in such manner as to result in obtaining private information about a person, otherwise than by way of an immediate response to circumstances.

Covert surveillance – RIPA advises that surveillance is covert only if it is carried out in such manner that the subjects are unaware that it is, or may be, taking place.

So, suppose you tell the subject that surveillance *will* be taking place and how it will be carried out. Let us suppose that when an abatement notice is served, the local authority write a covering letter advising that they will be carrying out monitoring to ensure compliance, and this monitoring will take the form of periodic visits, to be carried out at any time of day or night, by inspectors who will engage in non-intrusive monitoring. The 'target' could not complain that the surveillance was covert, since he not only knows it will be taking place, he has also been advised how it will be done and by whom. All he does not know is when. This knowledge is enough to take the surveillance out of the covert category and make it overt, and hence preclude the need for authorisation. It must be noted that this procedure has not been tested in court and may be subject to challenge.

It is essential for authorising officers to ensure that they give proper consideration to the reasons given for seeking authorisation to conduct directed surveillance. The authorising officer must be able to say why they believed what they were told about the complaint, what convinced them and what further inquiries, if any, they required to be made. Did they consider whether there was an alternative to directed surveillance and, if so, did they require that it was explored? If not, why not? It is essential that there is a clear paper trail that demonstrates the thinking of the authorising officer.

There is a higher threshold for directed surveillance with the exception of two trading standards offences. Local authority applications will only be granted for offences that attract a maximum custodial sentence of six months. Magistrates must approve any RIPA-regulated activity before it begins. This applies in England, Wales and Northern Ireland and to communications data in Scotland.

Home Office guidance says EHPs can continue to use covert techniques provided they have considered other less intrusive avenues first. Officers need to present magistrates with documentation and respond to questions by the officer with most knowledge of the case and show that covert surveillance is both proportionate and necessary. It is important to consider the questions that may be put by the magistrates – in other words, have less intrusive measures been used and subsequently failed?

Should communications data be needed then the following piece of legislation is relevant to EHPs.

AUTHORISATIONS FOR OBTAINING COMMUNICATIONS DATA

References

Investigatory Powers Act 2016, sections 73–75.

Extent

These provisions extend to England and Wales, Scotland and Northern Ireland.

Scope

A designated senior officer of a local authority may grant an authorisation for obtaining communications data only if it is necessary to obtain communications data for the purpose of preventing or detecting crime or of preventing disorder (section 73(3)).

The officer cannot grant an authorisation unless:

(a) the local authority is a party to a collaboration agreement (whether as a supplying authority or a subscribing authority or both), and
(b) that collaboration agreement is certified by the Secretary of State as being appropriate for the local authority (section 74(1)).

A designated senior officer of a local authority may only grant an authorisation to an officer of a relevant public authority which is a supplying authority under a collaboration agreement (or part of the agreement) to which the local authority is a party (section 74(2)–(4)).

The authority must apply to the court to confirm these authorisations by order (section 75(1)–(3)).

The local authority is not required to give notice of the application to any person to whom the authorisation relates, or that person's legal representatives (section 75(4)).

The relevant judicial authority may approve the authorisation if, and only if, the relevant judicial authority considers that:

(a) at the time of the grant, there were reasonable grounds for considering that the requirements were satisfied in relation to the authorisation, and
(b) at the time when the relevant judicial authority is considering the matter, there are reasonable grounds for considering that the requirements would be satisfied if an equivalent new authorisation were granted at that time (section 75(5)).

Where the relevant judicial authority refuses to approve the grant of the authorisation, it may make an order quashing the authorisation (section 75(6)).

Definitions

Designated senior officer, in relation to a local authority, means an individual who holds with the authority the position of director, head of service or service manager (or equivalent), or a higher position (section 73(2)).

Stake-outs and covert surveillance

The Court of Appeal in the case of R v Johnson[15] declared that there are two requirements for stake-outs.

First, the officer in charge of the inquiry must be in a position to give evidence that he visited the premises to be used and ascertained the attitude of the occupiers of the fact that the premises were to be used, and their attitude to the fact that a surveillance operation had been in place would have to be disclosed, which may then lead to the identification of both the premises and the occupants. He/she may also inform the court of any local difficulties in obtaining assistance from members of the public that may have caused the use of the particular premises chosen.

The second requirement is that a more senior officer should be in a position to give evidence that, immediately before the trial, he had visited the premises and ascertained whether the occupiers were still of the same view and, whether they were or not, what their attitude was to the possible disclosure of the use made of the premises and of facts which may lead to the identification of both premises and occupiers.

The lessons of R v Johnson for local authority officers are:

1. Make sure the local authority's enforcement policy takes account of the rules in R v Johnson; and
2. Be pragmatic, and use local authority premises or vehicles for surveillance where possible.

The Protection of Freedom Act 2012 includes provision for making codes of practice on surveillance, regulation of surveillance using CCTV and powers of the Secretary of State to amend powers of entry. Some are included below.

Covert Surveillance and Property Interference Code August 2018

This code of practice provides guidance on the use by public authorities of Part II of the Regulation of Investigatory Powers Act 2000 to authorise covert surveillance that is likely to result in the obtaining of private information about a person. The code provides guidance on when an application should be made for an authorisation and the procedures that must be followed before activity takes place. It also provides guidance on the handling of any information obtained by surveillance activity. It highlights differences between authorities in England and Wales, Scotland, and Northern Ireland.

Covert Human Intelligence Sources Revised Code of Practice August 2018

This code of practice provides guidance on the authorisation of the use or conduct of covert human intelligence sources ('CHIS') by public authorities under Part II of the Regulation of Investigatory Powers Act 2000. The code also provides guidance on the handling of any information obtained by use or conduct of a CHIS.

Surveillance Camera Code of Practice – June 2013 provides guidance on the appropriate and effective use of surveillance camera systems by relevant authorities in England and Wales who must have regard to the code when exercising any functions to which the code relates. Other operators and users of surveillance camera systems in England and Wales are encouraged to adopt the code voluntarily.

Obstruction

Inspecting officers often encounter being partially obstructed by being delayed just enough to mean that gaining access to a premises is pointless.

Delayed consent, however willingly given, does not constitute real consent, and taking up such an invitation to enter makes later pleading an allegation of obstruction in court difficult.

First approach

When immediate consent to enter is not given, officers could walk away on the grounds that delayed entry is pointless, and prosecute for obstruction, based on the facts of the initial refusal. The defendant did not allow immediate access and, as such, this is an offence.

Second approach

Stage 1 – If officers want to enter a premises and are being frustrated by being given only delayed consent, they could write to the owner and/or occupier telling them that they want to enter, that they have the power to do so and that when they ask to be allowed entry, access is required immediately. Owners etc. are then forewarned of what officers expect and can take legal advice should they want to.

Importantly, officers also have a paper trail that can be produced in evidence, showing the court that they have advised owners/occupiers of the position, and their claims not to know or to understand would therefore be groundless.

Stage 2 – Turn up unannounced and ask to be allowed in. If officers are delayed, even slightly, they should just go away. Do not wait. Then prosecute, alleging obstruction. Then repeat the process until the message gets through. Fines for obstruction can be significant, commonly in the order of £1,000. Being fined once may cause the defendants to consider their position; being fined several times will usually make a real difference.

Warrants

Obstruction of an officer in the course of their duties under legislation is an offence. Furthermore, the same Acts make provision for officers who have been refused entry, or who anticipate that entry will be refused, to make an application to the magistrates' court for a warrant authorising entry.

Obtaining a warrant is a straightforward procedure. An application is made to the magistrates' court; evidence as to why the warrant is required is given on oath, before the magistrates, usually in camera (in private); and subject to the magistrates being satisfied that the application has merit, the warrant will be issued. Where the premises in question is domestic, the property owner must be given 24 hours' notice of the hearing application, in order that they may attend, should they wish. This requirement does not apply to commercial premises. Where an EHP can make out a sound case for the issue of a warrant, it will be granted. To execute a warrant, an officer is required to show it to the owner of the premises. If the property owner still refuses entry in the face of a valid warrant, they are in contempt of court and should be prosecuted.

The requirement that must be satisfied when seeking a warrant is that the officer seeking to execute the warrant must have reasonable belief that an offence is being committed in the premises or on the land to which the warrant relates. The reasonableness of this view will be tested by the magistrate granting the warrant; therefore, it must be sufficiently robust to satisfy that inquisition.

Reasonable belief is that attributed to a reasonable man with reasonable judgement. Where there is dispute as to what is and is not reasonable, the judge in court will be asked to arbitrate, and it is for the judge to decide if, on the facts as known at a particular point in time, a prosecutor had reasonable grounds to believe that an offence had been committed.

Deciding to prosecute

Prosecutors must ask themselves the following:

1. On the evidence, is there a *prima facie* case to answer?
   And subject to the answer being yes:
2. Is it in the public interest to prosecute?
   The person making the decision to prosecute must be independent of the case, and must demonstrably be so.

Is there a prima facie case?

A lawyer will normally decide this – legal advice as to whether a certain series of facts constitutes an offence is definitive. A lawyer who is a member of an investigation team is not independent from the case, so a different, uninvolved lawyer, chief EHP or whoever is considered to be an appropriate person will have to review the file when it is completed and consider whether to prosecute. The independent case review must demonstrably be completely independent.

The Code for Crown Prosecutors 2018, which contains the Full Code Test, is a more sophisticated test, with guidance as to whether it is appropriate in a particular case to commence proceedings.

Is it in the public interest to prosecute?

This is for the officer in the case to consider. Sometimes there will be good reasons for prosecuting a case – for instance, the publicity generated by doing so may be dispropor-tionate to the offence itself but may influence businesses. However, it may also have a detrimental effect on the way the council is viewed by the public. Overall, it is the officer who must decide how best and most appropriately to deal with the offence.

Things to take into consideration

Courts will usually look at the overall offending when it imposes sentence. This will include the number and seriousness of offences; however, the court will not look favourably on a large number of informations.

Where the defendant has indicated that he will be pleading guilty, the prosecutor can pick out a couple of specimen offences and then ask if the defendant is prepared to ask the court to take the remaining counts into consideration. Where he is, they are tabulated, and after his plea is accepted, they are passed to the court to be taken into consideration when sentence is considered. The time and resource saved by this practice attracts discount off sentence, but the offences to be taken into consideration (TiCs) remain recorded against the defendant, and can be cited as part of his previous offending record in any future cases.

Where the defendant has pleaded guilty, he will usually accept the TiCs, which will have been put to him on his plea being indicated. It is an option for the prosecutor, even in the face of a not guilty plea, to list a number of TiCs and offer them to the defence on the basis that the defendant will accept them in the event of his being convicted against his plea. If he is not convicted, the main offences and the TiCs will fall, but where such an offer is made, the prosecutor will generally be confident that he will succeed at trial, and therefore the risk of major loss is minimal. This provision should be in every local authority's enforcement policy.

'Shall have regard to'

You must have regard to something when the circumstances which cause it to be relevant actually exist. Therefore, if the requirement is that, when considering how a potential prosecution should be disposed of – whether to instigate court action or to offer a formal caution – the officer making the decision may be required by an enforcement policy to take into consideration the character of the defendant, the officer need only consider the character of the defendant when making that decision.

Consideration is subjective – it requires free thought and an independent decision.

What the officer is required to do is to consider what he or she knows about the character of the defendant – good or bad – and to consider how that defendant will respond to whatever method of disposal is preferred.

The officer making the decision should record the reason for the decision being made, showing that he or she has taken the character of the defendant into consideration in its making the requirement.

You must be able to show that you have given it some weight, and that the consideration you gave the issue was true consideration that did impinge on the decision-making process, not just a token consideration because you had to do so.

Failure to take account of, consider or have regard to things that you are required to take account of etc. can cause your decision to be challenged through the mechanism of judicial review.

Any authority that does not consider all of the relevant issues, particularly those it is specifically required to consider, whether they are codes of practice, statutory guidelines, planning policy guidance, etc., can be accused of flawed decision making, have its decision overturned by the High Court and be returned to it for reconsideration.

Where directory words are used, officers should ensure that what is required is actually done, so that those affected by a decision, whilst perhaps not liking the decision itself, cannot challenge the process by which the decision was made.

Prompt action

Time between end of compliance period and check for completion of works

If an officer is trying to impress on the court the importance and urgency that necessitated tackling a statutory nuisance, the officer has also to demonstrate that he/she acted expeditiously throughout. This means that the officer went back to the site as soon as the compliance period ended to check for completion of works.

Anything less risks giving the impression that the matter had ceased to be important; in effect, the service of the notice rather than compliance with it resolved the issue. At the end of the compliance period, the site should be visited to assess compliance. If a matter is important enough to warrant action, it is important enough to follow through.

Time lapse between offence and bringing case to trial

Summary proceedings must be commenced by the laying of an information within six months of the prosecutor having sufficient evidence to make a *prima facie* case, and in either way or indictable matters, they should be laid within 12 months. Failure to comply may lead to a challenge based upon the case being out of time.

Officers must ensure that all proceedings are commenced as expeditiously as possible. Where there is the possibility of delay, for whatever reason, the reason for the delay and the manner in which the prosecutor has attempted to mitigate it should be documented in order that the court may be appraised, if necessary. In such scenarios, the defence will be to have the prosecution dismissed without consideration of the evidence on the basis of prejudice through delay, and breach of Article 6 of the European Convention on Human Rights.

It is highly unlikely that the court will be sympathetic where lack of progress is attributed to lack of time or resources, or pressure of other work. Valid reasons for unavoidable delay will have to be presented.

In some cases, commencement dates are prescribed within the legislation, to take account of the fact that covert illegal action may not be discovered for a significant period, e.g. the Food Safety Act 1990 – commencement must be either three years from the commission of the offence or within one year of its discovery by the prosecutor, whichever is the shorter (section 34).

What constitutes commencement? When is a case considered to be started? The date a case is commenced is the date on which the information making the allegation is laid.[16] For all practical purposes, this is the date on which the informations arrive at the magistrates' court. When they are delivered by hand, that date is clear.

When they are sent by post, certain rules apply to when delivery is assumed to have taken place. Similarly, it is now established that when informations are laid by electronic mail, they are laid on being sent, not on being received.[17]

When is an offence discovered? In some cases, it will be very clear – when an officer finds a breach of the legislation. Further, if an officer decides that they will not commence proceedings in respect of the breach but will serve a notice requiring remediation, it is not until they revisit at the end of the compliance period and discover that the requirements of the notice have not been satisfied that time begins to run on an allegation of non-compliance. In both cases, the date of discovery is clear and can be pinpointed – the date of the initial visit, or the date of the inspection to ascertain compliance.

The date of the discovery of an offence is the date upon which an investigating officer has reasonable ground to believe that an offence has been, or is being, committed.

Chapter 4

LEGAL PROCEEDINGS

PREPARING A CASE (PLAINTIFF OR APPELLANT)

Collecting evidence

Evidence is all the material on which either party seeks to rely on to prove their case, i.e. it is what the court sees or hears (e.g. testimony or oral evidence, documents, objects, videos, photographs).

You must first collect evidence of fact from witnesses, identify the reasons/causes of the breach or incident, then speak to those who may be responsible for the breach or incident, and finally gather interview information.

Gather all documents related to the cases so long as you can provide witnesses who can tell the court about their origin and content. Never write on original correspondence – using sticky notes that can be removed is preferable.

Appendix 3 shows a method of ensuring the relevant, admissible and fair evidence is collected for a particular offence. Each row should address an individual element of the offence that needs to be proved (first column). The second column shows the details of the offence element that you have collected. The remaining columns to the right are self-evident in that these are for the sources and references, ensuring that the evidence is well organised and can be used in court for efficient reference.

Exhibits

Give the court articles of real evidence, in the form of photographs, plans or maps of the location, and real 'things' that they can look at, weigh in their hands and scale against the rest of the scene. This is likely to influence the court more than witness statements.

Any physical evidence taken should be bagged and tagged, and then stored in a lockable evidence cupboard, to which access is restricted.

A register should be kept indicating:

- when the item of evidence was put in the cupboard;
- by whom it was put in the cupboard;
- if and when the item was taken out of the cupboard, by whom and for what purpose – for example, an item of food removed to be inspected by a company representative.

Photographs/videos

The role of the witness is to give evidence as to what was seen or heard; the court makes decisions of fact. Give short factual statements of what you saw to introduce a photograph or video. It is vital to let the photograph/video speak for itself. The rule is 'simply exhibit the photograph, and exhibit the photograph simply'.

Identification evidence

The leading case for identification is that of R v Turnbull.[18] In this case, the Court of Appeal laid down guidelines that should be applied where identification evidence is disputed.

Condensed, they are:

A   Amount of time the suspect was observed by the witness
D   Distance between suspect and witness
V   Visibility at the time of observation
O   Obstructions between witness and suspect
K   Known previously (i.e. does the witness know the suspect)?
A   Any significant reason for the witness to remember the suspect?
T   Time elapsed since witness saw the suspect and description written down
E   Error or discrepancies in the witness's description.

This is a list of all those points that a witness identifying a suspect *must* cover when making their identification, whether orally in the witness box or in a witness statement made as part of the case if their identification evidence is to be relied on. If one of these is not covered, the judge is obliged to warn the jury about the risks of relying on the identification evidence, and he may still do this even if all of the points are covered and the issue of identity is critical. It is a misdirection should he fail to do so and grounds for appeal.

Enforcement officers should subject their own identification evidence to this test to be sure that it will stand up in court.

Proving competence and authority

It is necessary to prove that the officer taking the case or who is the main witness is competent to do so. It needs to be shown that the officer holds a relevant qualification and is sufficiently competent in the particular field of the case. It may be necessary to show that knowledge and skills are held and practised, refresher training has been done and practical skills are up to date, as well as showing how knowledge is maintained.

Where newly qualified officers take a case, knowledge is the most up to date and skills, once obtained, can be practised often. Where officers practise within their field of expertise and keep their skills and knowledge up to date, it is very difficult to challenge competency successfully.

It is necessary to show that the officer is duly authorised by the local authority and was acting within their authority (see p. 41).

Witness statements

A witness statement is a document recording the evidence of a person, which is signed by that person to confirm that the contents of the statement are true. A statement should record what the witness saw, heard or felt. In the case of the EHP, this would include all of the exhibits that he collected in the course of the inquiry. You should rely on notes made at the time of the incident when making a witness statement. Lawyers expect to find a very close connection between the notes and the text of the witness statement.

Officers should consider producing a witness statement at timely intervals. Breaking the case up into manageable amounts makes the witness statement likely to be more reliable and therefore less vulnerable to challenge. Make sure in your witness statement you make it clear that you are familiar with the leading case, the current Code of Practice and the relevant guidance, i.e. the key legal background to the case. You should check all facts written in the statement are correct, with the correct date and supported by evidence. Never add or remove things from your statement. Describe behaviour and observable facts in the statement and avoid professional terms and jargon, using everyday language. Ensure any opinion stated is within your expertise and be aware of any inconsistencies. Witness statements can be made under two pieces of legislation:

- Section 9 of the Criminal Justice Act 1967 states that the contents of a written statement will be admissible, without the witness attending court to give oral evidence.
- Section 20(2)(j) of the Health and Safety at Work Act 1974 gives officers the power to require any person whom the officer has reasonable cause to believe will be able to provide information relevant to his examination or investigation to answer such questions thought fit to ask, and to sign a declaration of the truth of the answers.

Witness statements should be on prescribed forms. They should describe the sequence of events in a logical order, usually chronological. They should be certified by the witness that the evidence is true. They should be short and the information in it should be relevant to the offence.

Witness statements from lay people

It is important to impress on witnesses that they should only say what they know to be true, and thereafter to leave it to the court to determine effect.

The first step is to make sure that witnesses know that they should not discuss the matter with each other. Quite clearly you cannot stop this happening. All you can do is make sure you advise them, in simple and understandable terms, not to talk to anyone about their evidence. Stress the importance of it being their evidence and remaining uncontaminated. Make sure they understand your advice. The second step is to impress on a witness who is about to provide a witness statement the need to ensure that they can give their own evidence of what they are about to write by highlighting that their evidence will not be accepted per se, but will be challenged in cross-examination. If the witnesses come under pressure in the witness box and concede that they have chatted with other witnesses about their evidence, on their head be it. They will also have to concede that you told them not to, either in cross-examination or under re-examination, so at the very least you can demonstrate you did try to prevent evidence contamination.

Can you write the statement for the witness? You can write down the statement, at the dictation of the witness, but you must only record what you are told. You may not add bits, and you should not sanitise the use of English to make the statement more readable. A defence lawyer can usually tell whether a witness is using his own words or words are being used for him, and any suspicion that a statement has been doctored or sanitised will lead to the witness who made it being called and cross-examined on it.

The first thing you can do is talk the witness through their evidence. Get them to go through what they saw, heard, etc. in chronological order. You cannot impress on them what is important and what is not because to do so is to affect their judgement, but at least they will have had an opportunity to rehearse what they will say before committing it to paper. If the witness does not remember something vital, you can push very gently – e.g. 'Did you not say to me . . . that may be worth recording'. You cannot remind them; only they can remember.

Where a witness writes their own statement, the rule is that you can guide the pen but you may not push it. It is perfectly acceptable to suggest to the witness that he writes his statement in chronological order because it makes sense to record the story as it unfolded. Given that you will have talked it through as suggested above, this will ensure that the witness has a clear picture in his mind, recalls as much as possible and records it in admissible form. When the witness has finished writing, you can read over the statement and point out that he may have forgotten to mention something he has told you orally, giving him the chance to include it, provided the exchange to which reference is made actually took place.

How much assistance can you give to a witness who is about to appear and give oral evidence for the authority? Coaching is not allowed. Officers must be very vigilant to ensure that nothing that they do or say can be construed as witness coaching.

Sub judice

Information that is relevant to a case that has not yet come before the court is *sub judice*, literally 'in *the course of trial*', and must not be discussed in the public domain. It must be preserved intact, to be considered in court by a jury or lay bench who are hearing its contents for the first time, and can form a view as to its validity in context of the hearing and alongside all other relevant evidence.

The reason for the rule of *sub judice* is simple – it allows arbitrators to consider evidence on the basis of what they hear and see in court. Where information has been in the public domain, it is virtually impossible for an individual not to form a view about it themselves and for their own judgement not to be coloured by what they have heard from others.

Expert witnesses

Experts, and indeed EHPs, who are considering offering themselves as expert witnesses must be aware of the rules governing the conduct of expert witnesses and must abide by them. Failure to do so may damage the credibility of the expert witness in the eyes of the court, their client or others who may have considered engaging him. Expert witnesses hold a very privileged position in litigation, and with this comes responsibilities.

An expert witness may give the court evidence of his or her opinion, unlike a witness to fact, who is restricted to purely factual evidence. In order to give opinion evidence, the expert must know what each witness has said, and in order to facilitate him having this knowledge he is allowed to sit in the well of the court and listen to the evidence of the witnesses as to fact. He may then comment on that evidence when he gives evidence by, for example, indicating which scenario presented in the evidence is more likely, in his view, to be the case.

The expert witness has more freedom in court than the lay witness.

It is incumbent both on those who act as expert witnesses and on those who chose to offer expert witnesses in support of their cases to ensure that the restrictions of the role are understood.

Where an expert is asked a question or considers a proposition that takes him outside the core of his field of expertise, he should advise the court that he cannot give expert evidence on the point.

What is an expert witness?

An individual must be an expert in a particular and relevant field. Clear evidence of expertise in the form of qualifications, papers published, etc. must be provided to satisfy the court of expertise. The court would look for evidence that the expert was so regarded by his peers. The court will also consider whether her expertise is up to date – an individual who has not updated their knowledge will not be regarded as an expert. Equally, the court will want to know how much experience an expert has.

However, it is in the discretion of the court as to whether an expert witness is required.

The purpose of an expert witness is to help the court by providing an expert opinion based on the facts as presented by the lay witnesses, and if the court feels that it (or the jury) is capable of making a decision without such assistance, it will refuse to allow an expert witness to be called. If it does agree, the court also has to be persuaded that the person tendered as an expert witness is in fact someone who can be regarded as an expert.

Selecting an expert witness

It is important to strictly define that area in which the assistance of expert evidence is required.

It is necessary to be clear about the exact areas of evidence upon which expert opinion is required and to ensure that the expert chosen has sufficient expertise to consider all of them.

It should be remembered that a particular expert's field of expertise may not necessarily be as wide as the area upon which expertise is required, and in some cases more than one expert may be required.

Ensure that the court will accept the expert chosen:

- Is his expertise up to date?
- Is his expertise mainly theoretical or does he have field experience?
- Is he mainstream and does he speak as the majority in this field or not?

Consider whether you can understand your expert:

- Is what your expert has to say comprehensible to the court and the jury?
- Can your expert speak so that lay jurors or magistrates understand him and will be persuaded by what he says?

Take the advice of your expert:

- Does he agree with your evidence and your methods?
- Are there more tests he would recommend?
- Does he suggest another approach?
- Does his opinion conflict with your evidence?
- Does your expert know what is expected of him?

Expert rules

An expert is there to assist the court and his overriding duty is to the court, not to the party calling him. He must stay within his area of expertise and must inform the court if he speaks on matters outside that, when his evidence will not be treated as expert evidence.

It is incumbent on him to make it clear to the court when he speaks outside his area of expertise, since that is within his knowledge more than anyone else's.

The expert is required to refer to what facts and assumptions he has used and made when coming to his conclusions and also to state what material facts do not support his conclusions.

Where the expert has not been able to research the issue fully because, for example, insufficient data is available or because the theories relied upon may not be widely accepted, he should note that his view is provisional.

If, after exchange of statements, the expert changes his mind on a material matter, he should tell the other side and the court as soon as possible.

Photographs, plans, surveys and other documents relied on by the experts in coming to their stated views must be exchanged at the same time as the reports, so that each side can come to a fully advised view of the other's position.

There is an exception where the law says the relevant officer can act on an opinion. EHPs are lay witnesses and can only give evidence of fact unless the law says otherwise.

Importantly, the EHP will give opinion evidence on only one point: whether, in his or her opinion and based upon all of the factual evidence he or she gives as a lay witness, something is prejudicial to health or is a nuisance, for example.

However, EHPs can be regarded as expert witnesses, but only if they have specialised in a particular field and have reached a very high level in that field.

Professional witnesses

In cases where it is perceived that there may be some difficulty persuading lay persons to give evidence, local authorities can consider the use of professional witnesses.

Professional witnesses, usually from an investigation agency, undertake monitoring, as well as ensure that the evidence gathered will be admissible and available. In such cases, the cost of engaging the witnesses can legitimately be claimed as investigation costs when the matter goes to court.

Provided the witnesses are properly briefed as to what the investigation requires, and can be demonstrated to have no interest in the outcome of it, they are a more certain source of evidence than lay persons.

Interviews under caution (IUCs)

Before carrying out interviews, you should prepare well. Prepare your questions in advance. Ensure you have the ability to record the questions and answers in a notebook, interview record form or tape recorder.

The right to legal advice

An interview under caution is evidence and is admissible for the truth of its content in court.

When a person attends an interview voluntarily, he must be informed about his continuing right to consult privately with a solicitor and that free independent legal advice is available.

It is essential he is properly and appropriately advised by someone who not only understands the law but also understands the way in which it works in the context of an interview under caution.

If a local authority asks for a person to attend for interview, a solicitor may ask the local authority to disclose all materials involved in the case so he/she can advise the client whether to attend for interview. Interviewees must also be told why they are being asked to attend for interview so they have some idea what the interview is about.

However, you are not obliged to disclose every element of the offence or the defendant's role in it. You need to show enough so that a *prima facie* case is made.

Why carry out an interview under caution?

IUCs are governed by the requirements of PACE Code of Practice C.

A prosecutor carries out an IUC when he has *prima facie* evidence to suggest that an offence has been committed and that the interviewee is either the sole accused or has played a part in a multi-handed offence.

Interviews under cautions are not necessary if you already have enough evidence to establish a *prima facie* case against the proposed defendant. Conducted properly, an IUC should fill gaps in the inquiry by establishing the role of the accused, providing the interviewer with details not known at the time of the offence or, in some cases, allowing the interviewee to clear himself and eliminate himself from the inquiry. An IUC should not be done until the obvious ways of obtaining evidence have been exhausted.

Ending an interview under caution

The interview must cease when 'the officer in charge of the investigation . . . reasonably believes there is sufficient evidence to provide a realistic prospect of conviction for that offence'.

Interviewing officers conducting IUCs need to carefully consider every answer that they get and think about the cumulative effect of the answer. If that answer gives them enough evidence to believe that they have a reasonable prospect of a successful prosecution, they must stop the interview. There is no discretion to do otherwise and the case will not be assisted if they do.

At the end of the interview, invite the interviewee to sign the notes or the seal on the tape. Record the time the interview started and ended, any breaks taken and all persons present.

Evidence of previous condition and failure to comply

To introduce the previous nature of the premises in a prosecution is to open a witness statement with a sentence such as 'XYZ premises is rated as Risk Category 1 because of previous issues with . . .'. Risk assessment-based inspection is standard practice and there should be no problem in introducing these facts.

You can also include a potted inspection history to the statement as an exhibit. For example, 'This visit was the third within a period of x months. The inspection history of the premises is attached as Exhibit AB1', but no further detailed comment on the inspection history should be made.

Claiming costs

'Where ... any person convicted of an offence before the Crown Court, the Court may make such order as to the costs to be paid by the accused to the prosecutor as it considers just and reasonable' (Prosecution of Offences Act 1985 s.18(1)c).

Provided a local authority can show that its claim for costs is just and reasonable, there is no *prima facie* reason for the costs not being awarded. Costs start to run when the EHP enters the premises on a routine or investigatory visit. It is important when collating a costs application to accompany a prosecution that everything that can justifiably be claimed as costs of the investigation is included.

A complete schedule of costs should include:

- costs of the time spent by the EHP in the investigation;
- administrative staff costs in typing notices, making copies and transcribing interview notes;
- costs of photo processing and analysts' cost;
- costs incurred in service of documents by, for example, personal service or recorded delivery post;
- any conferences or meetings with the council's legal team should be costed, with the costs of both environmental health practitioner and lawyer being included in the claim;
- telephone calls relating to the matter, whether internally, with other officers of the council or with the defendant or his legal representative;
- expenses that are incurred in attending on witnesses and taking statements etc.;
- costs of environmental health practitioner, lawyers and witnesses in attending at the hearing.

Try, so far as you can, to estimate what the defendant may claim and cover it off.

For example, if he has had food hygiene training, advise your lawyer. He can then refer to the training and the knowledge that the defendant should have in consequence of it in his address to the court if there is a plea of guilty tendered or in cross-examination if the matter goes to trial.

Include the time spent advising and assisting the defendant. This has a dual purpose: it undermines any claim that he did not know and could not be expected to know what was required, and further rebuts any claim that the council has been unhelpful or uncooperative.

If you can show that he has made substantial profits from his activity, advise the prosecutor to give details of your calculation to the court – it will have to be either accepted or denied and if denied will have to be explained.

Give your lawyer as much background information as you can and make sure potential 'areas of dispute' are addressed either in the facts of the case or during cross-examination.

If the defence, having been sent a copy of the prosecution's application for costs, do not tell the prosecutor or the court that they propose to object to some or all of them before the verdict is delivered, they may not raise objection to any part of the costs application.

Local authorities are required by rules deriving from the *Associated Octel* judgement to serve their schedule of costs on the defendant at the earliest opportunity, in order that the defence may decide what parts, if any, they wish to challenge as unjust or unreasonable (R v Associated Octel[19]).

Defendant's previous convictions etc.

Find out about the defendant's previous convictions and evidence the offences through certificates of conviction from the court producing a schedule of convictions which are live having regard to the Rehabilitation of Offenders act 2012. Then serve the schedule on the offender and obtain agreement that the offences are those which he or she has been convicted.

Having done this you can now put the schedule of offences before the court after the defendant is convicted.

Laying informations

A prosecutor who wants the court to issue a summons must serve an information in writing on the court officer or, unless other legislation prohibits this, present an information orally to the court, with a written record of the allegation that it contains (Criminal Procedure Rules).

An information is a document which states in clear and unambiguous language the offence with which the accused is charged. There should be only one offence in the information. You should accurately describe the name and address of the accused and the specific offence with which the accused is charged in ordinary language, avoiding technical terms. It should include the appropriate section of the Act or regulation contravened. The information should also include the date of offence where possible. If the date is not accurately known, then the phrases 'on or about' or 'between date 1 and date 2' can be used. Finally, refer to the section where the penalty for the offence is contained.

Briefing solicitors or Counsel

Counsel is required to keep a distance from the client – he or she can only be contacted via the instructing solicitor. Solicitors also need to be briefed before they come to court. It is therefore important that the brief contains all of the necessary information, and copies of all of the relevant documents, as well as clear and unequivocal advice as to what the client wants.

Solicitors or Counsel may have no knowledge at all of the case or the facts surrounding it and have to be told everything, particularly those things that, as professionals, EHPs consider to be taken as read, or common knowledge, because for someone outside the field of environmental health they are not.

If you want Counsel to attend at court and represent the local authority, it is necessary to be clear and to say so. It is also necessary to underline what else you want – for example, that an application for costs (to include Counsel's reasonable fees) should be made, that a Confiscation Order should be sought, etc. and to advise what would be acceptable by way of plea bargain, should it be offered. If there is something non-negotiable, say so.

Where Counsel is asked to advise, it is essential to be clear about what advice you actually want. It is important to include some practical information which solicitors or Counsel would be keen to have, such as the fact that the lay witness is reluctant and very nervous, or that the defendant has a violent temper. This allows them to be ready for whatever may happen, and to plan accordingly.

Proving the case

In a court case, all that is necessary is to know what you have to prove and to show, by way of evidence, that you can do so.

It is very important to present all the relevant facts to the court. The courts want a presentation of all of the relevant facts – for example, in the case of environmental damage, the court should be made properly aware of environmental damage and understand how long the environment would take to recover from a particular incident, or the true cost of remediation. A plea of guilty should not prevent all of the relevant facts coming before the court, with sufficient detail, so that a bench of lay magistrates can understand the complexity of the damage caused by the incident and the scale and cost of remediation.

The information laid before the court has to be comprehensive and understandable, and, for example, in the case of environmental damage, it should take account of the following features, which demonstrate aggravating damage:

- adverse effects on human health, flora and fauna;
- noxious and persistent damage by pollutants;
- the need for extensive clean-up operations; and
- the effect of the incident on lawful activities.

Previous convictions of a like nature are also aggravating features.

Any financial gain to the defendant as a result of the incident should be demonstrated as it should be reflected in the fine, such that polluters cannot make enough from their activities to pay any fine and still be in pocket as the result of their activities.

The court wants:

- projections of clean-up costs and times, particularly since the cost of clean-up all too often falls to publicly funded bodies;
- evidence of the polluting incident, e.g. colour photographs or video footage can have a significant effect on those who do not have regular contact with such incidents;
- estimates of the financial benefit to the defendant (these are very valuable).

Prosecution file

The following list is an example of what could be contained in a prosecution file:

- file front sheet to include:
- defendant's name and address, date of birth, gender, occupation, nationality
- defendant's previous convictions
- alleged offences with details of offence and relevant legislation
- date of alleged offence
- date of laying information
- officer's authorisation details
- case summary
- evidence – witness statements, photographs and physical evidence
- witness reliability
- whether evidence is admissible, substantial and reliable
- whether there is a reasonable prospect of conviction
- whether it is in the public interest to prosecute
- initial witness assessment (whether witnesses are vulnerable and require support)
- witness non-availability
- witness statements
- exhibit list, including documents
- record of interviews
- unused material likely to undermine the case
- schedule of prosecution costs
- officer's contemporaneous notes.

If there is a not guilty plea, add:

- schedule of sensitive material
- schedule of non-sensitive unused material
- disclosure report
- update on witness non-availability.

Review of a legal file

If a local authority does not have a senior officer or manager, with appropriate expertise, who can review a case file, the authority could refer the file to another local authority that has such an officer. This ensures an independent review which does not prejudice the defendant or the prosecuting local authority.

Disclosable documents

Disclosure is the process of informing the defence of unused material which has been recorded or retained by the authority and not disclosed in evidence capable of undermining the prosecution case or assisting the defendant, e.g. notebooks, draft versions of witness statements if different to the final version, interview records, communication with experts and any material casting doubt on the reliability of the evidence or a witness.

Each party to a case is required to inform the other side of the existence of all disclosable documents that are or have been in their possession or control.

Always keep your notebook tidy and neat as the relevant pages of officers' notebooks are disclosable. If the book is untidy and disorganised, and the court sees it, it can be used to suggest a less than professional attitude.

Internal office documentation must also be disclosed – file notes or telephone messages written on whatever was to hand when the call was received are part of the case file, and are therefore disclosable.

Anything that relates to a prosecution, or goes onto a case file, has the potential to be disclosed; therefore officers should ensure that nothing other than relevant information is included.

The process of disclosure requires the prosecutor to provide to the defence a copy of all relevant material pertaining to a case. The material is divided into:

- material upon which reliance will be placed;
- material which is disclosed but upon which reliance will not be placed;
- other material which exists, is relevant, but will not be disclosed.

Into this third class goes material for which legal professional privilege is claimed. Advice from in-house lawyers needs to be kept separate from the case notes so that it attracts legal professional privilege.

Sentencing

It will help the court greatly, and shorten hearings, if a case summary can be supplied by the prosecutor at the time of the issue of proceedings. This document should set out the facts on which the prosecution rely and contain aggravating and mitigating features. These are known as Friskies Schedules and have their origin in a prosecution brought by the Health and Safety Executive.[20] In this case, the Court of Appeal made strong recommendations about presentation of health and safety cases, which should assist the judge to understand the aggravating and mitigating features of the case and to pass an appropriately tailored sentence.

The rules are:

1. The prosecutor should list, in writing, the facts of the case and the aggravating features that he says are present.
2. This list should be served on the defendant and the court.
3. In the event of the defendant pleading guilty, the defence should submit a similar document, but in their case outline those mitigating features they say are present.
4. If the prosecution and the defence agree on the aggravating and mitigating features, they should commit this to writing and put the agreement to the court. Where there is no agreement, the judge will hear argument on the disputed points, form a view and advise the parties of his view, which would be relevant in any appeal against the sentence handed down.

The Court of Appeal has also declared what constitutes aggravating and mitigating factors in health and safety cases, in the case of Howe and Son.[21]

The court stepped outside the facts of the case to list other features that might be relevant to sentence, but noted that, depending on the circumstances of the case, it would be for the court to decide whether they were aggravating or mitigating, and to what extent. These included degree of risk and extent of danger (i.e. whether death or injury was foreseeable), gravity and extent of breach, whether it was an isolated incident or continued over a long period, and the defendant's resources and the effect of a fine on its business, but it was noted that this list is not exhaustive.

Whilst both the cited cases were brought under health and safety legislation, it will assist the court to take a similar approach irrespective of the legislation under which proceedings are brought. Applying the same principle by providing information that will assist the magistrates or judge when considering sentence should ensure that the sentence is tailored to reflect the degree of criminality involved in the offence, which is in the interests of prosecution, the defendant and the general public's belief in the criminal justice system.

COURT PROCEDURE

The magistrates' court is the usual court in which local authorities take prosecutions. There are three lay magistrates sitting in court and they are assisted by the clerk of the court. The clerk of the court sits in front of the magistrates and gives them legal and procedural advice. The clerk also deals with general administrative work.

When in court, before the session starts, you should stand up when the magistrates or judge enter the court. The magistrates will normally bow and you should bow immediately afterwards. Sit down after the magistrates sit down.

The clerk will read the charge to the defendant and the defendant will plead guilty or not guilty.

Where the defendant pleads guilty, the prosecuting lawyer will read a statement of facts relating to the case. This includes submissions of the seriousness of the case, notice of the offence, description of the circumstances giving aggravating and mitigating circumstances, costs and compensation claims. The defendant's solicitor will then plead mitigation. The magistrates will then normally impose a sentence but they may ask for more information before passing sentence.

Where a not guilty plea is made, the prosecution will present its case, calling witnesses, and the defendant's lawyer will cross-examine those witnesses. Then the defendant's lawyer presents this defence, including witnesses. The prosecution lawyer will cross-examine the defendant's witness.

An EHP who has given evidence and sits in court may find the evidence of the defendant incredulous, or the comments of the defence solicitor in mitigation farcical, and may feel frustrated in the extreme that he or she cannot counter it. The temptation to roll your eyes or shake your head in disbelief is considerable, as it may be seen as the only way in which the evidence being advanced can be challenged. The temptation must be resisted because behaviour in court that may influence the magistrates or jury can form the ground for appeal. Be impassive to the antics of the defence.

Giving evidence

The prosecuting lawyer will ask the main witness, probably the EHP taking the case, questions about themselves and the details of the offence or offences based on the witness statement. When answering the lawyer, face the bench or the jury to ensure they understand what you are saying as they will be making the decision on the case. In the Crown Court, make eye contact with one or two of the jury. If they think the witness is looking at them, they concentrate. Crisp, sharp evidence engages juries.

If you are asked questions by the magistrates, address them collectively as 'your worships' and individually as 'Sir' or 'Ma'am'. Crown Court judges should be addressed as 'My Lord' or 'My Lady'. Appear serious, caring, flexible, well informed, fair and reasonable in giving evidence.

When giving oral evidence, it is absolutely critical that the witness says enough to establish that those things they are required to prove are indeed demonstrated, and that this is done sufficiently robustly to rebut any challenge that may be put in cross-examination.

When giving evidence in the magistrates' court, watch the clerk. If he or she is writing, they consider the point sufficiently important to record it. If they are not, they do not, so stop. Slow down if the clerk seems to be writing too fast and may be missing some important points. The same applies in the Crown Court and in other forms of administrative tribunal – if the judge or the inspector is writing, the point is important; carry on, slowly. If they are not, stop.

Watch your own lawyer. If they are nodding encouragingly, keep going. If they look as if they want to speak, let them. The rule is to keep it short, keep it simple and keep it relevant.

When it comes to language in court, the rule is no acronyms, no jargon and no unexplained use of technical terminology.

Juries like clear, unfussy information in understandable chunks and not embellished with unnecessary details. The witness needs to get the message across as one who speaks their language.

Opinion

Opinion is a judgement or belief not founded on certainty or proof, or an evaluation or judgement given by an expert. Opinion is the view you hold – the judgement you form, which need not be certain or proved to be true. If an officer can show he is properly informed and has considered his view, his opinion is valid, and he is entitled to act on it. Some legislation, such as the Health and Safety at Work Act 1974, allows officers to act on opinion. See expert witnesses and opinion (p. 64).

Role of the lawyer

The role of the local authority lawyer is to take the local authority's instructions and act so as to give effect to those instructions.

The lawyer must always advise the local authority of the likely outcome of the proposed action, and the likely outcome of any alternatives that he may propose, but at the end of the day it is always for the client (local authority/EHP) to decide which action should be taken.

The instructing department must be free to exercise their professional judgement as to the disposal of matters. The role of the legal department is to follow their clients' instructions. The role of the instructing department is to take legal advice, consider it and then make a decision, taking account of all of the relevant considerations.

Appeals

An appeal is not properly made if it was not made on the proper grounds for an appeal, it was not made in time or it was not paid for in full within the appeal period. Failing any one of these conditions is fatal to the appeal and the local authority does not need to defend against it.

FURTHER READING

This book is an interpretation of particular legislative procedures that are used in environmental health work. For a study of the wider legislative context in which these procedures are to be exercised, the reader is referred to:

1. *Legal Competence in Environmental Health*, Terence Moran, 2002, Taylor & Francis, London.
2. *Clay's Handbook of Environmental Health*, 21st edition, Stephen Battersby, Editor, 2016, Spon Press, London.
3. *Environmental Health and Housing, Issues for Public Health*, 2nd edition, Jill Stewart and Zena Lynch, 2018, Routledge, London.

NOTES

1. van Kamp I. and Davies H. 2008. Environmental noise and mental health: Five-year review and future directions. 9th International Congress on Noise as a Public Health Problem (ICBEN), Foxwoods, CT.
2. Thompson C.W., Roe J., Aspinall P., Mitchell R., Clow A. and Miller D. 2012. More green space is linked to less stress in deprived communities: evidence from salivary cortisol patterns. *Landscape Urban Plann.* 105, 221–229. doi: 10.1016/j.landurbplan.2011.12.015.
3. Beyer K.M., Kaltenbach A., Szabo A., Bogar S., Nieto F.J., and Malecki K.M. 2014. Exposure to neighborhood green space and mental health: evidence from the survey of the health of Wisconsin. *Int J Environ Res Public Health.* 11(3), 3453–3472.

4. Taylor A.F. and Kuo F.E. 2009. Children with attention deficits concentrate better after walk in the park. *J Atten Disord.* 12(5), 402–409.

5. Berman M.G., Kross E., Krpan K.M., Askren M.K., Burson A., Deldin P.J., Kaplan S., Sherdell L., Gotlib I.H. and Jonides J. 2012. Interacting with nature improves cognition and affect for individuals with depression. *J Affect Disord.* 140(3), 300–305.

6. Oudin A., Bråbäck L., Åström D.O. et al. 2016. Association between neighbourhood air pollution concentrations and dispensed medication for psychiatric disorders in a large longitudinal cohort of Swedish children and adolescents. *BMJ Open* 6, e010004. doi: 10.1136/bmjopen-2015-010004.

7. Elliott v Brighton [1980] 79 LGR 506.

8. Swansea City Council v Glass [1992] 3 WLR 123.

9. Kensington & Chelsea RLBC v Khan [2002] EWCA civ 279.

10. Graddage v Haringey LBC [1975] 1 WLR 241, and Sterling Homes (Midlands) Ltd v Birmingham CC [1996] Env. LR 121.

11. R v Mansfield DC ex parte Ashfield Nominees Ltd [1999] E.H.L.R. 290.

12. Swindon BC v Forefront Estates Ltd [2012] EWHC 231.

13. Paddington Borough Council v Finucane [1928] Ch567, Bristol CC v Virgin [1928] 2 KB 622 and Westminster CC v Haymarket Publishing Ltd 1[981] 1 WLR 677.

14. Stewart J. and Lynch Z. 2018. *Environmental Health and Housing: Issues for Public Health.* London: Routledge.

15. R v Johnson [1988] 1 WLR 1377.

16. Beardsley v Giddings (1904) 1KBD847.

17. R v Pontypridd Juvenile Court (unrep).

18. 1976 63 Cr Ap R 132 [1977] QB 224(CA).

19. R v Associated Octel Co Ltd. (Costs) TLR Nov 15 1996 646.

20. R v Friskies Petcare Ltd (2000) 2 CAR(S) 401.

21. R v Howe and Son (Engineering) Ltd [1999] 2 All ER 249.

APPENDICES

Local authority definitions

The definition of a local authority is very similar in all Acts of Parliament and regulations. The references for their definition are below.

- Animal Boarding Establishments Act – section 5(2)
- Animal Welfare (Licensing of Activities Involving Animals) (England) Regulations 2018 – regulation 2
- Anti-Social Behaviour and Policing Act 2014 – sections 57 and 74
- Building Act 1984 – sections 91 and 126
- Caravan Sites and Control of Development Act 1960 – section 29(1)
- Care Act 2014 – section 1(4)
- Clean Air Act 1993 – sections 55 and 64
- Control of Pollution Act 1974 – section 73(1)
- Criminal Justice and Public Order Act 1994 – section 77(6)
- Dangerous Wild Animals Act 1976 – section 7(4)
- Energy Efficiency (Private Rented Property) (England and Wales) Regulations 2015 – regulation 2(1)
- Environment Act 1995 – sections 91, 82 and 105
- Environmental Damage (Prevention and Remediation) Regulations 2009 – regulation 2
- Environmental Permitting (England and Wales) Regulations 2010 – regulation 6
- Environmental Protection Act 1990 – section 78A(9) – Contaminated land
- Environmental Protection Act 1990 – section 4 and section 30 (waste authorities)
- Food Safety Act 1990 – section 5, and Food Hygiene (England) Regulations 2006 – regulation 2
- Health and Safety at Work etc. Act 1974 – section 53
- Housing Act 1985 – section 1
- Housing Act 2004 – section 261
- Housing and Planning Act 2016 – section 56
- Local Government (Miscellaneous Provisions) Act 1982 – sections 13(11) 20(1) and 44(1) 29(4)
- Noise Act 1993 – section 11(1)
- Notification of condensers etc. – regulation 2
- Pet Animals Act – section 7(3)
- Prevention of Damage by Pests Act 1949 – section 1

- Protection against Eviction Act 1977 – section 6
- Public Health Act 1936 – sections 1(2) and 91
- Public Health (Control of Diseases) Act 1984 – section 1
- Refuse Disposal (Amenity) Act 1978 – section 11
- Riding Establishments Act – section 6(4)
- Scrap Metal Dealers Act 2013 – section 22(3)
- Sunday Trading Act 1994 – section 8
- Water Industry Act 1991 – section 219

APPENDIX 2

Service of notices references

- Animal Welfare (Licensing of Activities Involving Animals) (England) Regulations 2018 – regulation 23
- Anti-Social Behaviour and Policing Act 2014 – sections 55 and 79
- Building Act 1984 – sections 92 and 94
- Control of Pollution Act 1974 – section 105
- Energy Efficiency (Private Rented Property) (England and Wales) Regulations 2015 – regulation 3
- Environment Act 1995 – sections 123 and 124
- Environmental Permitting (England and Wales) Regulations 2010 – regulation 10
- Environmental Protection Act 1990 – section 160
- Food Safety Act 1990 – sections 49 and 50
- Health and Safety at Work etc. Act 1974 – section 46
- Housing Act 2004 – section 246
- Housing and Planning Act 2016 – section 42
- Local Government Act 1972 – sections 233 and 234
- Prevention of Damage by Pests Act 1949 – section 10
- Protection against Eviction Act 1977 –section 7
- Public Health Act 1936 – sections 283 to 285
- Public Health (Control of Diseases) Act 1984 – section 60
- Smoke and Carbon Monoxide Alarm (England) Regulations 2015
- Water Industry Act 1991 – section 216

APPENDIX 3

Powers of entry references

The powers of entry in Acts of Parliament relevant to the procedures in this book are provided for duly authorised officers as follows:

- Environment Act 1995, section 108.
- Environmental Protection Act 1990, section 115.
- Clean Air Act 1993, section 56.
- Noise Act 1996, section 10.
- Clean Neighbourhood and Environment Act 2005, sections 77 to 79.

- Control of Pollution Act 1974, sections 91–92.
- Food safety Act 1990, section 32.
- Food Safety and Hygiene (England) Regulations 2013, Regulation 16.
- Health and Safety at Work Act 1974, section 20.
- Housing Act 1985, section 260.
- Housing Act 2004, sections 40, 131, 239 to 242.
- Housing and Planning Act 2016, section 172.
- Building Act 1984, sections 95 and 96.
- Local Government (Miscellaneous Provisions) Act 1982, section 17.
- Energy Efficiency (Private Rented Property) (England and Wales) Regulations 2015.
- Smoke and Carbon Monoxide Alarm (England) Regulations 2015.
- Public Health Act 1936, sections 287–289.
- Caravan Sites and Control of Development Act 1960, section 26.
- Local Government and Housing Act 1989, section 97.
- Local Government (Miscellaneous Provisions) Act 1976, section 29.
- Refuse Disposal (Amenity) Act 1978, section 8.
- Animal Welfare Act 198, sections 60–66A.
- Prevention of Damage by Pests Act 1949, section 22.
- Public Health (Control of Disease) Act 1984, section 61.
- Animal Welfare Act 2006, section 53.
- Scrap Metal Dealers Act 2013, section 16.
- Local Government (Miscellaneous Provisions) Act 1982, section 17.
- Official Feed and Food Controls (England) Regulations 2009, Regulation 9 and 39.

APPENDIX 4

Evidence form – section <XX> Act <year>

Ref.	Offence element (point to prove)	Details of the offence element case law/definitions/ interpretation	Evidence of the offence element			
			Name of witness giving evidence of the element	Relevant exhibit no. (if any), witness initial and number	Page number and paragraph number of witness statement	Page number of interview under caution
1	Name					
2	Address					
3	Date and time of offences					
4						
5						
6						
7						
8						
9						
10						

You must ensure that there is sufficient evidence to prove that each element of the offence has been committed. The evidence must be referred to in a witness statement for it to be admissible in court. The exhibit(s) must be attached to witness statements for it to be admissible in court.

Taken together, the witness statements must cover every element of the offence – so some witnesses may prove some elements and others may prove other elements or reinforce the same elements where necessary.

PART 2

ENVIRONMENTAL PROTECTION

Chapter 1

AIR QUALITY

HEIGHT OF CHIMNEYS SERVING FURNACES

References

Clean Air Act 1993, sections 14 and 15
Clean Air (Height of Chimneys) (Exemption) Regulations 1969
Clean Air (Units of Measurement) Regulations 1992
Clean Air (Miscellaneous Provisions) (England) Regulations 2014
HMIP Technical Guidance Note D1 (01)
The third edition of the Chimney Heights Memorandum (HMSO, 1981)

Extent

These provisions apply in England, Scotland and Wales.

Scope

This procedure applies to chimneys attached to furnaces used to burn:

(a) pulverised fuel; or
(b) any other solid matter at a rate of 45.4 kilograms/h or more; or
(c) any liquid or gaseous matter at a rate of 366.4 kilowatts/h or more (section 14(1) and (2)).

Approval of chimney height

In relation to chimneys from these furnaces, the occupier of a furnace of a building or the person having possession of any fixed boiler or industrial plant outside of a building, must not use that furnace unless the height of its chimney has been approved by the local authority and any conditions of approval are complied with (sections 14(2)–(5)).

In relation to furnaces served by a chimney which was constructed before 1 April 1969, this requirement only applies where the combustion space of the furnace has been increased since that date or the furnace served by that chimney has been replaced by a larger one (schedule 5, paragraph 7).

Exemptions

The Clean Air (Miscellaneous Provisions) Regulations 2014 specify the following uses of boilers or industrial plant where chimney height approval is not required:

(a) temporarily replacing any other boiler or plant which is under inspection, maintenance or repair, or is being rebuilt or replaced;
(b) providing a temporary source of heat or power during building operations or work of engineering construction or for investigation and research;
(c) providing products of combustion to heat other plant to an operating temperature;
(d) providing heat or power for mobile or transportable agricultural plant (regulation 4(2)).

Criteria for decision

The local authority is required to reject the application unless it is satisfied that the height of the chimney will be sufficient to prevent, as far as practicable, the smoke, grit, dust, gases or fumes from becoming prejudicial to health or a nuisance having regard to:

(a) the purpose of the chimney;
(b) the position and description of nearby buildings;
(c) the levels of neighbouring ground;
(d) any other matters requiring consideration in the circumstances (section 15(2)).

Refusals and conditions

If the application is not approved or conditions are attached to approval, the local authority is required to give written notification of its reasons to the applicant within four weeks of the application. Conditions may relate to the rate and/or quality of the emissions from the chimney. Where the application is rejected, the local authority is also required to state the lowest height which it is prepared to accept conditionally, unconditionally or both (section 15(4) and (5)).

Similar requirements are placed upon the Secretary of State in relation to decisions following appeals (section 15(7) and (8)).

HEIGHT OF CHIMNEYS NOT SERVING FURNACES

References

Clean Air Act 1993, section 16
HMIP Technical Guidance Note D1 (01)

Extent

These provisions apply to England, Scotland and Wales.

FC3 Height of chimneys serving furnaces – sections 14 and 15 Clean Air Act 1993

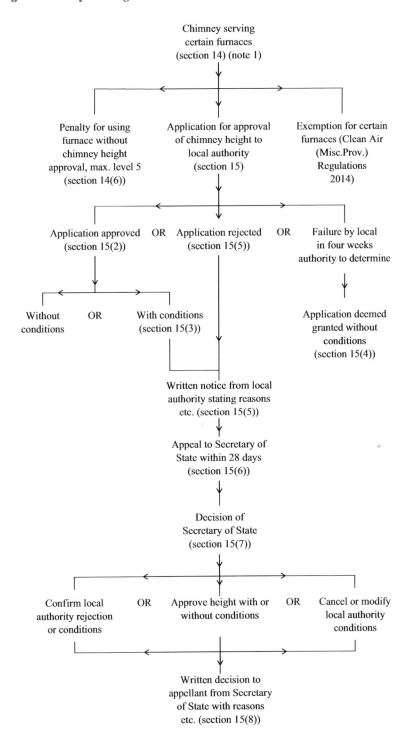

Note

1. For height of chimneys not serving furnaces, see FC4.

FC4 Height of chimneys not serving furnaces – section 16 Clean Air Act 1993

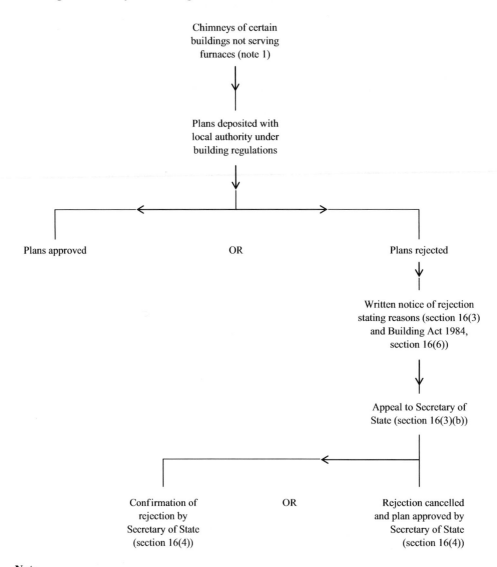

Chimneys of certain
buildings not serving
furnaces (note 1)

Plans deposited with
local authority under
building regulations

Plans approved   OR   Plans rejected

Written notice of rejection
stating reasons (section 16(3)
and Building Act 1984,
section 16(6))

Appeal to Secretary of
State (section 16(3)(b))

Confirmation of   OR   Rejection cancelled
rejection by   and plan approved by
Secretary of State   Secretary of State
(section 16(4))   (section 16(4))

Note

1. For height of chimneys serving furnaces, see FC3.

Scope

This procedure applies where plans deposited under building regulations for buildings other than:

(a) residences; or
(b) shops; or
(c) offices

show the proposed construction of a chimney, other than one serving a furnace for carrying smoke, grit, dust or gases from a building (section 16(1)).

Where plans are not deposited because the 'building notice' procedure is being used, the application of this chimney height approval procedure may be dealt with as a linked power.

Criteria for decision

The local authority is required to reject the plans unless it is satisfied that the height of the chimney will be sufficient to prevent, as far as practicable, the smoke, grit, dust or gases from becoming prejudicial to health or a nuisance having regard to:

(a) the purpose of the chimney;
(b) the position and description of buildings nearby;
(c) the levels of neighbouring ground;
(d) any other matters requiring consideration in the circumstances (section 16(2)).

CONTROL OF GRIT AND DUST FROM FURNACES

References

Clean Air Act 1993, sections 5 to 9
Clean Air (Arrestment Plant) (Exemption) Regulations 1969 – Scotland and Wales
Clean Air (Miscellaneous Provisions) (England) Regulations 2014
Clean Air (Emission of Grit and Dust from Furnaces) Regulations 1971
Clean Air (Units of Measurement) Regulations 1992

Extent

These provisions apply to England, Scotland and Wales.

Scope

The physical control of grit and dust from furnaces is part of the legislation and policy to protect and improve air quality and protect health of the local population. Arrestors are the current method of arresting dust and grit from various furnaces.

Requirement to fit arrestors

Furnaces, except those designed solely or mainly for domestic purposes and with a maximum heating capacity of less than 16.12 kilowatts, used in buildings to burn:

(a) pulverised fuel; or
(b) any other solid matter at a rate of 45.4 kilograms/h or more; or
(c) any liquid or gaseous matter at a rate of 366.4 kilowatts/h or more

must be provided with plant for arresting grit and dust which has been approved by the local authority and be properly maintained and used (section 6(1)).

Domestic furnaces burning pulverised fuel or solid fuel at a rate of 1.02 tonnes/hour or more are also required to be provided with grit arrestment plant approved by the local authority (section 8(1)).

Exemptions

Exemption from the requirement to fit arrestment plant may be by either:

(a) falling into the categories of exempted furnace set out in schedule 2 Clean Air (Miscellaneous Provisions) (England) Regulations 2014, which also set out the information to be provided in the application for exemption; or
(b) specific application for exemption to the local authority for furnaces falling outside the regulations in (a) above, in which case the local authority will need to be satisfied that the furnace can be operated without being prejudicial to health or a nuisance without arrestment plant (section 7).

Approval by local authority to proposed equipment

In situations where arrestment plant is required for either domestic or non-domestic furnaces, the plant must be approved by the local authority or installed in accordance with plans and specifications submitted to and approved by the local authority (sections 6(1) and 8(1)).

Emission of grit and dust

The occupier of any building containing a furnace, other than one designed solely or mainly for domestic purposes, with a maximum heating capacity less than 16.12 kilowatts/h which burns solid, liquid or gaseous matter must take all practicable means for minimising the emission of grit and dust from the chimney; otherwise he is committing an offence (section 5(5)).

This general requirement does not, however, apply to furnaces where maximum emission limits have been specified by the Clean Air (Emission of Grit and Dust from Furnaces) Regulations 1971.

Although there is no similar requirement for those domestic furnaces which are required to have arrestment plant, that plant must be properly maintained and used (section 8(1)).

Defences

In proceedings for an offence under the regulations, it is a defence to prove that the best practicable means have been used to minimise the emission (section 5(4)).

FC5 Control of grit and dust from furnaces – sections 6 to 9 Clean Air Act 1993

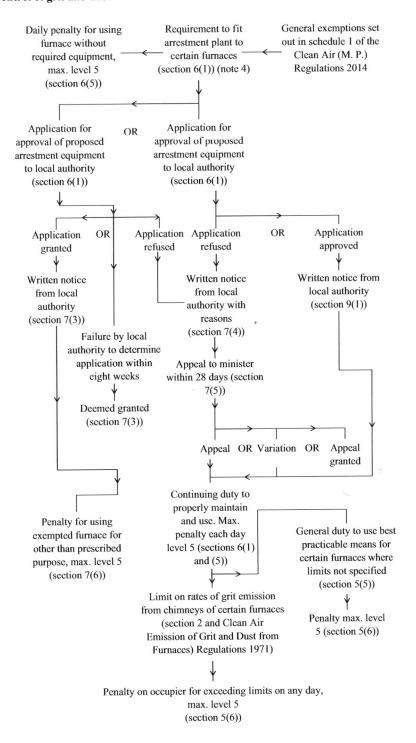

Notes
1. For grit, dust etc. emitted in smoke constituting statutory nuisance, see FC128.
2. For grit, dust etc. not emitted in smoke causing statutory nuisance, see FC128.
3. For measurement etc. of grit and dust, see FC6.
4. Local authority's powers to obtain information about furnaces and fuel under section 12 apply to this procedure.

Furnaces outside buildings

These provisions also apply to furnaces attached to a building or fixed to or installed on any land and in these cases the person responsible for compliance is the person having possession of the boiler or plant (section 13).

MEASUREMENT OF GRIT AND DUST FROM FURNACES

References

Clean Air Act 1993, sections 10 to 12
Clean Air (Measurement of Grit and Dust from Furnaces) Regulations 1971
Clean Air (Units of Measurement) Regulations 1992

Extent

These provisions apply to England, Wales and Scotland.

Scope

These provisions support the control of grit and dust provisions on p. 87.

Information

In connection with the provisions relating to grit arrestment plant and the measurement of grit and dust, local authorities are empowered to require any occupier to give them reasonable information about any furnaces used and the fuels burned following written notice giving not less than 14 days to do so (section 12(1)).

Measurement by occupiers

These provisions apply only to furnaces used to burn:

(a) pulverised fuel; or
(b) any other solid matter at a rate of 45.4 kilograms/h or more; or
(c) any liquid or gaseous matter at a rate of 366.4 kilowatts/h or more.

The regulations detailing measurement etc. are applied to an individual furnace by written notice from the local authority to the occupier (or, in relation to outside furnaces, the person having possession) (sections 10(1) and 13(2)).

Requirement for local authority to undertake measurement

This applies only to furnaces burning:

(a) solid matter, other than pulverised fuel, at a rate less than 1.02 tonnes/h; or
(b) liquid or gaseous matter at a rate less than 8.21 megawatts/h.

In these cases, the occupier by written notice may request the local authority to undertake the making and recording of measurements (section 11(2)).

FC6 Measurement of grit and dust from furnaces – sections 10 and 11 Clean Air Act 1993

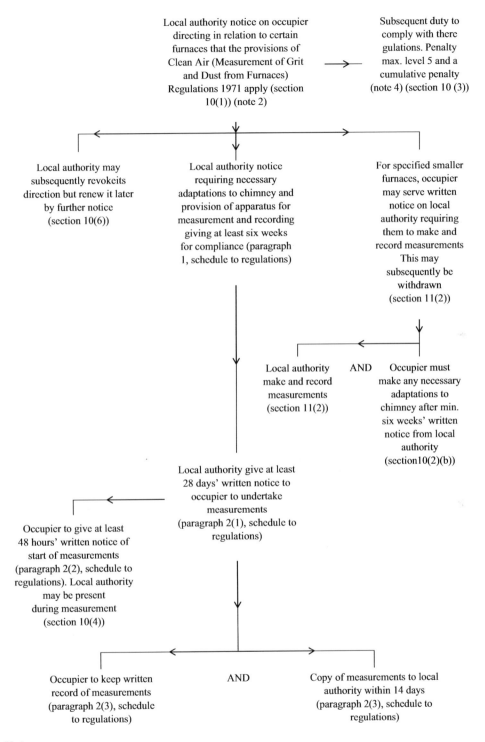

Local authority notice on occupier directing in relation to certain furnaces that the provisions of Clean Air (Measurement of Grit and Dust from Furnaces) Regulations 1971 apply (section 10(1)) (note 2)

→ Subsequent duty to comply with there gulations. Penalty max. level 5 and a cumulative penalty (note 4) (section 10 (3))

Local authority may subsequently revokeits direction but renew it later by further notice (section 10(6))

Local authority notice requiring necessary adaptations to chimney and provision of apparatus for measurement and recording giving at least six weeks for compliance (paragraph 1, schedule to regulations)

For specified smaller furnaces, occupier may serve written notice on local authority requiring them to make and record measurements This may subsequently be withdrawn (section 11(2))

Local authority make and record measurements (section 11(2)) AND Occupier must make any necessary adaptations to chimney after min. six weeks' written notice from local authority (section10(2)(b))

Local authority give at least 28 days' written notice to occupier to undertake measurements (paragraph 2(1), schedule to regulations)

Occupier to give at least 48 hours' written notice of start of measurements (paragraph 2(2), schedule to regulations). Local authority may be present during measurement (section 10(4))

Occupier to keep written record of measurements (paragraph 2(3), schedule to regulations)

AND

Copy of measurements to local authority within 14 days (paragraph 2(3), schedule to regulations)

Notes
1. For general powers of local authority regarding information about atmosphere pollution, see FC13.
2. There is no appeal against the local authority notice applying the provisions of the regulations.
3. These provisions also apply to outdoor furnaces, section 13.
4. In relation to offences which are a repetition or continuation of an earlier offence, there is a penalty of £50 max. daily.

The occupier remains responsible, however, for any necessary adaptations to the chimney to allow measurements to be taken.

Measurements etc.

The Clean Air (Measurement of Grit and Dust from Furnaces) Regulations 1971 lay down the detailed requirements. Although power is given to the Secretary of State in the Act to include the measurement of fumes, the regulations apply only to grit and dust.

The notice from the local authority requiring measurements may require them to be taken from time to time or at staged intervals, but the local authority cannot normally require measurements from a chimney in excess of once each three months unless the local authority thinks that this is necessary to obtain the true level of emission (paragraph 2(4) schedule to regulations).

The records to be kept of measurements must show:

(a) the date;
(b) the number of furnaces discharging into the chimney;
(c) results of measurements in pounds per hour grit and dust emitted and the percentage of grit in the solids emitted (paragraph 2(3) schedule to the 1971 regulations).

PROHIBITION OF DARK SMOKE ETC. FROM CHIMNEYS

References

Clean Air Act 1993, section 1
Dark Smoke (Permitted Periods) Regulations 1958
Dark Smoke (Permitted Periods) (Vessels) Regulations 1958 – (Scotland and Wales)
Clean Air (Miscellaneous Provisions) (England) Regulations 2014
The Clean Air (Emission of Dark Smoke) Regulations (Northern Ireland) 1981

Extent

These provisions apply in England, Scotland and Wales. Similar provisions occur in Northern Ireland.

Scope

It is an offence to emit dark smoke from the chimney of any building (including houses) on any one day and this includes chimneys serving the furnaces of boilers or industrial plant which are either attached to buildings or for the time being fixed to or installed on any land (section 1). Emissions lasting not longer than the following are, however, to be left out of account:

(a) Dark Smoke (Permitted Periods) Regulations 1958 – applies to emissions from chimneys other than on vessels:

No. of furnaces	Aggregate in any eight-hour period	
	Without soot blowing	With soot blowing
1	10 min.	14 min.
2	18 min.	25 min.
3	24 min.	34 min.
4 or more	29 min.	41 min.

FC7 Prohibition of dark smoke etc. from chimneys – section 1 Clean Air Act 1993

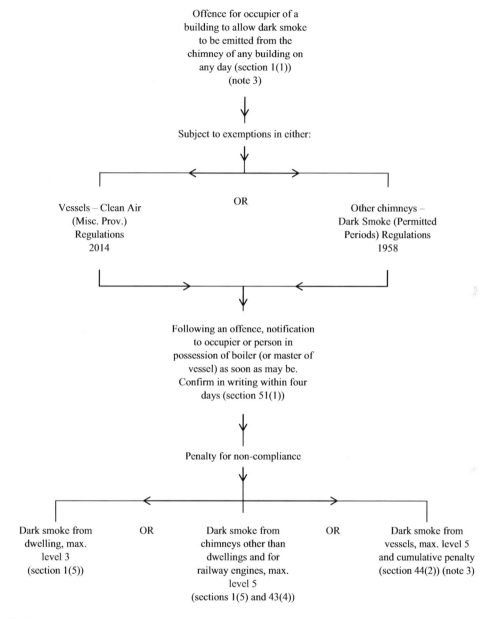

Offence for occupier of a
building to allow dark smoke
to be emitted from the
chimney of any building on
any day (section 1(1))
(note 3)

Subject to exemptions in either:

OR

Vessels – Clean Air
(Misc. Prov.)
Regulations
2014

Other chimneys –
Dark Smoke (Permitted
Periods) Regulations
1958

Following an offence, notification
to occupier or person in
possession of boiler (or master of
vessel) as soon as may be.
Confirm in writing within four
days (section 51(1))

Penalty for non-compliance

Dark smoke from
dwelling, max.
level 3
(section 1(5))

OR

Dark smoke from
chimneys other than
dwellings and for
railway engines, max.
level 5
(sections 1(5) and 43(4))

OR

Dark smoke from
vessels, max. level 5
and cumulative penalty
(section 44(2)) (note 3)

Notes

1. For dark smoke from industrial or trade premises (other than from chimneys), see section 2 of the Clean Air Act 1993 (FC8).
2. This procedure also applies to vessels and railway engines (Clean Air Act 1993, sections 43 and 44).
3. In relation to offences which are a repetition or continuation of an earlier offence, there is a penalty of max. £50 daily (section 50).

but in any event emissions must not exceed four minutes' continuous dark smoke (other than by soot blowing) or two minutes' black smoke in aggregate in any 30-minute period.

(c) Clean Air (Miscellaneous Provisions) (England) Regulations 2014 (regulation 2). These regulations apply only to vessels and details of the exemptions for the emission of dark smoke are set out in schedule 1 and vary depending upon the class of boiler. The exemptions do not permit the continuous emission of dark smoke for in excess of four minutes (ten minutes for certain natural-draught, oil-fired boilers) or the emission of black smoke for more than three minutes in any 30-minute period.

Defences

In any proceedings under section 1, it will be a defence to prove that the contravention was solely due to:

(a) lighting up from cold and all practicable steps were taken to prevent or minimise the emission; or

(b) failure etc. of the furnace that could not reasonably have been foreseen or prevented; or

(c) suitable fuel being unavailable and all practicable steps taken to prevent or minimise the emission; or

(d) any combination of (a)–(c) above; or

(e) there was a failure by the local authority to comply with the notification provisions (below) (sections 1(4) and 51(3)).

Person responsible

This is:

(a) for chimneys of buildings – the occupier of the building (section 1(1));

(b) for chimneys not attached to buildings – the person having possession of the boiler or plant (section 1(2));

(c) for vessels – the owner and master or other person in charge of the vessel (section 44(2));

(d) for railway engines – the owner (section 43(2)).

Notification of offences

Upon becoming aware that an offence under section 1 is being or has been committed, an authorised officer of the local authority must:

(a) notify the person responsible as soon as may be; and

(b) confirm the offence in writing before the end of the four days following the day on which the offence was committed (section 51).

PROHIBITION OF DARK SMOKE ETC. FROM INDUSTRIAL AND TRADE PREMISES (OTHER THAN FROM CHIMNEYS, E.G. TRADE BONFIRES)

References

Clean Air Act 1993 sections 2 and 33
Clean Air (Emission of Dark Smoke) (Exemption) Regulations 1969

Extent

These provisions apply in England, Wales and Scotland.

Scope

These provisions apply only to industrial and trade premises where the smoke is emitted other than from a chimney, e.g. bonfires etc. It is an offence for dark smoke to be omitted from these premises but this does not apply to dark smoke from a chimney covered by section 1 and to exempted material prescribed in regulations by the Secretary of State (sections 2(1) and (2)). Their main use is to control industrial bonfires.

Industrial or trade premises are defined as meaning premises used for any industrial or trade purposes or premises not so used on which matter is burnt in connection with any industrial or trade process (section 2(6)).

Where any material is burned on industrial and trade premises and this is likely to produce dark smoke, there is an assumption that dark smoke has been emitted unless the occupier or person who caused or permitted the emission can show that no dark smoke was emitted (section 2(3)).

Exemptions

These are specified in the Clean Air (Emission of Dark Smoke) (Exemption) Regulations 1969 and include:

- demolition waste (ABC);
- waste explosive (AC);
- matter burnt in connection with fire research and training (C);
- tar, pitch, asphalt and the like burnt in the laying of surfaces (C);
- carcases of diseased animals or poultry (AC);
- contaminated containers associated with veterinary or agricultural purposes (ABC).

Certain conditions must occur for these exemptions to apply. The conditions are listed below and the letters in brackets above must apply for the exemption to be valid.

- Condition A: that there is no other reasonably safe and practicable method of disposing of the matter.
- Condition B: that the burning is carried out in such a manner as to minimise the emission of dark smoke.
- Condition C: that the burning is carried out under the direct and continuous supervision of the occupier of the premises concerned or a person authorised to act on his behalf.

FC8 Prohibition of dark smoke etc. from industrial and trade premises (other than from chimneys) – section 2 Clean Air Act 1993

Offence for occupier of industrial or trade
premises to emit dark smoke on any day
(section 2) (note 1)

Exemptions provided by Clean Air
(Emission of Dark Smoke) (Exemption)
Regulations 1969

Notification of offence to occupier by local
authority confirmed in writing within four
days of offence (section 51(1))

On summary conviction there is a fine (section 2(5)) (note 3)

Notes

1. For procedure relating to dark smoke emitted from chimneys, see FC7.
2. For special offences relating to cable burning, see section 33.
3. Level 5 on the Standard Scale.

Defences

It is a defence in proceedings to prove that:

(a) the contravention was inadvertent and that all practicable steps had been taken to prevent or minimise the emission of dark smoke; or

(b) the notification procedure had not been followed (section 2(4)).

Notification of offences

The notes on FC8 are also applicable here (section 51).

There is a specific offence for cable burning. Section 33 of the Clean Air Act 1993 makes it an offence liable to a fine of level 5 unless it is subject to Part I of the Environmental Protection Act 1990 or section 2 of the Pollution Prevention and Control Act 1999.

SMOKE CONTROL AREAS

Although these provisions remain in force, those relating to the making of new smoke control areas (SCAs) are rarely used since the problem of smoke pollution from domestic chimneys has been largely overcome by earlier use of orders and the changing nature of domestic fuels. However, there is still the need to enforce existing orders and these procedures indicate how this may be done.

The current approach to the reduction of air pollution is through air quality management and the making of air quality management areas (see p. 113).

New provisions to be brought in to deal with the air pollution from wood burning stoves by 2022 may change enforcement provisions and support for the upgrading of existing stoves.

References

Clean Air Act 1993, sections 18 to 29

Smoke Control Areas (Exempted Fireplaces) (England) Order 2014. Similar regulations for Wales and Scotland

Smoke Control Areas (Authorised Fuels) (England) (No. 2) Regulations 2014. Similar regulations for Wales and Scotland

Extent

These provisions apply to England, Scotland and Wales. Similar provisions apply in Northern Ireland.

Directions by Secretary of State

Where the Secretary of State (or in Scotland the Scottish Environmental Protection Agency) considers it expedient to abate pollution by smoke and the local authority concerned has not exercised its powers to make SCAs sufficiently or at all, he may, after consulting the local authority, direct it to submit a programme to him. Following approval of its proposals by the Secretary of State or, in its default, following the determination of a programme by the

Secretary of State, the local authority is under a statutory obligation to comply (Clean Air Act 1993 section 19).

These powers have been used to secure compliance with the EC Air Quality Directive on sulphur dioxide and suspended particulates 80/779/EEC.

Smoke control areas

These may relate to the whole or part of a local authority area (section 18(1)) and in them the emission of smoke from the chimney of a building, or a chimney serving a fixed boiler or industrial plant is generally prohibited (sections 20(1) and (2)).

Exemptions

Exemptions from the general requirement not to emit smoke can come about in the following ways:

(a) The burning of an authorised fuel, i.e. one designated by one of the SCA (Authorised Fuels) Regulations, is a defence against proceedings (section 20(4) and (6)).

(b) By the use of one of a class of fireplaces exempted by a SCA (Exempted Fireplaces) Order following the Secretary of State being satisfied that the appliance is capable of burning other than authorised fuels without producing any or a substantial quantity of smoke (section 21).

(c) Exemption granted by the local authority in respect of a specified or class of building or specified or class of fireplaces, upon such conditions as the local authority may specify (section 18(2)).

(d) Exemptions granted by the local authority for the purpose of investigation and research (section 45).

Offences

Within a SCA, the following are offences:

(a) emitting smoke from the chimney of a building etc. (section 20(1) and (2)) subject to the exemptions listed above;

(b) acquiring solid fuel, other than authorised fuel for use in a building, fixed boiler or industrial plant in a SCA, unless intended for use in an exempted building or fireplace (section 23(1)(a) and (b));

(c) selling by retail solid fuel, other than authorised fuel, for delivery to a building or for any fixed boiler or industrial plant in a premises in a SCA. In this case the person cannot be convicted if he had reasonable grounds for believing:

    (i) that the building was exempted; or

    (ii) that the fireplace or plant for which the fuel was intended was exempted (section 23(1)(c) and (5)).

Upon becoming aware that an offence under section 20 relating to emissions of smoke in SCAs has been committed, an authorised officer of the local authority must:

(a) notify the person responsible as soon as may be; and

(b) confirm the offence in writing before the end of the four days following the day on which the offence was committed (section 51).

Adaptation of fireplaces in private dwellings

The notice procedure may be used where adaptations are required to avoid contraventions of section 20 and these may be served on the owner or occupier of the private dwelling either between the date of making the order and its operation, or after the date of operation. The period allowed for compliance must be reasonable and not less than the appeal period, i.e. 21 days. The provisions of Part 12 of the Public Health Act 1936 apply to the service and enforcement of these notices, but costs of carrying out work in default can be recovered only to an amount up to three-tenths of the expenditure by the local authority (section 24(3)).

Grants

The local authority is required to pay grants to owners or occupiers of private dwellings to assist with the adaptation of fireplaces where this is required to avoid contraventions of section 20, provided that the works are carried out to the local authority's satisfaction before the order becomes operative, unless the local authority has served a notice as above (section 25).

The normal grant is seven-tenths of necessary expenditure, but the local authority may pay more at its discretion. Payments to occupiers other than owners for move-able cooking or heating appliances are made at seven-twentieths initially and the remainder after two years, provided that the appliance is still present (schedule 2, paragraph 1). The local authority has a discretion to make up to 100 per cent grants for adaptations in churches, chapels or buildings used by charities (section 26).

Exchequer contributions towards the local authority's costs ceased with orders coming into operation from 31 March 1996 other than in exceptional cases where, for example, the local authority may breach 80/779/EEC and funding is not available from the local authority's own resources. The Secretary of State may fund the implementation of a Smoke Control Area (schedule 2 para. 4).

Meaning of 'adaptations'

In relation to the payment of grants and to notices, 'adaptations' include:

(a) adapting or converting a fireplace; or
(b) replacing any fireplace by another fireplace or by some other means of heating or cooking; or
(c) altering any chimney which serves any fireplace; or
(d) providing gas ignition, electric ignition or any other special means of ignition; or
(e) carrying out any operation incidental to any of the operations (a)–(d) above, and includes works both inside and outside that dwelling.

The works must be reasonably necessary in order to make what is, in all circumstances, suitable provision for heating and cooking without contravention of section 20 (section 27).

Designated appliances

Classes of heating appliance designated by a local authority for its area, or by the minister either generally or in the area of defined local authorities, which would impose undue strain on the fuel resources available, do not qualify for local authority grant (schedule 2, paragraph 2).

FC9 Smoke control areas: directions by the Secretary of State – section 19 Clean Air Act 1993

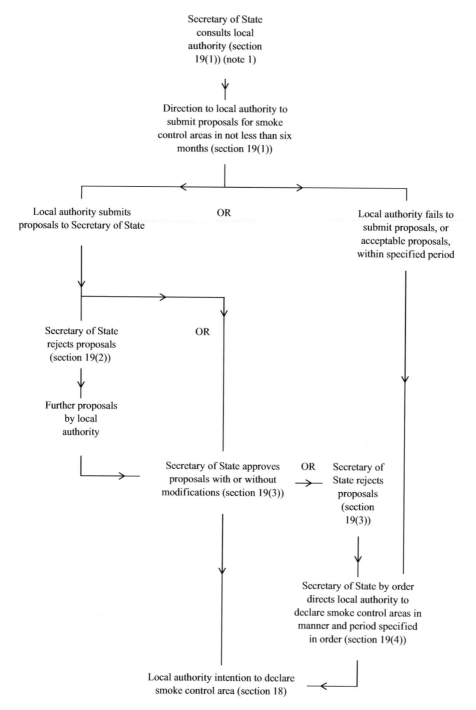

Note

1. In Scotland, the powers of the Secretary of State are exercised by the Scottish Environmental Protection Agency and the procedure chart needs to be read accordingly.

FC10 Smoke control areas: declaration by local authority – section 19 Clean Air Act 1993

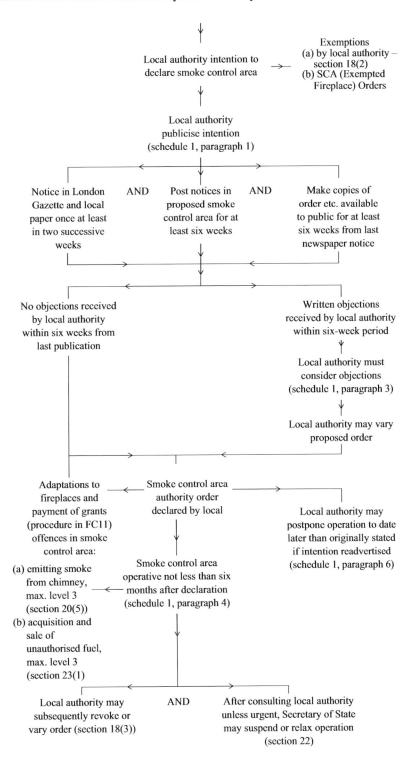

Local authority intention to
declare smoke control area

Exemptions
(a) by local authority –
section 18(2)
(b) SCA (Exempted
Fireplace) Orders

Local authority
publicise intention
(schedule 1, paragraph 1)

Notice in London
Gazette and local
paper once at least
in two successive
weeks

AND

Post notices in
proposed smoke
control area for at
least six weeks

AND

Make copies of
order etc. available
to public for at least
six weeks from last
newspaper notice

No objections received
by local authority
within six weeks from
last publication

Written objections
received by local authority
within six-week period

Local authority must
consider objections
(schedule 1, paragraph 3)

Local authority may vary
proposed order

Adaptations to
fireplaces and
payment of grants
(procedure in FC11)
offences in smoke
control area:

(a) emitting smoke
from chimney,
max. level 3
(section 20(5))
(b) acquisition and
sale of
unauthorised fuel,
max. level 3
(section 23(1)

Smoke control area
authority order
declared by local

Local authority may
postpone operation to date
later than originally stated
if intention readvertised
(schedule 1, paragraph 6)

Smoke control area
operative not less than six
months after declaration
(schedule 1, paragraph 4)

Local authority may
subsequently revoke or
vary order (section 18(3))

AND

After consulting local authority
unless urgent, Secretary of State
may suspend or relax operation
(section 22)

FC11 Smoke control areas: adaptations to fireplaces etc. – section 24 Clean Air Act 1993

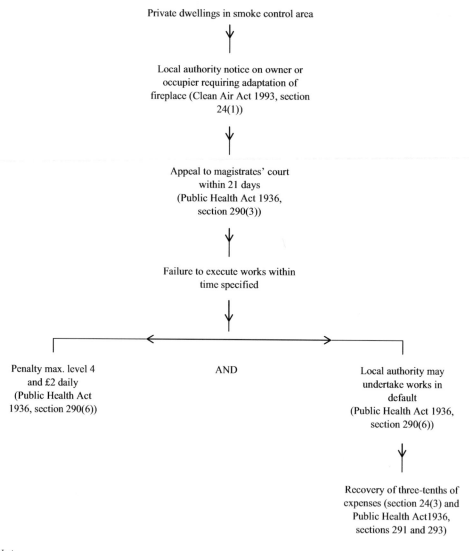

Private dwellings in smoke control area

Local authority notice on owner or
occupier requiring adaptation of
fireplace (Clean Air Act 1993, section
24(1))

Appeal to magistrates' court
within 21 days
(Public Health Act 1936,
section 290(3))

Failure to execute works within
time specified

Penalty max. level 4
and £2 daily
(Public Health Act
1936, section 290(6))

AND

Local authority may
undertake works in
default
(Public Health Act 1936,
section 290(6))

Recovery of three-tenths of
expenses (section 24(3) and
Public Health Act1936,
sections 291 and 293)

Note

1. The provisions of Part 12 Public Health Act 1936 relating to appeals against, and the enforcement of, notices
   requiring the execution of works apply to notices in this procedure (section 24(2)) (p. 21).

BURNING OF CROP RESIDUES

References

Environmental Protection Act 1990, section 152
The Crop Residues (Burning) Regulations 1993

Extent

The procedure applies in England, Wales and Scotland but not in Northern Ireland (section 164).

Scope

There is a prohibition on the burning of the following crop residues:

(a) cereal straw;
(b) cereal stubble;
(c) residues of:

    (i) oil-seed rape;
    (ii) field beans harvested dry; and
    (iii) peas harvested dry (regulation 4 and schedule 1).

Exemptions

The burning of the crop residues specified above is allowed, however, for the purposes of:

(a) education and research;
(b) disease control or the elimination of plant pests where a notice has been served under article 22 of the Plant Health (Great Britain) Order 1993;
(c) the disposal of straw stack remains or broken bales (regulation 4).

Following a review of the exemptions from these regulations in 1996, the government announced that the non-inclusion of linseed residues would continue. Farmers are, however, urged to follow the safety rules on allowable burning.

Restriction on allowable burning

Where any of the specified crop residues are to be burned, this must be carried out in strict compliance with the following restrictions and requirements:

1. No crop residue may be burned:

    (a) during the period between one hour before sunset and the following sunrise; or
    (b) on any Saturday, Sunday or bank holiday.

2. No crop residue may be burned if the area to be burned extends, in the case of cereal straw or cereal stubble, to more than ten hectares, and in any other case to more than 20 hectares.

3. No crop residue may be burned unless:

    (a) the area to be burned is surrounded by a fire-break, which borders on that area and which, in the case of cereal straw or cereal stubble, shall be at least ten metres wide and in any other case at least five metres wide;

    (b) any building, structure or other thing mentioned in paragraph 4(b) or (c) below which lies within the area to be burned is surrounded by a fire-break of the relevant width referred to in sub-paragraph (a) above which borders on any crop residues which are to be burned; and

    (c) in the case of any land ('intervening land') between a fire-break to be established in accordance with sub-paragraph (a) or (b) above and any other land or any building, structure or other thing mentioned in paragraph 4(b), (c) or (d) below which lies within the relevant distance there mentioned of the area to be burned or, as the case may be, of any crop residues within that area:

        (i) the intervening land is cleared of all crop residues; or

        (ii) all crop residues on the intervening land are incorporated into the soil before burning takes place.

4. No crop residue may be burned if the area to be burned:

    (a) is less than 150 metres from any other area in which crop residues are being burned;

    (b) in the case of cereal straw or cereal stubble, is less than 15 metres, and in any other case less than five metres from:

        (i) the trunk of any tree (including any tree in coppice or scrubland);

        (ii) any hedgerow;

        (iii) any fence not the property of the occupier of the land upon which the burning is carried out;

        (iv) any pole which is or may be used to carry telegraph or telephone wires;

        (v) any electricity pole, pylon or substation;

    (c) in the case of cereal straw or cereal stubble, is less than 50 metres, and in any other case less than 15 metres from:

        (i) any residential building;

        (ii) any structure having a thatched roof;

        (iii) any building, structure, fixed plant or machinery which could be set alight or damaged by heat from the fire;

        (iv) any scheduled monument which could be set alight by the fire;

        (v) any stack of hay or straw;

        (vi) any accumulation of combustible material other than crop residues removed in the making of a fire-break;

        (vii) any mature standing crop;

        (viii) any woodland or land managed as a nature reserve;

        (ix) any building or structure containing livestock;

        (x) any oil or gas installation on or above the surface of the ground; or

    (d) is less than 100 metres from:

        (i) any motorway;

        (ii) any dual carriageway;

        (iii) any A-road;

        (iv) any railway line.

5. No crop residue may be burned unless all persons concerned in the burning operation are familiar with the provisions of the regulations and, except where an emergency arising during the operation renders it impracticable, each area to be burned is supervised by at least two responsible adults, one of whom having experience of burning crop residues shall be in general control of the operation.

6. No crop residue may be burned unless there is available at the area being burned:

   (a) not less than 1,000 litres of water in one or more mobile containers, together with means of dispensing the water for fire-fighting purposes in a spray or jet at a rate of 100 litres per minute; and

   (b) not fewer than five implements suitable for use in fire-beating.

7. No crop residue may be burned unless every vehicle used in connection with the burning is equipped with a suitable and serviceable fire extinguisher.

8. No crop residue may be burned unless reasonable precautions have been taken to ensure that the fire will not cross a fire-break.

9. Ashes of burnt cereal straw or cereal stubble shall not, without reasonable excuse, be allowed to remain on the soil for longer than 24 hours after the time of commencement of the burning, but shall be incorporated into the soil:

   (a) within that period; or

   (b) in a case where to do so would be likely, having regard to wind conditions, to cause nuisance, as soon as conditions allow (schedule 2).

Notification of intended burning

Before any allowable burning takes place and so far as is reasonably practicable, notice must be given to:

   (a) the Environmental Health Department of the District Council where the burning is to take place;

   (b) occupiers of any adjacent premises; and

   (c) the air traffic control of any aerodrome with a perimeter fence within 800 metres of the area to be burned.

Such notice is to be given at least one hour or not more than 24 hours before the commencement of the burning (schedule 2, paragraph 6).

Definitions

Crop residue means straw or stubble or any other crop residue remaining on the land after harvesting of the crop grown thereon.

Fire-break means an area of ground of which the surface consists wholly or mainly of substances other than combustible material (regulation 2(1)).

CABLE BURNING

A person who burns insulation from a cable with a view to recovering metal from the cable shall be guilty of an offence unless the burning is part of a process subject to Part I of the Environmental Protection Act 1990 or an activity subject to regulations under section 2

FC12 Burning of crop residues – Crop Residues (Burning) Regulations 1993

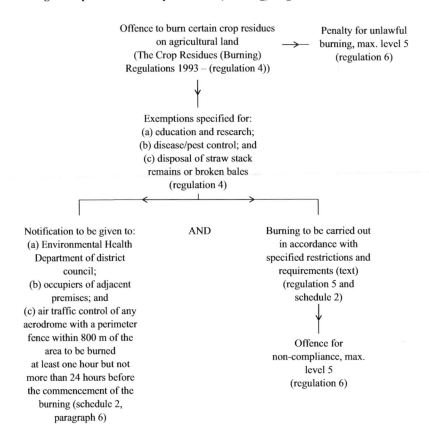

Offence to burn certain crop residues
on agricultural land
(The Crop Residues (Burning)
Regulations 1993 – (regulation 4))

Penalty for unlawful
burning, max. level 5
(regulation 6)

Exemptions specified for:
(a) education and research;
(b) disease/pest control; and
(c) disposal of straw stack
remains or broken bales
(regulation 4)

Notification to be given to:
(a) Environmental Health
Department of district
council;
(b) occupiers of adjacent
premises; and
(c) air traffic control of any
aerodrome with a perimeter
fence within 800 m of the
area to be burned
at least one hour but not
more than 24 hours before
the commencement of the
burning (schedule 2,
paragraph 6)

AND

Burning to be carried out
in accordance with
specified restrictions and
requirements (text)
(regulation 5 and
schedule 2)

Offence for
non-compliance, max.
level 5
(regulation 6)

of the Pollution Prevention and Control Act 1999. A person guilty of this offence is liable on summary conviction to a fine not exceeding level 5 on the standard scale (section 33 Clean Air Act 1993).

LOCAL AUTHORITY REVIEWS OF AIR QUALITY

References

Environment Act 1995, sections 80 to 91 and schedule 11
The Air Quality (England) Regulations 2000 (as amended)
The Air Quality (Scotland) Regulations 2000 (as amended)
The Air Quality (Wales) Regulations 2000 (as amended)
The Air Quality (Northern Ireland) Regulations 2003 (as amended)
The Air Quality (Standards) Regulations 2010
Air Quality Strategy for England, Scotland, Wales and Northern Ireland 2007
Clean Air Strategy 2019
DEFRA Guidance Notes:

    (a) LAQM. PG(09) Policy guidance, DEFRA, February 2009;
    (b) LAQM. PG(16) Policy guidance, DEFRA, April 2016;
    (c) LAQM. TG(09) Technical guidance, DEFRA, February 2009;
    (d) LAQM. TG(16) Technical guidance, DEFRA, February 2018.

Extent

These provisions apply to England, Wales and Scotland. Similar provisions apply in Northern Ireland.

Scope

This procedure deals with the statutory requirement on local authorities to review the quality, and the likely future quality, of air within the authority's area (section 82(1)).

Enforcing authorities

These are:

    (a) the Secretary of State;
    (b) the Environment Agency, the Natural Resources Body for Wales and the Scottish Environment Protection Agency;
    (c) a waste collection authority;
    (d) a local enforcing authority.

A local enforcing authority means:

    (a) a local enforcing authority within the meaning of Part 1 of the Environmental Protection Act 1990 (integrated pollution control and air pollution control by local authorities);
    (b) a local authority within the meaning of Part 2A of that Act (contaminated land);

(c) a local authority for the purposes of Part 4 of the Environment Act 1995 (air quality);

(d) a local authority under regulations made under section 2 of the Pollution Prevention and Control Act 1999 extending to England and Wales or regulations under section 61 of the Water Act 2014

In relation to the procedures in this chapter, the local authorities referred to above are:

(a) unitary authorities in England, Scotland and Wales, including London borough councils; and

(b) district councils that are not unitary councils (sections 91 and 108).

National Air Quality Strategy (NAQS)

The Secretary of State is required to prepare and publish a strategy containing policies with respect to the assessment and management of the quality of air.

Such a strategy takes account of UK obligations under the EC treaties and of international agreements and includes statements of:

(a) air quality standards – concentrations of pollutants in the atmosphere which can broadly be taken to achieve a certain level of environmental quality;

(b) air quality objectives – these provide a framework for determining the extent to which policies should aim to improve air quality and indicate the progress which can be made towards air quality standards; and

(c) measures to be taken by local authorities and others to achieve the objectives (section 80(1)–(5)).

The first strategy was published in March 1997 with revisions in 2000 and 2003 (CM 4548). The latest version was published in July 2007. It sets out certain standards and specific objectives for key air pollutants to be achieved in the long term (up to 2050).

A new Clean Air Strategy was published early in 2019 which promises new measures to be enacted during 2019 to support the process of developing and implementing clean air strategies.

The air quality objectives are set out in the Air Quality (England) Regulations 2000, as amended, which provide the statutory basis for the air quality objectives under local air quality management in England (see Table 1 in LAQM. PG (09)). Local authorities should note that they also have a duty to continue to work towards meeting the air quality objectives beyond the deadlines set in the regulations.

Note that there are separate air quality regulations for Wales, Scotland and Northern Ireland.

London Air Quality Strategy

Working within the NAQS, the Greater London Authority (GLA) is required to produce an Air Quality Strategy for London, which will explain how, working with London borough councils and other bodies, the national objectives will be achieved. The first strategy was issued by the GLA in September 2002, entitled 'Cleaning London's Air – the Mayor's Air Quality Strategy'. The latest update was in December 2010 and set targets for 2015 and 2020.

The Mayor's Air Quality Strategy does not replace the duties of the London borough councils, who will have to take account of it in undertaking their air quality management functions (section 362 Greater London Authority Act 1999).

Local air quality strategies

There is no statutory duty on local authorities to prepare a local strategy but they are urged to do so. It is recommended that all local authorities, particularly those that have not had to designate AQMAs or do not expect to designate an AQMA in the future, but who have areas at risk of exceedance, should consider drawing up an Air Quality Strategy. The authority should set up a steering group to develop the strategy (LAQM PG(16).

Local authority reviews and assessments

Each of the following local authorities (as enforcing authorities) is required to review periodically air quality in its area:

(a) Unitary authorities (in England, Wales and Scotland, including London borough councils);

(b) District councils which are not unitary authorities (but see 'functions' of county councils on p. 40) (sections 82(1) and 91).

LAQM. PG(09) and PG (16) are guidance issued by the Secretary of State under section 88(1) of the Environment Act 1995, and local authorities must have regard to it when carrying out their local air quality management duties. The government and the Mayor of London recommend that this policy guidance is made available to all local authority departments. The chapters covering transport and planning are relevant to those working in various local government departments, such as environmental health, land use planning, economic development and transport planning.

Every local authority shall review the air quality within its area, both at the present time and the likely future air quality (section 82). Local authorities should designate an air quality management area where air quality objectives are not being achieved, or are not likely to be achieved, within the relevant period, as set out in the Air Quality (England) Regulations 2000 (section 83). The local authority may from time to time revise their action plan (section 84).

Where local authorities do revise their air quality they should follow a two-step approach when carrying out review and assessment.

At Step 1, all authorities are required to undertake an updating and screening assessment. Where an authority identifies a risk that an air quality objective will be exceeded at a relevant location, the local authority is required to proceed to Step 2 – a detailed assessment.

To ensure continuity in the local air quality management process and fill the gaps between the three-yearly requirement to carry out a review and assessment, local authorities are required to prepare review and assessment progress reports in the years when they are not carrying out an updating and screening assessment. Where a detailed assessment is being undertaken, a short progress report should be provided within the detailed assessment for those areas not covered by the detailed assessment (LAQM (PG) (09) paragraph 1.7).

If DEFRA or the Mayor of London (as applicable) does not accept the conclusion contained within a local authority's report, the authority will be invited to provide written

FC13 Obtaining of information about atmospheric pollution – sections 35 to 40 Clean Air Act 1993

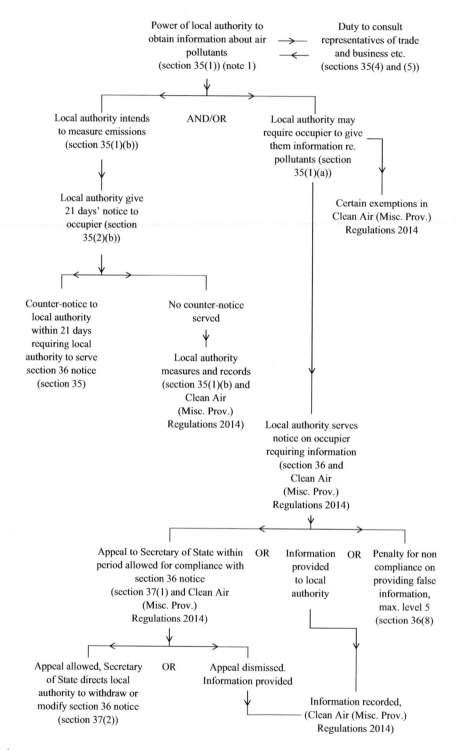

Note

1. For powers to require information about grit and dust, see Clean Air Act 1993, section 12.

FC14 Local authority reviews of air quality – sections 80 to 86 Environmental Protection Act 1990

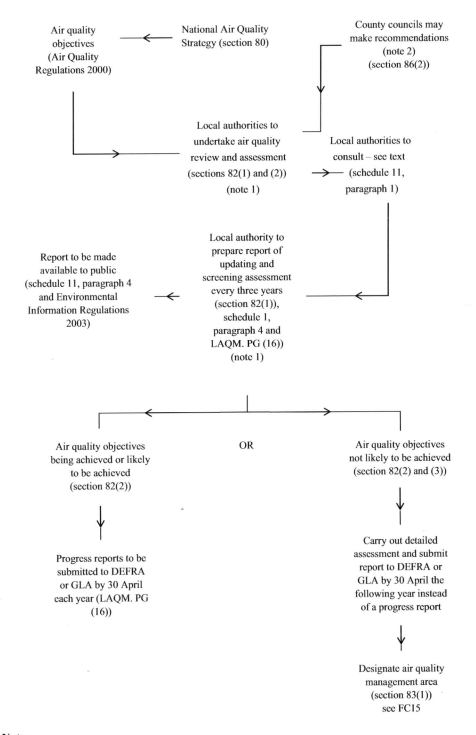

Notes

1. The Secretary of State or Scottish Environment Protection Agency has reserve powers in the event of local authority default (section 85).
2. The county council (where one exists) should be involved throughout the air quality management process.

comments justifying their decision within a specified deadline set out in the appraisal letter. This will involve a response within a short deadline in keeping with the need to complete the process as quickly as possible.

Local authorities are able to amend or revoke an existing air quality management area order at any time (section 83 (2)). In order to make a significant amendment or revoke an air quality management area, the local authority is required to submit a detailed assessment report clearly outlining the evidence for changes in the likelihood of exceedance of the objectives occurring and demonstrating the cause of these changes, such as changes to the source of the pollution and/or better monitoring/modelling information. DEFRA and the Mayor of London also expect the authority to consult all the relevant statutory consultees, local stakeholders, businesses and members of the public. Those local authorities should submit their further reports for appraisal showing the monitoring results and other evidence to justify their decision to take action.

Where it is accepted by DEFRA and/or the Mayor of London that the revocation or amendment is justified, local authorities will be expected to take the relevant action within four months following receipt of comments from DEFRA or the Mayor of London on detailed assessments or further assessments.

Where an air quality management area is revoked, local authorities should consider drawing up a local air quality strategy to ensure air quality issues maintain a high profile locally and to respond to any public expectations (chapter 2 LAQM PG 09).

Consultation

In undertaking an air quality review and assessment, the local authority is required to consult:

(a) the Secretary of State or the National Assembly in Wales;
(b) the Environment Agency (Scottish Environment Protection Agency in Scotland);
(c) in England and Wales, the Highways Authority;
(d) all neighbouring local authorities;
(e) the county council (where applicable);
(f) the National Park Authority (where applicable);
(g) any other public authority which the local authority considers appropriate; and
(h) bodies representing local business interests and such other bodies as the local authority considers appropriate (schedule 11, paragraph 1);
(i) and in Greater London, the London Borough Council or local authority contiguous with a London borough must consult the Mayor of Greater London (LAQM. PG (09)).

For the purposes of the 1995 Act, authorities must consult on their:

- air quality review and assessment;
- further air quality assessment in an air quality management area; and
- preparation or revision of an air quality action plan.

Local authorities are also expected to consult on the declaration, amendment or revocation of any air quality management areas.

Public access to information

A report of the results of the air quality review and a report of the results of any assessment must be made available to the public free of charge at all reasonable times (schedule 11, paragraph 4).

In any event, information which the local authority holds in relation to the quality of air needs to be made available (subject to exceptions) by the Environmental Information Regulations 2003.

Functions of county councils

It is expected that, while district councils will be the lead local authorities in the remaining two-tier areas of England, they will fully involve counties in the review and assessment process.

The county council may make recommendations to the district in relation to any review and assessment and the district must take account of them (section 86(2)).

In addition, schedule 11, paragraph 2 makes provision for the exchange of information between district and county councils throughout the air quality management process. LAQM. PG (09) and PG (16) deal with the integration, implementation and monitoring of local air quality plans into transport plans and planning.

Once an air quality management area has been designated, there is a further requirement that county councils should submit details of whatever actions they propose to the district council, for inclusion in an Action Plan within a relevant period (nine months) (section 86(3) and Air Quality (England) Regulations 2000, regulation 3).

AIR QUALITY MANAGEMENT AREAS (AQMAs)

References

Environment Act 1995, sections 83 and 84 and schedule 11
The Air Quality Regulations (England) (Scotland) and (Wales) 2000 (as amended)
The Air Quality Standards Regulations 2010
National Air Quality Strategy for England, Scotland, Wales and Northern Ireland CM 4548 (July 2007)
DEFRA Guidance Notes:

    (a) LAQM. PG(09) Policy guidance;
    (b) LAQM. PG(16) Policy guidance, DEFRA, April 2016;
    (c) LAQM. TG(09) Technical guidance, DEFRA, February 2009;
    (d) LAQM. TG(16) Technical guidance, DEFRA, February 2018.

Extent

These provisions apply to England and Wales. Similar provisions apply in Scotland and Northern Ireland as they implement EU legislation.

Scope

Where, following a review and assessment of air quality in its area, a local authority determines that prescribed air quality objectives are not likely to be achieved within the 'relevant period' as prescribed in the regulations, it must identify the areas where this is likely and, by order, designate them as an AQMA (section 83(1)).

Such orders may subsequently be varied or revoked by further orders following further assessment and, in the case of revocation, where the prescribed standards/objections are likely to be met throughout the 'relevant period' (section 83(2)).

While the order needs to designate the part/s of the local authority's area to which the AQMA status is to be attached, it is expected that delineation should make appropriate use of relevant physical and geographical boundaries (LAQM. PG(09) paragraph 3.2).

Where an AQMA comes into operation, the local authority should prepare, a written plan to achieve air quality standards and objectives in the designated area, using powers of the authority (section 84(2)).

Air quality action plans

These should include:

(i)   quantification of the source contributions to the predicted exceedances of the objectives. This will allow the action plan measures to be effectively targeted;
(ii) evidence that all available options have been considered;
(iii) how the local authority will use its powers and also work in conjunction with other organisations in pursuit of the air quality objectives;
(iv) clear timescales in which the authority and other organisations and agencies propose to implement the measures within its plan;
(v)   where possible, quantification of the expected impacts of the proposed measures and, where possible, an indication as to whether the measures will be sufficient to meet the air quality objectives. Where feasible, data on emissions could be included as well as data on concentrations where possible; and
(vi) how the local authority intends to monitor and evaluate the effectiveness of the plan. (LAQM (09) paragraph 4.2)

Air quality action plans and transport plans

Given the importance of vehicle pollution in the assessment of air quality and its improvement, it is necessary for the air quality action plan to be integrated with the transport plan. Guidance on this is given in chapter 6 of LAQM. PG(09).

Single vehicle emissions in AQMAs

The Road Traffic (Vehicle Emissions) (Fixed Penalty) (England) Regulations 2002 allow local authorities who have designated AQMAs to apply to the Secretary of State to be given powers to issue fixed penalty notices to users of vehicles who fail to meet the emission requirements in the Road Vehicles (Construction and Use) Regulations 1986.

Consultation

In undertaking the further air quality assessment and in preparing its action plan, the local authority is required to consult widely. The bodies concerned are the same as those for the initial review and assessment on p. 109 (schedule 11, paragraph 1).

Public access to information

The local authority is required to make available to the public free of charge copies of the order designating the AQMA, the report of the further assessment, the action plan and the proposals of the county council, where applicable (schedule 11, paragraph 4).

In any event, information which the local authority holds in relation to air quality must (with certain exceptions) be made available to the public under the Environmental Information Regulations 2004.

County councils

It is expected that, whilst the lead local authorities include district councils in the remaining two-tier areas of England, county councils will be fully involved by the district in the air quality management process generally, and schedule 11, paragraph 2 makes provision for the exchange of information between the two local authorities.

In relation to AQMAs, there is a specific requirement upon county councils that it should, within nine months from the date it is first consulted on the action plan, submit to the district details of its proposals for the action plan (sections 86(3) and (4) and regulation 3(1) of the Air Quality Regulations 2000). Disputes between districts and county councils about the content of action plans are to be settled by the Secretary of State or Scottish Environment Protection Agency (section 84(5)).

Secretary of State powers

The Act does provide for the Secretary of State (or SEPA in Scotland) to direct local authorities on the action it should take (including the designation of AQMAs) where prescribed air quality standards/objectives are unlikely to be achieved within the 'relevant periods', and the local authority has either failed to discharge a duty imposed on it or has taken inappropriate action to deal with the situation (sections 86(3) and (4)).

The Air Quality Standards Regulations 2010

These regulations implement European Community Directive 2008/50/EC on ambient air quality and cleaner air for Europe and Directive 2004/107/EC relating to arsenic, cadmium, mercury, nickel and polycyclic aromatic hydrocarbons in ambient air.

The regulations apply mainly in England.

The regulations give the Secretary of State duties in the following areas.

- assessment of sulphur dioxide, nitrogen dioxide and oxides of nitrogen, particulate matter, lead benzene and carbon monoxide;
- assessment of ozone;
- assessment of arsenic, cadmium, nickel, mercury, benzo(a)pyrene and other polycyclic aromatic hydrocarbons;
- limit values, target values, long-term objectives, information and alert thresholds and critical levels for the protection of vegetation;
- requirements in relation to $PM_{2.5}$ in addition to the limit value and target value for this pollutant and calculation of an average exposure indicator (AEI) for the UK, the calculation of a national exposure reduction target based on the AEI, attainment of the national exposure reduction target in the UK and compliance with a limit on the AEI for 2020;
- requirements to draw up air quality plans in relation to limit values and target values and short-term action plans in relation to alert thresholds.

FC15 Air quality management areas (AQMAs) – sections 83 and 84 Environment Act 1995

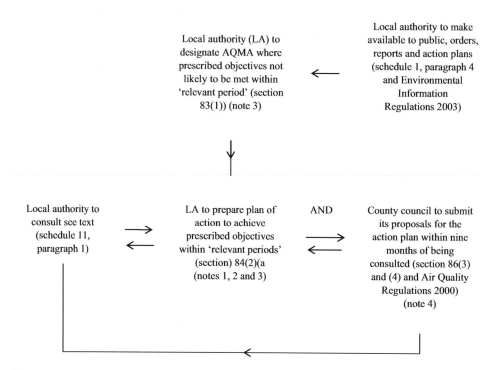

Notes

1. The local authority may revise the action plan periodically (section 84(4)) following reviews and assessments of air quality.
2. There is no statutory period for the preparation of an action plan but LAQM. PG(09) paragraph 4.3 suggests the action plan should be prepared within 12–18 months from the declaration of an AQMA.
3. The Secretary of State or Scottish Environment Protection Agency has powers of direction (text).
4. In the event of disagreement between district and county, the matter is determined by the Secretary of State (section 84(5)).

Chapter 2

ENVIRONMENTAL DAMAGE AND CONTAMINATED LAND

ENVIRONMENTAL DAMAGE

References

Environmental Damage (Prevention and Remediation) (England) Regulations 2015
Environmental Damage (Prevention and Remediation) (Wales) Regulations 2009

Extent

The Regulations apply in England and Wales and in specified marine waters and the seabed (regulations 1 and 6). This is EU derived legislation but retained as domestic law despite EU withdrawal.

Scope

The Regulations implement Directive 2004/35/EC and Article 38 of Directive 2013/30/EU on environmental liability with regard to the prevention and remedying of environmental damage.

They specify the types of damage to a protected species or natural habitat, a site of special scientific interest, water or land which constitute 'environmental damage' (regulation 4) and the types of activity causing environmental damage (regulation 5).

The regulations apply in relation to environmental damage if it is caused by an activity below:

- waste management operations;
- mining waste;
- discharges of all substances and pollutants into the inland surface and ground water that require prior authorisation;
- water abstraction and impoundment;
- dangerous substances, plant protection products and biocidal products;
- transport;
- genetically modified organisms;
- transboundary shipment of waste;
- operation of carbon dioxide storage sites (schedule 2)

The regulations also apply to environmental damage to protected species, natural habitats or a site of special scientific interest, and to environmental damage caused by any other activity if the operator intended to cause environmental damage or was negligent as to whether environmental damage would be caused (regulation 5).

There are some exemptions to the regulations, including acts of terrorism, an exceptional natural phenomenon, activities to protect from natural disasters, activities to serve national defence or international security, an oil pollution incident where compensation or liability applies certain activities involving radioactivity and damage caused in the course of some commercial sea fishing (regulation 8). There are some exemptions to damage to water – for example, damage caused by new modifications to the physical characteristics of a surface water body or an alteration to the level of a body of groundwater provided certain conditions are applied.

The conditions are:

(a) all practicable steps are taken to mitigate the adverse impact on the status of the body of water;

(b) the reasons for those modifications or alterations are specifically set out and explained in the river basin management plan;

(c) the reasons for those modifications or alterations are of overriding public interest, or the result of the damage is outweighed by the benefits of the new modifications or alterations to human health, to the maintenance of human safety or to sustainable development; and

(d) the beneficial objectives served by those modifications or alterations of the water body cannot for reasons of technical feasibility or disproportionate cost be achieved by other means.

Enforcing authorities under defined in the Environmental Permitting (England and Wales) Regulations 2016. Where the Environment Agency or the Natural Resources Body for Wales is responsible for granting the permit, if the damage is to marine waters in the Welsh zone, or to a natural habitat or protected species or a site of special scientific interest in those waters, by the Welsh Ministers and in any other case, the Environment Agency is responsible.

Where the local authority is responsible for granting the permit for Part 2; and Part 3 is to be enforced by the local authority, if the damage is to land; the Environment Agency, if the damage is to surface water or groundwater; the Marine Management Organisation, if the damage is to marine waters, or to a natural habitat or protected species or a site of special scientific interest within those waters, out to 12 nautical miles from the baselines in England; the Secretary of State, if the damage is to marine waters lying beyond 12 nautical miles from the baselines in England, other than any lying in the Welsh zone, or to a natural habitat or protected species or a site of special scientific interest in those waters; Natural England, if the damage is to a natural habitat or a protected species or a site of special scientific interest on land or in surface water or groundwater and the Welsh Ministers, if the damage is to marine waters in the Welsh zone, or to a natural habitat or protected species or a site of special scientific interest, in those waters (regulations 10 and 11).

If there is more than one type of damage, the regulations are enforced by any or all of the specified enforcing authorities. However, an enforcing authority may appoint any other enforcing authority to act on its behalf (regulation 12).

Preventing environmental damage

Where an operator of an activity that has caused, or causes, an imminent threat of environmental damage, or has caused damage where there are reasonable grounds to believe that the damage is or will become environmental damage, the enforcing authority may serve a notice on that operator that:

- describes the damage or threat;
- requires the operator to provide additional information on any damage that has occurred;
- specifies the measures required to prevent the damage or further damage; and
- requires the operator to take those measures, or measures at least equivalent to them, within the period specified in the notice.

Failure to comply with such a notice served is an offence (regulations 13 and 14).

The enforcing authority can take the actions which the operator of an activity should have taken in an emergency, if the operator cannot be ascertained, or if the operator fails to comply with a notice (regulation 15).

The operator may recover the costs when acting in accordance with the instructions of a public authority, and as a result causes or threatens to cause environmental damage, unless the instructions related to an emission or incident caused by the operator's own activities (regulation 16).

Remediation

Where damage has been caused, the enforcing authority must establish whether or not it is environmental damage (regulation 17).

If the enforcing authority decides that the damage is environmental damage, it must notify the operator of any activity or activities that caused the damage (the responsible operator) that:

(a) the damage is environmental damage;
(b) the responsible operator's activity was a cause of the environmental damage;
(c) the responsible operator must submit proposals, within a time specified, for measures that will achieve the remediation of the environmental damage; and
(d) the responsible operator has a right to appeal.

The enforcing authority may withdraw the notification if it is satisfied that the notification should not have been served or that an appeal against the notice is likely to succeed (regulation 18 and schedule 3).

Appeals

A person served with such a notification may notify the Secretary of State that they intend to appeal against that notification.

Notice of appeal must be within 28 days of service of the notification unless the time limit is extended by the Secretary of State.

The grounds of appeal are:

(a) the operator's activity was not a cause of the environmental damage;
(b) the enforcing authority has acted unreasonably in deciding that the damage is environmental damage;

(c) the environmental damage resulted from compliance with an instruction from a public authority (except an instruction relating to an emission or incident caused by the operator's own activities);

(d) the responsible operator was not at fault or negligent and the environmental damage was caused by an emission or event expressly authorised by, and fully in accordance with, the conditions of a permit;

(e) the responsible operator was not at fault or negligent and the environmental damage was caused by an emission or activity or any manner of using a product in the course of an activity that the operator demonstrates was not considered likely to cause environmental damage according to the state of scientific and technical knowledge at the time when the emission was released or the activity took place;

(f) the environmental damage was the result of an act of a third party and occurred despite the fact that the responsible operator took all appropriate safety measures.

The person deciding the appeal may confirm or quash the notice (regulation 19 and schedule 5).

Remediation notices

Where a responsible operator has received a notice from the enforcing authority, it must without delay identify potential remedial and submit them in writing to the enforcing authority for its approval.

On receiving the proposals from the responsible operator, the enforcing authority must consult any interested party and any person on whose land the remedial measures will be carried out and may consult any other person where necessary.

Following consultation, the enforcing authority must serve a notice on the responsible operator that specifies:

(a) the damage;

(b) the measures necessary for remediation of the damage, together with the reasons;

(c) the period within which those measures must be taken;

(d) any additional monitoring or investigative measures that the responsible operator must carry out during remediation; and

(e) the right of appeal against the notice.

Failure to comply with a remediation notice is an offence (regulation 20).

Appeal against the remediation notice

The responsible operator may notify the Secretary of State that they intend to appeal against the remediation notice on the grounds that its contents are unreasonable. Such an appeal may only be brought against those parts of the remediation notice that are different from proposals made by the responsible operator. Notice of appeal must be served within 28 days of service of the remediation notice unless the time limit is extended by the Secretary of State.

The Secretary of State may confirm, vary or quash the notice, and must give written notification of the final decision and the reasons for it. They may add further compensatory remediation requirements necessitated by the lapse of time since the remediation notice was served. The notice is frozen until the appeal is determined unless the person hearing the appeal directs otherwise (regulation 21 and schedule 5).

An enforcing authority may serve further remediation notices at any time while remediation is being carried out or, if remediation has not been achieved, at the end of the remediation period, requiring further or different remediation (regulation 22).

Works in default of notice

The enforcing authority may carry out any reasonable works:

(a) at any time if a responsible operator cannot be identified;
(b) if a responsible operator fails to comply with a remediation notice, whether or not an appeal is pending; or
(c) if the responsible operator is not required to remediate under these regulations (regulation 23).

Costs

The enforcing authority can recover justified costs (remediation and administration) when the enforcing authority acts instead of the operator. Relevant costs are:

- service of notices;
- assessing whether the damage is environmental damage;
- establishing who is the responsible operator;
- establishing what remediation is appropriate;
- carrying out necessary consultation; and
- monitoring the remediation, both during and after the work.

No proceedings for the recovery of costs may be commenced by the enforcing authority more than five years after the completion of the measures or the identification of the operator liable to carry out the measures, whichever is the later. Costs are recoverable from the owner as a charge on premises (regulations 24 to 27).

Requests for action by interested parties

Any person who is affected or likely to be affected by environmental damage, or who otherwise has a sufficient interest, may notify the appropriate enforcing authority of any environmental damage which is being, or has been, caused or of which there is an imminent threat. A notification must be accompanied by a statement explaining the way the notifier will be affected by the damage, or the reason that the notifier has a sufficient interest; and sufficient information to enable the enforcing authority to identify the location and nature of the incident.

The enforcing authority must consider the notification and inform the notifier as to the action, if any, that it intends to take.

Before taking any decision, the enforcing authority must, if practicable, notify the operator concerned of the notification and the accompanying information, and invite that operator to submit comments on them.

The enforcing authority need not act if:

(a) the notifier is not likely to be affected or does not have a sufficient interest;
(b) in the opinion of the enforcing authority, the information provided does not disclose any environmental damage or threat of environmental damage; or
(c) as a result of the urgency of the situation, it is not practicable for the enforcing authority to comply (regulation 29).

Powers of authorised persons

The powers in section 108 of the Environment Act 1995 apply in relation to these regulations, and the powers of persons authorised by the Environment Agency in that section are exercisable by persons authorised by any enforcing authority (regulation 31).

Provision of information to the enforcing authority

An enforcing authority may require an operator to provide such information as it may reasonably require enabling the enforcing authority to carry out its functions under the regulations. Failure to provide such information is an offence (regulation 32).

Enforcement

No enforcement action may be taken 30 years or more after the emission, event or incident concerned (regulation 33).

Penalties

A person guilty of an offence under these regulations is liable:

(a) on summary conviction, to a fine not exceeding the statutory maximum or to imprisonment for a term not exceeding three months, or both; or

(b) on conviction on indictment, to a fine or to imprisonment for a term not exceeding two years, or both.

Where a body corporate is guilty of an offence and that offence is proved to have been committed with the consent or connivance of, or to have been attributable to, any neglect on the part of:

(a) any director, manager, secretary or other similar person of the body corporate; or

(b) any person who was purporting to act in any such capacity,

that person is guilty of the offence as well as the body corporate (regulation 34).

The Secretary of State must carry out a review of these regulations from time to time and publish a report of the findings (regulation 35).

Definition

Environmental damage is damage to:

(a) protected species or natural habitats, or a site of special scientific interest;

(b) surface water or groundwater; or

(c) land (regulation 4(1)).

The rest of the regulation gives further description of the type of damage.

FC16 Request for action by interested parties – environmental damage – Regulation 29 Environmental Damage (Prevention and Remediation) Regulations 2009

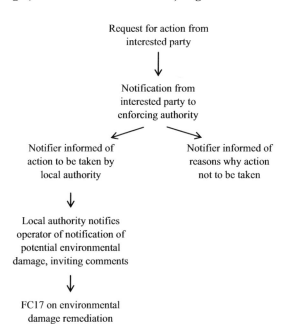

Request for action from
interested party

Notification from
interested party to
enforcing authority

Notifier informed of
action to be taken by
local authority

Notifier informed of
reasons why action
not to be taken

Local authority notifies
operator of notification of
potential environmental
damage, inviting comments

FC17 on environmental
damage remediation

FC17 Environmental damage – Regulations 13 to 23 Environmental Damage (Prevention and Remediation) Regulations 2009

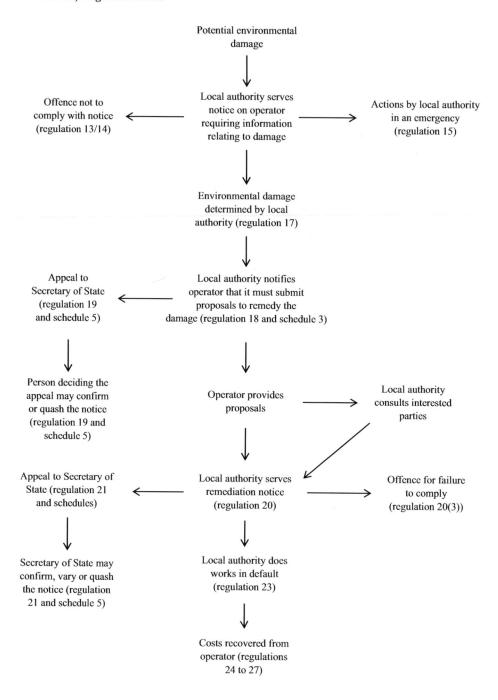

Potential environmental damage

Offence not to comply with notice (regulation 13/14)

Local authority serves notice on operator requiring information relating to damage

Actions by local authority in an emergency (regulation 15)

Environmental damage determined by local authority (regulation 17)

Appeal to Secretary of State (regulation 19 and schedule 5)

Local authority notifies operator that it must submit proposals to remedy the damage (regulation 18 and schedule 3)

Person deciding the appeal may confirm or quash the notice (regulation 19 and schedule 5)

Operator provides proposals

Local authority consults interested parties

Appeal to Secretary of State (regulation 21 and schedules)

Local authority serves remediation notice (regulation 20)

Offence for failure to comply (regulation 20(3))

Secretary of State may confirm, vary or quash the notice (regulation 21 and schedule 5)

Local authority does works in default (regulation 23)

Costs recovered from operator (regulations 24 to 27)

THE IDENTIFICATION AND REMEDIATION OF
CONTAMINATED LAND

Statutory nuisance provisions do not apply where the nuisance consists of or is caused by land being in a 'contaminated state', i.e. land where there are substances in, on or under that land such that harm is being caused, or there is a possibility of its being caused, or the pollution of controlled waters is being, or is likely to be, caused (section 79(1A) and (1B).

Contaminated land is dealt with by special procedures (see FC18 and FC19).

References

Environmental Protection Act 1990 Part 2A (as amended by the Environment Act 1995 and the Clean Neighbourhoods and Environment Act 2005)

The Contaminated Land (England) Regulations 2006 (as amended)

Waste and Contaminated Land (Northern Ireland) Order 1997

Environmental Protection Act 1990 Part 2A Contaminated Land Statutory Guidance (April 2012)

BS 10175:2001+A2:2017 Investigation of Potentially Contaminated Sites – Code of Practice

Contaminated land technical advice is available from the CIEH website at www.cieh.org/ policy – Guidance on Comparing Soil Contamination Data with a Critical Concentration

Model Procedures for the Management of Contaminated Land – CLR11 – DEFRA and the Environment Agency – September 2004

Extent

These provisions apply in England, Wales and Scotland with minor differences in Wales and Scotland. There are similar provisions in Northern Ireland.

The Act generally is applicable to Crown premises but does not create a criminal liability. The powers of entry given to local authority officers in respect of the various procedures may, however, be curtailed by the Secretary of State in the interests of national security in relation to specific premises and powers to be specified in a certificate (section 159).

Powers of entry

The general powers of entry for authorised officers for procedures under this part are either provided especially or are contained in powers relating to 'pollution control functions' in section 108 of the Environment Act 1995 (p. 42).

Scope

These procedures provide for the identification of contaminated land and formal determination by local authorities. They also provide for what is to be done by way of remediation, which may be required for any contaminated land where the local authority is the enforcing authority, i.e. all except special sites that are the responsibility of the Environment Agency. The statutory guidance is intended to explain how local authorities should implement the regime, including how they should go about deciding whether land is contaminated land. It also elaborates on the remediation provisions of Part 2A, such as the goals of remediation, and how regulators should ensure that remediation requirements are reasonable. It also explains specific aspects of the Part 2A liability arrangements, and the process by which the enforcing authority may recover the costs of remediation from liable parties in certain circumstances.

Role of local authorities

The role of the local authorities is to:

(a) prepare a written strategy regarding inspection of their area (section 78B(1) and section 2 of the statutory guidance);

(b) identify any potentially contaminated land (section 78B(1)) and any special sites (section 78C(1) and regulations 2 and 3);

(c) determine whether land is contaminated land (section 78A(2)–(5) and section 5 of the statutory guidance);

(d) establish who is the owner of the land (section 78A(9)), who appears to be in occupation, who is the appropriate person to bear responsibility for any necessary remediation action (sections 78E(1) and 78F);

(e) notify such persons and the Environment Agency that land is determined as contaminated and establish special sites (section 78B(3) and (4));

(f) require appropriate remediation defined in section 78A(7) and implement under sections 78E, 78H and 78N;

(g) maintain a register (section 78R, 78S, 78T, 78TA, 78TB and regulation 13).

In undertaking these roles, the local authorities are required to act in accordance with guidance issued by the Secretary of State (section 78B(2) and 78W) and to have regard to any site-specific advice given to it by the Environment Agency (Scottish Environment Protection Agency in Scotland) (section 78 V).

Contaminated land

This is defined as any land which appears to the local authority in whose area it is situated to be in such a condition by reason of substances in, on or under the land that:

(a) significant harm is being caused or there is a significant possibility of such harm being caused; or

(b) pollution of controlled waters is being, or there is a significant possibility of such pollution being caused (section 78A(2)).

In this context, 'harm' means harm to the health of living organisms or other interference with the ecological systems of which they form part and, in the case of man, includes harm to his property (section 78A(4)).

Section 78A(5) provides that the following are to be determined by the local authority in accordance with section 4 the statutory guidance:

(a) what is significant harm to human health (section 4.1);

(b) what is the significant possibility of significant harm to human health (section 4.2);

(c) what is significant harm and significant possibility of such harm (non-human receptors) (section 4.3);

(d) significant pollution of controlled waters and significant possibility of such pollution (section 4.4).

The Contaminated Land (England) Regulation 2006 and the Radioactive Contaminated Land (Modification of Enactments) (England) Regulation 2006 as amended, have brought within the scope of this definition land which is contaminated as a result of radioactive substances in or under the ground. Such sites are to be regarded as special sites (see below) and are to be dealt with by the Environment Agency.

Special sites

Wherever the local authority has determined land to be contaminated land, it must also decide whether it meets the description of special sites prescribed for the purposes of section 78C(8), regulations 2 and 3 and schedule 1 of the 2006 regulations and identify special sites for which the Environment Agency is the enforcing authority.

If the local authority considers that the land might be designated a special site, it should seek the advice of the Environment Agency (section 78C(3)). The Environment Agency also needs to consider whether any contaminated land should be designated as a special site. This might be based on information from its other pollution control functions. There is a duty on the Environment Agency to notify the local authority (section 78C(4)). Where the contaminated land meets one or more of the prescribed descriptions, the local authority must designate it a special site. Disputes between a local authority and the Environment Agency are to be settled by the Secretary of State.

Having made such a designation, the local authority must give written notice to:

(a) the Environment Agency;
(b) the owner of the land;
(c) any person who appears to be the occupier of all or part of the land; and
(d) each person who appears to be an appropriate person (see below).

Following this process, Environment Agency and not the local authority is responsible for securing any necessary remediation of special sites (sections 78C, 78D and 78E(1)).

The identification of contaminated land

Section 2 of the statutory guidance sets out the inspection duty. The local authority has sole responsibility for determining whether land is contaminated and it cannot delegate this except in accordance with section 101 of the Local Government Act 1972. The duty to inspect under section 78B(1) is extended by the statutory guidance to include a strategic approach. Every local authority must therefore have set out its approach to this duty in a written strategy (section 2.3 to 2.6). This approach should enable the local authority to identify, in a rational, ordered and efficient manner, the land which merits detailed individual inspection, identifying the most pressing and serious problems first and concentrating resources on the areas where contaminated land is most likely to be found.

Section 2.9 of the statutory guidance cover detailed inspections and use of powers of entry. If land has been determined as contaminated land and is likely to be designated a special site, the local authority should always make arrangements for the Environment Agency to carry out the inspection on its behalf. Section 108 of the Environment Act 1995 (see p. 122) can also be used to authorise Environment Agency staff.

'Contaminated Land Inspection Strategies' DETR (2001) is a Technical Advice Note for local authorities at this stage.

BS 10175:2001+A2:2017 provides guidance for local authorities on investigation techniques, sampling and on-site testing and laboratory analysis.

To enable a local authority to make a judgement regarding significant harm (section 78A(2)), scientific guidance is published by DEFRA:

- The Contaminated land exposure assessment (CLEA) tool;
- Land Contamination: risk management; Model Procedures for the Management of Land Contamination (CLR11
- Managing and reducing land contamination: guiding principles – GPLC2 – FAQs, technical information, detailed advice and references.
- Development of Category 4 Screening Levels for Assessment of Land Affected by Contamination. – DEFRA September 2014

Category 4 Screening Levels (C4SL) have been developed to complement the Statutory Guidance which sets out four levels of contaminated land and to assist enforcement authorities in assessing risk. Category 4 is the lowest level of contamination and comprises land where estimated levels of exposure to contaminants in soil are likely to form only a small proportion of what a receptor might be exposed to anyway through other sources of environmental exposure. Other indicators are known as General Assessment Criteria (GACs) or Soil Guidance Values (SGV's) and although these can be used with proper understanding of their derivation to assess contaminated land they should not be used as intervention levels (Statutory Guidance 2012).

Appropriate persons

These are the persons responsible for any remediation of contaminated land that the local authority may require. Part 2A of the Act defines two categories of 'appropriate persons' and sets out the circumstances in which they might be liable.

The first category is created by section 78F(2) – 'any person or any persons who caused or knowingly permitted the substances, or any of the substances, by which the contaminated land in question is such land, to be in, or under that land, is an appropriate person'. These are referred to as Class A persons by the statutory guidance. Such a person will be the appropriate person only in respect of any remediation, which is referable to particular substances, which he caused or knowingly permitted to be in, on or under the land (section 78F(3)).

The second category arises where it is not possible to find a Class A person. In these circumstances, the owner (see definitions) or occupier for the time being of the land is an appropriate person, referred to as a Class B person. Occupier is not defined but this would normally be the person in occupation, e.g. tenant or licensee. The enforcing authority may carry out necessary remediation works themselves where, after appropriate enquiry, it is not possible to identify appropriate person(s) (section 78N(3) – often referred to as the 'orphan sites' provision. The statutory guidance in section 7 guides local authorities in the circumstances where two or more appropriate persons are liable for remediation. It allows a local authority to determine who should bear the liability and identifies five distinct phases in this procedure.

Remediation (see definition on p. 132)

Having identified land as contaminated land, the local authority has a statutory duty to ensure that appropriate remediation is carried out (section 78E(1)(b)).

It is the government's intention that, as far as possible, remediation should be carried out by agreement rather than through the use of the formal notice procedures. In any event, the local authority is required to consult with 'appropriate persons' (see above), owners etc. about what is to be done by way of remediation before serving any remediation notices (section 78H(1)) and to produce a remediation scheme.

From a practical point of view, remediation is likely to be phased with different remediation actions being required at different times, e.g. assessment actions, remedial treatment actions and monitoring actions.

The local authority must have regard to the standard of remediation, i.e. it should result in land being 'suitable for use'. Section 6 of the statutory guidance give general advice on the various aspects of remediation and the circumstances where land is no longer contaminated.

The practicability, effectiveness and durability of remediation are governed by section 78E(5), and the guidance sets out the general criteria to meet these objectives. The guidance does not attempt to set out detailed technical procedures or working methods. In considering such matters, the enforcing authority may consult relevant technical documents (e.g. produced by the Environment Agency or other professional and technical organisations).

Section 78E(5)(c) requires that regard should be had to the reasonableness of remediation. Section 6(d) of the guidance sets out the criteria for this. Paragraphs 6.29 and 6.30 set out the matters which the local authority should take into account when considering the costs involved and compensation that is to be paid in accordance with section 78G(5) involving compliance with the remediation notice.

Urgent remediation action

Where there is imminent danger of serious harm or serious pollution of controlled water being caused, the local authority can either:

(a) serve an urgent remediation notice without going through the normal consultation and other procedural requirements (section 78H(4)); or

(b) where an urgent remediation notice would not result in remediation happening soon enough, carry out the urgent remediation itself and recover its costs where possible (section 78N).

Depending upon the circumstances, the local authority may take such action either before any remediation work has commenced by normal procedures or during the course of such work.

If the local authority carries out the work itself, it must produce a remediation statement describing the actions it has carried out (section 78H(7)).

'Imminent' and 'serious' are not defined in part 2A of the Act. The statutory guidance states that a local authority needs to judge each case on the normal meaning of the words and the facts of the case. Section 4 of the guidance helps local authorities with regard to the categories of harm which constitutes significant harm to human health and the significant possibility of it occurring.

Remediation notices

Where it has not been possible to secure a remediation scheme by agreement, the local authority is required to serve a remediation notice on each 'appropriate person' (see above) specifying what needs to be done by way of remediation and the time periods within which each action must be taken (section 78E(1) and regulation 4). The content of notices is specified in sections 78E(1) and (3) and in regulation 4.

Regulation 5 requires that copies of the notice are sent at the same time to those persons required to be consulted under sections 78G(3) and 78H(1) and to the Environment Agency.

Remediation notices may be served on different persons for action relating to different substances (section 78E(2)) and, where served on more than one person, the notice must state the proportion of the costs to be borne by each person (section 78E(3)).

In specifying the works to be undertaken, the local authority must consider any work to be reasonable having regard to:

(a) the costs; and
(b) the seriousness of the harm or of the pollution (section 78E(4)).

In determining what is to be done by way of remediation, the local authority must also have regard to the standard to which land is to be remediated and to what is regarded as reasonable (section 78E(5)). Remediation should not be required for the purpose of achieving any objectives other than those set out in paragraph 1.4 of the guidance; in particular, not for dealing with matters which do not themselves form part of a significant pollution linkage or making the land suitable for any uses other than its current use.

Notices may be modified by the authority under section 78L(2)(b) and are then subject to notification and appeal provisions (regulation 11).

Remediation declarations

Where the local authority has identified works which could be carried out but is precluded from including them on a remediation notice because of the 'cost' and 'seriousness' tests above, it must produce and publish a remediation declaration which indicates:

(a) the work in question;
(b) why the local authority would have otherwise specified that work; and
(c) the grounds on which the local authority feels justified in not specifying the work on the notice (section 78H(6)).

Such declarations must be included in the remediation register (section 78R(1)(c)).

Appeals

The grounds for appeal to the Secretary of State against a remediation notice are set out in regulation 7. The person has 21 days to appeal and, where it is duly made, the notice is suspended pending the decision. In these circumstances, the local authority needs to consider whether it should carry out urgent remediation itself using its powers under section 78N. The Secretary of State must hold a hearing or local enquiry (regulation 9).

Offences

Persons failing to comply with a remediation notice are liable, on summary conviction, to a fine not exceeding level 5 and to one-fifth of level 5 for each day on which the offence continues.

Where the land concerned is industrial, trade or business premises, the maximum fine is increased to £20,000 or one-tenth daily.

As an alternative to proceeding summarily, the local authority may go to the High Court when it feels the former would not secure an effectual remedy (section 78M).

In addition, the local authority may itself undertake the works in default and recover costs (section 78N).

Remediation statements

Where, without the service of a remediation notice, remediation is to take place, the person who is to carry it out (including the local authority where this is the case) is required to prepare and publish a remediation statement which records:

- (a) the things which have been, are being or are to be done;
- (b) the name and address of the person undertaking the works; and
- (c) the time which the work is expected to take (section 78H(7) and (8)).

These statements must be entered in the remediation register (section 78R(1)(c)).

Where the requirement for a remediation statement is not met by the person concerned, the local authority may produce it and recover its costs in so doing (section 78H(9)).

Remediation registers

Each local authority must maintain a remediation register which will contain details specified in regulation 13 and schedule 3 (section 78R(1) and regulation 13).

Certain information is excluded from this requirement where this is in the interest of national security or it is commercially confidential (sections 78S and 78T).

The register is available for inspection at the local authority's principal offices at all reasonable times (section 78R(8)(a)) and there must be a facility for members of the public to obtain copies of entries at a reasonable charge (section 78R(8)(b)).

Schedule 3 of the regulations prescribes full particulars of the following matters to be included in the register by the local authority:

- (a) remediation notices;
- (b) appeals against remediation notices;
- (c) remediation declarations;
- (d) appeals against charging notices;
- (e) designation of special sites;
- (f) notification of claimed remediation;
- (g) convictions for offences under section 78M;
- (h) guidance issued under section 78V(1); and
- (i) other environmental controls.

Definitions

Controlled waters are defined in section 78A(9) by reference to part 3, section 104 of the Water Resources Act 1991. This embraces territorial and coastal waters, inland fresh waters and groundwaters but does not include waters contained in underground strata but above the saturation zone.

Owner, in relation to any land in England and Wales, means a person (other than a mortgagee not in possession) who, whether in his or her own right or as trustee for any other person, is entitled to receive the rackrent of the land, or, where the land is not let at a rackrent, would be so entitled if it were so let.

Owner, in relation to any land in Scotland, means a person (other than a creditor in a heritable security not in possession of the security subjects) for the time being entitled to receive or who would, if the land were let, be entitled to receive the rents of the land in connection with which the work is used and includes a trustee, factor, guardian or curator and in the case of public or municipal land includes the persons to whom the management of the land is entrusted (section 78A(9)).

Remediation means:

(a) the doing of anything for the purpose of assessing the condition of:

   (i) the contaminated land in question;
   (ii) any controlled waters affected by that land; or
   (iii) any land adjoining or adjacent to that land;

(b) the doing of any works, the carrying out of any operations or the taking of any steps in relation to any such land or waters for the purpose:

   (i) of preventing or minimising, or remedying or mitigating the effects of, any significant harm, or any pollution of controlled waters, by reason of which the contaminated land is such land; or
   (ii) of restoring the land or waters to their former state; or

(c) the making of subsequent inspections from time to time for the purpose of keeping under review the condition of the land or waters (section 78A(7)).

FC18 Identification of contaminated land – Part 2A Environmental Protection Act 1990

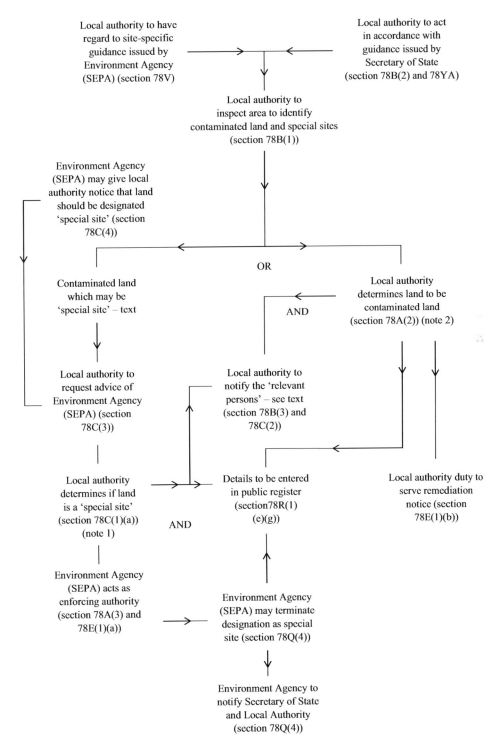

Local authority to have regard to site-specific guidance issued by Environment Agency (SEPA) (section 78V)

Local authority to act in accordance with guidance issued by Secretary of State (section 78B(2) and 78YA)

Local authority to inspect area to identify contaminated land and special sites (section 78B(1))

Environment Agency (SEPA) may give local authority notice that land should be designated 'special site' (section 78C(4))

OR

Contaminated land which may be 'special site' – text

Local authority determines land to be contaminated land (section 78A(2)) (note 2)

AND

Local authority to request advice of Environment Agency (SEPA) (section 78C(3))

Local authority to notify the 'relevant persons' – see text (section 78B(3) and 78C(2))

Local authority determines if land is a 'special site' (section 78C(1)(a)) (note 1)

Details to be entered in public register (section78R(1) (e)(g))

Local authority duty to serve remediation notice (section 78E(1)(b))

AND

Environment Agency (SEPA) acts as enforcing authority (section 78A(3) and 78E(1)(a))

Environment Agency (SEPA) may terminate designation as special site (section 78Q(4))

Environment Agency to notify Secretary of State and Local Authority (section 78Q(4))

Notes
1. Disputes between the local authority and Environment Agency (SEPA) about the designation of special sites are referred to the Secretary of State, whose decision must be notified to the relevant persons as well as to the parties involved (section 78D).
2. This information is subject to the Environmental Information Regulations 2004, subject to national security or commercial confidentiality.

FC19 Remediation of contaminated land – Part 2A Environmental Protection Act 1990

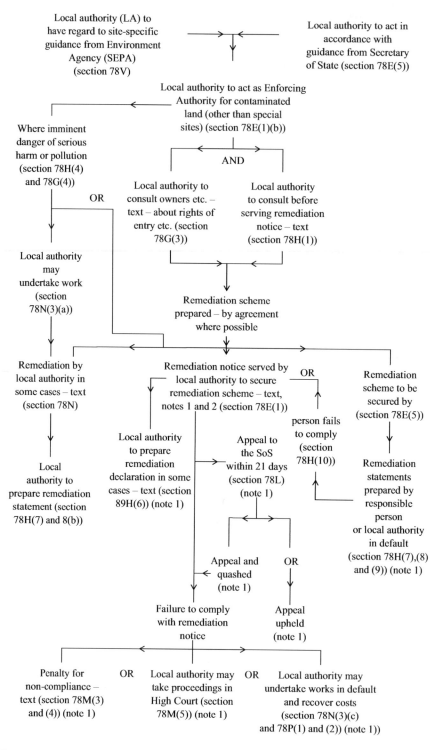

Notes

1. Details to be included in remediation register.
2. Remediation notices may be used several times on the same site if necessary to produce a staged result or to react to changing assessments and circumstances.

Chapter 3

NOISE

NOISE NUISANCES, NOISY PARTIES AND NOISE IN THE STREET

References

Control of Pollution Act 1974
Environmental Protection Act 1990
Noise and Statutory Nuisance Act 1993
Noise Act 1996
The Clean Neighbourhoods and Environment Act 2005
Antisocial Behaviour etc. (Scotland) Act 2004

Noise nuisances

Noise complaints have been dealt with under statutory nuisance provisions for a number of years and are still included in the statutory nuisances provisions of the Environmental Protection Act 1990 section 79 (see p. 573).

Noisy parties

Whilst certain types of these activities may be controlled through the Licensing Act 2003, many noisy parties are private, domestic, held in the home and for no private gain. In such cases, the use of the statutory noise nuisance legislation is one possible means of controlling noise.

It is unlikely that these provisions were drafted with this particular use in mind; nevertheless they can form an effective remedy, and DEFRA issues guidance on how to complain to a local authority about noise nuisances in its area. The CIEH have issued guidance to local authorities: 'Neighbourhood Noise Policies and Practice for Local Authorities – a Management Guide'. The way in which the statutory nuisance provisions can be used in this way is shown in FC20.

However, specific powers are now provided by the provisions of the Noise Act 1996, set out in FC24, to deal with night noise from certain premises.

Noise in streets

The Noise and Statutory Nuisance Act 1993 brought the definition of a new statutory nuisance from noise arising from a vehicle, machinery or equipment in streets. This was done to

FC20 Statutory nuisances – control over noisy parties – section 79 Environmental Protection Act 1990

Statutory noise
nuisance existing on
premises (section
79(1)(g)) (note 1)

↓

Authorised officer applies
to Justice of the Peace for
warrant (a) to enter in
emergency; and
(b) to effect a remedy
(schedule 3, paragraph
2(7))

↓

Authorised officer serves
abatement notice on either:
(a) person responsible (text); or
(b) owner or occupier of premises; or
(c) where the person cannot be named,
by addressing it to the occupier and
fixing it to the premises
(section 80(2))

←——————— ↓ ———————→

| Appeal to magistrates' court within 21 days but notice not suspended pending hearing (text) (note 3) | AND | Local authority may prosecute for non-compliance with notice (section 80(4)) (note 5) | AND/OR | Authorised officer may abate nuisance (section 81(3)), including confiscation and temporary removal of articles, e.g. audio equipment (schedule 3, paragraph 2A) (note 2) |

↓

See FC29 (note 4)

Notes

1. Anticipated noise nuisances can also be dealt with.
2. These powers must be exercised reasonably (section 81(3)).
3. The court may quash or vary the notice or dismiss the appeal (regulation 2(5)).
4. The procedure for dealing with the seizure of equipment is provided for in section 10 and the schedule to the Noise Act 1996 – see FC25.
5. For the adoptive procedure dealing with night noise from dwellings, see FC24.

FC21 Statutory nuisances – certain noise in the street – section 80 Environmental Protection Act 1990

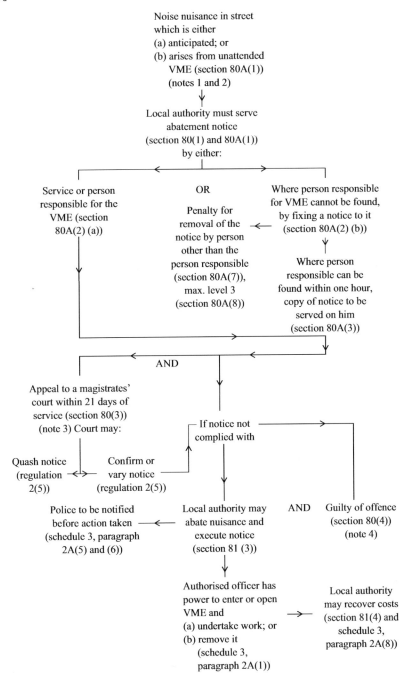

Noise nuisance in street which is either
(a) anticipated; or
(b) arises from unattended VME (section 80A(1))
(notes 1 and 2)

Local authority must serve abatement notice
(section 80(1) and 80A(1))
by either:

Service or person responsible for the VME (section 80A(2) (a))

OR

Penalty for removal of the notice by person other than the person responsible (section 80A(7)), max. level 3 (section 80A(8))

Where person responsible for VME cannot be found, by fixing a notice to it (section 80A(2) (b))

Where person responsible can be found within one hour, copy of notice to be served on him (section 80A(3))

AND

Appeal to a magistrates' court within 21 days of service (section 80(3)) (note 3) Court may:

If notice not complied with

Quash notice (regulation 2(5))

Confirm or vary notice (regulation 2(5))

Police to be notified before action taken (schedule 3, paragraph 2A(5) and (6))

Local authority may abate nuisance and execute notice (section 81 (3))

AND

Guilty of offence (section 80(4)) (note 4)

Authorised officer has power to enter or open VME and
(a) undertake work; or
(b) remove it
(schedule 3, paragraph 2A(1))

Local authority may recover costs (section 81(4) and schedule 3, paragraph 2A(8))

Notes

1. For other statutory noise nuisances in streets, see p. 135.
2. VME – vehicle, machinery or equipment.
3. The local authority will have indicated in the notice that it is to be suspended pending the hearing of any appeal (p. 143).
4. Details of penalties are given on p. 578.
5. For the control over loudspeakers in streets, see FC22.
6. A similar procedure for use in Scotland is set out in schedule 1 of the Noise and Statutory Nuisance Act 1993.
7. Regulation numbers refer to the Statutory Nuisance (Appeals) Regulations 1995.
8. See DoETR circ. 9/97 for guidance on the operation of this procedure.

allow local authorities to deal with problems that could give rise to considerable nuisance but which were not covered by previous legislation, which was restricted to noise from premises. In particular, this was the case with DIY car repairs, cooling engine noise from refrigerated lorries, misfiring of car alarms, buskers, etc.

The provisions also introduced a special procedure for dealing with noise which amounts to a statutory nuisance from vehicles, machinery or equipment in a street which are unattended and where the person responsible cannot be readily found.

The special procedure allows for the fixing of the abatement notice to the unattended vehicle but the authorised officer must spend up to one hour attempting to trace the person responsible (for appeals against these notices, see p. 576).

Where he is successful in so doing, that person must then comply with the notice, there being subsequent penalties for non-compliance and the local authority having default powers to take any steps necessary to abate the nuisance. To allow sufficient time for the person responsible, having been contacted, to abate the nuisance, the local authority may indicate on the notice attached to the unattended vehicle etc. that the time limit for compliance will be extended if it is possible for the person responsible to be served with a copy of the notice. A new notice can then be served with a different period allowed. Where the person responsible cannot be traced within one hour, the local authority may then enforce the notice. Before any action is taken, the police must be informed but the authorised officer has the power to enter or open the vehicle etc., if necessary by force, or remove it to a safe place. Having secured the abatement of the nuisance, the authorised officer must secure the vehicle etc., e.g. reset the vehicle alarm. The authorised officer must not cause any more damage than is necessary to execute the notice (section 80A).

Section 4 of the Clean Neighbourhoods and Environment Act 2005 now provides another remedy to some of the situations previously dealt with under this procedure. This provides that the person carrying on works of repair, maintenance, etc. to a motor vehicle on a road, with certain exceptions, is guilty of an offence. Section 6 of that Act provides a system of fixed penalty notices for this offence.

CONSENTS FOR THE USE OF LOUDSPEAKERS IN STREETS

References

Control of Pollution Act 1974, section 62
Noise and Statutory Nuisance Act 1993, section 8 and schedule 2
DEFRA Guidance on the Noise Act 1996 as amended – Guidance for Local Authorities, March 2008
Code of Practice on Noise from Ice-Cream Van Chimes etc. 2013

Extent

This provision applies in England, Wales and Scotland (section 109 and section 13).

Scope

This procedure allows a local authority to consent to the operation of a loudspeaker in a street between the hours of 9 pm and 8 am, being the period during which such an operation is generally prohibited under section 62(1) of the Control of Pollution Act 1974 (Noise and Statutory Nuisance Act 1993, schedule 2, paragraph 1).

Consents cannot be granted in connection with any election or for advertising any entertainment, trade or business (schedule 2, paragraph 1(2)). It is thought that local authorities may wish to use a consent system – for example, to allow charity events or street events to continue beyond 9 pm.

Adoption of scheme

The consent scheme operates only if the local authority has adopted its provisions by resolution of the council. Notice that such a resolution has been made must be published in a local newspaper in two consecutive weeks before becoming operative on a date specified, which cannot be less than one month from the date of the resolution.

The notice in the newspaper must set out the effect of the resolution and explain the procedure for applying for consents (section 8(1)–(4) Noise and Statutory Nuisance Act 1993).

Applications

These must be made to the local authority in writing and must contain the information requested by the local authority to enable it to reach a decision. The local authority may set a reasonable fee for each application (schedule 2, paragraphs 3 to 5 Noise and Statutory Nuisance Act 1993).

Considerations

The matters to be taken into account by the local authority are not specified but might reasonably include:

(a) location;
(b) adjacent activities;
(c) time;
(d) number of loudspeakers and their output power;
(e) methods of noise control;
(f) the planned route of any procession, etc.

Consents

Local authorities may decide not to issue a consent or issue one at their discretion, with or without conditions. There are no appeal provisions. Notification of the local authority's decision must be put in writing to the applicant within 21 days of the application being made, with details of any conditions which are to be attached (schedule 2, paragraph 4).

The conditions attached to any consent could include the specification of noise levels and of any route to be taken, but could cover any issue of legitimate concern to the local authority.

The local authority may (but is not obliged to) publicise details of any consents in a local newspaper (schedule 2, paragraph 6).

FC22 Consents for the use of loudspeakers in streets – section 62 Control of Pollution Act 1974 and section 8 Noise and Statutory Nuisance Act 1993

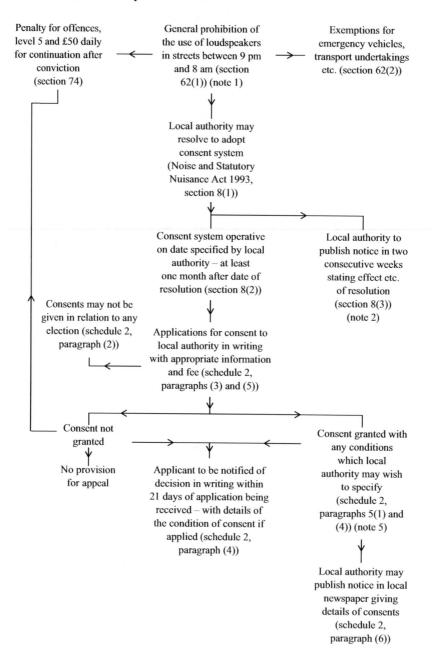

Notes

1. There is a total prohibition at any time on the use of loudspeakers in streets for advertising any entertainment, trade or business (there is an exception for advertising the presence of a vehicle selling perishable commodities, e.g. ice-cream, between noon and 7 pm). The consent procedure does not allow this prohibition to be breached.
2. There is no provision for the making or consideration of representations.
3. Where not otherwise indicated, section and schedule numbers refer to the Noise and Statutory Nuisance Act 1993.
4. For the control of statutory noise nuisances in streets, see FC21.
5. Operating a loudspeaker with a consent but in breach of its conditions would appear to create an offence under the Control of Pollution Act 1974, i.e. for breaching the general prohibition order (section 62(1)).

Consents and statutory nuisance

The issue of a consent does not affect the operation of procedures to deal with statutory noise nuisances in streets (FC21).

If during the operation of a consent such a nuisance is created, the local authority may use its powers under that procedure.

Information

An authority is provided with a general power to obtain information necessary to perform its functions under the Control of Pollution Act 1974 by the service of notice specifying the information required and giving a time period within which or time at which it must be supplied. The maximum penalty for non-compliance or wilfully giving false information is level 5 on the standard scale (section 93).

Definitions

Best practicable means – 'practicable' means reasonably practicable, having regard among other things to local conditions and circumstances, to the current state of technical knowledge and to the financial implications. 'Means' includes design, installation, maintenance and manner and periods of operation of plant and machinery, and the design, construction and maintenance of buildings and acoustic structures. The test of best practicable means is to apply only so far as is compatible with statutory duties imposed, with safety and safe working practices and with the exigencies of any emergency or unforeseen circumstances. Regard shall be had to any relevant provisions of an approved Code of Practice (section 72).

Local authority – see p. 40.

Noise includes vibration (section 73(1)).

Owner, except in Scotland, means the person for the time being receiving the rackrent of the premises in connection with which the word is used whether on his own account or as agent or trustee for another person, or who would so receive the rackrent if the premises were let at a rackrent (section 105(1)).

Premises includes land (section 105(1)).

Street means a highway and any other road, footway, square or court which is, for the time being, open to the public (Control of Pollution Act 1974 section 62(1)).

NOISE FROM CONSTRUCTION SITES

References

Control of Pollution Act 1974, sections 60 and 61

Control of Noise (Appeals) Regulations 1975

Control of Noise (Codes of Practice for Construction and Open Sites) (England) Order 2015

DEFRA Guidance on the Noise Act 1996 as amended – Guidance for Local Authorities, March 2008

Extent

These provisions apply to England, Wales and Scotland.

Scope

The first of these two procedures stems from an application for consent to the local authority, which, if operated in accordance with any conditions or with limitations, provides a defence against any proceedings taken by a local authority following the second procedure, i.e. a local authority notice specifying noise emission standards. The procedures could be operated together but only beneficially where, following application to the local authority, consent has been granted without conditions or limitations, or no consent was given.

The procedures are applicable to the following works:

(a) erection, construction, alteration, repair or maintenance of buildings, structures or roads;

(b) breaking up, opening or boring under any road or adjacent land in connection with the construction, inspection, maintenance or removal of works;

(c) demolition or dredging works;

(d) any other works of engineering construction (section 60(1)).

Criteria for consideration

In judging both applications for consent and whether or not to serve notices, local authorities must have regard to the following points:

(a) BS 5228 part 1 2009: Noise, part 2 2009: Vibration (Codes of Practice for Construction and Open Sites) (England) Order 2015);

(b) the need to ensure that best practicable means are employed to minimise noise;

(c) in considering the specification of any particular methods or plant etc., the interest of the applicant or recipient of the notice in regard to alternative methods etc. which might be as effective but more acceptable to them;

(d) the need to protect any persons in the locality from the effects of noise (section 60(4)).

Local authority notices

1. **Person responsible.** This is the person carrying out, or going to carry out, the works and any other persons who appear to the local authority to be responsible for, or have control over, those works. Notices may be served on all or any such persons (section 60(5)).

2. **Content.** Notices may include:

(a) specification of plant and machinery which is or is not to be used;

(b) the hours during which the works may be carried out;

(c) specification of noise levels from the premises, from any specified point on the premises or during specified hours;

(d) a provision for any change of circumstances (section 60(3)).

A notice may specify a time within which the notice should be complied with, which should not be less than the appeal period, i.e. 21 days, and may require the execution of works or other steps, but the local authority has no default powers (section 60(6)).

If the local authority intends that the notice should not be suspended in the event of an appeal, this must be stated in the notice (regulation 5 Control of Noise (Appeals) Regulations 1975).

3. **Appeals.** The grounds of appeal are set out in the Control of Noise (Appeals) Regulations 1975 as:

   (a) the notice is not justified;
   (b) there is some material informality, defect or error in the notice;
   (c) the local authority has refused unreasonably to accept compliance with alternative requirements or the requirements of the notice are unreasonable or unnecessary;
   (d) any time period allowed for compliance is not reasonably sufficient;
   (e) the notice should have been served on some other person in substitution for, or in addition to, the appellant;
   (f) the local authority has not had regard to the matters set out in section 60(4) (regulation 5).

   The court may dismiss the appeal, vary or quash the notice and/or make an order as to respective responsibilities for compliance and costs.

   In the event of an appeal, local authority notices are suspended until the appeal has been determined or abandoned where:

   (a) the noise is caused in the performance of a statutory duty imposed on the appellant; or
   (b) compliance would involve expenditure being incurred before the hearing, except in those cases where the local authority is of the opinion that:

      (a) the noise is injurious to health; or
      (b) the noise is likely to be of limited duration so that suspension would give no practical effect to the notice; or
      (c) the expenditure to be incurred would not be disproportionate to the public benefit from compliance, provided that this has been stated on the original local authority notice (regulation 10 Control of Noise (Appeals) Regulations 1975).

4. **Defence.** It is a defence in any proceedings for non-compliance with a local authority notice to prove that the alleged contravention amounted to the carrying out of the works in accordance with a consent under section 61 (section 61(8)).

Consents

1. **Applications.** These must be made at the same time as, or later than, any application for building regulations approval and may be made by anyone intending to carry out construction works. The application must contain particulars of:

   (a) the works proposed and the methods to be used to carry them out; and
   (b) the steps proposed to minimise noise (section 61(2) and (3)).

2. **Consent.** If the local authority considers that it would not serve a notice under section 60 if the works are carried out in accordance with the application, it must give consent and may:

    (a) attach conditions;
    (b) limit or qualify consent to allow for any changes in circumstances;
    (c) limit the duration of the consent (section 61(4) and (5)).

    A consent under this procedure does not of itself constitute a defence against proceedings for statutory nuisance by aggrieved persons under section 80 Environmental Protection Act 1990, and this must be stated in the consent. However, it will be a defence against proceedings by a local authority for failure to comply with an abatement notice under section 80 Environmental Protection Act 1990 to show that the alleged offence was covered by a notice under section 60 Control of Pollution Act or by a consent under sections 61 Control of Pollution Act. It will be a defence in proceedings following section 60 notices to show that the works were carried out in accordance with that consent (section 61(8) and (9) and section 80(9) Environmental Protection Act 1990).

3. **Appeals.** Grounds of appeal against no consent being granted or against attachment of conditions, limitations or qualifications are set out in Control of Noise (Appeals) Regulations 1975 as being:

    (a) that any condition, limitation or qualification is not justified;
    (b) that there is some material informality, defect or error in connection with the consent;
    (c) that the requirements of any condition are unreasonable in character or extent, or are unnecessary;
    (d) that the time within which any conditions etc. are to be complied with is not reasonably sufficient (regulation 6).

Publication

In order to bring either notices under section 60 or consents under section 61 to the attention of the public who might be affected, the local authority is able to publicise the contents of notices and consents in any ways which it thinks appropriate (sections 60(2) and 61(6)).

Definition

Work of engineering construction means the construction, structural alteration, maintenance or repair of any railway line or siding or any dock, harbour or inland navigation, tunnel, bridge, viaduct, waterworks, reservoir, pipeline, aqueduct, sewer, sewage works or gas holder (section 73(1)).

Practical issues

Some local authorities use the spirit of the legislation to work with contractors to limit noise complaints rather than serve notices, prior consents, etc. Responsible contractors approach local authorities for prior consent. They prefer an informal approach rather than notices and consents. Informal action works well with contacts in organisations through other areas of work, e.g. Integrated Pollution Prevention and Control (IPPC).

FC23 Noise from construction sites – sections 60 and 61 Control of Pollution Act 1974

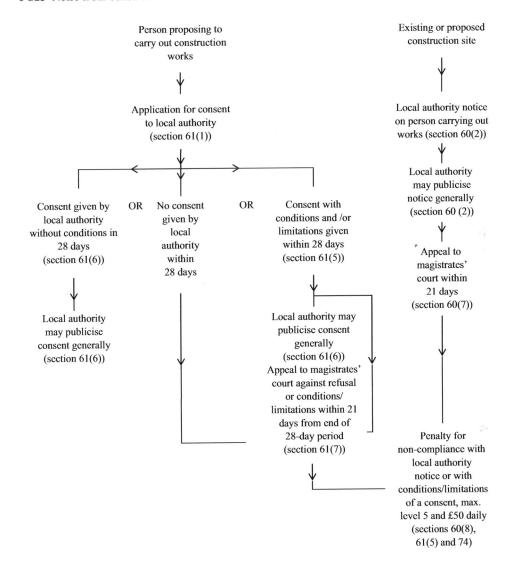

It is worth noting that local authorities need to coordinate their approaches to these organisations in licensing, planning, building control and noise control. Duplicating conditions in planning permission and in licensing may cause problems for licensing and noise control enforcement.

NOISE FROM CERTAIN PREMISES AT NIGHT

References

Noise Act 1996 as amended by the Antisocial Behaviour Act 2003 and the Clean Neighbourhoods and Environment Act 2005
The Environmental Noise Regulations (Northern Ireland) 2006
The Environmental Noise Regulations (Scotland) 2006
DEFRA Guidance on the CNEA 2005 – Noise 2006 and Fixed Penalty Notices 2006
Government guidance – Noise nuisances: how councils deal with complaints
DEFRA Circular NN/31/03/2004
CIEH/DEFRA Neighbourhood Noise Policies and Practice for Local Authorities – A Management Guide 2006

Extent

This procedure applies in England and Wales (section 1). Similar provisions apply in Scotland and Northern Ireland.

Scope

The procedure enables a local authority to deal with complaints from any individual present in a dwelling (definition below) during night hours that excessive noise is being emitted from another dwelling or a premises which is the subject of a premises licence or a temporary event notice under the Licensing Act 2003.

Night hours are defined as from 11 pm to 7 am the following morning (sections 2(2) and (6)).

NB. The general provisions of the Environmental Protection Act 1990 do not apply to this procedure.

Complaints

The local authority may investigate complaints of night noise which appear to be covered by the provisions. Such investigations must be undertaken by an officer of the local authority (section 2(1)).

Warning notices

Where upon investigation the officer is satisfied that:

(a) noise is being emitted from the offending premises during night hours; and
(b) the noise if measured within the complainants' dwelling would or might exceed the permitted level,

he may serve a warning notice (section 2(4)).

At this stage, actual measurement of the noise level is not obligatory. Noise above the permitted level does not necessarily create a statutory nuisance.

The permitted levels in relation to licenced premises and dwellings will be set by the Secretary of State in directions made under section 5. See Government Guidance – Noise nuisances: how councils deal with complaints.

Complaints made by a resident of one local authority area about noise arising from an adjoining local authority area may be pursued by the first local authority and warning notices served (section 2(7)).

In judging whether or not the noise complained about is, would or might exceed the permitted level, the officer has discretion as to whether or not measurements are taken and whether the assessment of the noise is from within or outside the complainants' dwelling (section 2(5)).

Content of warning notices

The notice must:

(a) state that the officer considers that there is night noise from the offending premises which exceeds, or may exceed, the permitted level as measured within the complainants' dwelling; and

(b) give warning that any person responsible for noise emitted above the permitted level in the period specified in the notice may be guilty of an offence (section 3(1)).

The period specified must begin not earlier than ten minutes from the notice being served and end at the following 7 am. Officers may use discretion as to how much 'grace' should be given before the notice comes into effect given the circumstances of the situation. The notice must also state the time at which it is served (section 3(2) and (4)).

Service of notices

Service of warning notices is effected either by delivering to any person present at or near the offending dwelling and appearing to be responsible for the noise or, where identification of such a person is not reasonably practicable, by leaving it at the offending dwelling (section 3(3)).

A person is deemed as responsible for the noise if he is a person by whose act, default or sufferance the emission of the noise is wholly or partly attributable (section 3(5)).

Offences

Once a warning notice has been served, if noise is emitted from the offending dwelling in excess of the permitted level as measured from within the complainants' dwelling, any person responsible (see above) is guilty of an offence and a fine not exceeding level 3 (section 4).

It is also an offence where noise from other premises exceeds permitted level measured from within the complainant's dwelling after service of a warning notice and the noise is emitted from the premises in the period specified in the notice, and the responsible person is liable on summary conviction to a fine not exceeding level 5 on the standard scale (section 4A).

It should be noted that, at this stage, the noise must be measured and the measurement taken from within the dwelling (as distinct from the situation at the investigative stage).

Evidence of measurements taken will not be admissible in court proceedings unless they were taken using a device approved by the Secretary of State and used in accordance with conditions prescribed by him (section 6).

There are detailed provisions about the way in which evidence must be gathered and presented (section 7).

It is a defence to show that there was a reasonable excuse for the act, default or sufferance in question (section 4(2)).

Fixed penalty notices

Where the authorised officer believes that an offence has been committed by the infringement of a warning notice, he may serve a fixed penalty notice as an alternative to prosecution.

The notice offers the person the opportunity of discharging liability to conviction by payment within 14 days. The notice is served either by giving it to the person or by leaving it, addressed to him, at the offending dwelling or licenced premises. In the latter case, the notice should be given to the most senior person present. The notice must state:

(a) details of the offence;
(b) the period during which proceedings will not be taken, i.e. 14 days;
(c) the amount of the fixed penalty (£100); and
(d) the ways in which the payment can be made (these are not 'on the spot' fines).

The fixed penalty in respect of dwellings is, at the discretion of the local authority, or £100. For other premises, the fine is fixed at £500 (sections 8 and 8A Noise Act 1996).

If an officer of a local authority gives a person a fixed penalty notice, the officer may require the person to give him his name and address. A person commits an offence if he fails to give his name and address when required to do so, or he gives a false or inaccurate name or address in response. A person guilty of this offence is liable on summary conviction to a fine not exceeding level 3 on the standard scale (section 8B).

Where a payment is not received during the period specified (if by post the date of payment is deemed to be the time at which the letter would be delivered in the ordinary course of post), the local authority may proceed by way of prosecution for the offence (section 8(3)).

Only one fixed penalty notice may be given to the same person for a particular premises the same night but can be prosecuted for subsequent offences (section 9(2)).

The local authority may use any sums it receives from penalty receipts for financing its qualifying functions (section 9(4)).

Equipment involved in these complaints may be seized under section 10 of the Noise Act 1996 (see below at p. 150).

Definitions

Dwelling means any building, or part of a building, used or intended for use as a dwelling and references to noise emitted from a dwelling include noise emitted from any garden, yard, outhouse or other appurtenance belonging to or enjoyed within the dwelling (section 11(2)).

FC24 Noise from certain premises at night – Noise Act 1996

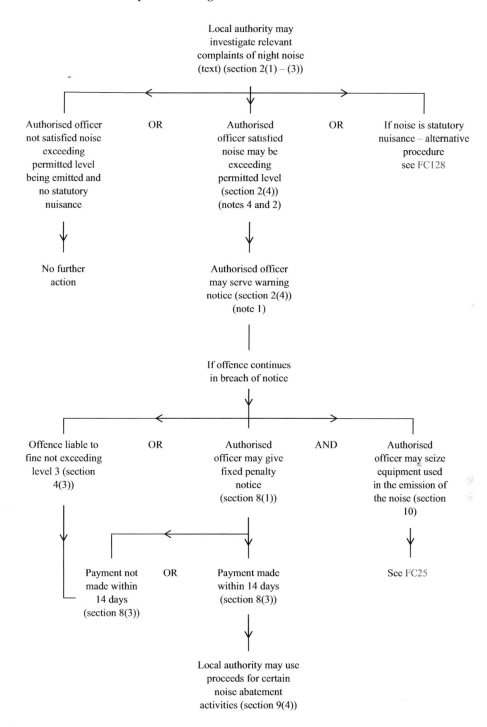

Notes
1. There is no prescribed form for a warning notice but see CIEH guidance.
2. Action may be taken against a premises within an adjoining local authority area (section 2(7)).
3. For noise which is a statutory nuisance, see FC128.
4. This may or may not be a statutory nuisance.

Permitted level of noise – the appropriate person (the Secretary of State for England and the National Assembly for Wales) may in writing determine the maximum level of noise which may be emitted during night hours from any dwelling or other premises (or vary it). The permitted level is to be a level applicable to noise as measured from within any other dwelling in the vicinity by an approved device used in accordance with any conditions subject to which the approval was given. Different permitted levels may be determined for different circumstances, and the permitted level may be determined partly by reference to other levels of noise. The appropriate person may from time to time vary his directions under this section by further directions in writing (section 5).

SEIZURE OF EQUIPMENT USED TO MAKE NOISE UNLAWFULLY

References

Noise Act 1996, section 10 and schedule 1
Environmental Protection Act 1990, sections 79(1)(g) and 81(3)
DEFRA Guidance on the Noise Act 1996 as amended – Guidance for Local Authorities, March 2008
CIEH Neighbourhood Noise Policies and Practice for Local Authorities – A Management Guide, 2006

Extent

In addition to England and Wales, the Act applies, with some minor amendments, in Northern Ireland but not in Scotland (section 14(4)).

Scope

This procedure is available to authorised officers of local authorities where:

(a) a warning notice under the Noise Act 1996 has been served (FC24) and noise emitted from the offending dwelling has exceeded the permitted level as measured from within the complainants' dwelling; (section 10) or
(b) the local authority is using its powers under section 81(3) of the Environmental Protection Act 1990 to itself abate or prevent a statutory noise nuisance under section 79(1)(g) following the service of an abatement notice (FC128).

Seizure of equipment

An officer of the local authority or someone authorised by the local authority has the power to enter the dwelling from which the noise is being or has been emitted and may seize and remove any equipment which is being or has been used to make the noise. Any person exercising these powers must produce his or her authority, if required to do so (sections 10(2) and (3)).

Note – if entry is in connection with enforcement under the 1990 Act, then 24 hours' notice of intention to enter must be given (see p. 578).

Warrants

Where a Justice of the Peace is satisfied on sworn information that:

(a) a warning notice has been served;
(b) noise has been emitted in breach of the requirements of the notice; and
(c) entry to the dwelling has been refused, such refusal is apprehended or a request for entry would defeat the object of admission,

the Justice of the Peace may by warrant authorise the local authority to enter, if necessary by force. Persons authorised by the local authority to effect the warrant may take with them any other person or equipment as may be necessary but must ensure when leaving that any unoccupied premises are left as effectively secured against trespass as they found them.

Warrants continue in force until the purpose for entry has been satisfied (sections 10(4)–(6)).

Offences

Any person wilfully obstructing someone exercising the powers of this procedure is liable to a fine not exceeding level 3 (section 10(8)).

Retention of equipment

Equipment seized under these powers may be retained by the local authority for 28 days unless it was:

(a) the subject to proceedings for a noise offence – in which case it may be retained until the proceedings have been dealt with; or
(b) equipment used to emit noise which has been the subject of a fixed penalty notice under section 8 – in which case it must be returned on request (schedule 1, paragraph 2).

Forfeiture order

When a person is convicted by a magistrates' court of a noise offence (either the breach of a warning notice under the Noise Act or an abatement notice for a statutory noise nuisance under the Environmental Protection Act), the court may order the forfeiture of any seized equipment. In these cases, third parties (e.g. hiring companies) may make application to the court within six months for its return. Persons subject to the forfeiture order have no further rights to the equipment named (paragraph 3 and 4).

The court may also give directions (where a forfeiture order is not made) as to the return, retention or disposal of the equipment (paragraph 5).

Return of seized equipment

Equipment which is claimed within the required periods or is ordered to be returned to its owner by a magistrates' court must be returned by the local authority once a reasonable charge has been paid to it (paragraph 6). If the equipment is sold then the proceeds must be paid to the owner after reasonable charges have been taken by the local authority.

FC25 Seizure of equipment used to make unlawful noise – section 10 Noise Act 1996

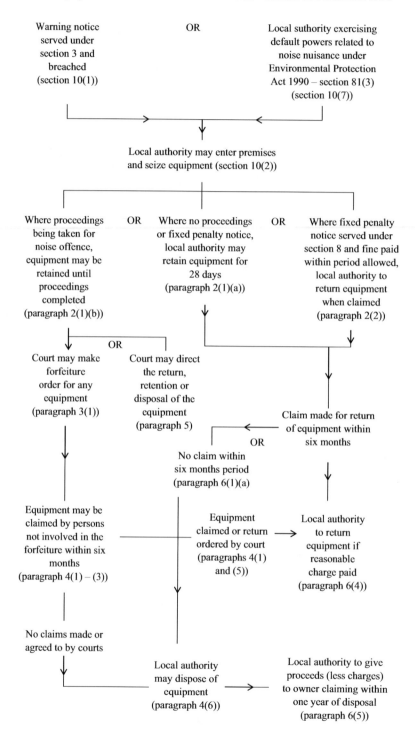

Note

1. Paragraph numbers refer to schedule 1 of the Noise Act 1996.

Notification of claim rights

Local authorities have a duty to take reasonable steps to bring to the attention of persons who may be entitled to do so their rights to make claims for the return of seized equipment (paragraphs 4(4) and 6(3)).

AUDIBLE ALARM NOTIFICATION AREAS

References

The Clean Neighbourhoods and Environment Act 2005, sections 69 to 81
Clean Neighbourhoods and Environment Act (Northern Ireland) 2011
The Environmental Offences (Fixed Penalties) (Miscellaneous Provisions) Regulations 2017
DEFRA Guidance on the Clean Neighbourhoods and Environment Act 2005 – Noise 2006 and Fixed Penalty Notices 2006
DEFRA Guidance on the Noise Act 1996 as amended – Guidance for Local Authorities, March 2008

Extent

This provision applies in England, Wales and Scotland. Similar provisions apply in Northern Ireland.

Scope

This is a procedure that enables local authorities to deal with annoyance caused by audible intruder alarms. It allows a local authority to designate any or all of its area as an alarm notification area (ANA). The effect is to require all intruder alarms, residential and non-residential, to be notified to the local authority with the identification of a key-holder (section 69(1)).

Designation

A notice of the proposal must be published in a local newspaper, stating:

(a) that representations can be made;
(b) the date by which these must be received being at least 28 days after the publication of the notice.

The local authority must consider all representations made within the time limit (section 69(2–5)).
If the local authority determines to go ahead, it must:

(i) arrange for the decision to be published in a local newspaper;
(ii) send a copy to all premises in the area (NB 'premises' does not include vehicles in this procedure);
(iii) specify a date of operation, which must be at least 28 days after the notice has been published (section 69(6–8)).

FC26 Audible alarm notification areas (ANAs) – sections 69 to 80 Clean Neighbourhoods and Environment Act 2005

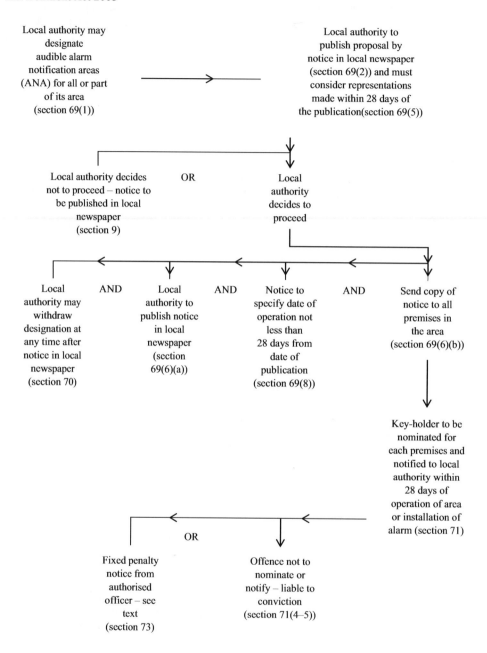

There is no requirement to notify newcomers to the area after the original designation although the local authority may wish to consider a mechanism for such notification.

If the local authority decides not to go ahead, it must publish a notice to that effect (section 69(9)).

Withdrawal of designation

The designation may be withdrawn by publishing the proposal with an implementation date and notifying all of the premises in the area (section 70).

Notification of nominated key-holders

In a designated area, the occupier, or owner where there is no occupier, must nominate a key-holder and then notify the local authority in writing within 28 days of the operation of the designation, or 28 days from installation, of that holder's name, address and telephone number.

A person eligible to be the key-holder:

(a) must hold sufficient keys to gain access;
(b) normally resides or is situated in the vicinity;
(c) has information that will enable him to silence the alarm;
(d) has agreed to be the nominated key-holder;
(e) for residential premises, is not the occupier or a key-holding company;
(f) for non-residential premises, is the responsible person or acts on their behalf or it is a key-holding company (sections 71 and 72(3–8)).

Offences

It is an offence to fail to nominate or notify details of a key-holder and the penalty on conviction is a fine not exceeding level 3 (section 71(4–5)).

Fixed penalties

Where it appears that an offence has been committed, an authorised officer may issue a fixed penalty notice and this, if paid within 14 days, removes the liability to conviction (section 73). The amount of the fixed penalty is set by the 2017 regulations at between £50 and £80 at the discretion of the local authority and with discounts for early payment. Sums collected by the local authority may be used to offset the costs of this procedure and other identified functions under the Noise Act 1996 and in dealing with statutory noise nuisances (section 75). Authorised officers are given the power to request the name and address of offenders where he proposes to issue a fixed penalty notice (section 76).

Power of entry

An authorised officer is able to enter a premises, other than by force, in order to silence an intruder alarm in any part of the district and not just in a designated area provided that:

• the alarm has been sounding for more than 20 minutes or intermittently for more than one hour; and

- the sounding of the alarm is likely to give reasonable cause for annoyance to persons living or working in the vicinity; and
- if the premises is in a notification area and reasonable steps have been taken to get the nominated key-holder to silence it (sections 77 and 79).

The authorised officer can also take necessary steps to silence the alarm (section 79(2)).

An authorised officer may apply to a Justice of the Peace for a warrant to enter a premises to silence an alarm but the Justice of the Peace must be satisfied that it has not been possible to enter without the use of force (section 78). Local authorities may recover costs of entry (section 79).

Chapter 4

POLLUTION PREVENTION

ENVIRONMENTAL PERMITTING

Introduction

The Environmental Permitting system implements the requirements of the Integrated Pollution Prevention and Control (IPPC) Directive, which applies to the more complex and potentially polluting industries and deals with any likely significant pollutant emission to air, water or land. It also addresses noise, energy efficiency, waste minimisation, prevention of accidental emissions and site protection and restoration.

The Environmental Permitting system is designed to allow regulators to focus on medium- and high-risk operations whilst continuing to protect the environment and human health.

The Environmental Permitting (England and Wales) Regulations 2016 provide industry, regulators and others with a permitting and compliance system, including systems for discharge, consenting, groundwater authorisations and radioactive substances regulation.

Larger industrial installations are subject to integrated controls on all likely significant emissions. Most are regulated by the Environment Agency and are often referred to as 'Part A(1)' activities. These are regulated for emissions to air, land, water and other environmental considerations.

Local authorities regulate premises with 'Part A(2)' (comparatively less polluting) installations for air, land, water and other environmental considerations and 'Part B' activities (smaller installations), which are generally limited to air pollution control.

The Part B regime is known as Local Air Pollution Prevention and Control (LAPPC) and the A2 regime as Local Authority Industrial Pollution Prevention and Control (LA-IPPC).

Part A(1) and A(2) installations

The IPPC system as enforced by both the Environment Agency and local authorities for the installations allocated to them applies an integrated approach to ensure a high level of protection to the environment as a whole and to prevent and, where that is not practicable, reduce emissions to acceptable levels. The conditions set in granting permits are to be based on the use of 'best available techniques' (BAT) – see below. The regulations also include provisions relating to energy efficiency, site restoration, noise, odour, waste minimisation, accident prevention and heat and vibrations.

Part B installations

Regulated by local authorities under LAPPC, these are also subject to conditions based on BAT but only so far as these are related to emissions to air.

Definitions

Activities 'are industrial activities forming part of an installation' and are defined as 'activities of any nature, whether:

(a) industrial or commercial or other activities; or
(b) carried on at particular premises or otherwise,

and includes (with or without other activities) the depositing, keeping or disposal of any substance'(Pollution Prevention and Control Act 1999 section 1(2)).
    Activity is defined more specifically in Schedule 1 Part 2 of the regulations.

Change in operation – a change in the nature or functioning, or an extension, of the installation which may have consequences for the environment. A change in operation could entail either technical alterations or modifications in operational or management practices, including changes to raw materials or fuels used and to the installation throughput (paragraph 5, schedule 5).

Emission means:

(a) in relation to a Part A installation, the direct or indirect release of substances, vibrations, heat or noise from individual or diffuse sources in the installation into the air, water or land;
(b) in relation to a Part B installation, the direct release of substances or heat from individual or diffuse sources in the installation into the air;
(c) in relation to a solvent emission activity, the direct or indirect release of substances from individual or diffuse sources in the regulated facility into the air;
(d) in relation to Part B mobile plant, the direct release of substances or heat from the mobile plant into the air;
(e) in relation to a waste operation, the direct or indirect release of substances, vibrations, heat or noise from individual or diffuse sources related to the operation into the air, water or land;
(f) in relation to a mining waste operation, the direct or indirect release of substances, vibrations, heat or noise from individual or diffuse sources related to the operation into the air, water or land;
(g) in relation to a radioactive substances activity, the direct or indirect release of radioactive material or radioactive waste;
(h) in relation to a small waste incineration plant, the direct or indirect release of substances from individual or diffuse sources in the regulated facility into the air or water;
(i) in relation to a medium combustion plant, the release of substances from the plant into the air;
(j) in relation to a specified generator, the release of substances from the plant into the air (regulation 2(1)).

Emission limit value means the mass, expressed in terms of specific parameters, concentration or level of an emission, which may not be exceeded during one or more periods of time.

Environmental pollution means pollution of the air, water or land which may give rise to any harm; and for the purposes of this definition (. . .)

(a) pollution includes pollution caused by noise, heat or vibrations or any other kind of release of energy; and

(b) air includes air within buildings and air within other natural or man-made structures above or below ground.

In this definition, harm means:

(a) harm to the health of human beings or other living organisms;

(b) harm to the quality of the environment, including:

  (i) harm to the quality of the environment taken as a whole;

  (ii) harm to the quality of the air, water or land; and

  (iii) other impairment of, or interference with, the ecological systems of which any living organisms form part;

(c) offence to the senses of human beings;

(d) damage to property; or

(e) impairment of, or interference with, amenities or other legitimate uses of the environment (expressions used in this paragraph having the same meaning as in Council Directive 96/61/EC) (Pollution Prevention and Control Act 1999 section 1(1)).

Installation means (except where used in the definition of 'excluded plant'):

(a) a stationary technical unit where one or more activities are carried on; and

(b) any other location on the same site where any other directly associated activities are carried on, and references to an installation include references to part of an installation.

Mobile plant means any of the following:

(a) Part B mobile plant;

(b) Waste mobile plant

(c) mobile medium combustion plant (regulation 2(1)).

Operator, in relation to a regulated facility, means:

(a) the person who has control over the operation of the regulated facility;

(b) if the regulated facility has not yet been put into operation, the person who will have control over the regulated facility when it is put into operation; or

(c) if a regulated facility authorised by an environmental permit ceases to be in operation, the person who holds the environmental permit (regulation 7).

Pollutant means any substance liable to cause pollution (regulation 2(1)).

Pollution, in relation to a water discharge activity or groundwater activity, means the direct or indirect introduction, as a result of human activity, of substances or heat into the air, water or land which may:

(a) be harmful to human health or the quality of aquatic ecosystems or terrestrial ecosystems directly depending on aquatic ecosystems;
(b) result in damage to material property; or
(c) impair or interfere with amenities or other legitimate uses of the environment.

Pollution, other than in relation to a water discharge activity or groundwater activity, means any emission as a result of human activity which may:

(a) be harmful to human health or the quality of the environment;
(b) cause offence to a human sense;
(c) result in damage to material property; or
(d) impair or interfere with amenities or other legitimate uses of the environment.

Regulated facility means any of the following:

(a) an installation;
(b) mobile plant;
(c) a waste operation;
(d) a mining waste operation;
(e) a radioactive substances activity;
(f) a water discharge activity;
(g) a groundwater activity (regulation 8).
(h) a small waste incineration plant;
(i) a solvent emission activity;
(j) a flood risk activity
(k) a medium combustion plant;
(l) a specified generator (regulation 8(1)).

Substantial change means a change in operation of an installation which in the regulator's opinion may have significant negative effects on human beings or the environment and includes:

(a) in relation to a Part A installation, a change in operation which in itself meets the thresholds, if any, set out in part 2 of schedule 1; and
(b) in relation to an incineration plant or co-incineration plant for non-hazardous waste, a change in operation which would involve the incineration or co-incineration of hazardous waste.

Best available technique (BAT)

Operators of Part A(1) installations are required to use BAT to achieve a high level of protection of the environment taken as a whole and, for Part B, in relation to emissions into the air. BAT is defined in the IPPC directive as:

the most effective and advanced stage in the development of activities and their methods of operation which indicates the practical suitability of particular techniques for providing in principle the basis for emission limit values designed to prevent and, where that is not practicable, generally to reduce emissions and the impact on the environment as a whole.

In essence, BAT is a technique that balances the costs to the operator with the benefits to the environment, with a consideration of local circumstances and provides the main basis for the setting of emission limit values – see definition on p. 160.

Guidance

1. *Integrated Pollution Prevention and Control – a Practical Guide, 4th ed. 2005 – DEFRA*
   This a general guide to the whole of the IPPC system and covers all aspects. It is primarily intended for Environment Agency regulators in relation to A(1) installations and persons affected by it.
2. Environmental permitting guidance relevant to the Local Authority Pollution Control regime, in particular the *LAPPC General Guidance Manual* and *Controlling Pollution from Industry: Regulation by Local Authorities – A Short Guide. April 2012*
   These three manuals are the principal guidance to local authorities in exercising their functions under the Environmental Permitting Regulations, i.e. LA-IPPC and LAPPC, and are applicable to all procedures in this chapter.
3. *Process Guidance Notes (PGNs) and Sector Guidance Notes (SGNs)*
   These are issued by the Secretary of State under regulation 64 and form statutory guidance on what constitutes BAT for LA-IPPC installations for each of the main sectors regulated.
4. *LAQM Notes*
   These are issued by DEFRA and form additional guidance to local authorities on a wide range of both technical and administrative issues.
5. *Explanatory Memorandum to Environmental Permitting (England and Wales) Regulations 2016*
6. *Environmental Permitting Guidance: The IPPC Directive Part A(1) Installations and Part A(1) Mobile Plant for the Environmental Permitting (England and Wales) Regulations 2010*
7. *Environmental Permitting Guidance: Core Guidance for the Environmental Permitting (England and Wales) Regulations 2010*
8. *Environmental Permitting General Guidance Manual on Policy and Procedures for A2 and B Installations Local Authority Integrated Pollution Prevention and Control (LA-IPPC) and Local Authority Pollution Prevention and Control (LAPPC). General Guidance Manual on Policy and Procedures for A2 and B Installations, Part B of Manual Annexes and Parts C and D. General guidance manual: application forms, specimen notices and consultation letter.*

All of the publications above are available on the DEFRA website at www.defra.gov.uk/industrial-emissions/las-regulations/guidance.

Instruments

An instrument means a notice, notification, certificate, direction or form under these regulations. An instrument must be in writing.

Instruments may be served on or given to a person by personal delivery, leaving it at the person's proper address or sending it by post or electronic means to the person's proper address.

In the case of a body corporate, an instrument may be served on or given to the secretary or clerk. In the case of a partnership, an instrument may be served on or given to a partner or a person having control or management of the partnership business.

The proper address is the last known address except:

(a) in the case of a body corporate or their secretary or clerk, the registered or principal office of that body, or the e-mail address of the secretary or clerk;

(b) in the case of a partnership or a partner or person having control or management of the partnership business, the principal office of the partnership, or the e-mail address of a partner or a person having that control or management (regulation 10).

Application to the Crown

Whilst the Crown is bound by the Environmental Permitting Regulations, it is not criminally responsible and cannot be prosecuted for non-compliance with notices etc. However, local authorities may apply to the High Court to have the Crown's actions, or lack of them, declared unlawful (regulation 11 and schedule 4).

Obtaining of information

By notice served on any person, local authorities may require to be given such information as is specified in the notice and in such form and time/period as is specified (regulation 61(1)).

PERMITTING OF ACTIVITIES BY LOCAL AUTHORITIES

References

Environment Act 1995
Pollution Prevention and Control Act 1999
Environmental Permitting Regulations (England and Wales) Regulations 2016
Environmental Permitting: General Guidance Manual on Policy and Procedures for A2 and B Installations, April 2012

Extent

These provisions apply to England and Wales. This is EU derived legislation but retained as domestic law despite EU withdrawal. Similar provisions occur in Scotland and Northern Ireland.

Scope

These procedures apply to activities, installations and mobile plant, prescribed in parts A(2) and B of schedule 1 to the Environmental Permitting Regulations 2016. These activities etc. are controlled and permitted by the local authority. In each case, the operator (definitions on p. 159) is required to obtain a permit from the local authority before operating the installation or plant.

Applications (FC27)

Applications are to be made in writing to the local authority, with the prescribed fee (see p. 167), and must contain all of the information specified in regulation 14 and paragraph 2 of part 1 of schedule 5. Application forms are in part C of the Secretary of State Guidance Manual.

For novel or complex installations, staged applications are possible by agreement between the operator and the local authority.

Local authorities are empowered to request by notice additional information sufficient for them to determine the application. If the information requested has not been provided within the time specified in the notice, the application is deemed to have been withdrawn (schedule 5, part 1, paragraph 4).

Consultations

Subject to the exceptions in schedules 2 and 4 (see FC33), the local authority must send copies of the application within the communication consultation period to the public consultees (a person whom the regulator considers is affected by, is likely to be affected by or has an interest in an application). The consultation period is 30 working days starting on the day the regulator receives a duly made application, unless national security or confidentiality is an issue, in which case the 30 days starts on the determination date.

The local authority must have regard to make representations from the public consultees before determining the application (schedule 5, paragraph 11).

The Environment Agency should always be a public consultee as should be the local authority when the Environment Agency is dealing with applications for part A(1). There is a specimen consultation letter in part C of the Guidance Manual.

Determination of applications

The general principles against which all applications, parts A and B, are to be considered are:

(i) that all appropriate preventative measures must be taken against pollution and in particular the application of BAT (see p. 160);
(ii) that no significant pollution will be caused.

The additional general principles to be considered for part A activities are:

(a) that satisfactory methods will be in place to avoid waste production and, where it is produced, for its recovery and disposal;
(b) that there is efficient use of energy;
(c) that measures are taken to prevent accidents and to limit their consequences;
(d) that, on the cessation of operation of the installation, satisfactory measures will be taken to avoid pollution and return the site to a satisfactory state.

The local authority must determine applications generally within three months of submission unless a longer period is agreed. In the case of transfer, it is two months and, where national security and confidentiality is an issue, it is four months. Failure to do so is deemed to be a refusal (paragraph 15, part 1, schedule 5).

Content and form of the permit

An environmental permit must specify the operator and the regulated facility whose operation it authorises. It can be in electronic form and must include a map, plan or other description of the site showing the geographical extent of the site of the regulated facility, but the map etc. does not apply to mobile plant or radioactive substances activities (regulation 14)

Conditions may be attached to the permit requiring the operator to carry out works or do other things on land which the operator is not entitled to do without obtaining the consent of another person (regulation 15).

A regulator cannot authorise the operation of more than one regulated facility under a single environmental permit but can authorise a single environmental permit for the operation by the same operator of more than one mobile plant, standard facility, regulated facility on the same site or more than one radioactive substances activity (regulation 17).

If the local authority decides to issue a permit, it may attach conditions (paragraph 12, part 1, schedule 5) appropriate to the following issues and in accordance with sectoral technical guidance. For all activities, these are:

(a) emission limit values (EMLs) or equivalent parameters for pollutants, in particular those likely to be emitted in significant quantities. These will normally be based on BAT (see p. 160), taking account of the particular characteristics and the local environment of the plant;
(b) best available technique for preventing or reducing emissions.

It is suggested that content and conditions should cover:

(a) long-distance and transboundary pollution;
(b) the protection of soil and groundwater and the management of waste;
(c) precautions to protect the environment when the installation is not operating normally, e.g. during start-up;
(d) site monitoring and remediation;
(e) the ongoing monitoring of emissions and the submission of reports to the local authority;
(f) notification procedures to deal with incidents or accidents;
(g) taking account of conditions for emissions to water specified by the Environment Agency; and
(h) avoiding conflict with other legislation prescribing release levels.

Standard rules

The Secretary of State is empowered to make standard rules (generally known as process and sector guidance notes) for certain types of installation, which can be used by a local authority instead of site-specific conditions. Standard rules will, by their nature, be suitable for industry sectors where installations share similar characteristics (regulations 26 to 30).

Refusals

A permit must be granted or refused following an application. The permit cannot be granted by the local authority unless it considers that the applicant will be the person who will have control after the permit is issued and will ensure that the installation or mobile plant is operated so as to comply with the environmental permit (paragraphs 12 and 13, part 1, schedule 5). This may be, for example, where there is reason to believe that the operator lacks the management systems or competence to run the installation according to the application or any permit conditions. There is an appeal procedure against refusal (see below).

Transfers (FC29)

Transfers of permits between operators must be the subject of a joint application to the local authority, who may agree to the transfer unless it considers that the conditions attached will not be complied with (regulation 21). There is an appeal against refusal (regulation 31) (see below).

Permit reviews

The local authority must periodically review the environmental permits and make periodic inspections of regulated facilities (regulation 34). It is suggested that the reviews be carried out where:

(a) the installation causes such significant pollution that the local authority must change the emission limit values;

(b) substantial changes in best available technique make it possible to reduce emissions significantly without excessive costs; and

(c) operators must change techniques for reasons of safety.

Variation of conditions (FC30)

Operators of permitted installations and mobile plant must make an application for variation to the local authority of their intention to make a 'substantial change' and the rules for consultation and public participation apply (form in part C of the General Guidance Manual) (schedule 5).

The local authority may vary the conditions at any time on its own initiative. The prescribed fee must accompany applications (see 'Charges' on p. 167). The application or local authority-led variation is subject to the same procedures as an initial application, including a period of public participation and consultation before determination.

There is an appeal against the change of conditions and against a refusal to grant an application for change (see below) (regulation 20).

Surrender of permits (FC31 and FC32)

An environmental permit may be surrendered by the operator of an activity where he intends to cease operation. The operator must notify the local authority about a Part B installation, except a waste operation, mobile plant or stand-alone water discharge activity or stand-alone groundwater activity. In all other cases, an application must be made to the local authority. The local authority may accept the surrender of the permit if it is satisfied that the steps to be taken to implement the closure are appropriate to avoid risk of pollution and will return the site to a satisfactory state.

In the case of partial surrenders, the regulator must serve a notice on the operator specifying that it is necessary to vary the conditions, the variation and the date the variation takes effect.

Applications must be determined in the same way as initial applications, variations and transfers and failure to do so is deemed to be a refusal. There is an appeal against refusal to accept surrender (regulations 24 and 25).

Enforcement notices (FC33)

If the local authority is of the opinion that an environmental permit has been is or is likely to be contravened, they may serve an enforcement notice on the operator. The notice must specify:

(a) the contravention;
(b) the steps necessary to remedy it; and
(c) the time period allowed. Non-compliance is an offence – see below (regulation 36).

Suspension notices

Where the local authority believes that the continued operation of the activities will involve a risk of serious pollution, it may serve a suspension notice on the operator. This applies whether or not the activities are permitted and may deal with a failure to comply with conditions of a permit. Alternatively, the situation may be dealt with by use of its powers under regulation 57 – see below. The suspension notice must:

(a) specify what the risk of serious pollution is;
(b) specify the steps to be taken;
(c) specify the period within which the steps must be taken;
(d) state that any permit shall cease to have effect to the extent specified in the notice until the notice is withdrawn;
(e) where the activities will continue in part, state any additional measures to (b) above that must be taken.

The local authority may withdraw the notice at any time and must do so when the risk has been removed (regulation 37).

Prevention or remedying of pollution (FC34)

Where the local authority believes that the operation of an activity where an environmental permit is in place involves a risk of serious pollution or there has been the commission of an offence by the operator that causes pollution, it may arrange itself for steps to be taken to remove that risk or pollution. Before taking any steps, the local authority must give the operator five days' notice.

The costs of the necessary works may be recovered from the operator unless the operator can show that there was no risk or the costs, or part of them, were incurred unnecessarily (regulation 57).

Revocation notices (FC35)

The local authority may revoke a permit at any time, in whole or in part, by the service of a revocation notice. This is a wide power that can be used whenever the local authority considers it to be appropriate. The local authority may vary the permit conditions to take into account the revocation.

The notice must be served on the operator and specify:

(a) the reasons for the revocation;
(b) in the case of a partial revocation:

    (i) the extent to which the permit is being revoked;
    (ii) any variation to the conditions of the permit; and

(d) the date on which the revocation will take place, at least 20 working days after the notice is served (regulations 22 and 23).

Unless the local authority withdraws a revocation notice, an environmental permit ceases to have effect on the date specified in the notice:

(a) in the case of a revocation in whole, entirely; or
(b) in the case of a partial revocation, to the extent of the part revoked.

Public registers

Local authorities must maintain public registers containing information on all Part A(2) and Part B permits issued by them for installations in their areas and also details of installations in their area regulated by the Environment Agency (regulation 46). The content of the register is prescribed by schedule 27, paragraph 1. The register may be in any form and the necessary content of registers is set out in annex 13 of the General Guidance Manual.

Commercial confidentiality

The local authority must exclude what it considers commercially confidential information or if it receives a notice from the operator that the information is confidential (see FC28 and FC36). Information is commercially confidential if the information is commercially or industrially confidential in relation to any person (regulation 48).

If the local authority considers that information may be confidential but has not received an objection notice, it must give notice of that view to the operator. The operator must, within 15 working days after the date of the notice, give consent to the regulator to include the information on the register or give an objection notice to the regulator (regulation 49).

Charges

The Secretary of State, under the powers of regulation 66, determines the type and level of charges that may be made by local authorities in relation to their activities for environmental permits. This scheme is reviewed annually and is set out in the local authority charging schemes: the Local Authority Permits for Part B Installations and Mobile Plant

(Fees and Charges) (England) Scheme 2017 and the Local Authority Permits for Part A(2) Installations and Mobile Plant (Fees and Charges) (England) Scheme 2017. The charges may be raised relating to:

(a) permit applications and 'substantial change';
(b) late applications;
(c) variations;
(d) transfers;
(e) surrender;
(f) subsistence charges on an annual basis.

The fees are based on a risk-based approach in line with the 'polluter pays' principle. All processes, except waste oil burners and petrol stations, are required to be risk-rated to establish the inspection frequency and hours that the local authority needs to spend on them.

Appeals

Unusually for environmental health enforcement work, rights of appeal are to the Secretary of State and not to a court of law. The operator may appeal against local authority decisions relating to:

(a) refusal of an application for a permit (regulation 13);
(b) refusal of an application for the variation of a permit (regulation 20);
(c) the service of a revocation, enforcement or suspension notice (regulations 22, 36 and 37);
(d) determination that information is not commercially confidential (regulation 53);
(e) refusal of an application to transfer or surrender a permit (regulations 21, 24 and 25);
(f) the service of a variation notice on the local authority's initiative (regulation 20).

The full procedure with timescales is set out in schedule 6 to the Environmental Permitting Regulations. The Secretary of State has the power to affirm or quash the local authority decisions and to alter the terms of any conditions (regulation 31(4) and schedule 6).

Offences

Offences committed against any of the requirements of the Environmental Permitting Regulations are punishable, on summary conviction, of a fine and/or up to 12 months' imprisonment. Conviction in the Crown Court may lead to a fine and/or to imprisonment for up to five years (regulations 38 and 39).

FC27 Permitting of scheduled activities by local authorities – Regulation 14 and Schedule 5 Environmental Permitting Regulations 2010

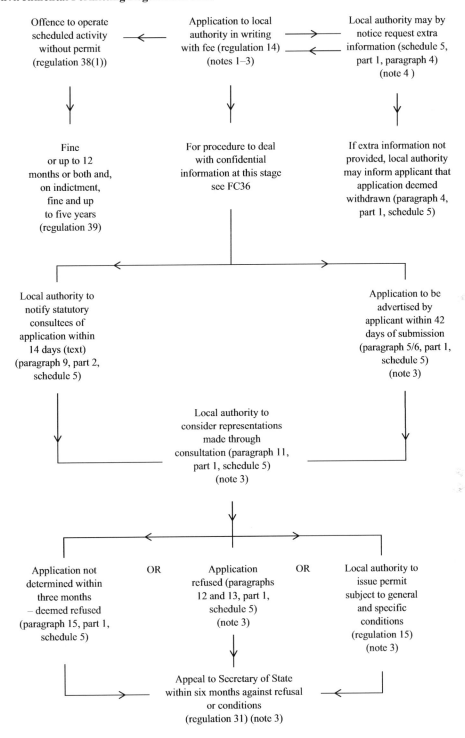

Notes

1. For details to be included, see regulation 14 and part 1, schedule 5.
2. For consolidation of a number of regulated facilities with a single permit (regulation 18).
3. Details of these events to be included in public register (regulation 46 and paragraph 1, schedule 27).
4. Request for information notice form in part D of General Manual.

FC28 Handling of commercially confidential and national security information by local authorities – Regulations 47 to 55 Environmental Permitting Regulations 2010

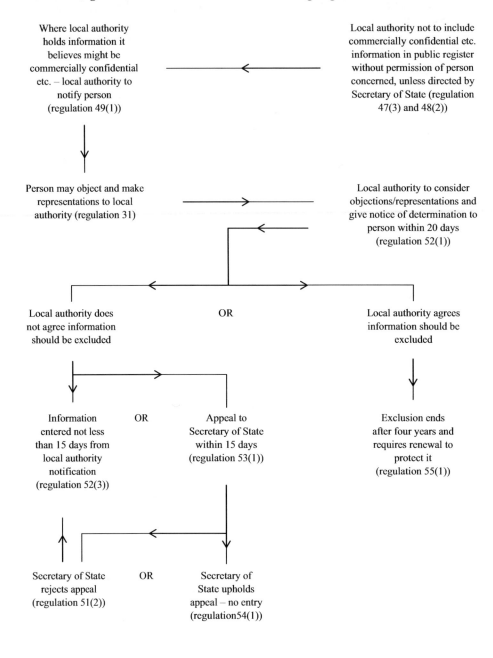

Where local authority holds information it believes might be commercially confidential etc. – local authority to notify person (regulation 49(1))

Local authority not to include commercially confidential etc. information in public register without permission of person concerned, unless directed by Secretary of State (regulation 47(3) and 48(2))

Person may object and make representations to local authority (regulation 31)

Local authority to consider objections/representations and give notice of determination to person within 20 days (regulation 52(1))

OR

Local authority does not agree information should be excluded

Local authority agrees information should be excluded

Information entered not less than 15 days from local authority notification (regulation 52(3))

OR

Appeal to Secretary of State within 15 days (regulation 53(1))

Exclusion ends after four years and requires renewal to protect it (regulation 55(1))

Secretary of State rejects appeal (regulation 51(2))

OR

Secretary of State upholds appeal – no entry (regulation54(1))

FC29 Transfer of permits – Regulation 21 and Schedule 5 Environmental Permitting Regulations 2010

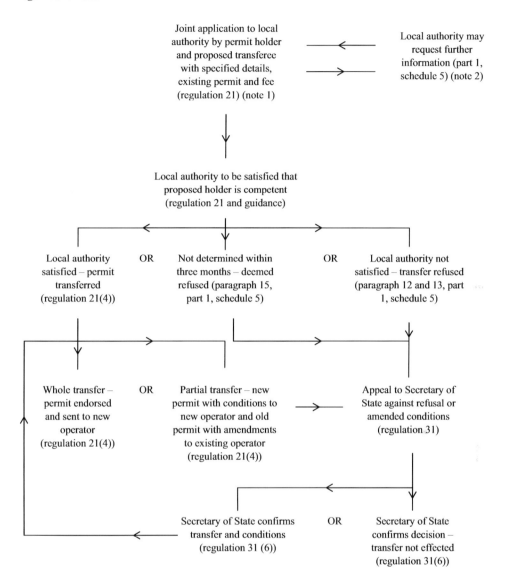

Joint application to local authority by permit holder and proposed transferee with specified details, existing permit and fee (regulation 21) (note 1)

Local authority may request further information (part 1, schedule 5) (note 2)

Local authority to be satisfied that proposed holder is competent (regulation 21 and guidance)

Local authority satisfied – permit transferred (regulation 21(4))

OR

Not determined within three months – deemed refused (paragraph 15, part 1, schedule 5)

OR

Local authority not satisfied – transfer refused (paragraph 12 and 13, part 1, schedule 5)

Whole transfer – permit endorsed and sent to new operator (regulation 21(4))

OR

Partial transfer – new permit with conditions to new operator and old permit with amendments to existing operator (regulation 21(4))

Appeal to Secretary of State against refusal or amended conditions (regulation 31)

Secretary of State confirms transfer and conditions (regulation 31 (6))

OR

Secretary of State confirms decision – transfer not effected (regulation 31(6))

Notes
1. Application form in part C of General Manual.
2. Specimen request for information notice in part D of the General Manual.
3. Information to be included in public register (regulation 46 and paragraph 1, schedule 24).

FC30 Variations to permit conditions – Regulation 20 and Schedule 5 Environmental Permitting Regulations 2010

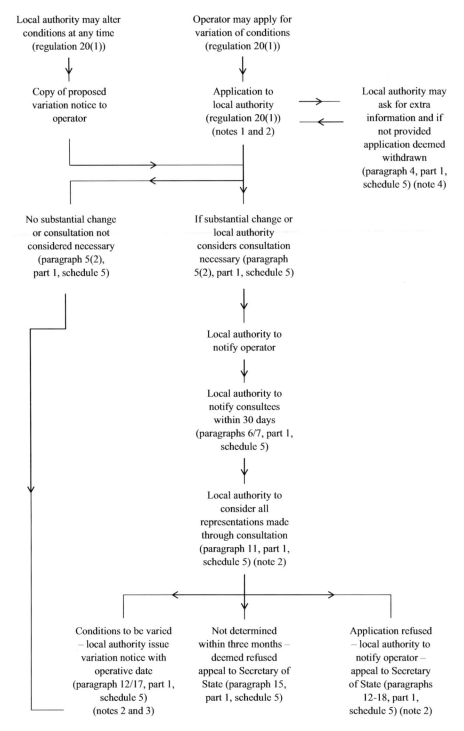

Notes

1. The details to be included in the application are set out in regulation 14 and paragraph 2 of part 1 of schedule 5.
2. Information to be included in public register (regulation 46 and paragraph 1, schedule 24).
3. Model form in part C of General Manual.
4. Specimen request for information notice in part D of General Manual.

FC31 Application for surrender of an environmental permit (Part A processes) – Regulation 25 Environmental Permitting Regulations 2010

Operator ceases or intends to cease operating whole
or part of an installation other than part B
installation, mobile plant, stand-alone water
discharge or groundwater activity (regulation 25(1))

Operator makes application to local authority
with specified details as if an initial application
is being made (text)
and a date of surrender not less than 20 days
from notification (regulation 24(3)) (note 1)

Permit ceases to be effective on date
specified in notification (regulation 24(4))

For partial surrenders local authority may
vary conditions for remaining permit
by serving a variation notice (see FC30)
(regulation 24(6))

Notes
1. Application form in part C of General Manual.
2. Part 1 of schedule 5 of the Environmental Permitting Regulations 2010 applies to surrender of an environmental permit.
3. The local authority can:

   - request more information (paragraph 4, part 1, schedule 5);
   - refuse an application (paragraph 12, part 1, schedule 5).

4. If application is refused or deemed to be refused, the operator can appeal to the Secretary of State (regulation 31).

FC32 Notification of surrender of an environmental permit (Part B processes) – Regulation 24 Environmental Permitting Regulations 2010

Operator ceases or intends to cease operating whole
or part of a part B installation, mobile plant,
stand-alone water discharge or groundwater activity
(regulation 24(1))

Operator to notify local authority with specified
details as if an initial application is being
made(text)
and a date of surrender not less than 20 days
from notification (regulation 24(3)) (note 1)

Permit ceases to be effective on date
specified in notification (regulation 24(4))

For partial surrenders local authority may
vary conditions for remaining permit
by serving a variation notice (see FC30)
(regulation 24(6))

Notes

1. Application form in part C of General Manual.
2. Part 1 of schedule 5 of the Environmental Permitting Regulations 2010 applies to surrender of an environmental permit.

FC33 Enforcement and suspension notices – Regulations 36 and 37 Environmental Permitting Regulations 2010

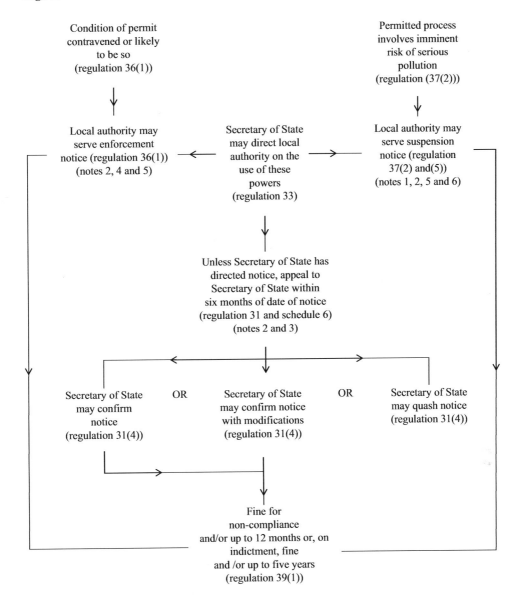

Notes

1. Where these circumstances exist, the local authority may serve a suspension notice unless taking action itself under regulation 57 – see FC35.
2. Details to be entered in public register (regulation 46 and paragraph 1, schedule 24).
3. The notices are not suspended during the appeal process (regulation 31(9)).
4. Instead of serving an enforcement notice, the local authority may prosecute for the contravention (regulation 42).
5. An enforcement or suspension notice may be withdrawn at any time (regulations 36(7) and 37(8)).
6. Specimen Suspension Notice in part D of the General Manual.

FC34 Local authority powers to prevent or remedy pollution – Regulation 57 Environmental Permitting Regulations 2010

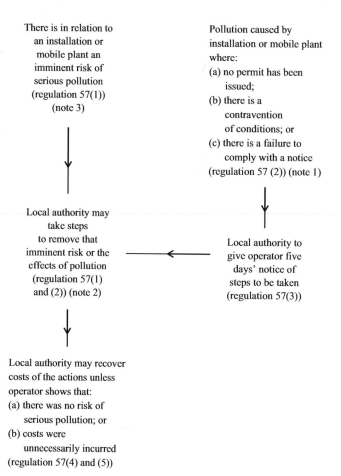

There is in relation to
an installation or
mobile plant an
imminent risk of
serious pollution
(regulation 57(1))
(note 3)

Pollution caused by
installation or mobile plant
where:
(a) no permit has been
    issued;
(b) there is a
    contravention
    of conditions; or
(c) there is a failure to
    comply with a notice
(regulation 57 (2)) (note 1)

Local authority may
take steps
to remove that
imminent risk or the
effects of pollution
(regulation 57(1)
and (2)) (note 2)

Local authority to
give operator five
days' notice of
steps to be taken
(regulation 57(3))

Local authority may recover
costs of the actions unless
operator shows that:
(a) there was no risk of
    serious pollution; or
(b) costs were
    unnecessarily incurred
(regulation 57(4) and (5))

Notes
1. This power may be implemented in addition to the prosecution of committed offences.
2. There is no provision for appeal against these actions, only a challenge to the recovery of costs.
3. Where these circumstances exist, the local authority may implement this procedure or serve a suspension notice – see FC33.

FC35 Revocation of permits by local authority – Regulations 22 and 23 Environmental Permitting Regulations 2010

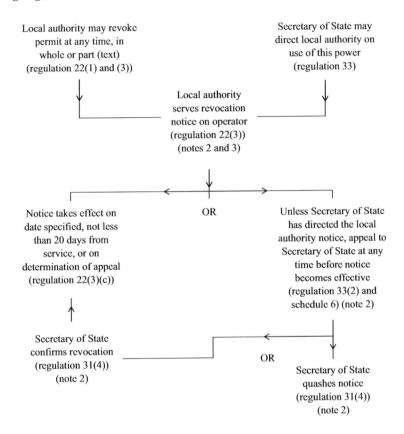

Local authority may revoke
permit at any time, in
whole or part (text)
(regulation 22(1) and (3))

Secretary of State may
direct local authority on
use of this power
(regulation 33)

Local authority
serves revocation
notice on operator
(regulation 22(3))
(notes 2 and 3)

OR

Notice takes effect on
date specified, not less
than 20 days from
service, or on
determination of appeal
(regulation 22(3)(c))

Unless Secretary of State
has directed the local
authority notice, appeal to
Secretary of State at any
time before notice
becomes effective
(regulation 33(2) and
schedule 6) (note 2)

Secretary of State
confirms revocation
(regulation 31(4))
(note 2)

OR

Secretary of State
quashes notice
(regulation 31(4))
(note 2)

Notes
1. The revocation is suspended pending determination of the appeal or withdrawal of the notice (regulation 31(9)(b)).
2. Details to be entered in public register (regulation 46 and paragraph 1, schedule 24).
3. Specimen Revocation Notice in part D of the General Manual.

FC36 Applications to local authority to exclude commercially confidential information from public register – Regulations 48 to 55 Environmental Permitting Regulations 2010

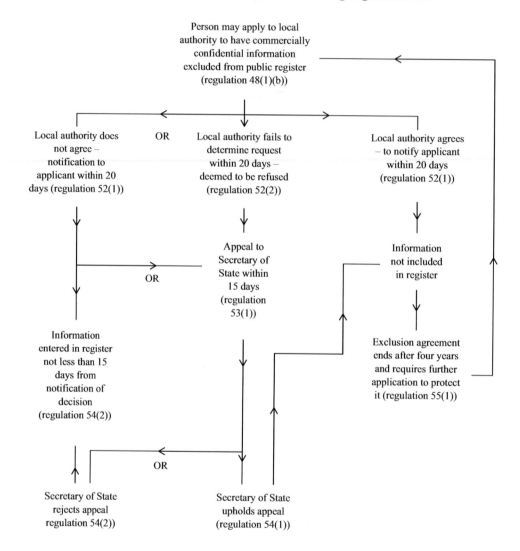

PART 3

FOOD SAFETY AND FOOD STANDARDS

FOOD CONTROL AND FOOD HYGIENE LEGISLATION

Legal structure

The current structure of food safety legislation for the UK took effect on 1 January 2006. This stems from two principal EC regulations:

(a) EC Regulation 852/2004 on the hygiene of foodstuffs; and
(b) EC Regulation 853/2004, laying down specific hygiene rules for food of animal origin.

Several others that contain important procedural matters relating to enforcement, including EC Regulation 854/2004, laying down specific rules for the organisation of official controls on products of animal origin intended for human consumption, support these. The UK decision to leave the European Union places serious doubt about what form food safety and food hygiene regime will be. The structure and detail of the legislation is very difficult to predict at the time of writing and no attempt will be made to predict the new regime. However, it is likely that the provisions will not change as we have to compete in a European and World market and we will need to maintain standards in order to do this.

These regulations have both a direct effect on UK enforcement, in that some of the provisions are directly enforceable, e.g. the registration and approval of food premises, and an indirect effect that has required new legislation in the UK to implement them, e.g. the use of hygiene improvement notices. In the light of UK devolution, the legislation dealing with these issues is a matter for the devolved governments. This work is based on the provisions in England but similar legislation is to be found in Northern Ireland, Scotland and Wales. The principal English regulations affecting this work are:

(a) the Food Safety and Hygiene (England) Regulations 2013 (FSH (E) Regulations); and
(b) the Official Feed and Food Controls (England) Regulations 2009 (OF and FC (E) Regulations).

The Food Safety Act 1990 is the main food safety legislation enacted in the UK and although it remains with some amendments, its enforcement provisions are likely to be used mainly in respect of food standards matters and not for food hygiene, where the provisions of the 2013 regulations take effect.

These Regulations implement Regulation (EC) No. 178/2002 of the European Parliament. They also implement a number of EU instruments:

Regulation (EC) No. 852/2004 on the hygiene of foodstuffs
Regulation (EC) No. 853/2004 laying down specific hygiene rules for food of animal origin
Regulation (EC) No. 854/2004 laying down specific rules for the organisation of official controls on products of animal origin intended for human consumption
Commission Regulation (EC) No. 2073/2005 on microbiological criteria for foodstuffs and
Commission Regulation (EC) No. 2075/2005 laying down specific rules on official controls for *Trichinella* in meat.

Codes of Practice (CoP) and Practice Guidance (PG) form a very important part of the procedural requirements.

The Food Law Code of Practice (England) March 2017 was issued by the Secretary of State under section 40 of the Food Safety Act 1990 and regulation 26 of the Food Safety and Hygiene (England) Regulations 2013. The Food Law Code of Practice gives instructions that local authorities must consider when enforcing food law. Local authorities need to follow and implement the relevant sections of the Code that apply. Food authorities are therefore required to follow and implement it, and food authorities who do not do so may find their decisions or actions successfully challenged or evidence ruled inadmissible.

The Food Law Practice Guidance (England) – Nov 2017 has a different status, being issued by the Food Standards Authority to assist food authorities with the enforcement of food law, and is non-statutory.

Authorised officers

An enforcement officer of the food authority enforces most of the procedures dealt with in this part. This is a person (whether or not an officer of the authority) who is authorised in writing, either generally or specifically, to act in matters arising under the Food Safety Act and the Food Safety and Hygiene (England) Regulations 2013.

There are specific requirements for the levels of qualifications and experience held by authorised officers carrying out inspections or other enforcement matters under food law. These are set out in the Code of Practice, chapter 4 and are noted in the procedures here.

Definitions

The following definitions are relevant to all procedures in this part:

1. Food

The definition in the Food Safety Act was changed by the Food Safety Act 1990 (Amendment) Regulations 2004 in order to bring it into line with EC Regulation 178/2002. It is as follows:

Food means any substance or product, whether processed, partially processed or unprocessed, intended to be, or reasonably expected to be, ingested by humans.
Food includes drink, chewing gum and any substance, including water, intentionally incorporated into the food during manufacture, preparation or treatment.

Food shall not include:

(a) feed;
(b) live animals unless they are prepared for placing on the market for human consumption;
(c) plants prior to harvesting;
(d) medicinal products within the meaning of Council Directives 65/65/ EEC and 92/73/EEC;
(e) cosmetics within the meaning of Council Directive 76/768/EEC;
(f) tobacco and tobacco products within the meaning of Council Directive 89/622/EEC;
(g) narcotic or psychotropic substances;
(h) residues and contaminants (section 1(1)).

2. Food business

Any business in the course of which commercial operations with respect to food or food sources are carried out (section1(1)).

This is 'any undertaking, whether for profit or not and whether public or private, carrying out any of the activities related to any stage of production, processing and distribution of food' (EC 178/2002, chapter 1, article 3).

Business includes the undertaking of a canteen, club, school, hospital or institution, whether carried on for profit or not, and any undertaking or activity carried on by a public or local authority (section1(1)).

3. Food business operator

This is 'the natural or legal person responsible for ensuring that the requirements of food law are met within the food business under their control' (EC 178/2002, chapter 1, article 3).

4. Specified EU provision

This means 'any provision of Regulation 178/2002 or the EU Hygiene Regulations that is listed in Column 1 of Schedule 2 and whose subject-matter is described in Column 2 of that Schedule' (Food Safety and Hygiene (England) Regulations 2013, regulation 2(1)).

5. The EU Hygiene Regulations

This means 'Regulation 852/2004, Regulation 853/2004, Regulation 854/2004, Regulation 2073/2005 and Regulation 2015/1375' (Food Safety and Hygiene (England) Regulations 2013, regulation 2(1)).

6. The Hygiene Regulations

This means the Food Safety and Hygiene (England) Regulations 2013) and the EU Hygiene Regulations' (Food Hygiene (England) Regulations 2013, regulation 2(1)).

General procedural provisions

1. Notices

In most cases, the forms to be used in the various procedures are specified in the Code of Practice at annexes 3, 4 and 5. The requirements relating to notices are:

    (a) all notices should be in writing;
    (b) they must be signed (including a facsimile signature) by the proper officer or any officer authorised in writing;
    (c) service may be effected by either:

        (i) delivering it to the person;
        (ii) by leaving it or sending it in a pre-paid letter addressed to him at the usual or last known residence;

(iii) for an incorporated company or body, by delivering it to the secretary or clerk at its registered office or by sending it in a pre-paid letter addressed to him there;

(iv) where it has not been possible to ascertain the name and address of an owner or occupier or if the premises are unoccupied, by addressing it 'the owner' or 'the occupier' and delivering it to someone on the premises or affixing it to the premises.

(Food Safety Act 1990, section 50 and similar provisions in the
Food Safety and Hygiene (England) Regulations 2013, regulation 30)

For some of the procedures here, the Code of Practice indicates acceptable, alternative methods of service, e.g. courier service, and these are indicated as appropriate.

2. Obstruction

Persons who obstruct anyone executing the 'Hygiene Regulations or Regulation 178/2002' or, without reasonable cause, fail to give assistance or information, or give false or misleading information are guilty of an offence (Food Safety Act 1990, section 33 and Food Safety and Hygiene (England) Regulations 2013, regulation 17).

3. Appeals

Where provision is made for appeals against the actions of authorised officers in these procedures, these are made to the magistrates' court within one month and may be followed by a further appeal to the Crown Court (Food Safety and Hygiene (England) Regulations 2013, regulations 22–24 and, in relation to approvals of premises dealing with foods of animal origin, the Official Feed and Food Control (England) Regulations 2009, regulations 12–13).

4. Due diligence

In any proceedings, it will be a defence to show that the person took all reasonable precautions and exercised due diligence to avoid the commission of the offence either by himself or by a person under his control (Food Safety Act 1990, section 21(1) and Food Safety and Hygiene (England) Regulations 2013, regulation 12).

5. Offences due to the fault of another person

Where the commission of an offence by a person is due to the act or default of another person, that other person is also guilty of an offence (Food Safety and Hygiene (England) Regulations 2013, regulation 11).

6. Time limit for prosecutions

No prosecution should take place under these procedures after the expiry of:

(a) three years from the commission of the offence; or
(b) one year from its discovery by the prosecutor,

whichever is the earlier (Food Safety Act 1990, section 34 and Food Safety and Hygiene (England) Regulations 2013, regulation 18).

7. Offences and penalties

Any person who contravenes any of the 'specified EU provisions' listed in schedule 2 of Regulation 178/2002 or the EU Hygiene Regulations is guilty of an offence. These provisions deal with the various relevant EU regulations. In addition, it is an offence not to comply with the 'Hygiene Regulations', which are the 2013 regulations and EC Regulations 852/2004, 853/2004, 854/2004, 2073/2005 and 2075/2005. The penalties for these offences are:

(a) on summary conviction, a fine not exceeding the statutory maximum; or

(b) on conviction on indictment, imprisonment for not exceeding two years, a fine or both (Food Safety and Hygiene (England) Regulations 2013, regulation 19).

A person guilty of obstruction is liable to a fine on the standard scale level 5, three months or both. There are other fines for specific offences.

Food Safety and Hygiene (England) Regulations 2013

The Regulations cover the following areas:

(a) create certain presumptions that specified food is intended for human consumption (regulation 3);

(b) provide that the Food Standards Agency is the competent authority except where it has delegated competence, and each food authority in its area or district (regulation 4);

(c) make provision for the execution of the Regulations, the EU Hygiene Regulations and Regulation 178/2002 (regulation 5);

(d) provide for the following enforcement measures to be available in respect of a food business operator –

(i) hygiene improvement notices (regulation 6);

(ii) hygiene prohibition orders (regulation 7);

(iii) hygiene emergency prohibition notices and orders (regulation 8);

(iv) remedial action notices (regulation 9), and

(v) detention notices (regulation 10);

(e) provide that where the commission of an offence is due to the act or default of another person, that other person commits the offence (regulation 11);

(f) provide that in any proceedings for an offence it is a defence for the accused to prove that they took all reasonable precautions and exercised all due diligence to avoid the commission of the offence (regulation 12);

(g) provide for defences in relation to food that is non-compliant with food law but is destined for export (regulation 13);

(h) provide for the procurement and analysis of samples (regulations 14 and 15);

(i) provide powers of entry for authorised officers of a food authority or the Food Standards Agency (regulation 16);

(j) create the offence of obstructing an officer (regulation 17);

(k) provide a time limit for bringing prosecutions (regulation 18);

(l) provide that anyone who contravenes or fails to comply with specified EU provisions commits an offence (regulation 19(1) and schedule 2);

(m) provide penalties for offences (regulation 19(2) and (3));

(n) provide that in relation to certain potential contraventions, no offence is committed provided certain conditions are met (regulation 19(4) to (8) and Schedules 3 and 7);

(o) provide that where an offence is committed by a corporate body or a Scottish partnership, officers of that body or partners of that partnership may be deemed to have also committed the offence (regulations 20 and 21);

(p) provide a right of appeal against a decision of an officer of an enforcement authority –

    (i) to serve a hygiene improvement notice or a remedial action notice, or

    (ii) to refuse to issue a certificate to the effect that the health risk condition no longer exists in relation to the food business concerned (regulation 22);

(q) provide for the application, for the purposes of section 9 of the Food Safety Act 1990, but with a specified modification (regulation 25);

(r) provide that the Secretary of State may issue codes of recommended practice to food authorities (regulation 26);

(s) provide for the protection of officers acting in good faith (regulation 27);

(t) provide for the revocation or suspension of the appointment or designation of specified officials (regulation 28);

(u) provide that when an authorised officer of an enforcement authority has certified that any food has not been produced, processed or distributed in accordance with these Regulations and the EU Hygiene Regulations, it is to be treated for the purposes of section 9 of the Food Safety Act 1990 as failing to comply with food safety requirements (regulation 29); and

(v) make provision for the service of documents (regulation 30).

REGISTRATION OF FOOD BUSINESS ESTABLISHMENTS

References

EC Regulation 852/2004 on the Hygiene of Foodstuffs, 29 April 2004
The Food Safety and Hygiene (England) Regulations 2013
Food Hygiene (Wales) Regulations 2006
Food Hygiene (Scotland) Regulations 2006
Food Hygiene (Northern Ireland) Regulations 2006
Food Law Code of Practice (England), March 2017
Approval of establishments – Guidance for local authority authorised officers – March 2016

Extent

These provisions apply in England only but there are similar provisions in Wales, Scotland and Northern Ireland. This is EU-derived legislation but retained as domestic law despite EU withdrawal.

Scope

EU Regulation 852/2004 requires that each food business operator must supply the competent authority, in this case the food authority, with details to enable each establishment used for any of the stages of production, processing and distribution of food to be registered (A6.2). An establishment is defined as including any unit of a food business and will therefore include those on land and water and those which are mobile (A2 paragraph 1(c)). The purpose is to allow the food authority to perform official controls efficiently (recital paragraph (19)). Each separate unit must be registered and details of those that require approval and those that are exempt are contained in the Approval of establishments – Guidance for local authority authorised officers – March 2016, p. 186.

Generally the exempt premises are:

The relevant exemptions from the requirements for approval under Regulation 853/2004 fall into three categories:

    (a) Direct supply of small quantities of primary products; Regulation 853/2004, Article 1(3)(c),(d) and (e)

    (b) Retail exemption; Regulation 853/2004, Article 1(5)(b)(ii)

    (c) Food containing both products of plant origin and processed products of animal origin (composite products); Regulation 853/2004, Article 1(2).

Local Authorities must be aware that establishments which are exempt from approval because they assemble, manufacture or handle composite products only, must demonstrate that the products of animal origin used to make the composite products were produced and handled in accordance with the specific requirements in Regulation (EC) 853/2004 as well as the requirements in Regulation (EC) 852/2004. This is specified in Article 1(2) of Regulation (EC) 853/2004 (Code of Practice 3.2.4).

Further detail on the types of processes that need approval together with exemptions are referred to in the Approval of Product-Specific Establishments section (p. 186).

Registration

Operators must register their premises with the food authority at least 28 days before food operations commence and it is an offence not to do so. The food authority is required to provide a registration form for this to be undertaken and a model form is contained in paragraph 3.3.4 of the Code of Practice (Code of Practice 3.2.3).

On receiving the form, the local authority must record the date of receipt on it and, if there are any activities noted that are outside its enforcement role, send it to the appropriate authority. If there are any parts not completed, the food authority should make contact with

FC37 Registration of food business establishments – EC Regulation 852/2004 on the Hygiene of Foodstuffs

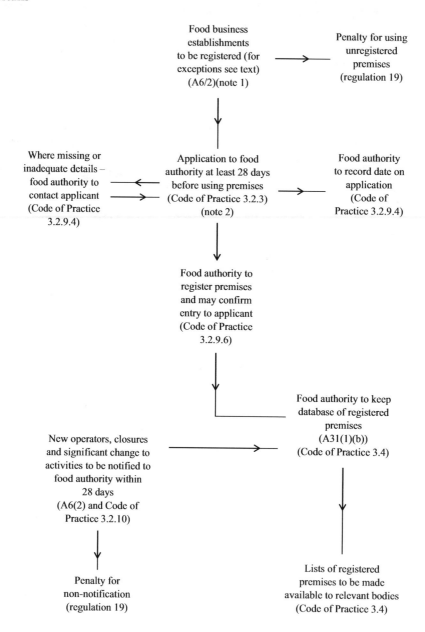

Notes

1. 'A' references are to EC Regulation 852/2004; regulation refers to the 2013 England regulations and Code of Practice to the 2016 Code of Practice.
2. A model form is at paragraph 3.3.4 of the Code of Practice and should be made available by the food authority.
3. For approval of product-specific establishments, see FC38.

the operator or, if the missing information is substantial, return it to him for final completion (Code of Practice 3.2.9.4).

Registration is then effected by entering the details on a database and keeping the form in the file for those premises. Certificates of registration are not to be issued to the operator but the food authority can confirm the receipt of the form and the entry of the premises in the database (Code of Practice 3.2.9.6).

Food authorities are asked to consider an inspection of the premises when the form is received but registration is not dependent upon the results of that inspection. The registration process is informatory (Code of Practice 5.2.2).

Database

Food authorities must draw up a list of food establishments registered with them and will include within it premises previously registered (EC Regulation 882/2004, A31 (1)(b)). The list is to be made available to the general public at all reasonable times and should contain the following details for each business:

(i) name of the food business;
(ii) address;
(iii) particulars and nature of the business.

Details of food premises records are to be made available to the Health Protection Agency and the Consultant in Communicable Disease Control to facilitate the investigation of outbreaks of disease (Code of Practice 3.4). These records are subject to both the Data Protection Act 1998 and the Freedom of Information Act 2000.

Changes to activities etc.

Food business operators are required to ensure that the food authority always has up-to-date information on establishments by notifying them of any significant changes in activities or of any closure (A6.2 of 852/2004 and Code of Practice 3.2.10). The new operator should make notification of a change of operator. The database should be updated accordingly and the details placed on the premises file (Code of Practice 3.2.10).

APPROVAL OF PRODUCT-SPECIFIC ESTABLISHMENTS

References

EC Regulation 852/2004 on the hygiene of foodstuffs, 29 April 2004
EC Regulation 853/2004 on specific hygiene rules for food of animal origin, 29 April 2004
EC Regulation 854/2004 laying down specific rules for the organisation of official controls on products of animal origin intended for human consumption
EC Regulation 882/2004 on the official controls performed to ensure the verification of compliance with feed and food law, animal health and animal welfare rules
The Food Safety and Hygiene (England) Regulations 2013
Food Hygiene (Wales) Regulations 2006
Food Hygiene (Scotland) Regulations 2006

Food Hygiene (Northern Ireland) Regulations 2006
The Official Feed and Food Controls (England) Regulations 2009
Food Law Code of Practice (England), March 2017 (Code of Practice)
Food Law Practice Guidance (England), March 2017 (Practice Guidance)

Extent

These provisions apply in England only but there are similar provisions in Wales, Scotland and Northern Ireland. This is EU-derived legislation but retained as domestic law despite EU withdrawal.

Scope

The competent authority must approve establishments handling products of animal origin. These are unprocessed and processed products of animal origin but not foods consisting partly of products of plant origin (A4 853/2004). 'Establishment' is defined in A2 paragraph 1(c) of EC852 as being 'any unit of a food business' and will therefore include both mobile units and those on water, including fishery vessels. The food authority is responsible for approvals of premises where any of the following products are produced alone or in combination:

(a) minced meat;
(b) meat preparations;
(c) mechanically separated meat;
(d) meat products;
(e) live bivalve molluscs, but not the classification of beds, which is done by the Food Standards Agency;
(f) fishery products;
(g) raw milk other than cows' milk;
(h) dairy products;
(i) eggs but not from primary production; (j) egg products;
(k) frogs' legs and snails;
(l) rendered animal fats and greaves;
(m) treated stomachs, bladders and intestines;
(n) gelatine and collagen – collection centres and tanneries supplying raw material for these premises do not require to be approved but do need authorisation by the food authority;
(o) certain cold stores and wholesale markets.

The competent authority for the purposes of the EU Hygiene Regulations is the Food Standards Agency except where it has delegated competences as provided for in those Regulations. The competent authorities for the purposes of Regulation 178/2002(3) are the Agency and each food authority in its area or district (regulation 4)

The Food Standards Agency is the enforcement authority for the approval of:

(i) raw cows' milk;
(ii) slaughterhouses;
(iii) game-handling establishments;

(iv) cutting plants putting fresh meat on the market; and

(v) activities co-located with these where approval would otherwise be dealt with by the food authority, e.g. meat products establishments (regulation 5, Food Safety and Hygiene (England) Regulations 2013).

Exemptions

The following are exempted from the need to seek approval:

1. Primary production for private domestic use;
2. Domestic preparation, handling or storage of food for private domestic consumption;
3. The direct supply by the producer of small quantities of primary products to the final consumer or to local retail establishments directly supplying the final consumer;
4. Hunters who supply small quantities of wild game or wild game meat directly to the final consumer or to local retail establishments supplying the final consumer;
5. Food containing both products of plant origin and processed products of animal origin;
6. The direct supply, by the producer, of small quantities of meat from poultry and lagomorphs slaughtered on the farm to the final consumer or to local retail establishments directly supplying such meat to the final consumer as fresh meat.
7. The production, transportation or storage of products not requiring temperature-controlled storage conditions (EC Regulation 853/2004 A1).
8. Activities such as the occasional preparation of food by individuals or groups for gatherings or for sale at charitable events are also exempt (recital 9 of 852/2004).
9. A detailed appraisal of these exemptions is given in the Code of Practice 3.3.3.

Applications for approval

The form of application is in annex 5 of the Code of Practice.

Having ensured that all the necessary information has been obtained, the local authority is required to undertake a primary inspection (A31(2)(b) of 882/2004 and 3.3.6 of Code of Practice). This is to check that the premises meets all of the requirements for that type of premises as set out in detail in annex 2 of 853/2004. A sectoral approach is taken in determining the specific hygiene provisions for each foodstuff but the food authority may grant special conditions to take account of traditional production methods.

Where the application is for a live bivalve mollusc purification centre, or a modification to an existing centre, a Code of Practice must be sent to the Food Standards Agency for consultation with the Centre for Environment, Fisheries and Aquaculture Science (CEFAS). The food authority cannot determine such applications until they have received a response from CEFAS and must include any operating conditions set by them in the approval document (Code of Practice, Chapter 8).

Approvals

Approval can only be given if the premises complies with all of the food law requirements relevant to that business (EC 882/2004 A31 (2)). Some conditions are specific for each type of food and are to be found in the annexes to EC Regulation 853/2004, whilst other general requirements are laid down in 852/2004. When approval is given, the food authority must notify the food business operator and the Food Standards Agency – see 3.3.16 of the Code of Practice. Detailed guidance to food authorities on each product dealt with by them is contained in Chapter 8 of the Practice Guidance.

Each establishment must be allocated a three-digit approval number unique to that authority, which should consist of the food authority's two-letter code followed by the approval number (A3 (3) 854/2004 and Code of Practice 3.3.14). The approval code is then to be incorporated into an identification mark to be applied to the products from that establishment, the form of which is contained in annex 2, section 1B of 853/2004 and 3.3.2 of the Practice Guidance.

Conditional approval

The food authority may grant a conditional approval where the premise does not fully comply but only if the infrastructure and equipment requirements are met. In such cases, a further inspection must be carried out within three months and the conditional approval may be extended for a further six months (A 31(2)(d) EC882/2004 and Code of Practice 3.3.11). When granting a conditional approval, the food business operator and the Food Standards Agency must be notified – see 3.3.16 of the Code of Practice. At the end of the time limit, the application must be granted or refused – 3.3.12 and 3.3.13 of the Code of Practice.

Refusals to approve

If the requirements of 853/2004 are not met and the food authority has decided not to issue a conditional approval, the food authority will refuse the application. The applicant must be notified in writing of the refusal with the reasons for that decision and the rights of appeal (Code of Practice 3.3.13).

Suspension and withdrawal of full and conditional approvals

Before deciding to use these options, the food authority must first have considered the use of the other enforcement options – see FCs 40–44. The withdrawal of an approval should only be implemented if the food authority is satisfied that it does not have a reasonable expectation that the deficiencies will be rectified and acceptable standards will be maintained in the future (EC Regulation 882/2004 A 31(2)(e)).

Decisions to suspend or withdraw approvals must be notified to the food business operator in writing together with the reasons for that decision and the rights of appeal. The Food Standards Agency must also be informed. Pending the outcome of an appeal, the establishment is able to continue business subject to any conditions imposed by the food authority to protect public health, although the food authority will still have available the other enforcement options in that period (Code of Practice 3.3.21).

Changes of details or activities

Food business operators must ensure that the food authority has up-to-date information and, in particular, notify the food authority of significant changes of operation and closures (A 6(2) 852/2004 and Code of Practice 3.3.17).

Appeals

Appeals against decisions of the food authority to refuse full or conditional approval or to withdraw or suspend approval are dealt with by regulations 12–13 of the Official Feed and Food Controls (England) Regulations 2009. They are made initially to the magistrates' court with further appeal rights to the Crown Court.

Database

Comprehensive details of premises subject to approvals and conditional approvals are to be kept by the food authority and this should be made available to the public (Code of Practice 3.3.23). Information on individual premises must also be supplied to the Health Protection Agency and the Consultant for Communicable Diseases Control to aid investigations of outbreaks of infectious diseases (Code of Practice 3.4). The database should be updated accordingly.

Penalties

Offences of operating non-approved premises and other offences related to this procedure are dealt with by regulation 19 of the Food Safety and Hygiene (England) Regulations 2013. On conviction, persons are subject to a fine not exceeding the statutory maximum on summary conviction and, on indictment, imprisonment for up to two years and/or a fine.

Further guidance

See www.food.gov.uk/enforcement

EU Commission Guidance Platform
https://ec.europa.eu/food/safety/biosafety/food_hygiene/guidance_en

Guidance document on the implementation of certain provisions of Regulation (EC) No 853/2004 on the hygiene of food of animal origin

https://ec.europa.eu/food/sites/food/files/safety/docs/biosafety_fh_legis_guidance_reg-2004-853_en.pdf

Guidance on food safety management systems for small food retailers (SFR-FSMS): the application of hazard identification, ranking and control in butcher, grocery, bakery, fish and ice cream shops

https://ec.europa.eu/food/sites/food/files/safety/docs/biosafety_fh_legis_guidance_mngt-systems-sfr-fsms_en.pdf

Key questions related to import requirements and the new rules on food hygiene and official food controls

https://ec.europa.eu/food/sites/food/files/safety/docs/biosafety_fh_legis_guidance_interpretation_imports.pdf

Guidance document on the implementation of certain provisions of Regulation (EC) No 852/2004 On the hygiene of foodstuffs

https://ec.europa.eu/food/sites/food/files/safety/docs/biosafety_fh_legis_guidance_reg-2004-852_en.pdf

FC38 Approval of product-specific establishments by food authorities – EC Regulation 882/2004

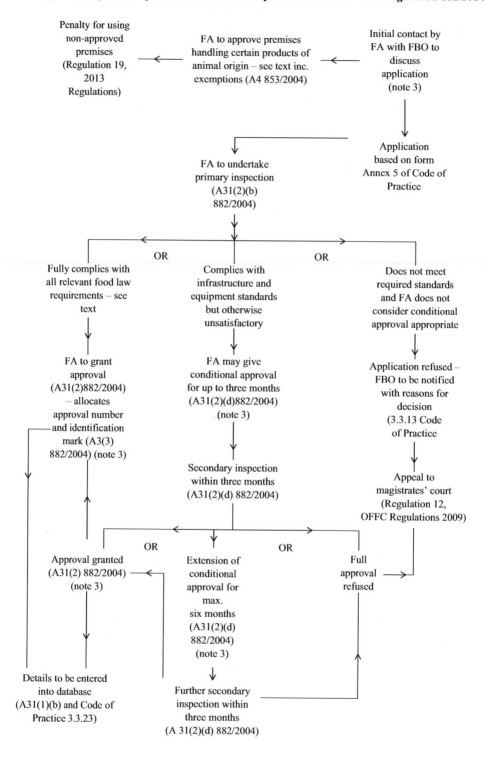

Notes
1. FBO = food business operator.
2. FA = food authority.
3. FSA to be notified of these decisions.
4. OFFC Regulations = Official Feed and Food Controls Regulations 2009.

Commission Notice on the implementation of food safety management systems covering prerequisite programs (PRPs) and procedures based on the HACCP principles, including the facilitation/flexibility of the implementation in certain food businesses

https://eur-lex.europa.eu/legal-content/EN/TXT/PDF/?uri=OJ:C:2016:278:FULL&from=EN

Community Guide to the Principles of Good Practice for the Microbiological Classification and Monitoring of Bivalve Mollusc Production and Relaying Areas

https://cc.curopa.cu/food/sites/food/files/safety/docs/biosafety_fh_guidance_community_ guide_bivalve_mollusc_monitoring_en.pdf

REGISTRATION AND APPROVAL OF ANIMAL BY-PRODUCT PLANTS

References

Animal By-Products (Enforcement) (England) Regulations 2013
Animal By-Products (Enforcement) (Scotland) Regulations 2013
Animal By-Products (Enforcement) (Wales) Regulations 2013
Animal By-Products (Enforcement) (Northern Ireland) Regulations 2015

Extent

These provisions apply to England. Similar provisions apply to Wales, Scotland and Northern Ireland.

Scope

Animal by-products plants and establishments are required to register with a competent authority and seek approval to operate.

Procedure for registration of plants and establishments

A notification must be made in writing to the competent authority where it is made with a view to registration in accordance with Article 23(1) (registration of operators, establishments or plants) of the EU Control Regulation; or to inform the authority of changes in accordance with Article 23(2) of that Regulation (regulation 11).

The competent authority must give notice in writing to –

(a) the operator who makes notification, of the registration of the operator; or the decision not to register the operator;

(b) a registered operator, of:

(i) a prohibition made under Article 46(2) (prohibition on operations) of the EU Control Regulation;

(ii) a requirement to comply with Article 23(1)(b) or (2) of the EU Control Regulation (information on activities and up-to-date information); or

(iii) the amendment of the registration or the ending of the registration where an operator has notified the competent authority of the closure of an establishment in accordance with Article 23(2) (up-to-date information) of the EU Control Regulation (regulation 12).

Procedure for approval

Operators to whom Article 24(1) (approval of establishments or plants) of the EU Control Regulation applies, must apply in writing to the competent authority for approval, including approval after the grant of temporary approval where Article 33 of the EU Implementing Regulation (re-approval of plants and establishments after the grant of temporary approval) applies (regulation 13).

Notification of decisions on approval

The competent authority must give notice in writing to –

  (a) the applicant for approval, of the:

   (i) grant of approval in accordance with Articles 24 (approval) and 44 (procedure for approval) of the EU Control Regulation;
   (ii) grant of conditional approval in accordance with Articles 24 and 44 of the EU Control Regulation, or the extension of such approval in accordance with Article 44; or
   (iii) refusal to grant approval in respect of an initial application or extension;

  (b) the operator of a plant or establishment subject to conditional approval granted in accordance with Articles 24 and 44 of the EU Control Regulation, of the:

   (i) grant of full approval;
   (ii) extension of such approval;
   (iii) imposition of conditions (suspensions, withdrawals and prohibitions on operators) of the EU Control Regulation;
   (iv) suspension of such approval;
   (v) withdrawal of such approval;
   (vi) making of a prohibition; or
   (vii) refusal to extend or grant full approval;

  (c) the operator of an approved plant or establishment, of the –

   (i) imposition of conditions;
   (ii) suspension of such approval;
   (iii) making of a prohibition; or
   (iv) withdrawal of such approval (regulation 14).

Reasons for decisions

Where a decision is made by the competent authority and notified, the competent authority must give reasons in writing for that decision (regulation 15).

Appeals procedure

Where the competent authority has made a notification to which regulation 15(1) applies, a person may appeal against it by making written representations, within 21 days of the issuing of notification of that decision, to a person appointed by the Secretary of State (regulation 16(1)).

The competent authority may also make written representations to the appointed person concerning the decision (regulation 16(2)).

The appointed person must then report in writing to the Secretary of State (regulation 16(3)).

The Secretary of State must give the applicant written notification of the final determination and the reasons for it (regulation 16(4)).

Offences and penalties

A person who fails to comply with an animal by-product requirement commits an offence (regulation 17).

It is an offence –

(a) intentionally to obstruct an authorised person;
(b) without reasonable cause, to fail to give to an authorised person any information or assistance or to provide any facilities that such person may reasonably require;
(c) knowingly or recklessly to give false or misleading information to an authorised person; or
(d) to fail to produce a record or document when required to do so by an authorised person (regulation 18).

A person guilty of an offence under these Regulations is liable on summary conviction, to a fine not exceeding the statutory maximum or to imprisonment not exceeding three months or both; or on conviction on indictment, to a fine or to imprisonment for a term not exceeding two years, or both (regulation 20).

Power to disclose information for enforcement purposes

Where information received by an enforcement authority or an authorised person in the course of enforcing these Regulations, that person may disclose the information to any comparable enforcement authority or authorised person (appointed elsewhere within the United Kingdom to enforce the EU Control Regulation and the EU Implementing Regulation) for the purposes of their enforcement role (regulation 26).

Definition

Enforcement authority

Regulation 10 is enforced by –

(a) in relation to any slaughterhouse, cutting plant or game-handling establishment, the Food Standards Agency; and
(b) in relation to any other premises, the Food Standards Agency or the food authority in whose area the premises are situated (regulation 21(1)).

Otherwise these Regulations are enforced by –

(a) the relevant local authority;
(b) the port health authority (regulation 21(2)).

FC39 Registration and approval of animal by-product plants – Animal By-Products (Enforcement) (England) Regulations 2013

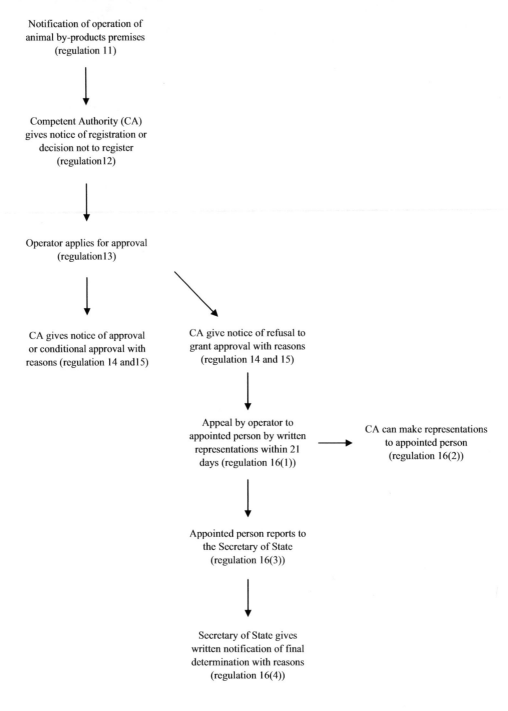

Notification of operation of
animal by-products premises
(regulation 11)

Competent Authority (CA)
gives notice of registration or
decision not to register
(regulation12)

Operator applies for approval
(regulation13)

CA gives notice of approval
or conditional approval with
reasons (regulation 14 and15)

CA give notice of refusal to
grant approval with reasons
(regulation 14 and 15)

Appeal by operator to
appointed person by written
representations within 21
days (regulation 16(1))

CA can make representations
to appointed person
(regulation 16(2))

Appointed person reports to
the Secretary of State
(regulation 16(3))

Secretary of State gives
written notification of final
determination with reasons
(regulation 16(4))

HYGIENE IMPROVEMENT NOTICES

References

The Food Safety and Hygiene (England) Regulations 2013
Food Hygiene (Wales) Regulations 2006
Food Hygiene (Scotland) Regulations 2006
Food Hygiene (Northern Ireland) Regulations 2006
Food Law Code of Practice (England), March 2017, DEFRA (CoP)
Food Law Practice Guidance (England), March 2017, FSA (PG)

Extent

These provisions apply in England only but there are similar provisions in Wales, Scotland and Northern Ireland. This is EU-derived legislation but retained as domestic law despite EU withdrawal.

Scope

If an authorised officer of the food authority has reasonable grounds for believing that a food business operator is failing to comply with the 'Hygiene Regulations', he may serve a hygiene improvement notice on that person (regulation 6(1)). The 'Hygiene Regulations' are the 2013 Regulations and Community Regulations 852/2004, 853/2004, 854/2004, 2073/2005 and 2015/1375 (regulation 2(1)).

The service of a hygiene improvement notice is considered to be appropriate where:

(a) formal action is proportionate to the risk to public health; and/or
(b) there is a record of non-compliance with breaches of the food hygiene regulations; and/ or
(c) the authorised officer has reason to believe that an informal approach will not be successful (Code of Practice 7.2.1.1).

They are not considered to be appropriate where:

(a) the contravention may be a continuing one;
(b) in transient situations, swift enforcement action is needed;
(c) there is a breach of good hygiene practice but no breach of regulations (Code of Practice 7.2.1.2).

The hygiene improvement notice may only be signed by an officer authorised to do so by the food authority, who must be either:

(i) an environmental health officer enforcing food hygiene or food processing regulations;
(ii) a holder of the Higher Certificate in Food Premises Inspection, who is authorised to carry out inspections; or
(iii) a holder of the Ordinary Certificate in Food Premises Inspection in relation to the premises they are authorised to inspect (Code of Practice, Chapter 4 and 7.2.2).

FC40 Hygiene improvement notice (HIN) – Regulation 6 Food Safety and Hygiene (England) Regulations 2013

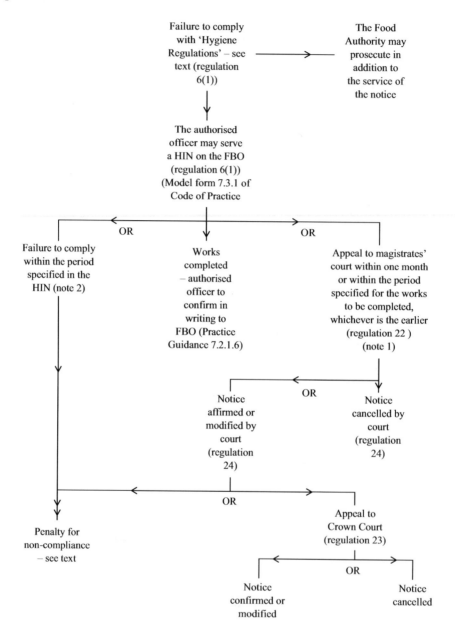

Failure to comply with 'Hygiene Regulations' – see text (regulation 6(1)) ⟶ The Food Authority may prosecute in addition to the service of the notice

↓

The authorised officer may serve a HIN on the FBO (regulation 6(1)) (Model form 7.3.1 of Code of Practice

OR — OR

Failure to comply within the period specified in the HIN (note 2)

Works completed – authorised officer to confirm in writing to FBO (Practice Guidance 7.2.1.6)

Appeal to magistrates' court within one month or within the period specified for the works to be completed, whichever is the earlier (regulation 22) (note 1)

OR

Notice affirmed or modified by court (regulation 24)

Notice cancelled by court (regulation 24)

OR

Penalty for non-compliance – see text

Appeal to Crown Court (regulation 23)

OR

Notice confirmed or modified

Notice cancelled

Notes

1. An appeal suspends the period for compliance until it has been disposed of (regulation 24).
2. On application from the FBO the authorised officer can extend the period where the Code of Practice considers it reasonable to do so in appropriate cases (Practice Guidance 7.2.1.4).
3. FBO = food business operator.

Content

The hygiene improvement notice should:

(a) state the officer's grounds for believing that the food business operator is failing to comply with the Hygiene Regulations;
(b) specify the matters which constitute that failure;
(c) specify the measures, in the officer's opinion, to be taken to secure compliance; and
(d) specify a period of at least 14 days during which the measures must be taken (regulation 6(1)).

A model form of notice is found at 7.3.1 of the Code of Practice.

Time limits

The hygiene improvement notice should indicate that, should he wish to do so, the food business operator may apply in writing for the time limit to be extended. The authorised officer will consider such request and, if feeling that an extension is justified, should withdraw the hygiene improvement notice and serve a new one with a different time limit (Practice Guidance 7.2.1.4).

Appeals

The recipient of the hygiene improvement notice may appeal to a magistrates' court within one month of service or, if it is shorter, the period specified in the notice for compliance (regulation 22). The court may cancel or affirm the notice with or without amendment. The notice is not enforceable until the appeal is disposed of or withdrawn (regulation 24). There is provision for a subsequent appeal to the Crown Court (regulation 23).

Penalty

A food business operator failing to comply with the notice is guilty of an offence (regulation 6(2)) and is liable to a penalty on summary conviction of a fine up to the statutory maximum or, on indictment, a fine and/or imprisonment for up to two years (regulation 19(2)). The prosecutor must commence action for prosecution before three years from the commission of the offence or one year from its discovery (regulation 18).

HYGIENE PROHIBITION ORDERS

References

Food Safety and Hygiene (England) Regulations 2013
Food Hygiene (Wales) Regulations 2006
Food Hygiene (Scotland) Regulations 2006
Food Hygiene (Northern Ireland) Regulations 2006
Food Law Code of Practice, March 2017, DEFRA (CoP)
Food Law Practice Guidance, March 2017, FSA (PG)

FC41 Hygiene prohibition orders (HPOs) – Regulation 7 Food Safety and Hygiene (England) Regulations 2013

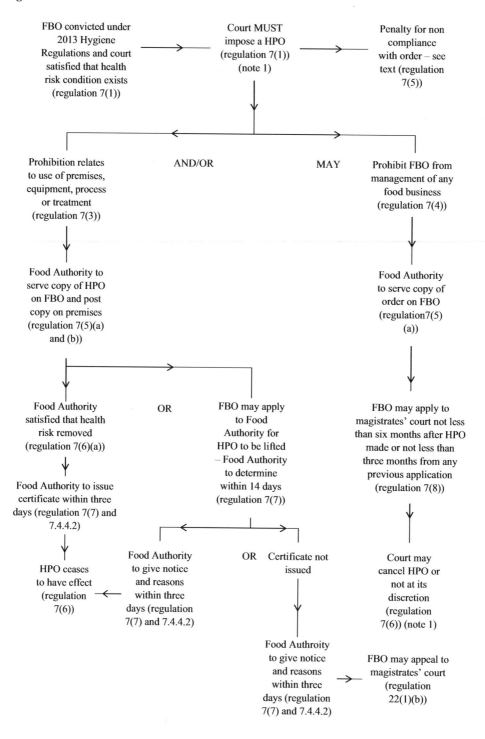

Notes

1. The food authority should notify other food authorities through the Chartered Institute of Environmental Health of a Prohibition order against a person (Code of Practice, paragraph 7.2.6.1).
2. FBO = food business operator.

Extent

These provisions apply in England only but there are similar provisions in Wales, Scotland and Northern Ireland. This is EU-derived legislation but retained as domestic law despite EU withdrawal.

Scope

Hygiene prohibition orders are made by a magistrates' court following the conviction of the food business operator of offences under the 2013 regulations (regulation 7(1)). There is no requirement for application to be made by the food authority, although they may bring the procedure to the attention of the court.

Health risk condition

Before making an order, the court must be satisfied that there is a risk of injury to health from:

(a) the use of any process or treatment;
(b) the construction of the premises or the use of equipment; or
(c) the condition of the premises or equipment (regulation 7(2)).

Where any of these conditions exist and the food business operator is convicted of an offence under the 2013 regulations, the court must make a hygiene prohibition order, although the prohibition of a person from the management of a business is discretionary (regulation 7(4)).

Types of prohibition

The type of prohibition specified in the order must be appropriate to the circumstances and may either:

(a) prohibit the use of a process or treatment for the purposes of the business;
(b) prohibit the use of the premises or equipment for the food business or similar food businesses;
(c) prohibit the use of the premises or equipment for the purposes of any food business (regulation 7(3)).

In the case of the prohibition of a person, the food business operator may be prohibited from participation in the management of any food business or businesses of a particular type (regulation 7(4)). This type of prohibition may also be applied to any other person convicted as a result of the commission of offences by the food business operator (regulation 7(10)).

Guidance on the appropriate use of these prohibitions against persons is given in the Code of Practice at paragraph 7.2.5.3.

Determination of hygiene prohibition orders

Orders prohibiting a person from the management of premises are determined by the court following application by the person. The food authority deals with all other hygiene prohibition orders made with other prohibitions, either on its own initiative or on application by the food business operator. Orders should be posted on the premises at the time of service and it is an offence of obstruction if the notice is removed or tampered with.

Appeals

Decisions of the food authority not to determine the hygiene prohibition order are to be communicated to the food business operator together with reasons for that decision and are subject to appeal to the magistrates' court (regulation 7(7) (b) and regulation 22). Details of appeals are to be given with the notice of reasons for the decision.

Penalties

A food business operator failing to comply with a hygiene prohibition order is guilty of an offence (regulation 7(5)) and is liable to a penalty on summary conviction of a fine up to the statutory maximum or, on indictment, a fine and/or imprisonment for up to two years (regulation 19(2)).

HYGIENE EMERGENCY PROHIBITION NOTICES AND ORDERS

References

Food Safety and Hygiene (England) Regulations 2013
Food Hygiene (Wales) Regulations 2006
Food Hygiene (Scotland) Regulations 2006
Food Hygiene (Northern Ireland) Regulations 2006
Food Safety (Improvement and Prohibition – Prescribed Forms) Regulations 1991
Food Law Code of Practice (England), March 2017, DEFRA (CoP)
Food Law Practice Guidance (England), March 2017, (FSA)

Extent

These provisions apply in England only but similar provisions apply in Wales, Scotland and Northern Ireland. This is EU-derived legislation but retained as domestic law despite EU withdrawal.

Scope

Unless the use of a voluntary procedure is appropriate (see below), an authorised officer may instigate emergency action to secure a hygiene emergency prohibition order (HEPO) where there is an *imminent* risk of injury to health in respect of any food business (regulation 8(1) and (2)).

Hygiene emergency prohibition notices (HEPNs)

This is served by an authorised officer on the food business operator where there is an *imminent* risk of injury to health. The parameters for the determination of this health risk condition are the same as for HEPOs (see below), except that the risk to health must be *imminent* (regulation 7(2) and 8(1)). Guidance on the health risk condition is given in the Code of Practice at 7.2.4.2.

The effect of the service of a HEPN is to immediately close the premises, or prevent the use of equipment or a process/treatment. Unlike a HEPO, this procedure may not be used to prohibit a person from management of a business.

HEPNs may only be signed by environmental health officers who are adequately trained, competent and duly authorised (Code of Practice 4.8.2 and Practice Guidance, Chapter 4).

Hygiene emergency prohibition orders (HEPOs)

The HEPN ceases to have effect after three days (or before if the conditions which led to it are removed) unless the authorised officer applies to the magistrates' court for the making of an HEPO. The food business operator must be given at least one day's notice of the intention to apply for an order (regulation 8(3), (7) and (8)). If the court is satisfied that the imminent health risk condition exists, it must make an HEPO (regulation 8(2)).

As soon as possible after the order is made, an authorised officer must fix a copy of it on the premises and serve a copy on the food business operator (regulation 8(6)).

Determination of HEPOs

The food authority effects this when satisfied that the health risk condition has been removed and it may be undertaken on its own initiative or on written application from the food business operator (regulation 8(8) and (9)). In the latter case, the decision must be made within 14 days (regulation 8(9)). A certificate that the HEPO has been removed must be issued within three days of the determination but, if refused, the food business operator must be notified with reasons for the decision (regulation 8(9)(b)). There is a right of appeal to the magistrates' court, which must be made within one month of the notice of the decision (regulation 22).

Compensation

The food authority must compensate the food business operator for losses arising from the service of an HEPN if an HEPO is not applied for within three days or if the court, in dealing with the application, considers that the health risk condition did not exist at the time the notice was served (regulation 8(10)).

Offences

Persons contravening an HEPN or an HEPO are guilty of an offence (regulation 8(6)) and are subject to a penalty on summary conviction of a fine not exceeding the statutory maximum or, on indictment, a fine and/or imprisonment for up to two years (regulation 19(2)).

Voluntary procedure

The removal of a situation where an imminent health risk condition exists may be dealt with voluntarily by the closure of the premise, where the food business operator agrees that the condition exists. Such agreements are to be made in writing (Code of Practice 7.2.5).

FC42 Hygiene emergency prohibition notices and orders (HEPNs and HEPOs) – Regulation 8 Food Safety and Hygiene (England) Regulations 2013

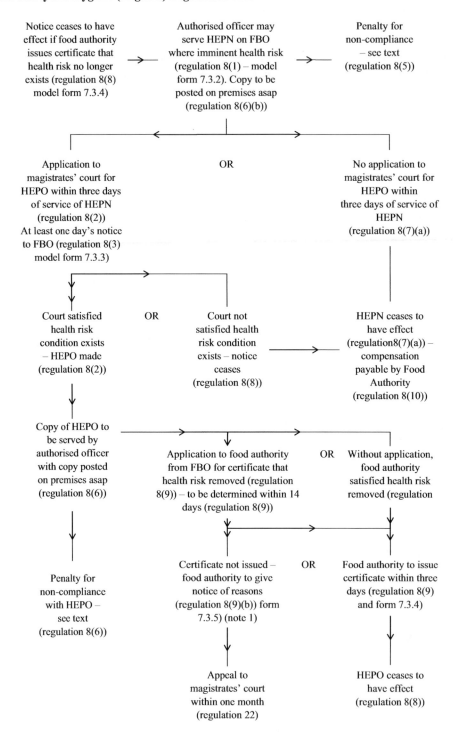

Notes

1. No time period is indicated within which this must be done.
2. The model forms are found in 7.3.2 of the Code of Practice.
3. FBO = food business operator.

REMEDIAL ACTION NOTICES AND DETENTION NOTICES

References

EC Regulation 853/2004 on specific hygiene rules for food of animal origin, 29 April 2004
The Food Safety and Hygiene (England) Regulations 2013
Food Hygiene (Wales) Regulations 2006
Food Hygiene (Scotland) Regulations 2006
Food Hygiene (Northern Ireland) Regulations 2006
Food Law Code of Practice, March 2017, DEFRA (CoP)

Extent

These provisions apply in England only but there are similar provisions in Wales, Scotland and Northern Ireland. This is EU-derived legislation but retained as domestic law despite EU withdrawal.

Scope

This procedure allows an authorised officer of the food authority to deal with a situation at premises handling food of animal origin and approved by them (FC38) where:

(a) any of the 'Hygiene Regulations' are being breached; or
(b) inspection under those regulations is being hampered (regulation 9(1)).

The 'Hygiene Regulations' are the 2013 Regulations and Community Regulations 852/2004, 853/2004, 854/2004, 2015/1375 and (regulation 2(1)).
The Code of Practice indicates that they should only be used in the following circumstances:

- at failure of any equipment or part of an establishment to comply with the requirements of the 'Hygiene Regulations' as defined by regulation 2 of the Food Safety and Hygiene (England) Regulations 2013
- where there is a need to impose conditions upon or the prohibition of the carrying on of any process breaching the requirements of the Regulations or hampering adequate health inspection in accordance with the Regulations; and
- where the rate of operation of the business is detrimental to its ability to comply with the Regulations (Code of Practice 7.2.12.1).

This procedure can be applied to businesses that are not actually approved by the authority but are required to be approved (Regulation 9, para. 1 of the Food Safety and Hygiene England Regulations).

Remedial action notices (RANs)

Where it is considered to be the most appropriate form of action, an authorised officer may serve a remedial action notice in writing on the food business operator or on his duly authorised representative. The notice may:

(i) prohibit the use of any equipment or any part of the establishment;
(ii) impose conditions on or prohibit the carrying out of any process;
(iii) require the rate of operation to be reduced to such an extent as specified or to be stopped completely.

FC43 Remedial action notices (RANs) and detention notices (DNs) – Regulation 9 Food Safety and Hygiene (England) Regulations 2013

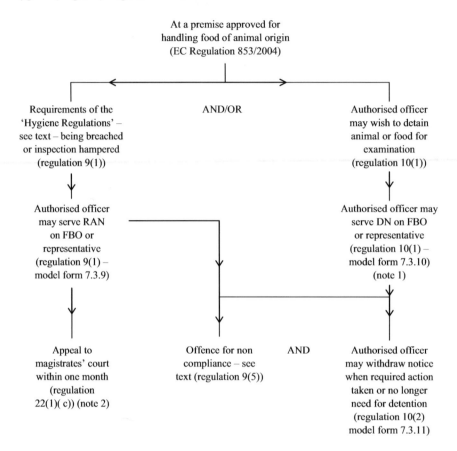

Notes

1. There is no provision for appeal against a detention notice.
2. The court may cancel, affirm or modify the notice (regulation 22(1)).
3. FBO = food business operator.

The notice must be served as soon as is practicable and must state why it is being served. If served under (i) above, it must state the breach and the action needed to be taken (regulation 9(1–3)).

Detention notices (DNs)

These notices may be served by an authorised officer where, at approved premises, he wishes to detain any animal or food for examination, including the taking of samples (regulation 10). The action may be taken either alone or in combination with the service of a remedial action notice.

Withdrawal of RANs and DNs

As soon as such action as required has been taken to remedy the issue or there is no longer a need for the animal or food to be detained, an authorised officer of the same food authority may withdraw the notice in writing served on the food business operator or his authorised representative (regulation 9(4) and 10(2)).

Qualifications of authorised officers for the service of RANs and DNs

These may be served only by environmental health officers who are properly trained, competent and duly authorised (Code of Practice Chapter 4 and 4.8.2).

Appeals

There is a right of appeal to a magistrates' court against a remedial action notice but not a detention notice. The appeal must be brought within one month of service. The remedial action notice is not suspended pending appeal. The court may cancel, approve or modify the notice and there is a further right of appeal to the Crown Court (regulations 22, 23 and 24).

Penalties

It is an offence not to comply with a remedial action notice or a detention notice (regulation 9(5) and 10(3)), the penalty for which on summary conviction is a fine not exceeding the statutory maximum or, on indictment, a fine and/or imprisonment for up to two years (regulation 19(2)).

DETENTION AND SEIZURE OF FOOD

References

The Food Safety and Hygiene (England) Regulations 2013
Food Hygiene (Wales) Regulations 2006
Food Hygiene (Scotland) Regulations 2006
Food Hygiene (Northern Ireland) Regulations 2006
Food Safety Act 1990, section 9
Food Safety (Northern Ireland) Order 1991

Detention of Food (Prescribed Forms) Regulations 1990
The Meat (Enhanced Enforcement Powers) Regulations 2000
Food Law Code of Practice March 2017, DEFRA (Code of Practice)
Food Law Practice Guidance March 2017, FSA (Practice Guidance)

Extent

These provisions apply in England, Wales and Scotland. Similar provisions apply in Northern Ireland. This procedure is based on both the Food Safety Act 1990 and the Food Safety and Hygiene (England) Regulations 2013.

Power of inspection of food

An authorised officer of a food authority may at all reasonable times inspect any food intended for human consumption which:

(a) has been sold;
(b) is offered or exposed for sale;
(c) is in the possession of, has been deposited with or consigned to any person for the purpose of sale or preparation for sale (section 9(1)).

Whilst the 1990 Act defines 'human consumption' as 'includes use in the preparation of food for human consumption' (section 53(1)), the phrase 'intended for human consumption' is defined in greater detail in regulation 3 of the Food Safety and Hygiene (England) Regulations 2013 in relation to procedures under those regulations as:

(a) any food commonly used for human consumption which is found on premises used for the preparation, storage, or placing on the market of that food; and
(b) any article or substance commonly used in the manufacture of food for human consumption which is found on premises used for the preparation, storage or placing on the market of that food,
       shall be presumed, until the contrary is proved, to be intended for placing on the market, or for manufacturing food for placing on the market, for human consumption.
(c) Any article or substance capable of being used in the composition or preparation of any food commonly used for human consumption which is found on premises on which that food is prepared shall, until the contrary is proved, be presumed to be intended for such use.

For the definition of 'food', see p. 182.

Voluntary procedures

It is possible for an authorised officer to use a voluntary procedure to remove food not suitable for human consumption from the food chain, at the instigation of either the authorised officer or the owner. However, such a procedure must be well documented in accordance with Code of Practice, paragraph 7.2.10.7.

Detention and seizure of food

Where either:

(a) on inspection, food is found to fail to comply with food safety requirements; or
(b) it appears to the authorised officer that any food is likely to cause food poisoning or any disease communicable to humans,

the authorised officer may either:

(i) detain the food by notice; or
(ii) seize it and remove it in order to have it dealt with by a Justice of the Peace (section 9(1)–(3)).

Food is deemed not to comply with food safety requirements if:

(a) it is unfit for human consumption; or
(b) it has been rendered injurious to health by the addition of articles or substances, by the use of an article or substance as an ingredient in its preparation, by the abstraction of any constituent or by the application of any process or treatment; or
(c) it is so contaminated (by extraneous matters or otherwise) that it would not be reasonable to expect it to be used for human consumption in that state (section 8(2));
(d) an authorised officer certifies that the food has not been produced, processed or distributed in compliance with the 'Hygiene Regulations' – see p. 185. Such certification should use the model form at 7.3.6 of the Practice Guidance (regulations 25 and 29).

The Practice Guidance requires that the procedures dealing with the inspection, detention and seizure of food under section 9 should only be exercised by authorised officers (paragraph 7.2.5).

Detention of food notice

The detention of the food is effected by a notice (form prescribed by the Detention of Food (Prescribed Forms) Regulations 1990) to be served on the person in charge of the food requiring that:

(a) the food is not to be used for human consumption; and
(b) it is not to be removed, except to a place specified in the notice.

Failure to comply with the notice is an offence for which the penalty on indictment is an unlimited fine and/or imprisonment for up to two years or on summary conviction to a fine up to the statutory maximum and/or imprisonment for not exceeding six months (sections 9(3) and 35(2)).

Having served such notice, the authorised officer is required to determine whether or not the food complies with food safety requirements as soon as is reasonably practicable and in any event within 21 days. If satisfied of compliance, the food is released and the notice withdrawn using a withdrawal of detention of food notice (PF No. 2); if satisfied of non-compliance, the food must be seized to have t dealt with by a Justice of the Peace (section 9(4)).

Seizure

Seizure of food, with or without prior detention, is effected by the authorised officer removing it to have it dealt with by a Justice of the Peace. The person in charge of the food must be informed by the service of a Food Condemnation Warning (practice guidance 7.2.5.8) and any person who may be liable to prosecution for the condition of the food may attend before the Justice of the Peace and is entitled to be heard and call witnesses (section 9(5)).

If the Justice of the Peace considers, on the basis of the evidence, that the food fails to comply with food safety requirements or is likely to cause food poisoning or a human communicable disease, he must condemn the food and order it to be destroyed or disposed of so as to prevent it being used for human consumption. In this case, the owner of the food is required to defray the costs incurred in the destruction or disposal of the food (section 9(6)).

The Code of Practice (paragraph 7.2.10.3) requires that, if possible, food which is seized should be presented to a Justice of the Peace within two days and, where the food is highly perishable, as soon as possible.

Compensation

If the Justice of the Peace is not satisfied that the food fails to comply, the food authority is required to compensate the owner for any depreciation in its value. This is also the case if a detention notice is withdrawn by the authorised officer (section 9(7)).

Disposal of food

Once a Justice of the Peace has condemned food, it is the responsibility of the food authority to ensure its adequate destruction by supervision throughout the process. If necessary, the food should be rendered unusable for a return to the food chain (Code of Practice 7.2.10.8).

Dealing with batches, lots or consignments (Code of Practice 7.2.10.6)

Article 14(6) of EC Regulation 178/2002 states that where

> any food that is unsafe forms part of a batch, lot or consignment of food of the same class or description, it shall be presumed that all the food in that batch etc. is also unsafe, unless following a detailed assessment there is no evidence that the rest of the batch etc. is unsafe.

This definition is applicable to the detention and seizure of food using these procedures (Code of Practice 7.2.10.6).

FC44 Detention and seizure of food notices – section 9 Food Safety Act 1990

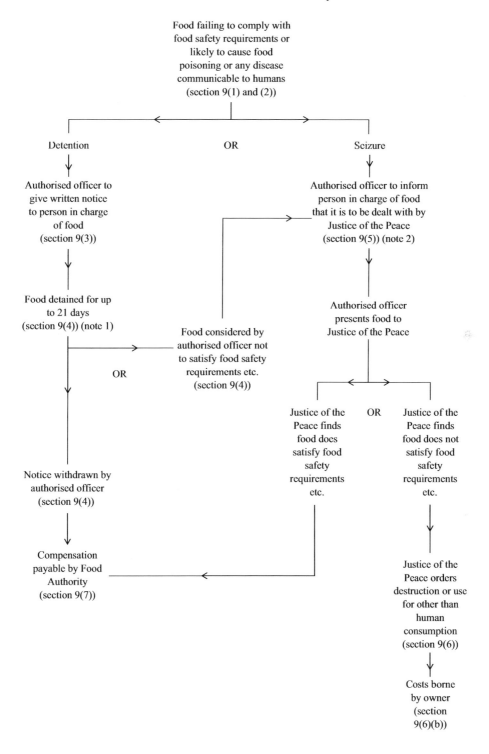

Notes

1. It is an offence to breach the requirements of the notice (section 9(3)).
2. Persons liable to prosecution have a right to attend the hearing and present evidence and witnesses (section 9(5)(a)).
3. The prescribed forms are detailed in the Detention of Food (Prescribed Forms) Regulation 1990.

IMPORTED FOOD – OFFICIAL CONTROLS ON FEED AND FOOD OF NON-ANIMAL ORIGIN FROM THIRD COUNTRIES

References

Official Feed and Food Controls (England) Regulations 2009 (in this procedure, regulation numbers refer to these regulations)
Official Controls (Animals, Feed and Food) (England) Regulations 2006
Official Feed and Food Controls (Wales) Regulations 2009
Official Feed and Food Controls (Scotland) Regulations 2009
Official Feed and Food Controls (Northern Ireland) Regulations 2009
Official Controls (Animals, Feed and Food) (Wales) Regulations 2007
Official Controls (Animals, Feed and Food) (Scotland) Regulations 2007
Official Controls (Animals, Feed and Food) (Northern Ireland) Regulations 2007
Food Law Code of Practice, March 2017, DEFRA (CoP)
Food Law Practice Guidance March 2017, FSA (Practice Guidance)

Extent

These provisions apply in England but similar provisions apply in Wales, Scotland and Northern Ireland. This is EU-derived legislation but retained as domestic law despite EU withdrawal.

Enforcement authorities

Port health authorities are local authority bodies at seaports set up under the Public Health (Control of Diseases) Act 1984. At UK airports and some seaports, the local authority in whose area the port is located is responsible for enforcing the food import controls.

Outside of port areas, local authorities ('inland' local authorities) are responsible for enforcing food safety controls on foodstuffs, including imported food.

Deferred execution and enforcement

The enforcement authority which is the ultimate destination of imported food becomes responsible for enforcing and executing the import provisions with respect to that product once it arrives there, in the following circumstances:

- a product from a third country has entered England;
- customs examination of that product has been completed or has been deferred until it reaches its place of destination elsewhere in the United Kingdom;
- an authorised officer of the enforcement authority for the place of entry has on reasonable grounds issued an authorisation confirming that:

  (i) examination of the product for the purposes of the import provisions should be deferred until the product arrives at its destination elsewhere in England; or

  (ii) such examination should take place when the product arrives at its destination elsewhere in the United Kingdom under legislation with respect to imported products in force there; and

- a person importing the product gives that authorised officer an undertaking in writing which states:

  (i) the destination of the product; and

  (ii) confirms that the container containing the product has been sealed and will not be opened until it has reached that destination, the opening of the container has been authorised by the enforcement authority for the place in which the destination is located, and the container will be available at that destination for examination under the import provisions.

Where an authorised officer issues an authorisation, he shall notify the enforcement authority where the imported food is destined for that the product has not been examined under the import provisions, if customs examination of the product has been deferred and a copy of any undertaking (regulation 27).

Checks on products

The importer must allow an authorised officer to carry out checks in relation to the product. The importer must provide the facilities and assistance which the authorised officer reasonably requires to carry them out at a specified place (regulation 29).

Checks include a systematic documentary check, a random identity check and, as appropriate, a physical check (Article 16 of Regulation 882/2004).

An enforcement authority has the powers of detention, destruction, special treatment, re-dispatch and other appropriate measures and charging for costs (regulation 31).

Notices under Articles 18 and 19 of Regulation 882/2004 (imports of feed and food from third countries)

An authorised officer must serve a notice on the feed or food business operator responsible for it, where he proposes:

- to place a consignment of feed or food under official detention;
- to destroy the feed or food;
- to subject it to special treatment;
- other appropriate measures;
- where already on the market, to monitor, recall or withdraw it before taking the measures above (regulation 32).

This notice can only be used to deal with food not of animal origin. Enforcement officers must use The Trade in Animals and Related Products Regulations 2011 (regulation 19) to control illegally imported products of animal origin.

Right of appeal

The person receiving this notice can appeal against the notice to the magistrates' court by way of complaint for an order within one month from the date of the notice. Where the magistrates' court determines that the decision of the authorised officer is incorrect, the authority must follow the instructions of the court (regulation 33). A person aggrieved by the dismissal of an appeal by a magistrates' court may appeal to the Crown Court (regulation 34).

Costs and fees

The enforcement authority may claim costs for its actions from the feed or food business operator or its representative (regulation 36).

Procurement by authorised officers of samples with regard to food

An authorised officer may:

(a) purchase a sample of any food, or any substance capable of being used in the preparation of food;

(b) take a sample of any food, or any such substance, which:

   (i) appears to him to be intended for placing on the market or to have been placed on the market, for human consumption; or

   (ii) is found by him on or in any premises which he is authorised to enter;

(c) take a sample from any food source, or a sample of any contact material, which is found by him on or in any such premises; and

(d) take a sample of any article or substance which is found by him on or in any such premises and which he has reason to believe may be required as evidence in proceedings (regulation 37).

Analysis etc. of samples

An authorised officer can submit the sample for analysis by a public analyst or examination by a food examiner. A certificate specifying the result of the analysis or examination must be provided and signed by the food analyst or examiner (regulation 38).

Obstruction etc. of officers

It is an offence to obstruct any person acting in the execution of the import provisions or to fail to give any assistance or information which that person may reasonably require of him for the performance of his functions. It is also an offence to recklessly give information that is known to be false or misleading in a material particular (regulation 40).

Offences and penalties

It is an offence to contravene or fail to comply with any of the specified import provisions and, if a person is guilty of an offence, they are liable on summary conviction to a fine not exceeding the statutory maximum or, on conviction on indictment, to imprisonment for a term not exceeding two years, to a fine or to both.

A person guilty of obstruction is liable on summary conviction to a fine not exceeding level 5 on the standard scale or to imprisonment for a term not exceeding three months, or to both (regulation 41).

Time limit for prosecutions

Prosecutions must be taken within three years of the offence or one year from its discovery by the prosecutor, whichever is the earlier (regulation 42).

FC45 Imported food – Regulations 27 to 42 Official Food and Feed Control (England) Regulations 2009

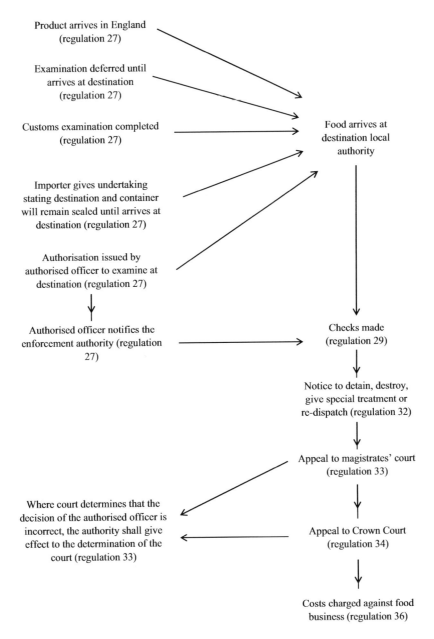

SAMPLING OF FOOD FOR ANALYSIS AND EXAMINATION

References

Food Safety Act 1990, sections 29–31
Food Safety (Sampling and Qualifications) Regulations 2013
Food Safety and Hygiene (England) Regulations 2013
Food Safety (Sampling and Qualifications) (Wales) Regulations 2013
Food Hygiene (Wales) Regulations 2006
Food Safety (Sampling and Qualifications) (Scotland) Regulations 2013
Food Hygiene (Scotland) Regulations 2006
Food Safety (Sampling and Qualifications) (Northern Ireland) Regulations 2013
Food Hygiene (Northern Ireland) Regulations 2006
Quick-Frozen Foodstuffs (England) Regulations 2007
Materials and Articles in Contact with Food (England) Regulations 2012
Food Law Code of Practice, March 2017, DEFRA (CoP)
Food Law Practice Guidance, March 2017, FSA (PG)
Guidance for local enforcement sampling officers on priorities for FSA National Coordinated
    Sampling Programme
Guidance on Food Sampling for Microbiological Examination, January 2006, LACORS
EU Directive 92/2 – sampling procedure and community method of analysis for the official
    control of temperatures of quick-frozen foods intended for human consumption

Extent

Most provisions apply in England, Wales and Scotland. Similar provisions apply in Northern Ireland. This is EU-derived legislation but retained as domestic law despite EU withdrawal.

Scope

Powers of sampling for analysis (chemical) and examination (microbiological) by an authorised officer are provided in both the Food Safety Act 1990, section 29 and the 2013 regulations (regulation 14). In practical terms, examination for microbiological condition will relate to the hygienic conditions in a premises and should therefore be undertaken using the 2013 regulations. The powers under the 1990 Act should be used to check food standards requirements (chemical).

Both sampling regimes are, however, covered by the Food Safety (Sampling and Qualifications) Regulations 2013.

Generally, the authorised officer may purchase or take (without payment) a sample of food or substance capable of being used in food that appears to be intended for human consumption or is found on any premises. Samples may also be taken from any food sources or of contact materials including the use of the sample as evidence in proceedings (regulation 14).

Sampling policy

The formal sampling procedures as identified here should form only part of a sampling policy, which itself should be an integral part of the food authority's enforcement policy. The sampling policy should be drawn up in liaison with the public analyst and/or the food examiner and should take account of the home and primary authority arrangements (Code of Practice 6.1.1).

Analysis and examination

The Food Safety Act 1990 recognises that the microbiological examination of food is a separate activity from its analysis.

Analysis: 'includes microbiological assay and any technique for establishing the composition of the food' (section 53(1)).

Examination is defined as: 'a microbiological examination' (section 28(2)).

Samples taken by the authorised officer are to be submitted either to a public analyst for analysis or, if for examination, to a food examiner. The necessary qualifications for each are specified in the Food Safety (Sampling and Qualifications) Regulations 2013 (section 30(1) and regulations 4 and 5). Code of Practice, paragraph 6.1.4 requires that all samples for analysis must be submitted to a laboratory accredited for that purpose and which appears on the list of official food control laboratories.

Food authorities, other than remaining two-tier district councils in England, who are not generally involved in matters of food composition, are required to appoint one or more public analysts for their areas (section 27).

Food authorities including district councils may provide their own facilities for microbiological examination (section 28) or, alternatively, may submit samples to a food examiner.

Analysts' and examiners' certificates

Following analysis or examination, the analyst or examiner must submit a certificate to the authorised officer specifying the result (section 30(6) and regulation 10). Each certificate must be signed (section 30(7)). The form of certificate to be used is specified in schedule 3 of the Food Safety (Sampling and Qualifications) Regulations 2013.

In any legal proceedings taken following the analysis or examination, the certificate or a copy of it is sufficient evidence of the facts stated therein, unless either party wishes the food analyst or examiner to be called as a witness (section 30(8)).

Sampling procedure

Following sampling, the sample must be dealt with in accordance with the Food Safety (Sampling and Qualifications) Regulations 2013 and the procedure is indicated in FC46. More detailed guidance will be found in the Code of Practice, chapter 6 and the Practice Guidance, chapter 6.

This procedure is *not* applicable to sampling under the following food regulations, in which particular sampling procedures are laid down:

- Poultrymeat (England) Regulations 2011 (minor application changes);
- Natural Mineral Water, Spring Water and Bottled Drinking Water (England)
- Regulations 2007 (regulations 17, 18 and 19);
- The Containments in Food (England) Regulations 2013 (regulation 8(2)).
- The Materials and Articles in Contact with Food Regulations 2012 (see below\);
- The Animals and Animal Products (Examinations for Residues and Maximum Residue Limits) (England and Scotland) Regulations 2015 (Part 3) (schedule 1 of the regulations).

The sampling procedure detailed in FC46 covers the sampling of foods in all other situations.

FC46 Sampling of food for analysis and examination – Food Safety Act 1990 and Food Safety (Sampling and Qualifications) Regulations 1990

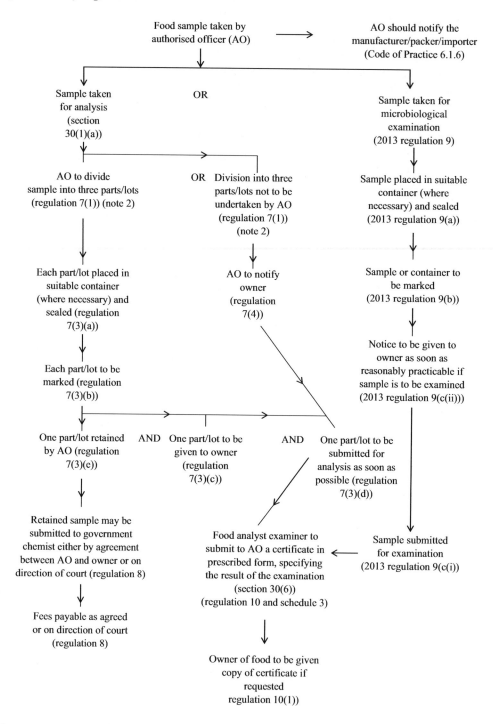

Food sample taken by authorised officer (AO) → AO should notify the manufacturer/packer/importer (Code of Practice 6.1.6)

OR

Sample taken for analysis (section 30(1)(a))

Sample taken for microbiological examination (2013 regulation 9)

AO to divide sample into three parts/lots (regulation 7(1)) (note 2)

OR Division into three parts/lots not to be undertaken by AO (regulation 7(1)) (note 2)

Sample placed in suitable container (where necessary) and sealed (2013 regulation 9(a))

Each part/lot placed in suitable container (where necessary) and sealed (regulation 7(3)(a))

AO to notify owner (regulation 7(4))

Sample or container to be marked (2013 regulation 9(b))

Notice to be given to owner as soon as reasonably practicable if sample is to be examined (2013 regulation 9(c(ii)))

Each part/lot to be marked (regulation 7(3)(b))

One part/lot retained by AO (regulation 7(3)(e))

AND One part/lot to be given to owner (regulation 7(3)(c))

AND One part/lot to be submitted for analysis as soon as possible (regulation 7(3)(d))

Retained sample may be submitted to government chemist either by agreement between AO and owner or on direction of court (regulation 8)

Food analyst examiner to submit to AO a certificate in prescribed form, specifying the result of the examination (section 30(6)) (regulation 10 and schedule 3)

Sample submitted for examination (2013 regulation 9(c(i)))

Fees payable as agreed or on direction of court (regulation 8)

Owner of food to be given copy of certificate if requested regulation 10(1)

Notes

1. Regulation numbers refer to the Food Safety (Sampling and Qualifications) Regulations 2013, except where 2013 is shown, in which case they refer to the Food Safety and Hygiene (England) Regulation 2013.
2. Division into three parts or lots is not required if this is not reasonably practicable or may impede analysis (regulation 7(1)).
3. This sampling procedure does not apply to the enforcement of certain food regulations (p. 219).
4. Sometimes there are difficulties in establishing who the 'owner' is in certain multiple retailer situations – in these cases, the registered head office of the company should be notified.

Division of samples

Samples intended for analysis (but not for examination) are to be divided into three parts unless either:

(a) this is not reasonably practicable; or
(b) it is likely to impede analysis (regulation 7(1) and (4)).

If the sample itself consists of sealed containers and the authorised officer believes that the opening of them would impede proper analysis, he may proceed by dividing the sample into parts by putting the containers into three lots (regulation 7(2)).

Sampling and analysis under other provisions

Quick-frozen foodstuffs (England) Regulations 2007 (regulation 7)

If, on inspection, an authorised officer has reasonable grounds to believe that the temperatures that are being or have been maintained in respect of any quick-frozen foodstuff are not the temperatures prescribed, he must inspect the quick-frozen foodstuff and their temperatures in accordance with the provisions of EU Directive 92/2.

Materials and articles in contact with food (England) Regulations 2012

An authorised officer who has procured a sample of materials or articles in contact with food under the Food Safety Act 1990 and who considers it should be analysed shall divide the sample into three parts.

If the sample consists of sealed containers and opening them would, in the opinion of the authorised officer, impede a proper analysis, the authorised officer shall divide the sample into parts by putting the containers into three lots, and each lot shall be treated as being a part. The authorised officer shall:

(a) if necessary, place each part in a suitable container and seal it;
(b) mark each part or container;
(c) as soon as is reasonably practicable, give one part to the owner and notify the owner in writing that the sample will be analysed;
(d) submit one part for analysis; and
(e) retain one part for future submission (regulation 25).

Secondary analysis by the government chemist

Where a sample has been retained under these regulations and proceedings are intended to be or have been commenced against a person for an offence under the regulations, and the prosecution intends to present as evidence the result of the analysis, the authorised officer should send the retained part of the sample to the government chemist for analysis:

- of the officer's own volition;
- if requested by the prosecutor (if a person other than the authorised officer);

- if the court so orders; or
- if requested by the accused.

Where a request is made by the accused, the authorised officer may give notice in writing to the accused requesting payment of a fee specified in the notice to defray some or all of the government chemist's charges for performing the analysis and, in the absence of agreement by the accused to pay the fee specified in the notice, the authorised officer may refuse to comply with the request (regulation 26(6)).

The government chemist shall analyse the part and send the authorised officer a certificate specifying the results of the analysis.

The authorised officer shall immediately on receipt supply the prosecutor and the accused with a copy of the government chemist's certificate of analysis (regulation 26).

Live molluscs

There is detailed guidance on sampling live molluscs in the Practice guidance at 8.1.12.

CLOSURE NOTICES ON MOLLUSC HARVESTING AREAS

References

EC Regulation 852/2004. EC Regulation 853/2004. EC Regulation 854/2004
Food Safety and Hygiene (England) Regulations 2013
Food Hygiene (Wales) Regulations 2006
Food Hygiene (Scotland) Regulations 2006
Food Hygiene (Northern Ireland) Regulations 2006
Food Law Code of Practice, March 2017, DEFRA (CoP)
Food Law Practice Guidance, March 2017, FSA (PG)

Extent

These provisions are made under EC regulations and apply in England. Similar provisions apply in Wales, Scotland and Northern Ireland. This is EU-derived legislation but retained as domestic law despite EU withdrawal.

Scope

Where, based on sampling, either:

(a) the health standards set out in Regulation 853/2004, section 7, chapter 5 are exceeded (these standards incorporate the microbiological standards in EC Regulation 852/2004); or
(b) there is otherwise a risk to human health,

the food authority must either:

(a) close the production area by the issue of a closure notice (CN); or
(b) if appropriate, re-classify the area as class B or C if it meets the criteria in part A of annex 2 of EC854/2004 (see below).

Consultation

The food authority should liaise with the Food Standards Agency in considering its actions, including on possible action to withdraw any molluscs from that area which may have been distributed (Code of Practice 8.6). They should also consider consulting appropriate experts, such as the Consultant Communicable Disease Control or the consultant microbiologist at the Health Protection Agency (Practice Guidance 8.1.16).

Re-classification

Classification of harvesting areas is normally undertaken by the Food Standards Agency but, in the urgent situation faced by this procedure, re-classification is by the food authority, in consultation with the food business operator and the Food Standards Agency. The classifications are laid out in EC 854/2004 at annex 2, chapter 2A and are:

(a)   Class A areas from which molluscs may be collected for direct human consumption – these must meet the standards in annex 3, section 7, chapter 5 of EC 853/2004;
(b)   Class B areas where molluscs may be collected but only placed on the market after treatment in a purification centre or after relaying so as to meet the standards in (a) above;
(c)   Class C from where molluscs may only be placed on the market after relaying for a long period so as to meet the standards in (a) above.

Closure notices

The issue of these notices by the food authority has the effect of temporarily closing the harvesting area, although no time limit is set. The form of notice to be used is in the Practice Guidance 8.1.19.

Information

The food authority must immediately inform all interested parties, such as producers, gatherers and operators of purification and dispatch centres, of any re-classification or issue of a closure notice (854/2004, annex 2, chapter 2 E(b)).

The Code of Practice suggests that this is best done in relation to closure notices by sending them copies. Where there is re-classification, notices should also be posted prominently around the area (Code of Practice 8.6).

Re-opening of production areas

The food authority should remove the closure notice immediately upon being satisfied that harvesting in accordance with the EC regulations may be resumed. This will follow a period of sampling (854/2004, annex 2, chapter 2 C2 and Code of Practice 5.3.5). Again, interested parties must be informed as above.

Offences

Non-compliance with a closure notice constitutes an offence (regulation 19(1), Food Safety and Hygiene (England) Regulations 2013).

FC47 Closure notices on mollusc harvesting areas

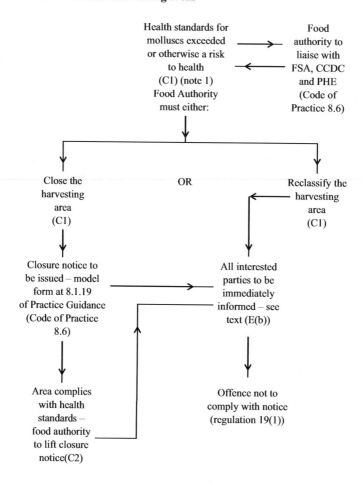

Notes

1. C and E numbers refer to EC Regulation 854/2004, annex 2, chapter 2.
2. FSA = Food Standards Agency.
3. CCDC = Consultant in Communicable Disease Control.
4. PHE = Public Health England.

FOOD STANDARDS

References

Food Safety Act 1990
Food Safety (Northern Ireland) Order 1991

Extent

These provisions apply to England, Wales and Scotland. Similar provisions apply in Northern Ireland.

Scope

These procedures include inspection and seizure of suspected food, Improvement Notices, Prohibition Orders and Emergency Prohibition Orders. They are mainly used to control food condition, information and labelling. They are distinct from similar powers which are used to deal with food hygiene conditions and practices in premises.

Inspection and seizure of suspected food

An authorised officer of a food authority can at all reasonable times inspect any food intended for human consumption which has been sold or is offered or exposed for sale; is in the possession of, or has been deposited with or consigned to, any person for the purpose of sale or of preparation for sale; or is otherwise placed on the market within the meaning of Regulation (EC) No. 178/2002 (section 9(1)).

If it appears to an authorised officer that any food is likely to cause food poisoning or any disease communicable to human beings (section 9(2)) the officer may either:

(a) give notice to the person in charge of the food that, until the notice is withdrawn, the food or any specified portion of it is not to be used for human consumption; and either is not to be removed or is not to be removed except to some place specified in the notice; or

(b) seize the food and remove it in order to have it dealt with by a Justice of the Peace;

and any person who knowingly contravenes the requirements of a notice is guilty of an offence (section 9(3)).

Where the authorised officer gives notice food is not to be used for human consumption, he shall, as soon as is reasonably practicable and in any event within 21 days, determine whether or not he is satisfied that the food complies with food safety requirements and:

(a) if satisfied, must withdraw the notice;

(b) if not satisfied, must seize the food and remove it in order to have it dealt with by a Justice of the Peace (section 9(4)).

Where an officer seizes the food he must inform the person in charge of the food of his intention to have it dealt with by a Justice of the Peace (section 9(5)).

If it appears to a Justice of the Peace, on the basis of evidence as he considers appropriate in the circumstances, that any food fails to comply with food safety requirements, he must condemn the food and order:

(a) the food to be destroyed or to be so disposed of as to prevent it from being used for human consumption; and

(b) any expenses reasonably incurred in connection with the destruction or disposal to be defrayed by the owner of the food (section 9(6)).

If a notice is withdrawn, or the Justice of the Peace refuses to condemn it, the food authority, must compensate the owner of the food for any depreciation in its value resulting from the action taken by the authorised officer (section 9(7)).

Any disputed question as to the right to or the amount of any compensation payable will be determined by arbitration (section 9(8)).

Improvement notices

If an authorised officer of an enforcement authority has reasonable grounds for believing that the proprietor of a food business is failing to comply with any relevant regulations, he may, by a notice served on that proprietor:

(a) state the officer's grounds for believing that the proprietor is failing to comply with the regulations;

(b) specify the matters which constitute the proprietor's failure so to comply;

(c) specify the measures which, in the officer's opinion, the proprietor must take in order to secure compliance; and

(d) require the proprietor to take those measures, or measures which are at least equivalent to them, within such period (not being less than 14 days) as may be specified in the notice (section 10(1)).

Any person who fails to comply with an improvement notice shall be guilty of an offence (section 10(2)).

Prohibition orders

If the proprietor of a food business is convicted of an offence under any relevant regulations; and the court by or before which he is so convicted is satisfied that the health risk condition is fulfilled with respect to that business, the court must by an order impose the appropriate prohibition (section 11(1)).

The health risk condition involves risk of injury to health from food in the following activities:

(a) the use for the purposes of the business of any process or treatment;

(b) the construction of any premises used for the purposes of the business, or the use for those purposes of any equipment; and

(c) the state or condition of any premises or equipment used for the purposes of the business (section 11(2)).

The appropriate prohibition is:

(a) a prohibition on the use of the process or treatment for the purposes of the business;
(b) a prohibition on the use of the premises or equipment for the purposes of the business or any other food business of the same class or description;
(c) a prohibition on the use of the premises or equipment for the purposes of any food business (section 11(3)).

If the proprietor or manager of a food business is convicted of an offence and the court by or before which he is so convicted thinks it proper to do so in all the circumstances of the case, the court may, by an order, impose a prohibition on the proprietor participating in the management of any food business, or any food business of a class or description specified in the order (section 11(4)).

As soon as practicable after the making of a prohibition order the enforcement authority must serve a copy of the order on the proprietor of the business; and affix a copy of the order in a conspicuous position on the premises and any person who knowingly contravenes such an order is guilty of an offence (section 11(5)).

A prohibition order ceases to have effect on the issue by the enforcement authority of a certificate to the effect that they are satisfied that the proprietor has taken sufficient measures to secure that the health risk condition is no longer fulfilled with respect to the business or where the court gives that direction (section 11(6)).

The enforcement authority must issue such a certificate within three days of their being satisfied and on an application by the proprietor for such a certificate, the authority must determine, as soon as is reasonably practicable and in any event within 14 days, whether or not they are so satisfied; and if they determine that they are not satisfied, give notice to the proprietor of the reasons for that determination (section 11(7)).

The court must give a direction on an application by the proprietor, where the court thinks it proper to do so having regard to all the circumstances of the case, including in particular the conduct of the proprietor since the making of the order; but no such application shall be entertained if it is made within six months after the making of the prohibition order; or within three months after the making by the proprietor of a previous application for such a direction (section 11(8)).

The relevant regulations for improvement and prohibition notices are those:

(a) for requiring, prohibiting or regulating the use of any process or treatment in the preparation of food; or
(b) for securing the observance of hygienic conditions and practices in connection with the carrying out of commercial operations with respect to food or food sources (section 10(3)).

Emergency prohibition notices and orders

Emergency prohibition notices are available in similar circumstances to a prohibition order (section 12(1) and (2)).

The officer must serve a notice on the proprietor of the business of his intention to apply for the order one day before doing so (section 12(3)).

The risk of injury to health must involve an imminent risk of injury (section 12(4)).

The same requirements for posting notice on the premises apply as does the relevant offence.

An emergency prohibition notice ceases to have effect:

(a) if no application for an emergency prohibition order is made within three days of the service of the notice;

(b) if such an application is made, on the determination or abandonment of the application (section 12(7)).

The same provisions apply to the issue of a certificate and compensation where the enforcement authority are satisfied that the proprietor has taken sufficient measures to secure that the health risk condition is no longer fulfilled with respect to the business (section 12(8), (9) and (10)).

Offences

Rendering food injurious to health

Any person who renders any food injurious to health by:

(a) adding any article or substance to the food;

(b) using any article or substance as an ingredient in the preparation of the food;

(c) abstracting any constituent from the food; and

(d) subjecting the food to any other process or treatment,

with intent that it shall be sold for human consumption, is guilty of an offence (section 7).

Selling food not complying with food safety requirements

Food fails to comply with food safety requirements if it is unsafe within the meaning of Article 14 of Regulation (EC) No. 178/2002

1. Food shall not be placed on the market if it is unsafe.
2. Food shall be deemed to be unsafe if it is considered to be:

(a) injurious to health;

(b) unfit for human consumption.

3. In determining whether any food is unsafe, regard shall be had:

(a) to the normal conditions of use of the food by the consumer and at each stage of production, processing and distribution, and

(b) to the information provided to the consumer, including information on the label, or other information generally available to the consumer concerning the avoidance of specific adverse health effects from a particular food or category of foods.

4. In determining whether any food is injurious to health, regard shall be had:

(a) not only to the probable immediate and/or short-term and/or long-term effects of that food on the health of a person consuming it, but also on subsequent generations;

(b) to the probable cumulative toxic effects;

(c) to the particular health sensitivities of a specific category of consumers where the food is intended for that category of consumers.

5. In determining whether any food is unfit for human consumption, regard shall be had to whether the food is unacceptable for human consumption according to its intended use, for reasons of contamination, whether by extraneous matter or otherwise, or through putrefaction, deterioration or decay.

6. Where any food which is unsafe is part of a batch, lot or consignment of food of the same class or description, it shall be presumed that all the food in that batch, lot or consignment is also unsafe, unless following a detailed assessment there is no evidence that the rest of the batch, lot or consignment is unsafe.

7. Food that complies with specific Community provisions governing food safety shall be deemed to be safe insofar as the aspects covered by the specific Community provisions are concerned.

8. Conformity of a food with specific provisions applicable to that food shall not bar the competent authorities from taking appropriate measures to impose restrictions on it being placed on the market or to require its withdrawal from the market where there are reasons to suspect that, despite such conformity, the food is unsafe.

9. Where there are no specific Community provisions, food shall be deemed to be safe when it conforms to the specific provisions of national food law of the Member State in whose territory the food is marketed, such provisions being drawn up and applied without prejudice to the Treaty, in particular Articles 28 and 30 thereof.

Any part of, or product derived wholly or partly from, an animal –

   (a) which has been slaughtered in a knacker's yard, or of which the carcase has been brought into a knacker's yard; or

   (b) in Scotland, which has been slaughtered otherwise than in a slaughterhouse,

is deemed to be unfit for human consumption (section 8).

Selling food not of the nature or substance or quality demanded

Any person who sells to the purchaser's prejudice any food which is not of the nature or substance or quality demanded by the purchaser is guilty of an offence (section 14(1)).

FOOD INFORMATION

References

Food Safety Act 1990
Food Information Regulations 2014
Food Information Regulations (Wales) 2014
Food Information Regulations (Scotland) 2014
Food Information Regulations (Northern Ireland) 2014

Extent

These provisions apply to England. Similar provisions apply to Wales, Scotland and Northern Ireland. This is EU-derived legislation but retained as domestic law despite EU withdrawal.

Scope

Regulation 12 of the Food Information Regulations (England) 2014 applies section 10 of the Food Safety Act 1990 to enable Improvement Notices to be served for a contravention of food information legislation.

Falsely describing or presenting food

Any person who falsely presents or describes the food through a label or advertisement the food or is likely to mislead as to the nature or substance or quality of the food, is guilty of an offence (section 15(1), (2) and (3)).

The fact that a label or advertisement in respect of which the offence is alleged to have been committed contained an accurate statement of the composition of the food shall not preclude the court from finding that the offence was committed (section 15(4)).

England and Wales (note: not yet in force)

Food information means information concerning a food and made available to the final consumer by means of a label, other accompanying material, or any other means including modern technology tools or verbal communication (Regulation (EU) No. 1169/2011) and (section 15A).

Contravention of food information law: seizure of food etc.

Where it appears to an authorised officer of a food authority, on an inspection, that food information law is being, or has been, contravened in relation to any food intended for human consumption the officer may:

(a) give notice that, until the notice is withdrawn –

   (i) the food, or any specified portion of it, is not to be used for human consumption; and

   (ii) the food, or any specified portion of it, and any related food information, or any specified part of it, is not to be removed (or is not to be removed except to some place specified in the notice); or

(b) seize the food and remove it in order to have it dealt with by a Justice of the Peace (section 15B(1) and (2)).

Notice is to be given to the person in charge of the food; and the owner of the food (where not the person in charge of the food) (section 15B(3)).

But notice need not be given if the officer, after making reasonable inquiries, does not know who owns the food (section 15B(4)).

Any person who knowingly contravenes the requirements of a notice commits an offence (section 15B(5)).

An officer who gives a notice must, as soon as is reasonably practicable and in any event within 21 days, determine whether or not food information law has been contravened in relation to the food in respect of which the notice was given (section 15B(6)).

After making a determination, the officer must:

(a) if satisfied that food information law has not been contravened, forthwith withdraw the notice; or
(b) if not so satisfied, seize the food and remove it in order to have it dealt with by the Justice of the Peace (section 15B(7)).

An officer who seizes and removes food may also:

(a) copy, make extracts of or take away any food information relating to the food that has been seized;
(b) where any such food information is in electronic form, require the information to be produced in a legible form in which it may be copied or taken away (section 15B(8)).

An officer who seizes and removes food must also inform the person in charge of the food and the owner of the food (where not the person in charge of the food) of the officer's intention to have it dealt with by the Justice of the Peace (section 15B(9)).

But the owner of the food need not be informed if the authorised officer, after making reasonable inquiries, does not know who owns the food (section 15B(10)).

Any person who might be liable to a prosecution for contravening food information law in relation to any food seized and removed is, if the person attends before the Justice of the Peace, entitled to be heard and to call witnesses (section 15B(11)).

If it appears to the Justice of the Peace (JP) that food information law has been contravened in relation to any food seized and removed, the JP may make such order as he considers appropriate in respect of the food and any food information relating to it (section 15B(12)).

An order made above may, in particular, order –

(a) that the food be destroyed or otherwise disposed of so as to prevent it from being used for human consumption;
(b) that any food information relating to the food be modified, destroyed or otherwise disposed of;
(c) that any food which is fit for human consumption (and any related food information, modified as the sheriff considers appropriate) be –

 (i) returned to the person who was in charge of the food; or
 (ii) distributed to such other person as the sheriff may determine (section 15B(13)).

Such an order:

(a) must, where the owner of the food is known, require the owner to meet any expenses reasonably incurred in connection with any destruction, modification, disposal, return or distribution of any food or food information which is carried out in pursuance of the order; and
(b) may require the owner of the food to meet any expenses reasonably incurred by the food authority in connection with any action taken by the authorised officer, or otherwise by or on behalf of the authority, in respect of any food or food information to which the order relates (section 15B(14)).

Where a notice is withdrawn; or the Justice of the Peace refuses to make an order the food authority must compensate the owner of the food for any depreciation in its value resulting from the action taken by the authorised officer (section 15B(15) and (16)). Any disputed question as to the right to or the amount of any compensation payable is to be submitted to arbitration for resolution (section 15B(17)).

Duty to report non-compliance with food information law

A food business operator must as soon as reasonably practicable inform the Food Standards Agency if the food business operator –

(a) is, or has been, in charge of any food which is intended for human consumption and has been placed on the market; and
(b) considers or has reason to believe that food information law is being contravened in relation to the food (section 15C(1)).

Any person who fails to comply shall be guilty of an offence (section 15C(2)).

A food business operator is to be treated as being, or having been, in charge of any food which it has received; imported; produced; processed; manufactured; distributed; or otherwise placed on the market (section 15C(3)).

Power to obtain information

Where a food business operator has informed the Food Standards Agency about food information law being contravened the operator must as soon as reasonably practicable provide any further information which is reasonably required by the Food Standards Agency which relates to the food (and any food information relating to it) and the circumstances which led the food business operator to inform the agency (section 15D(2)). Any person who fails to comply with this requirement is guilty of an offence (section 15D(3)).

Food Information Regulations Improvement Notices (Code of Practice 7.2.3)

Competent Authorities should deal with breaches of the Food Information Regulations 2014 by using the enforcement powers provided by those Regulations. Improvement notices should be used in line with the enforcement policy of a Competent Authority and must be considered as part of the escalation of enforcement action in line with the hierarchy of enforcement.

Competent Authorities should deal with breaches of the Food Safety and Hygiene (England) Regulations 2013 by using the enforcement powers provided by those Regulations (such as hygiene improvement notices under regulation 6). However, where labelling legislation made under the Food Safety Act 1990 is involved such as the Fish Labelling Regulations 2013, authorised officers should issue an improvement notice under section 10 of that Act. (Code of Practice 7.2.2.1)

Improvement Notices served under section 10 of the Food Safety Act 1990 must only be signed by officers who have been authorised to do so by the Competent Authority and meet the competency requirements set out in Chapter 4 of the Code of Practice. The officer who signs the notice must have witnessed the contravention and be satisfied that it constitutes a breach of the relevant legislation.

Competent Authorities must continue to use the prescribed forms set out in the Food Safety (Improvement and Prohibition – Prescribed Forms) Regulations 1991³³ when using powers under section 10 of the Food Safety Act 1990.

An improvement notice may be served on a person (food businesses operator) requiring the person to comply with the following:

- those listed in schedule 5, (the main provisions of 1169/2011), except as they relate to net quantity (section 10(1A)(a) to (c)); and
- those in section 10(1A)(d). These relate to:

  o the national requirements for non-prepacked foods requiring meat QUID labelling for foods containing meat (regulation 7(1), (4) and (5)); and
  o food irradiation labelling (the provisions of regulation 8(1) and (3)).
  o the national requirements under regulation 6 to provide name of food for non-prepacked foods (section 10(1A)(d)(ii) (Code of Practice 7.2.3.1)

A Food Information Regulations improvement notice would be inappropriate where breaches exist in respect of food hygiene, or where breaches exist in respect of food standards which pose a potential and imminent risk of injury to health and it is considered that swift enforcement action is needed.

It is an offence to fail to comply with certain allergen labelling and information requirements (these are listed in Regulation 10 of the Regulations) but, where there is a failure to comply with those provisions, enforcement officers will need to choose, based on the circumstances, between taking a criminal prosecution in relation to the contravention or serving an improvement notice or both (Code of Practice 7.2.3.2)

Food Authorities must have regard to the Framework Agreement on Local Authority Food Law Enforcement (the Framework Agreement), which reflects the requirements of the Code of Practice. The Framework Agreement is also consistent with the principles of the Regulators Code.

FOOD ADDITIVES

References

Food Additives, Flavourings, Enzymes and Extraction Solvents (England) Regulations 2013

Food Additives, Flavourings, Enzymes and Extraction Solvents (Wales) Regulations 2013

Food Additives, Flavourings, Enzymes and Extraction Solvents (Scotland) Regulations 2013

Food Additives, Flavourings, Enzymes and Extraction Solvents (Northern Ireland) Regulations 2013

Extent

These provisions apply in England. Similar provisions apply in Wales, Scotland and Northern Ireland.

Scope

For certain types of non-compliance with labelling requirements, an authorised officer can serve a compliance notice requiring specified steps to be taken, failing which an offence will be committed.

Compliance notices

If an authorised officer has reasonable grounds for believing that any person has not complied with, is not complying with, or is not likely to comply with regulations concerning food additives, flavourings, enzymes and extraction solvents (schedules 1, 2, 3 and 4 and regulation 13(1)), the officer may serve a compliance notice on that person (regulation 7(1)).

A compliance notice must state –

(a) the steps the person must take;
(b) the date and, if appropriate, the time by which each step must be taken;
(c) the reason for the service of the notice and for the steps required to be taken;
(d) that a failure to comply with the notice is an offence; and
(e) the details of the right to appeal against the notice (regulation 7(2)).

An authorised officer may serve a notice on a person withdrawing, varying or suspending a compliance notice (regulation 7(3)).

Any person who fails to comply with a compliance notice served on them commits an offence (regulation 7(4)).

Appeal against a compliance notice

Any person served with a compliance notice may appeal against that notice to a magistrates' court (regulation 8(1)).

The procedure on appeal to a magistrates' court shall be by way of complaint for an order (regulation 8(2)).

The period within which an appeal may be brought shall be one month from the date on which the compliance notice was served on the person wishing to appeal (regulation 8(3)).

A compliance notice is not suspended pending an appeal unless an authorised officer suspends it; or the court directs that it be suspended (regulation 8(4)).

The court may confirm, vary or evoke the notice or any requirement contained in it (regulation 8(5)).

Condemnation of food

Where any food is certified by a food analyst as being food which it is an offence to place on the market, that food may be seized and destroyed under an order of a Justice of the Peace as failing to comply with food safety requirement (section 9 of the Food Safety Act 1990 – p. 210) (regulation 18).

PART 4

HEALTH AND SAFETY

GENERAL PROCEDURAL PROVISIONS

Unless otherwise indicated, the following provisions are applicable to procedures in this Part.

References

Health and Safety at Work etc Act 1974 as amended

Health and Safety at Work (Northern Ireland) Order 1978

The Health and Safety (Enforcing Authority) Regulations 1998

Health and Safety (Enforcing Authority) Regulations (Northern Ireland) 1999

Extent and application

The procedures in this Part apply in Wales and Scotland but similar legislation applies in Northern Ireland (section 84).

Enforcing authority

Section 18(7)(a) of the Health and Safety at Work etc. Act 1974 defines enforcing authority as 'the Executive (Health and Safety Executive (HSE)) or any other authority which is by relevant statutory provision or by regulations . . . made responsible for the enforcement of any of the provisions to any extent'.

The Health and Safety (Enforcing Authority) Regulations 1998 make local authorities (as defined below) responsible for enforcement where the following are the main activities undertaken at any non-domestic premises.

1. The sale or storage of goods for retail or wholesale distribution except:
   - (a) at container depots where the main activity is the storage of goods in the course of transit to or from dock premises, an airport or a railway;
   - (b) where the main activity is the sale or storage for wholesale distribution of any dangerous substance or dangerous preparation;
   - (ba) where the main activity is the sale or storage for wholesale distribution of any hazardous substance or mixture;
   - (c) where the main activity is the sale or storage of water or sewage or their by-products or natural or town gas;

     and for the purposes of this paragraph where the main activity carried on in premises is the sale and fitting of motor car tyres, exhausts, windscreens or sunroofs, the main activity shall be deemed to be the sale of goods.

2. The display or demonstration of goods at an exhibition for the purposes of offer or advertisement for sale.
3. Office activities.
4. Catering services.

5. The provision of permanent or temporary residential accommodation, including the provision of a site for caravans or campers.
6. Consumer services provided in a shop, except dry-cleaning or radio and television repairs, and in this paragraph 'consumer services' means services of a type ordinarily supplied to persons who receive them otherwise than in the course of a trade, business or other undertaking carried on by them (whether for profit or not).
7. Cleaning (wet or dry) in coin-operated units in launderettes and similar premises.
8. The use of a bath, sauna or solarium, massaging, hair transplanting, skin piercing, manicuring or other cosmetic services and therapeutic treatments, except where they are carried out under the supervision or control of a registered medical practitioner, a dentist registered under the Dentists Act 1984, a physiotherapist, an osteopath or a chiropractor.
9. The practice or presentation of the arts, sports, games, entertainment or other cultural or recreational activities, except where the main activity is the exhibition of a cave to the public.
10. The hiring out of pleasure craft for use on inland waters.
11. The care, treatment, accommodation or exhibition of animals, birds or other creatures, except where the main activity is horse breeding or horse training at a stable, or is an agricultural activity or veterinary surgery.
12. The activities of an undertaker, except where the main activity is embalming or the making of coffins.
13. Church worship or religious meetings.
14. The provision of car parking facilities within the perimeter of an airport.
15. The provision of childcare, or playgroup or nursery facilities (schedule 1).

Schedule 2 of the Health and Safety (Enforcing Authority) Regulations 1998 allocate the following premises to the HSE for enforcement:

1. Any activity in a mine or quarry other than a quarry in respect of which notice of abandonment has been given under section 139(2) of the Mines and Quarries Act 1954.
2. Any activity in a fairground.
3. Any activity in premises occupied by a radio, television or film undertaking in which the activity of broadcasting, recording or filming is carried on, and the activity of broadcasting, recording or filming wherever carried on and, for this purpose, 'film' includes video.
4. The following activities carried on at any premises by persons who do not normally work in the premises:

   (a) construction work if:

      (i) the project which includes the work is notifiable within the meaning of regulation 6(1) of the Construction (Design and Management) Regulations 2015; or;
      (ii) the whole or part of the work contracted to be undertaken by the contractor at the premises is to the external fabric or other external part of a building or structure; or
      (iii) it is carried out in a physically segregated area of the premises, the activities normally carried out in that area have been suspended for the purpose of enabling the construction work to be carried out, the contractor has authority to exclude from that area persons who are not attending in connection with the carrying out of the work and the work is not the maintenance of insulation on pipes, boilers or other parts of heating or water systems or its removal from them;

(b) the installation, maintenance or repair of any gas system, or any work in relation to a gas fitting;

(c) the installation, maintenance or repair of electricity systems;

(d) work with ionising radiations except work in one or more of the categories set out in schedule 1 to the Ionising Radiations Regulations 2017.

5. The use of ionising radiations for medical exposure (within the meaning of regulation 2(1) of the Ionising Radiations Regulations 2017).

6. Any activity in premises occupied by a radiography undertaking in which there is carried on any work with ionising radiations.

7. Agricultural activities, and any activity at an agricultural show which involves the handling of livestock or the working of agricultural equipment.

8. Any activity on board a sea-going ship.

9. Any activity in relation to a ski slope, ski lift, ski tow or cable car.

10. Fish maggot and game breeding, except in a zoo.

11. Any activity in relation to a pipeline within the meaning of regulation 3 of the Pipelines Safety Act 1996.

12. The operation of –

(a) a guided bus system; or

(b) any other system of guided transport, other than a railway, that employs vehicles which for some or all of the time when they are in operation travel along roads.

13. The operation of a trolley vehicle system

14. Any activity regulated by the Acetylene Safety (England and Wales and Scotland) Regulations 2014 (schedule 2).

The 'main activity' test does not apply to the following premises, all of which are allocated to the HSE:

(a) the tunnel system within the meaning of section 1(7) of the Channel Tunnel Act 1987;

(b) an offshore installation within the meaning of regulation 3 of the Offshore Installations and Pipeline Works (Management and Administration) Regulations 1995;

(c) a building or construction site, that is to say, premises where the only activities being undertaken are construction work and activities for the purpose of or in connection with such work;

(d) the campus of a university, polytechnic, college, school or similar educational establishment;

(e) a hospital (regulation 3(5)).

In addition, all premises occupied by the following are reserved for HSE enforcement:

(a) local authority;

(b) parish and community councils;

(c) police authorities;

(d) fire and rescue authorities;

(e) organisations under the International Headquarters and Defence Organisations Act 1964, including authorities of visiting forces;

(f) United Kingdom Atomic Energy Authority;

(g) the Crown.

Where premises are mainly occupied by one of these bodies but also occupied by someone else, the Health and Safety Executive assumes responsibility for the whole premises (regulation 4).

Regulation 5 allows for a transfer of enforcement responsibility between Health and Safety Executive and local authorities in either direction by agreement.

Guidance on the Enforcing Authority Regulations is given on the HSE website ESite www.hse.gov.uk/foi/internalops/og/og-00073.htm.

Appointment of inspectors

An enforcing authority may appoint persons having suitable qualifications as inspectors. Each appointment must be in writing and specify which of the powers given to inspectors by the Act (below) are to be exercisable. In exercising these duties, the inspector must be able to produce if required his instrument of appointment or a duly authenticated copy of it (section 19).

Powers of inspectors

An enforcing authority may confer on an inspector all or any of the following powers in connection with his appointment:

(a) enter any premises at any reasonable time or, where a situation may be dangerous, at any time;

(b) take with him a police constable;

(c) take with him any other person duly authorised by the enforcing authority and any equipment or materials he may require;

(d) make such examination and investigation as necessary;

(e) direct that a premises or part of it remain undisturbed for as long as reasonably necessary for examination or investigation;

(f) take such measurements, photographs or records as necessary;

(g) take samples of articles, or substances in the premises or the atmosphere in, or in the vicinity of, the premises;

(h) dismantle, treat or test any article or substance likely to cause danger to health or safety, if requested, in the presence of a responsible person at the premises;

(i) take possession of or detain an article or substance likely to cause danger to health or safety in order to examine it, prevent it being tampered with or to keep it available for evidence, having previously taken a sample, where possible, and left notice of his action in the premises;

(j) require any person to answer and sign a declaration of truth;

(k) require the production of, inspect and take copies of any books or documents;

(l) require such facilities or assistance as necessary;

(m) exercise any other power necessary to carry out his responsibilities (section 20).

Notices

(a) **Form.** Notices should be in writing.

(b) **Authentication.** Since improvement and prohibition notices are served by an inspector, they should be signed by that person.

(c) **Service.** Notices to be served by an inspector may:

(i) be served by delivering it to the person, or leaving it at his proper address or sending it by post to him at that address;

(ii) in the case of a body corporate, be served or given to the secretary or clerk of that body;

(iii) in the case of a partnership, be served on or given to a partner or person having control or management of the partnership.

The proper address is:

(a) his last known address; or

(b) in the case of a body corporate on their secretary or clerk, at the registered or principal office; or

(c) in the case of a partnership, the principal office of the partnership.

If a person has specified an address in the UK, other than his proper address, to which notices should be sent, this will be acceptable as proper service. Notices to be served on an owner or occupier of any premises may be served by sending it by post to him at that premises or by addressing it to him by name but delivering it to some responsible person, resident or employee on the premises. Where, after reasonable enquiry, the name and address of the owner or occupier cannot be found, notice may be served by addressing it to 'the owner' or 'the occupier' of the premises (describing them) and delivering it to a responsible person, resident or employee at the premises, or, if no such person is available, by fixing it, or a copy, to some conspicuous part of the premises (section 46).

Offences

Failure to discharge a duty, to contravene any of the provisions of the act or its regulations or to obstruct an inspector will result on summary conviction in a penalty of imprisonment for a term not exceeding 12 months, or a fine, or both. The penalty on conviction on indictment is imprisonment for a term not exceeding two years, or a fine, or both (section 33 and schedule 3A).

Definitions

Unless otherwise indicated, the following definitions are applicable to procedures in this Part.

Employee means an individual who works under a contract of employment.

Personal injury includes any disease and any impairment of a person's physical or mental condition.

Premises includes any place and, in particular, includes:

(a) any vehicle, vessel, aircraft or hovercraft;

(b) any installation on land (including the foreshore and other land intermittently covered by water), any offshore installation, and any other installation (whether floating, or resting on the seabed or the subsoil thereof, or resting on other land covered with water or the subsoil thereof); and

(c) any tent or moveable structure.

Substance means any natural or artificial substance, including micro-organisms, whether in solid or liquid form or in the form of a gas or vapour (section 53).

Work means work as an employee or as a self-employed person. The employee is at work throughout the time when he is in the course of his employment, but not otherwise and a self-employed person is at work throughout such time as he devotes to work as a self-employed person (section 52(1)).

The meaning of 'work' has been extended by regulation to include any activity involving genetic manipulation or dangerous pathogens and the duty of care under section 3(2) of the Act extends to such an activity undertaken by research students etc. 'Work' has also been extended to include participants on government training schemes, school-age pupils on work experience and college students on 'sandwich course' external training (Health and Safety (Training for Employment) Regulations 1990).

In relation to premises outside of Great Britain but controlled under the Health and Safety at Work etc. Act 1974, e.g. oil rigs, the definition of 'at work' has also been extended so that an employee or a self-employed person is deemed to be at work throughout the time that he is present at the premises (regulation 23(2), the Management of Health and Safety Regulations 1999).

IMPROVEMENT AND PROHIBITION NOTICES

References

Health and Safety at Work etc. Act 1974, sections 21 to 24 inclusive
Health and Safety at Work (Northern Ireland) Order 1978
Employment Tribunals (Constitution and Rules of Procedure) Regulations 2013
Enforcement Policy Statement HSE
HELA Circular LAC 22/1 Rev., September 2000 – Choice of Appropriate Enforcement Procedure
Enforcement Management Model (EMM) General guidance on Application to Health Risks
Guidance for other people to accompany enforcement officers on site LAC: 22/2
Guidance on the appointment of Local Authority Inspectors to enforce the Health and Safety at Work etc. Act 1974 LAC 22/8
LA Enforcement in premises in which they may have an interest LAC: 22/10
Incident Selection Criteria Guidance LAC: 22/13
Formal cautions – HSE guidance LAC: 22/19
Guidance to LAs 2001 (Rev.), which includes at Annex I the Statement of Enforcement Policy, January 2002
National Local Authority Enforcement Code – Health and Safety at Work – England, Scotland & Wales

Extent

These provisions apply in England, Wales and Scotland. Similar provisions apply in Northern Ireland.

Scope

Where breaches of health and safety occur appointed inspectors can serve enforcement notices on the person(s) responsible to require them to carry out improvement works or to stop doing a particular activity.

Inspector

Improvement and prohibition notices may be served by persons appointed by an enforcing authority as an inspector under the Act and no further authority is required for them to exercise these powers (section 19).

Relevant statutory provisions

Both notice procedures relate to the application of 'relevant statutory provisions', which are:

(a) Part 1 Health and Safety at Work etc. Act 1974;
(b) health and safety regulations; and
(c) enactments and regulations made under them in force before the passing of the Health and Safety at Work etc. Act 1974 and now listed in schedule 1 of that Act (section 53(1)).

Enforcement procedures

In using the powers to serve improvement and prohibition notices, local authorities must have regard to the detailed guidance from the Health and Safety Executive in their Enforcement Policy Statement which local authorities are expected to comply with. Guidance is also to be found in several documents listed in the references above. One aspect of this is to be sure that the service of an improvement notice or prohibition notice is the most appropriate form of action and, in this context, attention is drawn to the enforcement option of a formal caution.

Improvement notices

(a) **Scope.** Improvement notices may be served when a relevant statutory provision is either being contravened or, having been contravened, likely to be continued or repeated (section 21).
(b) **Content.** The notice must include:

   (i)   the inspector's opinion concerning the contravention;
   (ii) specification of the provision or provisions being contravened;
   (iii) reasons for him being of that opinion;
   (iv) a requirement to remedy the contravention; and
   (v)   a period within which the contravention must be remedied, which must be not less than 21 days (section 21).

The notice need not specify the actual measures to be taken to remedy the contravention but, if it does, it may refer to any approved Code of Practice and allow a choice of remedies (section 23(2)).

Prohibition notices

(a) **Scope.** These notices may be used by an inspector where there is a risk of serious personal injury from an activity which is either being carried on or is likely to be carried on, and to which a relevant statutory provision applies (section 22(1) and (2)).

(b) **Content.** The notice must:

  (i) state the inspector's opinion relating to the risk of serious personal injury;

  (ii) specify the matters which give rise to that risk;

  (iii) where a relevant statutory provision is being contravened, specify the provision concerned and give reasons for that opinion;

  (iv) direct that the activity concerned shall not be carried on unless the matters specified in the notice have been remedied.

The notice will normally be given immediate effect if the inspector is of the opinion that there is a risk of serious personal injury. Alternatively, a period must be specified following which the notice comes into effect. There is no minimum period to be allowed and, since in the event of an appeal the notice is not automatically suspended, the period allowed may be shorter than the appeal period. The notice can only be suspended on application to the tribunal and the direction of the tribunal (sections 22 and 24).

The notice need not specify the actual measure to be taken but, if it does, it may refer to any approved Code of Practice and allow a choice of remedies (section 23(2)).

Standard of requirements

Unless a relevant statutory provision lays down a specific requirement, requirements in improvement notices relating to buildings must be no more onerous than that necessary to secure compliance with any building regulations in force as if the building was being newly erected (section 23(3)).

Person responsible

Improvement notices are to be served on the person who is contravening the relevant statutory provision or who has contravened it and is likely to continue or repeat it (section 21). Prohibition notices must be served on the person by or under the control of whom the particular activity is being or is about to be carried on (section 22(2)).

Appeals

Appeal against either notice must be made to an employment tribunal within 21 days of the date of service, although the tribunal may allow an extension of this period on application from the appellant, either before or after the expiration of the 21 days, if satisfied that it was not reasonably practicable to have brought the appeal within 21 days.

There are no specific grounds in either the Act or the regulations for lodging an appeal. The procedure to be adopted by the employment tribunal for receiving and hearing appeals is laid down in schedules 1–3 to the regulations (section 24 and Employment Tribunals (Constitution and Rules of Procedure) Regulations 2013).

Where an appeal is made against an improvement notice, the notice stands suspended until the appeal is determined or the appeal is withdrawn but appeals against

FC48 Improvement notices – section 21 Health and Safety at Work etc. Act 1974

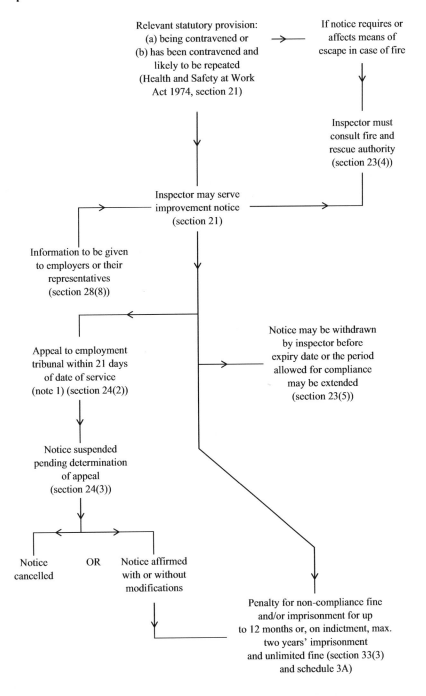

Notes
1. The period allowed for appeal may be extended by an employment tribunal (Employment Tribunals (Constitution and Rules of Procedure) Regulations 2013).
2. This procedure cannot be applied to the Crown (Health and Safety at Work Act 1974, section 48).
3. Unless the notice imposes requirements solely relating to the protection of persons at work, it must be included in a public register under the provisions of the Environment and Safety Information Act 1988.

FC49 Prohibition notices – section 22 Health and Safety at Work Act 1974

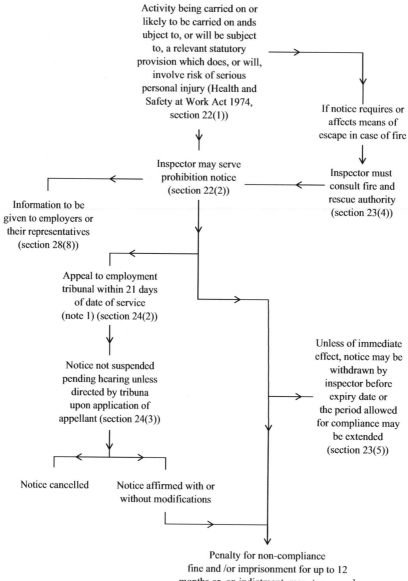

Activity being carried on or likely to be carried on ands ubject to, or will be subject to, a relevant statutory provision which does, or will, involve risk of serious personal injury (Health and Safety at Work Act 1974, section 22(1))

If notice requires or affects means of escape in case of fire

Inspector may serve prohibition notice (section 22(2))

Inspector must consult fire and rescue authority (section 23(4))

Information to be given to employers or their representatives (section 28(8))

Appeal to employment tribunal within 21 days of date of service (note 1) (section 24(2))

Unless of immediate effect, notice may be withdrawn by inspector before expiry date or the period allowed for compliance may be extended (section 23(5))

Notice not suspended pending hearing unless directed by tribuna upon application of appellant (section 24(3))

Notice cancelled   Notice affirmed with or without modifications

Penalty for non-compliance fine and /or imprisonment for up to 12 months or, on indictment, max. two years' imprisonment and unlimited fine (section 33(3) and schedule 3A)

Notes

1. The period allowed for appeal may be extended by an employment tribunal (Employment Tribunals (Constitution and Rules of Procedure) Regulations 2013).
2. This procedure cannot be applied to the Crown (Health and Safety at Work Act 1974, section 48).
3. Unless the prohibition notice imposes restrictions solely relating to the protection of persons at work, it must be included in a public register under the provisions of the Environment and Safety Information Act 1988.

prohibition notices do not automatically suspend the notice. Any suspension of a prohibition notice may be ordered only by the tribunal and then only from the date of the direction (section 24(3)).

Note – some employment tribunals call themselves industrial tribunals, particularly in Northern Ireland.

Defences

Where the duty imposed on a person by a relevant statutory provision is qualified by either:

(a) so far as is practicable; or
(b) as far as is reasonably practicable; or
(c) to use the best practicable means,

it is for the defendant to prove that it was not practicable or not reasonably practicable to do more than was in fact done to satisfy the duty or requirement, or that there was no better practicable means than was in fact used to satisfy the duty or requirement (section 40).

REPORTING OF INJURIES, DISEASES AND DANGEROUS OCCURRENCES

References

The Reporting of Injuries, Diseases and Dangerous Occurrences Regulations 2013
Reporting accidents and incidents at work: A brief guide to the Reporting of Injuries, Diseases and Dangerous Occurrences Regulations 2013 (RIDDOR) INDG 453

Extent

These regulations apply to England, Scotland, Wales and Northern Ireland.

Scope

These provisions require the keeping of records and reporting of certain accidents and injuries that occur in the workplace.

Non-fatal injuries to workers

Where any person at work, as a result of a work-related accident, suffers the following injuries:

(a) any bone fracture diagnosed by a registered medical practitioner, other than to a finger, thumb or toe;
(b) amputation of an arm, hand, finger, thumb, leg, foot or toe;
(c) any injury diagnosed by a registered medical practitioner as being likely to cause permanent blinding or reduction in sight in one or both eyes;

(d) any crush injury to the head or torso causing damage to the brain or internal organs in the chest or abdomen;

(e) any burn injury (including scalding) which –

   (i) covers more than 10% of the whole body's total surface area; or
   (ii) causes significant damage to the eyes, respiratory system or other vital organs;

(f) any degree of scalping requiring hospital treatment;

(g) loss of consciousness caused by head injury or asphyxia; or

(h) any other injury arising from working in an enclosed space which –

   (i) leads to hypothermia or heat-induced illness; or
   (ii) requires resuscitation or admittance to hospital for more than 24 hours,

the responsible person must follow the reporting procedure (see below p. 251 (regulation 4(1)).

Where any person at work is incapacitated for routine work for more than seven consecutive days (excluding the day of the accident) because of an injury resulting from an accident arising out of or in connection with that work, the responsible person must send a report to the relevant enforcing authority in an approved manner as soon as practicable and in any event within 15 days of the accident (regulation 4(2)).

Where any person not at work, as a result of a work-related accident, suffers:

(a) an injury, and that person is taken from the site of the accident to a hospital for treatment in respect of that injury; or

(b) a specified injury on hospital premises,

the responsible person must follow the reporting procedure (regulation 5).

Work-related fatalities

Where any person dies as a result of a work-related accident or occupational exposure to a biological agent, the responsible person must follow the reporting procedure (regulation 6(1) and (2)).

Where an employee has suffered a reportable injury which is a cause of his death within one year of the date of the accident, the employer must notify the relevant enforcing authority of the death in an approved manner without delay, whether or not the injury has been reported (regulation 6(3)).

This regulation does not apply to a self-employed person who suffers a fatal accident or fatal exposure on premises controlled by that self-employed person (regulation 6(4)).

Dangerous occurrences

Where there is a dangerous occurrence, the responsible person must follow the reporting procedure (regulation 7). Dangerous occurrences as detailed in **schedule 2** are notifiable.

Dangerous occurrences are certain, specified 'near-miss' events (incidents with the potential to cause harm.) Not all such events require reporting. There are 27 categories of dangerous occurrences that are relevant to most workplaces. For example:

- the collapse, overturning or failure of load-bearing parts of lifts and lifting equipment;
- plant or equipment coming into contact with overhead power lines;
- explosions or fires causing work to be stopped for more than 24 hours.

Certain additional categories of dangerous occurrences apply to mines, quarries, offshore workplaces and certain transport systems (railways etc). For a full, detailed list, refer to the online guidance at: www.hse.gov.uk/riddor and schedule 2 of the regulations.

Diseases

Where a person at work suffers from any of the listed occupational diseases and his work involves a corresponding work activity, that situation is reportable (regulation 5(1)). To become reportable, the disease must have been confirmed by a registered medical practitioner (regulation 5(2)).

The diseases and activities are set out in **schedule 3** to the regulations and include conditions due to physical agents and the physical demands of work such as:

- Ionising radiation
- Repetitive work injuries
- Infections due to biological agents
- Conditions due to harmful substances including poisonings.

Gas incidents

Death or major injury arising out of gas being distributed, filled, imported or supplied are notifiable to the Health and Safety Executive (regulation 6).

Defence

In proceedings against any person for failing to comply with a requirement of these Regulations, it is a defence for that person to prove that they were not aware of the circumstances which gave rise to that requirement, so long as that person had taken all reasonable steps to be made aware, in sufficient time, of such circumstances (regulation 16).

Recording and record-keeping

The responsible person must keep a record of any:

(a) reportable incident
(b) reportable diagnosis
(c) injury to a person at work resulting from an accident arising out of or in connection with that work, incapacitating that person for routine work for more than three consecutive days (excluding the day of the accident), and
(d) other particulars approved by the Executive or the Office of Road and Rail (regulation 12(1)).

An entry in the record must be kept for at least three years from the date on which it was made, and the record must be:

(a) kept at the place where the work to which it relates is carried on, or at the usual place of business of the responsible person; and

(b) in the case of a mine or quarry, available for inspection by any nominated person and workmen's inspectors (excluding any health record of an identifiable individual). (regulation 12(2)).

The responsible person must send to the relevant enforcing authority such extracts from the record as that enforcing authority may require (regulation 12(3)).

Particulars to be kept in records of any reportable incident

The date and time of the accident or dangerous occurrence.
   In respect of an accident injuring a person at work, that person's –

(a) full name;
(b) occupation;
(c) injury.

In respect of an accident injuring a person not at work, that person's –

(a) full name;
(b) status (for example 'passenger', 'customer', 'visitor' or 'bystander'); and
(c) injury,

unless these are not known and it is not reasonably practicable to ascertain them.
   The place where the accident or dangerous occurrence happened.
   A brief description of the circumstances in which the accident or dangerous occurrence happened.
   The date on which the accident or dangerous occurrence was first notified or reported to the relevant enforcing authority.
   The method by which the accident or dangerous occurrence was first notified or reported.
   The date of diagnosis of the disease.
   The name of the person affected.
   The occupation of the person affected.
   The name or nature of the disease.
   The date on which the disease was first reported to the relevant enforcing authority.
   The method by which the disease was reported.
   The date and time of the accident.
   The following particulars of the injured person –

(a) full name;
(b) occupation;
(c) injury.

The place where the accident happened.
   A brief description of the circumstances in which the accident happened (schedule 1).

Reporting procedure

Injuries, fatalities and dangerous occurrences

Where required to follow the reporting procedure (except in relation to a mine or quarry), the responsible person must –

(a) notify the relevant enforcing authority of the reportable incident by the quickest practicable means without delay; and
(b) send a report of that incident in an approved manner to the relevant enforcing authority within 10 days of the incident or, in the case of a dangerous occurrence, within 10 working days of the incident.

This does not apply to a self-employed person who is injured at premises owned or occupied by that self-employed person, and it is sufficient for a self-employed person to make arrangements for the report to be sent to the relevant enforcing authority by some other person. It also does not apply to dangerous occurrences offshore.

Diseases

Where required to follow the reporting procedure, the responsible person must send a report of the diagnosis in an approved manner to the relevant enforcing authority without delay.

It is sufficient for a self-employed person to make arrangements for the report to be sent to the relevant enforcing authority by some other person.

Carcinogens, mutagens and biological agents

Where required to follow the reporting procedure the responsible person must notify the relevant enforcing authority in an approved manner.

Mines and quarries

Where required to follow the reporting procedure in the case of a mine or quarry, the responsible person must:

(a) notify the relevant enforcing authority and any nominated person of the reportable incident by the quickest practicable means without delay; and
(b) send a report of that incident in an approved manner –

    (i) to any nominated person within seven days of the incident; and
    (ii) to the relevant enforcing authority within 10 days of the incident.

Where the responsible person becomes aware of a person subsequently dying as the result of an accident which gave rise to an injury reported, the responsible person must notify any nominated person of the death.

How to report

Online – www.hse.gov.uk/riddor and complete the appropriate online report form. The form will then be submitted directly to the RIDDOR database. The responsible person will receive a copy for records.

Telephone – All incidents can be reported online but a telephone service remains for reporting fatal and specified injuries only. Call the Incident Contact Centre on 0845 300 9923 (opening hours Monday to Friday 8.30 am to 5 pm).

HSE has an out-of-hours duty officer. Circumstances where HSE may need to respond out of hours include:

- a work-related death or situation where there is a strong likelihood of death following an incident at, or connected with, work;
- a serious accident at a workplace so that HSE can gather details of physical evidence that would be lost with time; and
- following a major incident at a workplace where the severity of the incident, or the degree of public concern, requires an immediate public statement from either HSE or government ministers. (INDG453(rev1))

Definitions

Accident includes an act of non-consensual physical violence done to a person at work (regulation 2).

Dangerous occurrence means an occurrence which arises out of or in connection with work and is of a class specified in schedule 2 (regulation 2).

Disease includes a medical condition (regulation 2).

Reportable incident means an incident giving rise to a notification or reporting requirement under these Regulations (regulation 2).

Reporting procedure means, in relation to –

    (a) an injury, death or dangerous occurrence (except at a mine or quarry), the procedure described in Schedule 1;

    (b) an occupational disease or a disease offshore, the procedure described in Schedule 1;

    (c) exposure to a carcinogen, mutagen or biological agent, the procedure described in Schedule 1; or

    (d) an injury, death or dangerous occurrence at a mine or quarry, the procedure described in Schedule 1 (regulation 2).

Responsible person –

    (a) in relation to an injury, death or dangerous occurrence reportable involving –

        (i) an employee, that employee's employer; or

        (ii) a person not at work or a self-employed person, or in relation to any other dangerous occurrence, the person who by means of their carrying on any undertaking was in control of the premises where the reportable or recordable incident happened, at the time it happened; or

    (b) in relation to a diagnosis reportable in respect of –

        (i) an employee, that employee's employer; or

        (ii) a self-employed person (regulation 3).

There are slight variations with regards to mines.

FC50 The reporting of injuries, diseases and dangerous occurrences – RIDDOR 2013

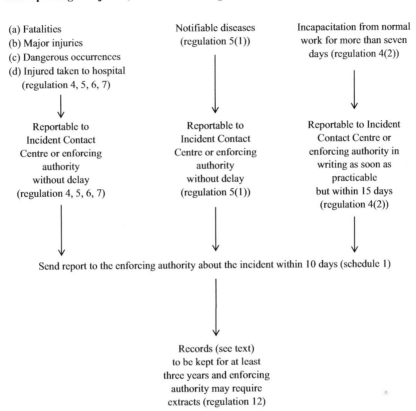

(a) Fatalities
(b) Major injuries
(c) Dangerous occurrences
(d) Injured taken to hospital
(regulation 4, 5, 6, 7)

Notifiable diseases
(regulation 5(1))

Incapacitation from normal
work for more than seven
days (regulation 4(2))

Reportable to
Incident Contact
Centre or enforcing
authority
without delay
(regulation 4, 5, 6, 7)

Reportable to
Incident Contact
Centre or enforcing
authority
without delay
(regulation 5(1))

Reportable to Incident
Contact Centre or
enforcing authority in
writing as soon as
practicable
but within 15 days
(regulation 4(2))

Send report to the enforcing authority about the incident within 10 days (schedule 1)

Records (see text)
to be kept for at least
three years and enforcing
authority may require
extracts (regulation 12)

Specified injury means any injury or non-fatal injuries specified in regulation 4(1)(a) to (h) above (regulation 2).

Work-related accident means an accident arising out of or in connection with work (regulation 2).

NOTIFICATION OF COOLING TOWERS AND EVAPORATIVE CONDENSERS

References

The Notification of Cooling Towers and Evaporative Condensers Regulations 1992

The Notification of Cooling Towers and Evaporative Condensers Regulations (Northern Ireland) 1994

Approved Code of Practice and Guidance – L8. Legionnaires' disease: the control of legionella bacteria in water systems, HSE 2013

Control of legionella: Inspection of evaporative cooling systems and investigation of outbreaks of Legionnaires' disease – HSE website

Technical Guidance HSG 274:

Part 1: The control of legionella bacteria in evaporative cooling systems.
Part 2: The control of legionella bacteria in hot and cold water systems.
Part 3: The control of legionella bacteria in other risk systems.

Extent

These provisions apply in England, Wales and Scotland. Similar provisions apply in Northern Ireland.

Scope

Notification is required to be made to the local authority in whose area the premises are situated in respect of all cooling towers and evaporative condensers except:

   (a)   where it contains no water exposed to air;
   (b)   its water supply is not connected; and
   (c)   its electrical supply is not connected (regulation 2 and 3(1)),

and where the device is situated in premises other than domestic premises, which are used for, or in connection with, the carrying on of a trade, business or undertaking (whether for profit or not).

The purpose of the notification is to allow the local authority to monitor the environmental control of the devices with particular reference to the avoidance of Legionnaire's disease and to have an awareness of the location of them in order to investigate outbreaks of such disease.

Person responsible

Notification is to be made by each person who has, to any extent, control of the premises and must be made before the device is situated there. Where the premises fall to any extent under

FC51 Notification of cooling towers and evaporative condensers – Notification of cooling towers and evaporative condensers Regulations 1992

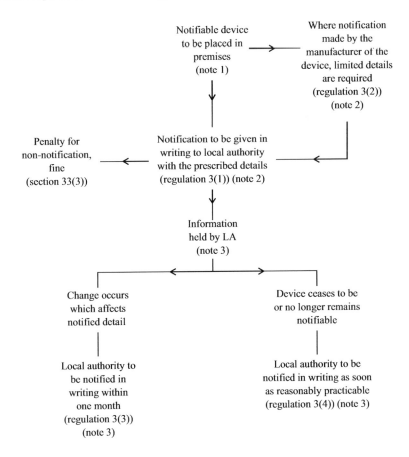

Notes
1. Notification must be made before the device is placed in the premises, otherwise an offence is committed (regulation 3(1)).
2. The form to be used is prescribed by the Health and Safety Executive.
3. Local authorities should make available to the Health and Safety Executive information from notifications relating to equipment on premises for which the Health and Safety Executive has enforcement responsibility.

the control of the manufacturer of the device, e.g. during installation, that person is required to effect the notification (regulation 3(1) and (2)).

Notification

This is to be done in writing using the form prescribed by the Health and Safety Executive for the purpose. The detail to be supplied is:

(a) the address of the premises where the notifiable device is to be situated;
(b) the name, address and telephone number of a person who has, to any extent, control of the premises referred to in (a) above;
(c) the number of notifiable devices at the premises referred to in (a) above;
(d) the location on the premises of each notifiable device referred to in (c) above.

Where the notification is being made by the manufacturer, the details required relate only to (a) and (b) above (regulation 3(1) and (2)).

Subsequent change

Any changes to the notified details must be notified to the local authority by either the person having control of the premises or the manufacturer in writing within one month of the change (regulation 3(3)).

The local authority must be informed in writing as soon as reasonably practicable where the device ceases to be used, other than for seasonal shutdown or maintenance (regulation 3(4)).

Definitions

Cooling tower means a device whose main purpose is to cool water by direct contact between that water and a stream of air.

Evaporative condenser means a device whose main purpose is to cool a fluid by passing that fluid through a heat exchanger, which is itself cooled by contact with water passing through a stream of air.

Heat exchanger means a device for transferring heat between fluids which are not in direct contact with one another.

SEIZURE OF DANGEROUS ARTICLES OR SUBSTANCES

Reference

Health and Safety at Work etc. Act 1974, section 25

Extent

This provision applies to England, Scotland, Wales and Northern Ireland.

Scope

An inspector appointed under the Act by an enforcing authority may seize any article or substance on any premises which he has power to enter if he believes that, in the circumstances in which he found it, the article or substance is a cause of imminent danger of serious personal injury to either employees or other persons (section 25(1)).

Customs officers are given the power to assist inspectors by seizing and detaining for up to two working days any imported article or substance (section 25A).

Sampling

Before seizing either:

(a) any article which forms part of a batch of similar articles; or
(b) any substance,

the inspector is required, where it is practicable for him to do so, to take a sample and give a portion of it, marked in such a way as to identify it, to a responsible person at the premises where it was found (section 25(2)).

Action following seizure

Having seized the article or substance, the inspector is to render it harmless and this may be by any method, including destruction (section 25(1)).

Having dealt appropriately with the article or substance, the inspector must:

(a) prepare and sign a written report of the matter, including the circumstances in which the article or substance was seized and dealt with; and
(b) give a copy of that report to a responsible person at the premises where the article or substance was found; and
(c) serve a copy of the notice on the owner of the article or substance where that is a different person from (b) above.

If, after reasonable enquiry, the inspector is unable to ascertain the name and address of the owner, a copy may be served on the owner by giving it to the responsible person (section 25(3)).

FC52 Seizure of dangerous articles or substances – section 25 Health and Safety at Work etc. Act 1974

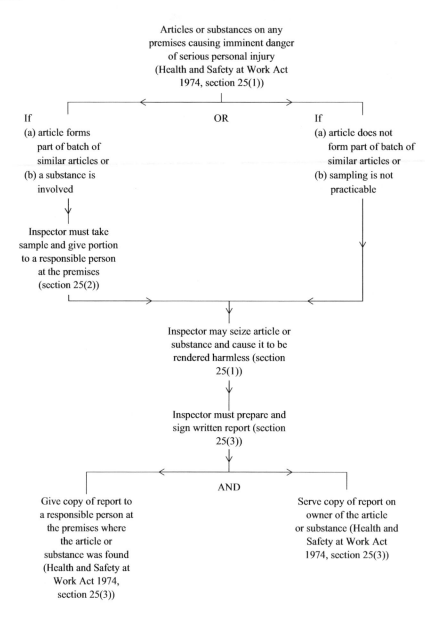

Articles or substances on any
premises causing imminent danger
of serious personal injury
(Health and Safety at Work Act
1974, section 25(1))

OR

If
(a) article forms
part of batch of
similar articles or
(b) a substance is
involved

If
(a) article does not
form part of batch of
similar articles or
(b) sampling is not
practicable

Inspector must take
sample and give portion
to a responsible person
at the premises
(section 25(2))

Inspector may seize article or
substance and cause it to be
rendered harmless (section
25(1))

Inspector must prepare and
sign written report (section
25(3))

AND

Give copy of report to
a responsible person at
the premises where
the article or
substance was found
(Health and Safety at
Work Act 1974,
section 25(3))

Serve copy of report on
owner of the article
or substance (Health and
Safety at Work Act
1974, section 25(3))

Note

1. These procedures cannot be applied to Crown premises (Health and Safety at Work Act 1974, section 48).

PART 5

HOUSING AND HEALTH

Chapter 1

STRUCTURE OF THE LAW DEALING
WITH HOUSING STANDARDS
AND THEIR ENFORCEMENT

This section outlines the key definitions of terms, power to obtain information with regard to enforcement, service of notice powers, powers of entry, appeal provisions and charging for enforcement powers. This framework is key to how housing powers are enforced.

The main provisions are found in the Housing Act 2004, which contains:

(a) a system for assessing housing conditions based on the use of the Housing Health and Safety Rating System (HHSRS);
(b) enforcement options; and
(c) licensing of certain houses in multiple occupation and selective licensing for other residential accommodation.

The Housing Act 1985 provisions relating to demolition orders and clearance areas together with those relating to overcrowding continue in force with amendments. The ability for local authorities to declare Renewal Areas is found in the Local Government and Housing Act 1989, and the Housing Grants, Construction, Development and Regeneration Act 1996 and the Regulatory Reform (Housing Assistance) (England and Wales) Order 2002 have provisions for grants for the renewal of private sector housing.

GENERAL PROCEDURAL PROVISIONS

Extent

The provisions in this part apply to England and Wales

- Housing Act 1985, section 625(3),
- Housing Grants, Construction and Regeneration Act 1996, section 148
- Housing Act 2004, section 270
- Housing and Planning Act 2016, section 215.

but, whilst the power to make orders and regulations under the Housing Act 2004 lies with the Secretary of State, he must consult the National Assembly for Wales before acting in relation to any residential properties in Wales.

Definitions

Appropriate Tribunal means:

    (a) in relation to a building in England, the First-Tier Tribunal or, where determined by or under Tribunal Procedure Rules, the Upper Tribunal; and

    (b) in relation to a building in Wales, a residential property tribunal.

Local housing authorities

In relation to England, these are:

- unitary councils;
- district councils;
- London borough councils;
- the common Council of the City of London;
- the Sub-Treasurer of the Inner Temple and the Under-Treasurer of the Middle Temple; and
- the Council of the Isles of Scilly.

For Wales, they are:

- county councils or county borough councils (Housing Act 2004, section 261). Throughout this part, local housing authority is abbreviated to local authority.

Residential premises

In relation to **Part 1 of the Housing Act 2004**, this means:

    (a) a dwelling;

    (b) a house in multiple occupation;

    (c) unoccupied accommodation;

    (d) any common parts of a building containing one or more flats (section 1(4)).

This definition is further clarified:

    (a) 'a building containing one or more flats' (in (d) above) does not include a house in multiple occupation;

    (b) 'common parts' in relation to a building containing one or more flats includes the structure and exterior of the building and common facilities provided (whether or not in the building) for persons who include the occupiers of one or more of the flats;

    (c) 'dwelling' means a building or part of a building occupied or intended to be occupied as a separate dwelling;

    (d) 'external common parts' in relation to a building containing one or more flats means common parts of the building which are outside it;

    (e) 'flat' means a separate set of premises (whether or not on the same floor) which forms part of a building, is constructed or is adapted for use for the purposes of a dwelling and either the whole or a material part of which lies above or below some other part of the building (section 1(5));

    (f) house in multiple occupation is as defined in sections 254–259 – see p. 360.

'Unoccupied house in multiple occupation accommodation' means a building or part of a building constructed or adapted for use as a house in multiple occupation but for the time being is either unoccupied or only occupied by persons who form a single household.

References to a dwelling, house in multiple occupation or a building containing one or more flats include (where the context permits) any yard, garden, outhouses and appurtenances belonging to or usually enjoyed with the dwelling, house in multiple occupation or building (or any part of it) (section 1(6)).

Person having control

This is the person who receives, either directly or as agent or trustee, the market rent from the tenants or is entitled to receive the rent if the premises were let, i.e. an owner (section 263).

Person managing

This is a person being an owner or lessee of the premises who receives the rent directly, or through an agent or trustee, from the occupier. Rent includes ground rent, so this person may be a managing agent (section 263).

Obtaining information

The local authority may require by notice the production of documentation that may be necessary to carry out its enforcement functions. Such notices may be served on the following 'relevant persons':

- holders of house in multiple occupation licences or those proposing to hold such a licence;
- any person with an estate or interest in the premises;
- a person who is or is proposing to be managing or having control;
- anyone otherwise involved, or so proposing, in the management of the premises; and
- an occupier.

Persons failing to comply with the notice are guilty of an offence and subject to a fine not exceeding level 5 (sections 235 and 236).

Service of notices

The form, authentication and methods of service of notices or other documents arising from the Housing Act 2004 are as set out in sections 233–234 of the Local Government Act 1972 and the provisions of section 16 of the Local Government (Miscellaneous Provisions) Act 1976 relating to the requisition of information regarding ownership etc. are also made available for enforcement of this Act (see p. 46) (Housing Act 2004, section 246).

Where the recipient has given prior approval, a notice or licence may be served electronically and detailed provisions for this are to be set out in the regulations (sections 247–248).

The Secretary of State has powers by regulations to dispense with the need for notices etc. where there is no prescribed form (section 244 and 245).

Appeals

Appeals against the various actions of the local authority in these procedures are to be made to the **Appropriate Tribunal**. These are the First-Tier Tribunal and Upper Tribunals in dealing with these procedures. The powers of the Tribunals are contained in section 230 and these include a general power to give such directions as it considers necessary or desirable for securing the just, expeditious and economical disposal of the proceedings. The more detailed provisions for appeals are noted in each procedure below.

With the approval of the First-Tier Tribunal, appeals against its decisions are to the Upper Tribunal within 28 days of notification of the decision (section 231).

Detailed rules for the operation of these Tribunals are found at www.justice.gov.uk/tribunals/residential-property.

Charging for enforcement action

Local authorities are given the power to make a reasonable charge for recovering certain costs in connection with:

1. the service of improvement notices;
2. making prohibition orders;
3. serving hazard awareness notices;
4. taking emergency remedial action;
5. making emergency prohibition orders;
6. making demolition orders.

The costs that may be recovered relate to:

1. inspection of the premises;
2. consideration of the action to be taken;
3. the service of the notice (section 49).

The local authorities should take account of personal circumstances of the person to be charged (Enforcement Guidance, paragraph 5.52) (p. 364).

There is an appeal to the First-Tier Tribunal against the charges raised. The demand for payment becomes operative 21 days after service and is, until recovered, a charge on the property and is a local land charge. After one month, the debt may be recovered by the appointment of a receiver (section 50).

HOUSING STANDARDS: THE HOUSING HEALTH AND SAFETY RATING SYSTEM (HHSRS)

References

Housing Act 2004

Housing Health and Safety Rating System **Operating Guidance**, Housing Act 2004 Guidance about inspections and assessment of hazards given under section 9, Department for Communities and Local Government (DCLG), 2006

Housing health and safety rating system (HHSRS) **enforcement guidance**: housing conditions: Guidance for local housing authorities about their duties and powers under part 1 of the Housing Act 2004

Background

The Housing Health and Safety Rating System of assessing housing conditions was introduced in the Housing Act 2004 and replaced the former system based on the test of fitness for human habitation. The system operates by reference to category 1 and category 2 hazards in residential premises. The system, and the way in which it should be operated, are described in Housing Health and Safety Rating System Operating Guidance Housing Act 2004, Guidance about inspections and assessment of hazards given under section 9, Department for Communities and Local Government (DCLG), 2006. The system and the options for remedial actions apply to residential premises (definition on p. 262).

Assessment process

The assessment process is:

(a) full inspection of the residential premises to identify any deficiencies;
(b) inspectors judge whether the deficiencies mean that there are any hazards that are significantly worse than the average for residential premises of that age and type; and if so:

    (i) the likelihood of an occurrence that could cause harm over the next 12 months; and
    (ii) the severity of the outcomes from such an occurrence.

The Housing Health and Safety Rating System Operating Guidance issued by the DCLG in February 2006 under section 9 of the Act gives detailed technical guidance on the way in which the system is to be used, and local authorities must have regard to this in carrying out their functions.

Review of housing conditions and Inspection

Local authorities are required to keep under review the housing conditions in their area with a view to identifying what courses of action are necessary under the Housing Acts 1985–2004 (section 3 Housing Act 2004 and Enforcement guidance 2.3).

Where the local authority considers, from information gained from its duties under section 3, or for any other reason, that it would be appropriate for a residential premises to be inspected with a view to determining if category 1 or 2 hazards exist, the local authority must arrange for an inspection. An inspection obligation also follows an official complaint about the condition of the premises from a Justice of the Peace or from a parish or community council (section 4).

Inspections are to be carried out in accordance with regulation 5 of the Housing Health and Safety Rating System (England) Regulations 2005 and an inspector must:

- have regard to any guidance issued under section 9 (Enforcement Guidance 3.2);
- inspect with a view to preparing an accurate record of the condition of the property; and
- prepare and keep such a record in written or electronic form.

Where an inspection reveals that category 1 or 2 hazards exist, the proper officer must report in writing to the local authority, who must consider the report as soon as possible (section 4(6–7)).

Hazards

A hazard is a situation where a risk of harm (see below) is associated with any of the matters or circumstances listed in schedule 1 of the Housing Health and Safety Rating System (England) Regulations 2005, being as follows:

Damp and mould growth

  1. Exposure to house dust mites, damp, mould or fungal growths.

Excess cold

  2. Exposure to low temperatures.

Excess heat

  3. Exposure to high temperatures.

Asbestos and MMF

  4. Exposure to asbestos fibres or manufactured mineral fibres.

Biocides

  5. Exposure to chemicals used to treat timber and mould growth.

Carbon monoxide and fuel combustion products

  6. Exposure to:

      (a) carbon monoxide;
      (b) nitrogen dioxide;
      (c) sulphur dioxide and smoke.

Lead

  7. The ingestion of lead.

Radiation

  8. Exposure to radiation.

Uncombusted fuel gas

  9. Exposure to uncombusted fuel gas.

Volatile organic compounds

10. Exposure to volatile organic compounds.

Crowding and space

11. A lack of adequate space for living and sleeping.

Entry by intruders

12. Difficulties in keeping the dwelling or house in multiple occupation secure against unauthorised entry.

Lighting

13. A lack of adequate lighting.

Noise

14. Exposure to noise.

Domestic hygiene, pests and refuse

15. (a) Poor design, layout or construction such that the dwelling or house in multiple occupation cannot readily be kept clean.
    (b) Exposure to pests.
    (c) An inadequate provision for the hygienic storage and disposal of household waste.

Food safety

16. An inadequate provision of facilities for the storage, preparation and cooking of food.

Personal hygiene, sanitation and drainage

17. An inadequate provision of:

    (a) facilities for maintaining good personal hygiene;
    (b) sanitation and drainage.

Water supply

18. An inadequate supply of water free from contamination, for drinking and other domestic purposes.

Falls associated with baths etc.

19. Falls associated with toilets, baths, showers or other washing facilities.

Falling on level surfaces etc.

20. Falling on any level surface or falling between surfaces where the change in level is less than 300 millimetres.

Falling on stairs etc.

21. Falling on stairs, steps or ramps where the change in level is 300 millimetres or more.

Falling between levels

22. Falling between levels where the difference in levels is 300 millimetres or more.

Electrical hazards

23. Exposure to electricity.

Fire

24. Exposure to uncontrolled fire and associated smoke.

Flames, hot surfaces, etc.

25. Contact with:

   (a) controlled fire or flames;
   (b) hot objects, liquid or vapours.

Collision and entrapment

26. Collision with, or entrapment of body parts in doors, windows or other architectural features.

Explosions

27. An explosion at the dwelling or house in multiple occupation.

Position and operability of amenities etc.

28. The position, location and operability of amenities, fittings and equipment.

Structural collapse and falling elements

29. The collapse of the whole or part of the dwelling or house in multiple occupation.

Assessment of hazards

The assessment of hazards identified is a two-stage process, which first considers the likelihood of an occurrence and second the range of harm outcomes. The calculations are based on the risk to the most vulnerable potential occupant, whether or not anyone is resident at the time of inspection. The system relates poor housing conditions to the kinds of harm attributable to such conditions. It does not try to assess a specific health outcome in relation to the current occupant.

'Likeliness' is considered within a period of 12 months, beginning on the date of the assessment, during which the relevant occupier may suffer harm from identified hazards. Likelihood is expressed as ratio from Table 1 in regulation 6 of the 2005 regulations.

In this connection, 'harm' is defined by the 2005 regulations by reference to four classes contained in schedule 2. These classes are grouped on harm as is reasonably foreseeable as a result of the hazard and being extreme, severe, serious or moderate harm. Each class has a list of specific conditions that may be suffered by the occupiers. The inspector must assess which of the four classes of harm a relevant occupier is most likely to suffer during the 12-month period.

To allow for the comparison of the significance of the widely differing hazards, the Housing Health and Safety Rating System uses a formula to generate hazard scores. This formula uses the likelihood expressed as a ratio in Table 1 of regulation 6, the spread of possible harm outcomes expressed as a percentage in Table 2 (regulation 6) and the weightings given to the four Housing Health and Safety Rating System Classes of Harm (reflecting the degree of incapacity suffered by each class). The hazard score is:

	Class of harm weighting	*Likelihood*	*Spread of harm (%)*
S1 =	10,000 x	1/L x	O1
S2 =	1,000 x	1/L x	O2
S3 =	300 x	1/L x	O3
S4 =	10 x	1/L x	O4

For the purposes of this formula:

L is the representative scale point in column 2 of Table 1 of regulation 6.
O1 is the representative scale point of the percentage range (RSPPR) recorded under paragraph (4) in relation to Class 1 harm.
O2 is the RSPPR for Class 2 harm.
O3 is the RSPPR for Class 3 harm.
O4 is the RSPPR for Class 4 harm.

The numerical hazard scores are banded into ten bands in Table 3 of regulation 7 and these are:

Band	Numerical score range
A	5,000 or more
B	2,000–4,999
C	1,000–1,999
D	500–999
E	200–499
F	100–199
G	50–99
H	20–49
I	10–19
J	9 or less

Categories of hazard

For the enforcement of the procedures to remedy unsatisfactory conditions, hazards are classified as either category 1 or category 2. Category 1 are those hazards banded as A, B or C in the table above and category 2 are those banded as D–J (regulation 8).

Application of Housing Health and Safety Rating System in houses in multiple occupation

The Part 1 functions are available for houses in multiple occupation but are not part of the licensing procedure for those premises – see below. There is a requirement in section 55(5)(c) and 55(6) that, not later than five years after an application for a house in multiple occupation licence, there should be no Part 1 enforcement functions (see below) that ought to be carried out by the local authority.

In Scotland, there is a Tolerable Standard in the Housing (Scotland) Act 1987.

In Northern Ireland, there is a Fitness Standard in the Housing (NI) Order 1992.

HOUSING STANDARDS FOR ACCOMMODATION OF THE HOMELESS

Reference

Homelessness (Suitability of Accommodation) (England) Order 2012

Extent

These standards apply in England only but there are similar provisions in Wales.

Scope

Where a local authority wishes to secure accommodation for homeless the accommodation must comply with suitable standards.

Article 2 of the order makes provision for the matters to be taken into account in determining whether accommodation is suitable for a person.

Article 3 sets out circumstances where accommodation is not to be regarded as suitable.

Matters to be taken into account in determining whether accommodation is suitable for a person.

In determining whether accommodation is suitable for a person, the local authority must take into account:

- the location of the accommodation;
- any disruption which would be caused by the location of the accommodation to the employment, caring responsibilities or education of the person or members of the person's household;
- the proximity and accessibility of the accommodation to medical facilities and other support which are currently used by or provided to the person or members of the person's household, and are essential to the well-being of the person or members of the person's household; and
- the proximity and accessibility of the accommodation to local services, amenities and transport (article 2).

Circumstances in which accommodation is not to be regarded as suitable for a person

Accommodation in the private rented sector shall not be regarded as suitable where in the view of the local authority, one or more of the following apply –

(a) the accommodation is not in a reasonable physical condition;
(b) any electrical equipment supplied with the accommodation is unsuitable
(c) the landlord has not taken reasonable fire safety precautions with the accommodation and any furnishings supplied with it;
(d) the landlord has not taken reasonable precautions to prevent the possibility of carbon monoxide poisoning in the accommodation;
(e) that the landlord is not a fit and proper person to act in the capacity of landlord, having considered if the person has committed certain sexual, discrimination or housing related offences,
(f) the accommodation an unlicensed house in multiple occupation,
(g) the accommodation does not have a valid energy performance certificate
(h) the accommodation does not have a current gas safety record
(i) the landlord has not provided the local authority a written tenancy agreement, which the landlord proposes to use, and the local authority considers to be adequate (article 3).

FITNESS FOR HUMAN HABITATION

Reference

Landlord and Tenant Act 1985, sections 8–10 (as amended by the Homes (Fitness for Human Habitation) Act 2019)

Extent

These provisions apply to England and Wales.

Scope

These standards apply to tenancies under the Landlord and Tenant Act 1985.

Implied terms as to fitness for human habitation

In a tenancy contract or lease for the letting of a house for human habitation there is implied, even though there may be a stipulation to the contrary:

(a) a condition that the house is fit for human habitation at the commencement of the tenancy, and
(b) an undertaking that the house will be kept by the landlord fit for human habitation during the tenancy (section 8(1) and 9A(1)).

It applies if the rent limit and the letting is not on such terms as to the tenant's responsibility (sections 8(3) and (4)). There are slight differences in the application of these provisions for agricultural workers (section 9).

There are detailed provisions about the responsibility of the lessor and lessee in tenancy agreements in sections 9A, 9B and 9C.

Fitness for human habitation

In determining whether a house or dwelling is unfit for human habitation, regard shall be had to its condition in respect of the following matters:

- repair,
- stability,
- freedom from damp,
- internal arrangement,
- natural lighting,
- ventilation,
- water supply,
- drainage and sanitary conveniences,
- facilities for preparation and cooking of food and for the disposal of waste water; in relation to a dwelling in England, any prescribed hazard;

and the house shall be regarded as unfit for human habitation if, and only if, it is so far defective in one or more of those matters that it is not reasonably suitable for occupation in that condition (section 10). This is very similar to the definition of fitness for human habitation before the HHSRS standard was introduced.

ENERGY EFFICIENCY IN PRIVATE RENTED PROPERTY

Reference

Energy Efficiency (Private Rented Property) (England and Wales) Regulations 2015

Extent

These provisions apply to England and Wales.

Scope

The regulations establish a minimum level of energy efficiency for privately rented property in England and Wales. Landlords of privately rented domestic and non-domestic property must ensure that their properties reach at least an energy performance certificate (EPC) rating of E before granting a new tenancy to new or existing tenants.

A tenant can request relevant energy efficiency improvements be carried out on the rented property by the landlord. It is the landlord's duty not to unreasonably refuse a tenant's request. The landlord must make an initial and full response to tenant's request. The landlord may make a counter proposal. There are certain exemptions that apply to the energy efficiency improvements. The Secretary of State must establish and maintain a PRS Exemptions Register (regulation 36). If a landlord believes that an EPC F or G rated property they let qualifies for an exemption from the minimum energy efficiency standard, that exemption must be registered on the PRS Exemptions Register – a self-certification database.

The tenant may appeal to the First-Tier Tribunal if not satisfied with the landlord's proposals or failure to act.

The First-Tier Tribunal must determine:

(a) the landlord's, or the superior landlord's, refusal of consent,
(b) the landlord's initial response,
(c) the landlord's full response,
(d) the counter proposal, or
(e) the landlord's failure to make energy efficiency improvements specified in a counter proposal, failed to comply with these Regulations (regulation 18(1)).

If the First-Tier Tribunal determines that the landlord, has failed to comply with these Regulations, the First-Tier Tribunal may by Order consent to the making of any relevant energy efficiency improvement specified in the tenant's request (regulation 18(2)).

Compliance notices

An enforcement authority may serve a notice on the landlord (L) where L appears to it to be, or to have been at any time within the 12 months preceding the date of service of the compliance notice, in breach of one or more of the requirements of these regulations requesting information as it considers necessary to enable it to monitor compliance (regulation 37(1)).

A notice may request L to produce for inspection originals, or copies, of the following:

(a) the energy performance certificate for the property which was valid at the time the property was let,
(b) any other energy performance certificate for the property in L's possession,
(c) any current tenancy agreement under which the property is let,
(d) any qualifying assessment in relation to the property,
(e) any other document which the enforcement authority considers necessary to enable it to carry out its functions,

and may request L to register copies of any of them on the PRS Exemptions Register (regulation 37(2)).

A notice must specify the name and address of the person to whom the documents or other information required must be provided, and the date by which they must be provided within one month from when the compliance notice is served (regulation 37(3)).

L must comply with the notice, and allow the enforcement authority to take copies of any original document produced (regulation 37(4)).

A notice may be varied or revoked in writing at any time by the enforcement authority that issued it (regulation 37(5)).

An enforcement authority may take into account any information held by it, whether or not provided to it in determining whether L has complied (regulation 37(6)).

Penalty notices

An enforcement authority may serve a notice on L where it is satisfied that L is, or has been at any time in the 18 months preceding the date of service of the penalty notice, in breach of one or more of the requirements of these regulations imposing a financial penalty, a publication penalty, or both (regulation 38(1)).

A penalty notice must

(a) specify the provision of these Regulations which the enforcement authority believes L has breached,
(b) give such particulars as the enforcement authority considers necessary to identify the matters constituting the breach,
(c) specify:

　(i) any action the enforcement authority requires L to take to remedy the breach,
　(ii) the period within which such action must be taken,

(d) specify:

　(i) the amount of any financial penalty imposed and, where applicable, how it has been calculated,
　(ii) whether the publication penalty has been imposed,

(e) require L to pay any financial penalty within a period specified in the notice,

(f) specify the name and address of the person to whom any financial penalty must be paid and the method by which payment may be made,

(g) state the effect of the appeal provisions, and

(h) specify:

  (i) the name and address of the person to whom a notice requesting a review may be sent (and to whom any representations relating to the review must be addressed), and

  (ii) the period within which such a notice may be sent (regulation 38(2)).

Each of the periods specified must not be less than one month, beginning the day on which the penalty notice is served (regulation 38(3)).

Where L fails to take the action required by a penalty notice within the period specified, the enforcement authority may issue a further penalty notice (regulation 38(4)).

Publication penalty

The publication penalty means publication on the PRS Exemptions Register of such of the following information in relation to a penalty notice as the enforcement authority decides:

(a) where L is not an individual, L's name,

(b) details of the breach of these Regulations in respect of which the penalty notice has been issued,

(c) the address of the property in relation to which the breach has occurred, and

(d) the amount of any financial penalty imposed (regulation 39(1)).

This information must be published for a minimum period of 12 months, and may be published for a longer period as the enforcement authority decides (regulation 39(2)).

Breaches in relation to domestic PR property

These penalties apply where L is, or was, the landlord of a domestic PR property (regulation 40(1)). Regulation 40 details the penalties that apply in specific situations.

Breaches in relation to non-domestic PR property

Regulation 41 details the penalties that apply where L is, or was, the landlord of a non-domestic PR property.

Reviews, waiving and modification of penalties

L may, within the period specified, serve notice on the enforcement authority requesting a review of its decision to serve a penalty notice (regulation 42(1)).

Where L gives notice or where the enforcement authority decides to review its decision to serve a penalty notice in any other case, the enforcement authority must:

(a) consider any representations made by L and all other circumstances of the case,

(b) confirm or withdraw the penalty notice, and

(c) serve notice of its decision to L (regulation 42(2)).

If, on a review, the enforcement authority:

(a)   ceases to be satisfied that L committed the breach specified in the notice,
(b)   is satisfied that L took all reasonable steps and exercised all due diligence to avoid committing the breach specified in the notice, or
(c)   decides that in the circumstances of the case it was not appropriate for a notice to be served on L,

the enforcement authority must serve a further notice on L withdrawing the notice (regulation 42(3)).

A notice confirming the penalty notice must state the appeal provisions (regulation 42(4)).

On a review, the enforcement authority may:

(a)   waive a penalty,
(b)   allow L additional time to pay any financial penalty,
(c)   substitute a lower financial penalty where one has already been imposed, or
(d)   modify the application of a publication penalty (regulation 42(5)).

Appeals

If, after a review, a penalty notice is confirmed by the enforcement authority, L may appeal to the First-Tier Tribunal on the grounds that:

(a)   the issue of the notice was based on an error of fact or an error of law
(b)   the notice does not comply with a requirement imposed by these Regulations, or
(c)   in the circumstances of the case it was inappropriate for the notice to be served on L (regulation 43).

Effect and determination of appeal

The bringing of an appeal suspends the penalty notice being appealed taking effect, pending determination or withdrawal of the appeal (regulation 44(1)).

The First-Tier Tribunal may quash the notice, or affirm the notice, whether in its original form or with such modification as it sees fit (regulation 44(2)).

If the notice is quashed, the enforcement authority must repay any amount paid as a financial penalty in pursuance of the notice (regulation 44(3)).

Recovery of financial penalty

The amount of an unpaid financial penalty is recoverable from L as a debt owed to the enforcement authority unless the notice has been withdrawn or quashed (regulation 45(1)).

Regulation 45 details the proceedings for the recovery of the financial penalty.

Effect of an improvement notice (see p. 290 on housing improvement notices)

Where a landlord served with a copy of a tenant's request or with an intended counter proposal, has also been served with an improvement notice in relation to the property, or

FC53 Energy efficiency compliance notices – Energy Efficiency (Private Rented Property) (England and Wales) Regulations 2015

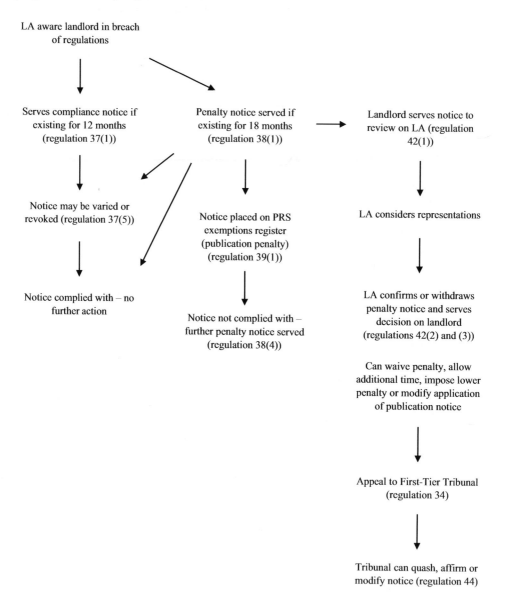

the part of the building, the landlord must, as soon as reasonably practicable after the date of service of the tenant's request or the intended counter proposal:

(a) provide the landlord with a copy of the improvement notice, and
(b) specify the works which the superior landlord intends to carry out to comply with the improvement notice, and the date by which the superior landlord proposes to carry out those works (regulation 14(1)).

Where a tenant is served with a copy of an improvement notice at any time before the landlord's full response is served, the tenant's request ceases to have effect (regulation 14(3)).

Definitions

Energy efficiency improvement, in relation to a property, means –

(a) a measure for improving efficiency in the use of energy in the property, or
(b) where indicated, a measure installed for the purposes of enabling the supply of gas through a service pipe to the property in any case where the property is not fuelled by mains gas, and is situated within 23 metres from a relevant main of a gas transporter.

Enforcement authority in relation to a domestic PR property means a local authority, in relation to a non-domestic PR property means a local weights and measures authority.

An enforcement authority must enforce compliance with these requirements in properties in its area (regulation 34).

Improvement notice means a notice served under section 11 or section 12 of the Housing Act 2004 where the time for compliance with the notice has not expired, or the notice has been appealed, and the appeal has not been determined.

PRESCRIBED SMOKE AND CARBON MONOXIDE ALARMS

Reference

Smoke and Carbon Monoxide Alarm (England) Regulations 2015
Housing (Scotland) Act 2014

Extent

These provisions apply to England only. Similar provisions apply in Scotland.

Scope

Rented dwellings must be provided with a smoke alarm and carbon monoxide alarm with a solid fuel burning combustion appliance equipped on each storey of the premises on which there is a room used wholly or partly as living accommodation. Checks must be made by or on behalf of the landlord to ensure that each prescribed alarm is in proper working order on the day the tenancy begins, if it is a new tenancy (regulation 4(1)).

Duty of local housing authority to serve a remedial notice

Where a local housing authority has reasonable grounds to believe that, a landlord is in breach of one or more of the above duties, the authority must serve a remedial notice on the landlord (regulation 5(1)).

A remedial notice must:

(a) specify the premises;
(b) specify the duty or duties that the local housing authority considers the landlord is failing or has failed to comply with;
(c) specify the remedial action the authority considers should be taken;
(d) require the landlord to take that action within 28 days of the notice being served;
(e) explain that the landlord is entitled to make written representations against the notice within 28 days of the notice being served;
(f) specify the person to whom, and the address (including email address) at which, any representations may be sent; and
(g) explain the effect of regulations 6, 7 and 8, including the maximum penalty charge which a local housing authority may impose (regulation 5(2)).

The local housing authority must serve a remedial notice within 21 days of the authority deciding it has reasonable grounds (regulation 5(3)).

Duty of relevant landlord to comply with a remedial notice

Where a remedial notice is served on a landlord who is in breach of one or more of the duties, the landlord must take the remedial action specified in the notice within the period specified (regulation 6(1)).

A landlord is not to be taken to be in breach of the duty if the landlord can show he, she or it has taken all reasonable steps, other than legal proceedings, to comply with the duty (regulation 6(2)).

Duty of local housing authority to arrange remedial action

Where a local housing authority is satisfied that a landlord is in breach of their duty, the authority must, if the consent of the occupier is given, arrange for an authorised person to take the remedial action specified in the remedial notice (regulation 7(1)).

The remedial action must be taken within 28 days of the date the landlord is required to take remedial action (regulation 7(2)).

An authorised person must give not less than 48 hours' notice of the remedial action to the occupier of the premises, and if required to do so by or on behalf of the landlord or occupier, produce evidence of identity and authority (regulation 7(3)).

A local housing authority is not to be taken to be in breach of its duty where the authority can show it has taken all reasonable steps to comply with the duty (regulation 7(5)).

Penalty for breach of the duty

Where a local housing authority is satisfied that a landlord on whom it has served a remedial notice is in breach of their duty, the authority may require the landlord to pay a penalty charge (regulation 8(1)). The amount of the penalty charge must not exceed £5,000 (regulation 8(2)).

Where a local housing authority decides to impose a penalty charge, the authority must serve 'a penalty charge notice' on the landlord within six weeks of the date landlord is required to take remedial action (regulation 8(3)).

Content of penalty charge notice

A penalty charge notice must state:

(a) the reasons for imposing the penalty charge;
(b) the premises to which the penalty charge relates;
(c) the number and type of prescribed alarms which an authorised person has installed at the premises;
(d) the amount of the penalty charge;
(e) that the landlord is required, within a period specified (less than 28 days – regulation 9(3)) in the notice:

   (i) to pay the penalty charge, or
   (ii) to give written notice to the local housing authority that the landlord wishes the authority to review the penalty charge notice (regulation 9(2));

(f) how payment of the penalty charge must be made; and
(g) the person to whom, and the address (including email address) at which, a notice requesting a review may be sent and to which any representations relating to the review may be addressed (regulation 9(1)).

Review of penalty charge notice

If the landlord serves a notice on the local housing authority requesting a review, the authority must consider any representations made by the landlord and decide whether to confirm, vary or withdraw the penalty charge notice and serve notice of its decision to the landlord (regulation 10(1) and (2)).

Appeals

A landlord who is served with a notice confirming or varying a penalty charge notice may appeal to the First-Tier Tribunal against the local housing authority's decision (regulation 11(1)).

The grounds for appeal are that –

(a) the decision to confirm or vary the penalty charge notice was based on an error of fact;
(b) the decision was wrong in law;
(c) the amount of the penalty charge is unreasonable;
(d) the decision was unreasonable for any other reason (regulation 11(1)).

Where a landlord appeals to the Tribunal, the operation of the penalty charge notice is suspended until the appeal is finally determined or withdrawn (regulation 11(3)).

The Tribunal may quash, confirm or vary the penalty charge notice, but cannot increase the amount of the penalty charge (regulation 11(4)).

Recovery of penalty charge

The local housing authority may recover the penalty charge on the order of a court, as if payable under a court order (regulation 12(1)).

Proceedings for the recovery of the penalty charge cannot be started within 28 days of the period for payment (regulation 12(2)).

If, within that period, the landlord gives notice to the local housing authority that the landlord wishes the authority to review the penalty charge notice, proceedings for the recovery of the penalty charge are suspended (regulation 12(3) and (4)).

In proceedings for the recovery of the penalty charge a certificate which is signed by the local housing authority's chief finance officer, and states that the penalty charge has not been received by a date specified in that certificate, is conclusive evidence of that fact (regulation 12(5)).

Sums received by a local housing authority under a penalty charge may be used by the authority for any of its functions (regulation 12(6)).

Information to be published by local housing authority

A local housing authority must prepare and publish a statement of principles which it proposes to follow in determining the amount of a penalty charge (regulation 13(1)).

It may revise its statement of principles and it must publish the revised statement (regulation 13(2)).

In determining the amount of a penalty charge, it must have regard to the statement of principles which was most recently prepared and published at the time when the breach in question occurred (regulation 13(3)).

Definitions

New tenancy means a tenancy granted on or after 1 October 2015, but does not include –

(a) a tenancy granted before that date;
(b) a periodic shorthold tenancy on the coming to an end of a fixed term shorthold tenancy;
(c) a tenancy which comes into being on the coming to an end of an earlier tenancy, under which, on its coming into being –

  (i)   the landlord and tenant are the same as under the earlier tenancy as at its coming to an end; and
  (ii) the premises let are the same or substantially the same as those let under the earlier tenancy as at that time (regulation 4(4)).

A tenancy begins on the day on which, under the terms of the tenancy, the tenant is entitled to possession (regulation 4(3)).

Premises does not include vehicles or vessels, a licensed HMO or a house where a licence is required (regulation 2).

Remedial action means action to install or repair a prescribed alarm; or to check a prescribed alarm is in proper working order (regulation 2).

FC54 Prescribed smoke and carbon monoxide alarms – Smoke and Carbon Monoxide Alarm (England) Regulations 2015

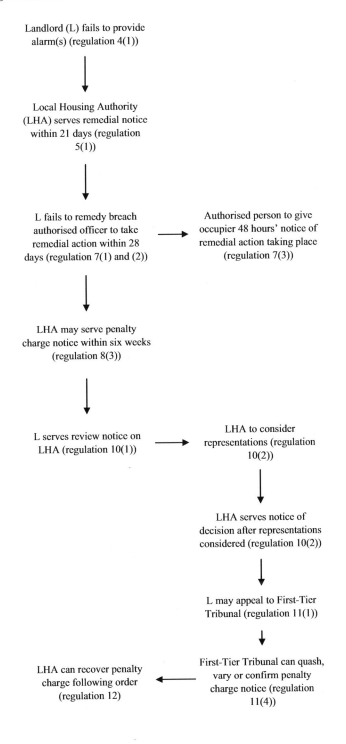

Chapter 2

ENFORCEMENT OPTIONS

GENERAL APPROACH TO ENFORCEMENT

The Enforcement Guidance Office of the Deputy Prime Minister (ODPM), February 2006, Part 2 sets out the policy context in which it is expected that local authorities will conduct their enforcement responsibilities. The outline of this is as follows:

- Local authorities are required to consider the housing conditions in the district and consider what action under the 2004 Act needs to be taken.
- Local authorities need to take a view of the spread of hazards and prioritise their actions on those with the more serious impact on health or safety.
- Where practicable, local authorities should consult adjoining local authorities about actions in estates that straddle boundaries.
- Enforcement policies should take account of the circumstances and views of tenants, landlords and owners and should provide for consultation with social services, tenancy support, housing needs and housing management officers.
- Local authorities are encouraged to adopt the Regulators Code (p. 4), which provides a basis for fair, practicable and consistent enforcement. The essence of this is that anyone likely to be the subject of enforcement action should receive clear explanations of what they are required to do and have an opportunity to resolve difficulties before formal action is taken. This allows local authorities to take informal action where that appears to be the most satisfactory course of action in the circumstances.

Part 1 enforcement options

The process of identifying and assessing hazards is the first part of the process that may lead to enforcement action. The Enforcement Guidance indicates a three-stage process:

1. The hazard score as calculated above.
2. The local authority judgement as to the most appropriate action to deal with the hazard in all the circumstances, including taking account of both potential and actual vulnerable occupants.
3. Whether the local authority, in the light of the score, has a discretion or duty to act.

The enforcement provisions that follow the Housing Health and Safety Rating System assessments are contained in Part 1 of the 2004 Act and are known as the Part 1 functions.

In reaching a decision in each case on the appropriate action to be taken, the local authority must have regard to the Enforcement Guidance 2006.

Section 5 of the Housing Act 2004 *imposes a duty* on a local authority to take appropriate (best course) enforcement action where there is a category 1 hazard. The options available to local authorities in respect of category 1 hazards are:

- improvement notices (Housing Act 2004, section 11 – FC56);
- prohibition orders (Housing Act 2004, section 20 – FC57);
- hazard awareness notices (Housing Act 2004, section 28 – FC65);
- emergency remedial action notices (Housing Act 2004, section 40 – FC59);
- emergency prohibition orders (Housing Act 2004, section 43 – FC60);
- demolition orders (Housing Act 1985, section 265(1) or (2) – FC70);
- clearance area declarations (Housing Act 1985, section 289(2ZB) – FC72); and
- in relation to prohibition and demolition orders, local authorities also have the alternative of making a determination to purchase under section 300(1) or (2) of the Housing Act 1985.

Where two or more courses of action are available to deal with a category 1 hazard, the local authority must take that which they consider to be the most appropriate (section 5(4)).

Section 7 lists the discretionary enforcement powers of local authorities where a category 2 hazard exists and these are:

- improvement notices (Housing Act 2004 section 12 – FC56);
- prohibition orders (Housing Act 2004, section 21 – FC57);
- hazard awareness notices (Housing Act 2004, section 29 – FC58);
- demolition orders (Housing Act 1985, section 265(3) or (4) but only in prescribed circumstances – FC70); and
- clearance area declarations (Housing Act 1985, section 289(2ZB) but only in prescribed circumstances – FC72).

There are other options in other legislation which are very specific which can deal with issues such as empty buildings, buildings beyond repair, energy efficiency and safety in dwellings. Those procedures are included in this Part. Part 7 includes procedures which deal with public health issues related to dwellings and conditions which can be dealt with efficiently through specific legislation.

Reasons for decision

In each case where the local authority decides to take a Part 1 action, whether in relation to category 1 or category 2 hazards, a statement of reasons as to why the particular course of action is being taken in preference to others must be prepared. This is to be served, along with the notice or order, on those in receipt of the notice or order or a copy of them. In addition, statements that relate to the inclusion of premises in a clearance area must be published as soon as possible after the resolution declaring the clearance area has been passed (section 8).

FIXED PENALTY NOTICES

References

Housing Act 2004 Schedule 13A

Housing and Planning act 2016 – sections 238 and 126 and Schedule 1 and 9

Civil penalties under the Housing and Planning Act 2016 – Guidance for Local Housing Authorities

Housing (Amendment) Act (Northern Ireland) 2011

Extent

These provisions apply to England and Wales. Similar provisions apply in Northern Ireland.

Scope

A power to impose a civil penalty as an alternative to prosecution for offences was introduced by sections 238 and 126 and schedule 9 of the Housing and Planning Act 2016.

Local housing authorities will be able to impose a civil penalty as an alternative to prosecution for the following offences under the Housing Act 2004 and Housing and Planning Act 2016:

- Failure to comply with an Improvement Notice (section 30 of the Housing Act 2004)
- Offences in relation to licensing of Houses in Multiple Occupation (section 72 of the Housing Act 2004)
- Offences in relation to licensing of houses under Part 3 (section 95 of the Housing Act 2004)
- Offences of contravention of an overcrowding notice (section 139 of the Housing Act 2004)
- Failure to comply with management regulations in respect of Houses in Multiple Occupation (section 234 of the Housing Act 2004),
- Breach of a banning order (section 21 of the Housing and Planning Act 2016)

Where a landlord breaches a prohibition order, local housing authorities can now seek a rent repayment order in addition to prosecuting the landlord.

The maximum penalty is £30,000. The amount of penalty is to be determined by the local housing authority in each case. In determining an appropriate level of penalty, local housing authorities should have regard to the guidance at paragraph 3.5 which sets out the factors to take into account when deciding on the appropriate level of penalty. Only one penalty can be imposed in respect of the same offence.

Local housing authorities are expected to develop and document their own policy on when to prosecute and when to issue a civil penalty and should decide which option it wishes to pursue on a case-by-case basis in line with that policy.

Authorities should use their existing powers to, as far as possible, make an assessment of a landlord's assets and any income they receive (not just rental income) when determining an appropriate penalty. They should consider the following factors to help ensure that the civil penalty is set at an appropriate level:

(a) Severity of the offence.
(b) Culpability and track record of the offender.
(c) The harm caused to the tenant.
(d) Punishment of the offender.
(e) Deter the offender from repeating the offence.
(f) Deter others from committing similar offences.
(g) Remove any financial benefit the offender may have obtained as a result of committing the offence.

Procedure

The procedure for imposing a civil penalty is set out in Schedule 13A of the Housing Act 2004 and Schedule 1 of the Housing and Planning Act 2016 and summarised below.

The local housing authority must give the person a 'notice of intent' to impose a financial penalty.

The notice must set out:

- the amount of the proposed financial penalty;
- the reasons for proposing to impose the penalty; and
- information about the right of the landlord to make representations.

The notice must be given no later than six months after the authority has sufficient evidence of the conduct to which the penalty relates, or at any time when the conduct is continuing.

A person who is given a notice of intent may make written representations to the local housing authority about the intention to impose a financial penalty.

Any representations must be made within 28 days from the date the notice being given.

After the end of the period for representations, the local housing authority must decide whether to impose a penalty and, if so, the amount of the penalty.

If the authority decides to impose a financial penalty, it must give the person a 'final notice' requiring that the penalty to be paid within 28 days.

The final notice must set out:

- the amount of the financial penalty;
- the reasons for imposing the penalty;
- information about how to pay the penalty;
- the period for payment of the penalty (28 days);
- information about rights of appeal; and
- the consequences of failure to comply with the notice.

The local housing authority may at any time withdraw a notice of intent or final notice; or reduce the amount specified in the notice.

On receipt of a final notice imposing a financial penalty, a landlord can appeal to the First-Tier Tribunal against the decision to impose a penalty and/or the amount of the penalty. The appeal must be made within 28 days of the date the final notice was issued. The final notice is suspended until the appeal is determined or withdrawn.

FC55 Fixed penalty notices – Housing Act 2004 Schedule 13A and Housing and Planning Act 2016 – sections 238 and 126 and schedule 1 and 9

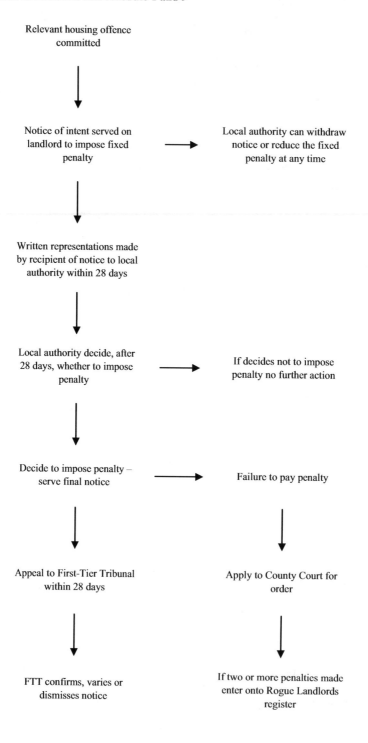

Appeals

This procedure is set out in schedule 13A, paragraph 10 of the Housing Act 2004 and schedule 1, paragraph 10 of the Housing and Planning Act 2016.

The First-Tier Tribunal has the power to confirm, vary (increase or reduce) the size of the civil penalty imposed by the local housing authority, or to cancel the civil penalty. If the Tribunal decides to increase the penalty, it may only do so up to a maximum of £30,000.

While the Tribunal is not bound by it, it will have regard to the guidance.

The Tribunal can dismiss an appeal if it is satisfied that the appeal is frivolous, vexatious or an abuse of process, or has no reasonable prospect of success.

Enforcement

Where the landlord or property agent fails to pay a civil penalty, the local housing authority should refer the case to the County Court for court order.

A certificate signed by the chief finance officer of the local housing authority which states that the amount due had not been received by a specified date will be treated by the courts as conclusive evidence of that fact.

Section 31 and Schedule 3 of the Housing Act 2004 relating to works in default continue to operate.

Where a landlord receives two or more civil penalties over a 12-month period, local housing authorities may include that person's details in the database of rogue landlords and property agents. While it is not compulsory, local housing authorities are strongly encouraged to do so. This will help ensure that other local housing authorities are made aware that formal action has been taken against the landlord.

If a landlord receives a civil penalty, that fact can be taken into account if considering whether the landlord is a fit and proper person to be the licence holder for a House in Multiple Occupation or any other property subject to licensing.

Income received from a civil penalty can be retained by the local housing authority provided that it is used to further the local housing authority's statutory functions in relation to their enforcement activities covering the private rented sector, as specified in Regulations.

RENEWAL AREAS

References

Local Government and Housing Act 1989
Housing (Scotland) Act 2006

Extent

These provisions apply to England and Wales. Similar provisions apply in Scotland.

Scope

Renewal areas are a tool to deliver a geographical area of housing improvement over a period of time with sufficient funding.

Declaration of renewal area

Where a local housing authority consider a report about the housing conditions below, and are satisfied:

(a) that the living conditions in an area within their district consisting primarily of housing accommodation are unsatisfactory, and

(b) that those conditions can most effectively be dealt with by declaring the area to be a renewal area,

they may declare a renewal area for the period specified in the declaration (section 89(1)).

The housing conditions contained in the report referred to above are:

(a) the living conditions in the area concerned;

(b) the ways in which those conditions may be improved (whether by the declaration of a renewal area or otherwise);

(c) the powers available to the authority if the area is declared to be a renewal area;

(d) the authority's detailed proposals for the exercise of those powers during the period that the area will be a renewal area (if declared);

(e) the cost of those proposals;

(f) the financial resources available, or likely to be available, to the authority (from whatever source) for implementing those proposals; and

(g) the representations (if any) made to the authority in relation to those proposals.

The report must contain a recommendation, with reasons, as to whether a renewal area should be declared and, if so, the period for which the area should be a renewal area. (section 89(3)).

A renewal area is such until the end of the period specified in the declaration, and can be extended by the authority (section 89(4)).

In considering whether to declare a renewal area, or to extend the period of its existence, an authority must have regard to guidance given by the Secretary of State (section 89(5)).

An authority must take steps to secure the following before declaring a renewal area:

(a) that the detailed proposals referred to the report or, where the authority are considering the extension of the period, those proposals that remain to be implemented, are brought to the attention of persons residing or owning property in the area; and

(b) that those persons are informed of the name and address of the person to whom should be addressed inquiries and representations concerning those proposals (section 89(6) and (7)).

Steps to be taken after declaration or extension and duty to publish information

As soon as possible after declaring a renewal area or extending the period for a renewal area, a local authority must take the steps best designed to secure:

(a) that the resolution to which the declaration, or extension (or further extension) of the period, relates is brought to the attention of persons residing or owning property in the area; and

(b) that those persons are informed of the name and address of the person to whom should be addressed inquiries and representations concerning action to be taken with respect to the renewal area (section 91).

They must also from time to time publish information about:

(a) the action they propose to take in relation to the area,
(b) the action they have taken in relation to the area, and
(c) the assistance available for the carrying out of works in the area (section 92).

General powers of local housing authority

The authority may acquire by agreement, or be authorised by the Secretary of State to acquire compulsorily, any land in the area on which there are premises consisting of or including housing accommodation or which forms part of the curtilage of any such premises; and the authority may provide housing accommodation on land acquired (section 93(2)).

The objectives of these powers are:

(a) the improvement or repair of the premises, either by the authority or by a person to whom they propose to dispose of the premises;
(b) the proper and effective management and use of the housing accommodation, either by the authority or by a person to whom they propose to dispose of the premises comprising the accommodation; and
(c) the well-being of the persons for the time being residing in the area (section 93(3)).

For the purpose of effecting or assisting the improvement of the amenities in the area, the authority may acquire by agreement, or be authorised by the Secretary of State to acquire compulsorily, any land in the area (including land which the authority propose to dispose of to another person who intends to effect or assist the improvement of those amenities) (section 93(4)).

The authority may carry out works (including works of demolition) on land owned by the authority in the area (whether or not that land was acquired (section 93(5)).

The authority may enter into an agreement with a housing association or other person under which, in accordance with the terms of the agreement (section 93(6)).

Power to apply for orders extinguishing right to use vehicles on highway

An authority who have declared a renewal area may exercise the powers to extinguish the right to use vehicles on certain highways (section 94(1)). The authority cannot make an application to Secretary of State to make or revoke order extinguishing right to use vehicles except with the consent of the local planning authority (section 94(2)). Any application must firstly be sent to the highway authority who must send it to the Secretary of State (section 94(3)). Compensation for loss of access to highway is payable by them instead of by the local planning authority (section 94(4)).

Exclusion of land from, or termination of, a renewal area

The local housing authority may by resolution exclude land from a renewal area, or declare that an area shall cease to be a renewal area (section 95). Similar requirements of publicity and notification apply.

Penalty for obstruction

A person who commits an offence under subsection (3) above is liable on conviction to a fine not exceeding level 3 on the standard scale (section 97).

IMPROVEMENT NOTICES

References

Housing Act 2004, sections 11–18 and schedules 1 and 2
Enforcement Guidance, HHSRS, ODPM, February 2006, part 5, paragraphs 5.4–5.14
Housing (Scotland) Act 1987
Housing (Northern Ireland) Order 1981

Extent

These provisions apply to England and Wales. Similar provisions apply in Scotland and Northern Ireland.

Scope

Improvement notices are a possible response where the local authority considers that this action is the most appropriate in all the circumstances to deal with one or more category 1 and/or category 2 hazards at residential premises (sections 11 and 12). They are notices that require the person on whom they are served to take the specified remedial action to remove the hazard.

Decision to serve an improvement notice

The local authority must be satisfied that the service of an improvement notice is the appropriate action having regard to the Enforcement Guidance. In addition, the circumstances necessary for the service of such a notice are:

(a) that the local authority is satisfied that a category 1 or category 2 hazard exists;
(b) that there is no interim or final management order relating to the premises (sections 11 and 12);
(c) that, if redevelopment of the land by the owner has been approved and is being proceeded with to a specified time limit, an improvement notice may not be used (section 39(5)).

Persons to be served with an improvement notice

The improvement notice is to be served on the person responsible. Identifying this depends on the type of premises (section 18 and schedule 1).

(a) In the case of houses in multiple occupation etc. licensed under Part 2 or 3 of the 2004 Act – the holder of the licence.

(b) In the case of premises not licensed under either Part 2 or 3:

    (i) where the premises is not a flat – the person having control;

    (ii) where the premises is not a flat, but is a house in multiple occupation – the person having control or the person managing the house in multiple occupation;

    (iii) where the premises is a flat and is a dwelling and not a house in multiple occupation – an owner who, in the local authority's opinion, ought to take the action specified in the notice;

    (iv) where the premises is a flat and is a house in multiple occupation – an owner who, in the local authority's opinion, ought to take the action specified in the notice, or the person managing the flat.

(c) In the case of common parts or any non-residential part of a building containing flats – an owner who, in the local authority's opinion, ought to take the action specified in the notice (schedule 1).

In addition to the person responsible, the local authority is required to serve a copy of the improvement notice on the occupier of the premises, and on every person who, to the knowledge of the local authority, is a freeholder, mortgagee or lessee of the premises (schedule 1, paragraph 5).

Content of improvement notice

An improvement notice can deal with one or more hazards of both categories (section 12(5)) and must, in relation to each hazard, specify:

(a) whether the notice has been served under section 11 or 12;

(b) the nature of the hazard and the residential premises to which the notice relates;

(c) the deficiency giving rise to the hazard;

(d) the premises in relation to which remedial action is to be taken and the nature of the remedial action;

(e) the date by which that remedial action is to be started (which cannot be earlier than the twenty-eighth day after the date the notice was served);

(f) the period within which the remedial action is to be completed, or the periods within which each part of that work is to be completed;

(g) information on the right to appeal under Part 3 of schedule 1;

(h) the period within which an appeal may be made.

In relation to category 1 hazards, the notice must require at least the removal of that hazard but may go further. Requirements must be reasonable (section 13).

Accompanying the improvement notice must be a statement of reasons (section 8(3) and (4)). This statement must explain why the local authority decided that the appropriate action was the service of an improvement notice rather than any of the other enforcement options (section 8(3) and (4)).

Operation of improvement notice

An improvement notice becomes operative 21 days from the date on which it was served. Where an appeal is brought, the notice does not become operative until the appeal is determined (or withdrawn) and confirmed and the period for further appeal expires (section 15).

Suspension of an improvement notice

The local authority has the discretion to suspend the operation of an improvement notice either:

(a) until a time, which may be when a person of a particular description begins or ceases to occupy the premises;
(b) until a specified event occurs, which may be a breach of an undertaking, which breach has been made known to the person on whom the notice was served (section 14).

The local authority has power to review a suspended improvement notice at any time, but must review it not later than one year after the date of service and on the anniversary of that date. The local authority's decision on review must be served on all those on whom the notice and copies were served (section 17).

The time or events triggering an end to suspension may be varied by the local authority at their own volition or as a result of the application of the person on whom it was served (section 16(4) and (8)).

Where an improvement notice has been suspended, it becomes operative again on the twenty-first day after the suspension ends and the minimum 28 days for starting work becomes 21 days (section 14(5)).

Revocation and variation of an improvement notice

Review of whether an improvement notice should or could be revoked or varied can be at the local authority's own initiative or on the application of the person on whom the notice was served (section 16(8)).

The local authority must revoke the notice where they are satisfied the requirements have been complied with (section 16(1)). This applies whether the notice relates to one or to more than one hazard.

The local authority is given a power to revoke a notice where:

(a) in the case of a notice served in respect of a category 1 hazard, they consider that there are special circumstances making revocation appropriate; or
(b) in the case of a notice served in respect of a category 2 hazard, they consider revocation appropriate (section 16(2)).

Where the notice relates to more than one hazard, the local authority has the power to revoke a part of the notice; they also have the power to vary the remainder of the notice (section 16(3)).

An improvement notice may be varied with the agreement of the person on whom it was served. It may also be varied without that agreement to alter the time or events triggering the end of a suspension (section 16(4)).

Change of person liable to comply

Where the person served with an improvement notice transfers their interest to a third party while the notice is still active, then the new person is deemed to be in the same position as the original person. However, the new person is not held subject to any liability incurred by the original recipient of the notice (section 19). It is important to note that such a transfer of interest has no effect on the period given for compliance in the notice.

Appeals

The persons on whom the improvement notice was served are given a general right to appeal against the notice to the Appropriate Tribunal within 21 days from the date of service. In addition to the open right of appeal, the schedule specifies the following grounds:

(i) that one or more other persons, as owner(s) of the premises, ought to take the specified action;
(ii) that one or more other persons, as owner(s) of the premises, ought to pay all or part of the cost of taking the specified action;
(iii) that, rather than an improvement notice, a hazard awareness notice, a prohibition order or a demolition order would have been the best course of action.

There is also an appeal process against decisions of the local authority relating to revocation and variation within 21 days from service of that notice.

The appeal is by way of a re-hearing, and the Appropriate Tribunal may have regard to matters of which the local authority was unaware. On hearing the appeal, the Tribunal may confirm, vary or quash the notice (schedule 1, part 3).

Prosecution for non-compliance with an improvement notice

It is a criminal offence to fail to comply with the requirements of an improvement notice within the period or period(s) specified. The local authority may take proceedings against the person on whom the notice was served in the magistrates' court. The maximum fine on summary conviction is not to exceed level 5 on the standard scale. It is a defence for the person to satisfy the court that there was a reasonable excuse for failing to comply with the notice. Non-compliance is deemed to be not carrying out the specified actions within the periods specified in the notice or the actions and period(s) as determined on appeal. It should be noted that the obligation to comply with the notice continues even when the period for completion has expired (section 30).

FC56 Improvement notices (INs) – sections 11 to 19 Housing Act 2004

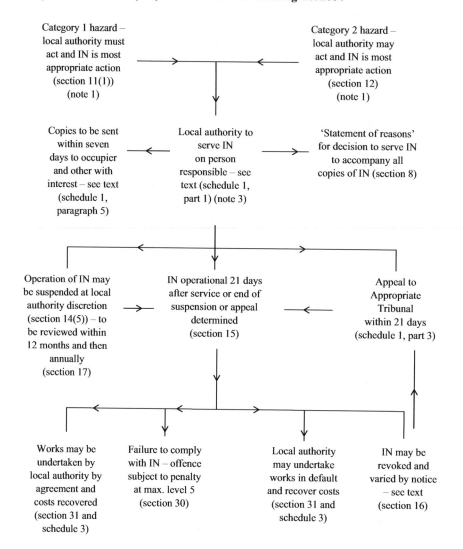

Notes

1. Where there are multiple hazards including category 2, the IN can require action to deal with both category 1 and category 2.
2. For emergency remedial action, see FC59.
3. The IN must be registered as a local land charge (section 37).

Default powers

The Act enables a local authority to carry out the actions specified in a notice in the following circumstances:

(a) with the agreement of the person on whom the notice was served;

(b) where, in the local authority's opinion, insufficient progress is made to complete the actions required, even though the period(s) specified in the notice have not expired;

(c) where the notice has not been complied with.

Where the local authority agrees to carry out the action, or decides to do so, they have all the rights that the person served with the notice had. Seven days before invoking their right to take action, the local authority must serve notice on the person on whom the improvement notice was served and serve a copy on the occupier and on any owner of the relevant premises.

Once the local authority has taken on the right to carry out the actions required, it is an offence for the person on whom the notice was served, or for any workman or contractor of that person, to be on the premises for the purpose of carrying out any works.

Any expenses, including administrative expenses, incurred by the local authority in taking action without agreement are recoverable. Such expenses are a charge on the premises and the local authority has the powers under the Law of Property Act 1925 as if the local authority were mortgagee by deed.

Where the local authority considers that the expense incurred (and any interest) is not likely to be recovered, and that rents or other payments are being made to a person, then the local authority may apply to the Appropriate Tribunal for an order that that person makes such payments as the Tribunal thinks just.

There is a right of appeal against a demand for payment of expenses (and any interest) made by the local authority. The appeal is to the Appropriate Tribunal and must be made within 21 days of the demand (section 31 and schedule 3).

PROHIBITION ORDERS

References

Housing Act 2004, sections 20–27 and schedule 2
Enforcement Guidance, HHSRS, Part 1, ODPM, February 2006, 5.15–5.30

Extent

These provisions apply to England and Wales.

Scope

Prohibition orders are a possible response available to the local authority to perform its duty to deal with category 1 hazards and is a discretionary power which may be used to deal with category 2 hazards, in both cases where it is considered to be the most appropriate form of action (section 20(1) and 21(1)). They may not be used when the premises is the subject of a management order – see FC80 and FC81.

A prohibition order may prohibit the use of all or part of a residential premises (definition on p. 262) (sections 20(1) and 21(1)). The non-residential part of premises may also be prohibited if the defect giving rise to a hazard is located there and the prohibition is necessary for the health and safety of actual or potential occupants (section 20(4)).

The Enforcement Guidance gives examples of when the use of a prohibition order might be appropriate.

These include:

- where the conditions present a serious threat to health or safety and remedial action would be unreasonable or impracticable;
- to specify maximum numbers of occupants where it is too small for the household's needs;
- to control the number of occupants where there are insufficient facilities;
- to prohibit use by a specified group or where conditions are hazardous to some people but relatively safe for others;
- in a house in multiple occupation, to prohibit the use of specified dwelling units or common parts.

The Enforcement Guidance also indicates additional matters to be considered before deciding upon the use of a prohibition order:

- Have regard to the exclusion of vulnerable people.
- Consider if the building is listed or protected.
- Take account of the relationship of the building to neighbouring buildings, which may mean that prohibition is preferable to, for example, demolition.
- Consider alternative uses other than those proposed by the owner.
- Take account of the existence of a conservation or renewal area.
- Consider effects of complete prohibition on the community and the appearance of the area.
- Consider the availability of local accommodation for rehousing displaced occupants.
- Consider appropriate financial advice or assistance (Enforcement Guidance, paragraphs 5.15–5.23).

Content of prohibition orders

A prohibition order can relate to one or more hazards and may include both category 1 and category 2 hazards, either in a separate order or combined in the same prohibition order (sections 20(5) and 21(4)). Orders under both sections 20 and 21 must state for each hazard:

1. whether the hazard is considered to be category 1 or category 2;
2. the nature of the hazard and the premises where it exists;
3. the deficiency giving rise to the hazard;
4. the premises to which the prohibition relates;
5. any remedial action considered necessary for the prohibition order to be revoked;
6. the rights of appeal (section 22(2) and (6)).

The order may prohibit:

1. use for some or all purposes unless approved by the local authority;
2. occupation by a particular number of households or occupants or particular descriptions of people, e.g. by age (section 22 (3)–(5)).

Service of prohibition orders

Copies of a prohibition order are to be served within seven days of the order being made, with the first day being the day on which it is made. They are to be served on:

(a) an owner or occupier of the whole or part of the premises;
(b) anyone authorised to permit persons to occupy;
(c) a mortgagee.

The requirement relating to occupiers will be met by fixing a copy to a conspicuous part of the premises (schedule 2, part 1).

Operation of prohibition order

Unless there is an appeal or suspension, prohibition orders come into effect 28 days after they are made (section 24).

Approvals of use by local authority

Where the prohibition relates to the use of the whole or part of the premises without the approval of the local authority, that approval must not be unreasonably withheld. Where use is not approved, the local authority must notify the person who made application within seven days and include the reasons for the decision together with their rights of appeal. There is an appeal against refusal to the Appropriate Tribunal within 28 days (section 22(7)–(9)).

Suspension of prohibition order and reviews

The local authority may suspend the operation of a prohibition order at its discretion and may:

- specify an event, such as the breach of an undertaking, as ending the suspension; or
- specify a time at which the suspension ends (section 23).
- The local authority must review a suspended prohibition order within a year of its being made and at least annually thereafter (section 26).

Revocation and variation

Revocation and variation may take place either on application or on the initiative of the local authority.

The local authority must revoke the prohibition order if it is satisfied that the hazard that necessitated the order no longer exists. Where there are multiple hazards, the revocation applies individually to each and, where the revocation is for less than them all, the prohibition order will need to be varied. There is then a discretionary power to revoke in relation to a category 1 hazard if satisfied that there are special circumstances. Revocation orders made in agreement come into force on the day they are made, otherwise 28 days after the decision.

The local authority may vary a prohibition order either with the agreement of every person on whom a copy has been served or, for a suspended order, to alter the time or events to which the suspension is to end (section 25).

The service and content of notices are prescribed in schedule 2, part 2.

Appeals

There is provision for appeals against the making of a prohibition order, refusals to approve use and decisions relating to revocation and variation. Such appeals are to be made to the Appropriate Tribunal within 28 days of the making of the order or decision.

The grounds of appeal include that the prohibition order was not the most appropriate course of action, which would have been an improvement notice, a hazard awareness notice or a demolition order (schedule 2, part 3).

Enforcement of prohibition order

Failure to comply with a prohibition order is an offence punishable by a fine of up to level 5 and £20 a day whilst the order is breached (section 32).

Determinations to purchase

In circumstances where the local authority would be required to make a prohibition order and it appears to them that the building can be rendered capable of providing accommodation that is adequate for the time being, the local authority may purchase the premises instead of making a prohibition order (Housing Act 1985, section 300).

Recovery of possession

Where it is necessary for the owner to take possession of the premises or part of it in order to comply with the order, this may be done by a notice to quit. The restrictions on the recovery of possession in the Rent Act 1977, the Rent (Agriculture) Act 1976 and part 1 of the Housing Act 1988 will not apply (section 33).

Compensation

Compensation is payable by the local authority to every owner of premises subject to a prohibition order (Housing Act 1985, section 584A) but is repayable where the prohibition order is revoked under section 25 (Housing Act 1985, section 584B).

FC57 Prohibition orders (POs) – sections 20 to 27 Housing Act 2004

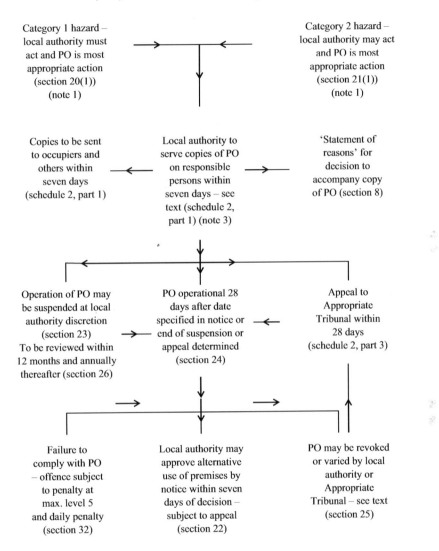

Notes
1. Where there are multiple hazards including both category 1 and category 2, the prohibition order may include both.
2. For emergency prohibition orders, see FC60.
3. The prohibition order is to be registered as a local land charge (section 37).

HAZARD AWARENESS NOTICES

References

Housing Act 2004, sections 28 and 29 and schedule 1, part 1
Enforcement Guidance, HHSRS, ODPM, February 2006, paras 5.38–5.43

Extent

These provisions apply to England and Wales.

Scope

Hazard awareness notices may be served where they are considered to be the most appropriate action in response to both the local authority duty for category 1 (section 28(1)) and its discretionary powers for category 2 (section 29(1)) hazards, in both cases where there is no management order in place – see FC80 and FC81.

The Enforcement Guidance considers that section 29 would be a reasonable response to a less serious hazard, whilst section 28 might be used where works of improvement or prohibition of use may not be practicable or reasonable. It also suggests that a hazard awareness notice might be used instead of an improvement notice where the owner or landlord has agreed to take remedial action and the local authority is confident that the work will be completed within a reasonable time (Enforcement Guidance, paragraphs 5.38–5.39 and 5.43).

These notices are not enforceable and require no further action from the recipient. There is therefore no provision for appeal. They are simply advisory, although the Enforcement Guidance suggests that the local authority should consider monitoring the position at that premises (Enforcement Guidance, paragraph 5.41). However, the service of a hazard awareness notice does not prevent the use of a formal notice later if an unacceptable hazard remains.

Content of hazard awareness notice

The notice must specify:

1. the nature of the hazard and the premises on which it exists;
2. the deficiency giving rise to the hazard;
3. the premises on which the deficiency exists;
4. the reasons for deciding upon the action as the most appropriate course (section 8);
5. details of the remedial action required which the local authority considers would be practicable and reasonable (Enforcement Guidance, paragraph 5.40).

A hazard awareness notice served under section 28 may only contain category 1 hazards but, where there are also category 2 hazards, the notices under sections 28 and 29 may be combined (section 29(6)).

Service of hazard awareness notice

The provisions relating to the service etc. of hazard awareness notices is the same as for improvement notices in schedule 1, part 1 – see below.

FC58 Hazard awareness notices (HANs) – sections 28 and 29 Housing Act 2009

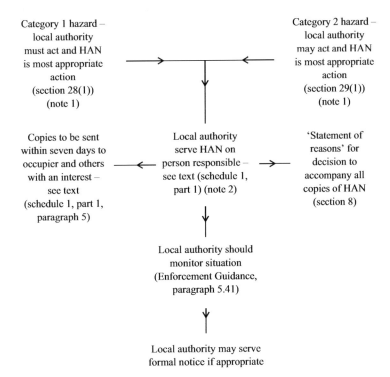

Category 1 hazard –
local authority
must act and HAN
is most appropriate
action
(section 28(1))
(note 1)

Category 2 hazard –
local authority
may act and HAN
is most appropriate
action
(section 29(1))
(note 1)

Copies to be sent
within seven days to
occupier and others
with an interest –
see text
(schedule 1, part 1,
paragraph 5)

Local authority
serve HAN on
person responsible –
see text (schedule 1,
part 1) (note 2)

'Statement of
reasons' for
decision to
accompany all
copies of HAN
(section 8)

Local authority should
monitor situation
(Enforcement Guidance,
paragraph 5.41)

Local authority may serve
formal notice if appropriate

Notes

1. Where both category 1 and category 2 hazards exist, the notices under sections 28 and 29 may be combined (section 29(6)).
2. There is no appeal since there is no enforcement process.
3. There is no requirement for the HAN to be registered as a local land charge.

NOTICE OF EMERGENCY REMEDIAL ACTION

References

Housing Act 2004, sections 40–42 and 45 and schedule 3
Enforcement Guidance, HHSRS, ODPM, February 2006, part 5, paragraphs 5.31–5.34

Extent

These provisions apply to England and Wales.

Scope

Where the local authority is satisfied that:

1. a category 1 hazard exists in a residential premises;
2. the hazard involves an imminent risk of serious harm to the health or safety of any of the occupants of that or other residential premises;
3. no management order is in force; and
4. action could be taken by way of an improvement notice under section 11,

they may take emergency remedial action as one of the options available to comply with their duty to act (section 40(1)–(3)).

'Emergency remedial action' means such remedial action in respect of the hazard concerned as the authority considers immediately necessary in order to remove the imminent risk of serious harm to the health or safety of any of the occupants (section 40(2)). Action may be taken in respect of more than one category 1 hazard on the same premises or in the same building containing one or more flats (section 40(4)).

Commencement of work

In relation to such hazards, the local authority may commence remedial work immediately without waiting for a notice of emergency remedial action to be served (paragraphs 3 to 6 of schedule 3).

Notice of emergency remedial action

Within seven days of commencing work, the local authority must serve a notice of emergency remedial action and copies of it on the same persons as would be the case for an improvement notice – see p. 364 (section 40(7)). The notice must specify:

- the nature of the hazard;
- the residential premises where it exists;
- the deficiency giving rise to the hazard;
- the premises where the remedial action has been or is to be taken;
- the nature of the action;
- the power under which the action is taken;
- the date on which the action was or is to be taken; and
- information about rights of appeal (section 41(2)).

FC59 Emergency remedial action (ERA) notices – sections 40 to 42 and 45 Housing Act 2004

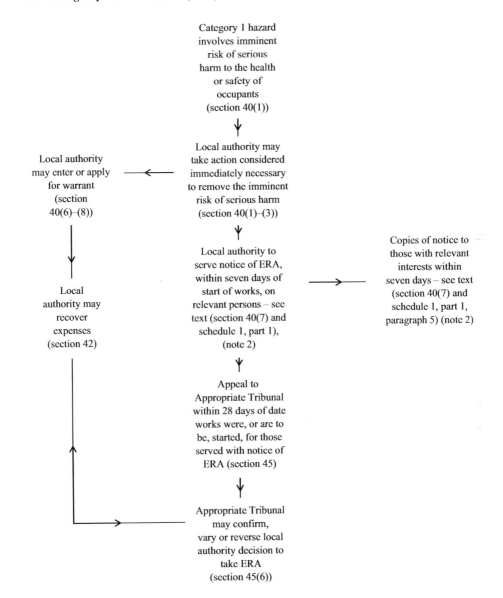

Category 1 hazard
involves imminent
risk of serious
harm to the health
or safety of
occupants
(section 40(1))

↓

Local authority
may enter or apply
for warrant
(section
40(6)–(8))

←

Local authority may
take action considered
immediately necessary
to remove the imminent
risk of serious harm
(section 40(1)–(3))

↓

Local
authority may
recover
expenses
(section 42)

Local authority to
serve notice of ERA,
within seven days of
start of works, on
relevant persons – see
text (section 40(7) and
schedule 1, part 1),
(note 2)

→

Copies of notice to
those with relevant
interests within
seven days – see text
(section 40(7) and
schedule 1, part 1,
paragraph 5) (note 2)

↓

Appeal to
Appropriate Tribunal
within 28 days of date
works were, or are to
be, started, for those
served with notice of
ERA (section 45)

↓

Appropriate Tribunal
may confirm,
vary or reverse local
authority decision to
take ERA
(section 45(6))

Notes
1. For the procedure for using emergency prohibition orders, see FC60.
2. Notice and copies to be accompanied by a 'statement of reasons' for the decision (section 8).

Appeals

Persons on whom the notice of emergency remedial action was served may appeal to a Appropriate Tribunal within 28 days of the date specified in the notice when the work was, or was to be, started. The tribunal may allow a longer period for appeal where it feels there is good reason to do so. An appeal will not prevent the action from being put into effect and this may have been started and finished before a notice has been served. There are no specified grounds of appeal. The tribunal may confirm, reverse or vary the local authority decision to take the action (section 45). Since works may have been completed at this time, the effect of such a decision would relate to the local authority's ability to recover expenses.

Recovery of expenses

The provisions of paragraphs 6–14 of schedule 3, which relate to enforcement of improvement notices, apply to the recovery of costs under this (section 42).

EMERGENCY PROHIBITION ORDERS

References

Housing Act 2004, sections 43–45 and schedule 1, part 1
Enforcement Guidance, HHSRSD, ODPM, February 2006, part 5, paragraphs 5.35–5.37

Extent

These provisions apply to England and Wales.

Scope

Where the local authority is satisfied that:

1. a category 1 hazard exists on a residential premises;
2. the hazard involves an imminent risk of serious harm to the health or safety of any of the occupants of that or other residential premises;
3. no management order is in force,

they may make an emergency prohibition order as one of the options available to comply with their duty to act (section 43(1)–(3)).

An emergency prohibition order is an order imposing, with immediate effect, prohibition/s on the use of any premises specified in the order (section 43(2)).

The same supplementary provisions as those applying to prohibition orders apply to emergency prohibition orders, i.e. approvals for use, revocation and variation, enforcement, recovery of possession, appeals and compensation (see below).

The order may prohibit:

- use for some or all purposes unless approved by the local authority;
- occupation by a particular number of households or occupants or particular descriptions of people, e.g. by age (section 43(3) and section 22(3)–(5)).

Contents of emergency prohibition order

The emergency prohibition order must specify:

- the hazard/s and the residential premises on which it exists;
- the deficiency giving rise to the hazard;
- the premises on which the prohibition is to operate;
- any remedial action which would cause the local authority to revoke the order.

The notice must also contain the rights of appeal and the period within which this may be done (section 44). The provisions of section 22(3)–(5) and (7)–(9) relating to the content of the notice also apply to emergency prohibition orders as they apply to prohibition orders – p. 295 (section 43(3)).

Service of emergency prohibition orders

Copies of an emergency prohibition order are to be served on the day that the order is made. They are to be served on:

- an owner or occupier of the whole or part of the premises;
- anyone authorised to permit persons to occupy;
- a mortgagee.

The requirement relating to occupiers will be met by fixing a copy to a conspicuous part of the premises (section 43(4) and schedule 2, part 1).

Operation of emergency prohibition order

The emergency prohibition order has immediate effect on its making (section 43(2)).

Approvals of use

Where the emergency prohibition relates to the use of the whole or part of the premises without the approval of the local authority, that approval must not be unreasonably withheld. Where use is not approved, the local authority must notify the person who made application within seven days and include the reasons for the decision together with their rights of appeal. There is an appeal provision against refusal to the Appropriate Tribunal within 28 days (section 43(5) and section 22(7)–(9)).

Revocation and variation

Revocation and variation may take place either on application or on the initiative of the local authority. The local authority must revoke the emergency prohibition order if it is satisfied that the hazard that necessitated the order no longer exists. Where there are multiple hazards, the revocation applies individually to each and, where the revocation is for less than them all, the emergency prohibition order will need to be varied. There is then a discretionary power to revoke in relation to a category 1 hazard if satisfied that there are special circumstances. Revocation orders made in agreement come into force on the day they are made, otherwise 28 days after the decision.

FC60 Emergency prohibition orders (EPOs) – sections 43 to 45 Housing Act 2004

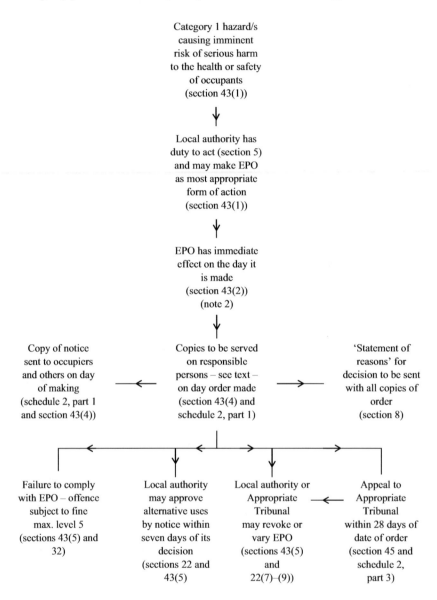

Category 1 hazard/s
causing imminent
risk of serious harm
to the health or safety
of occupants
(section 43(1))

Local authority has
duty to act (section 5)
and may make EPO
as most appropriate
form of action
(section 43(1))

EPO has immediate
effect on the day it
is made
(section 43(2))
(note 2)

Copy of notice
sent to occupiers
and others on day
of making
(schedule 2, part 1
and section 43(4))

Copies to be served
on responsible
persons – see text –
on day order made
(section 43(4) and
schedule 2, part 1)

'Statement of
reasons' for
decision to be sent
with all copies of
order
(section 8)

Failure to comply
with EPO – offence
subject to fine
max. level 5
(sections 43(5) and
32)

Local authority
may approve
alternative uses
by notice within
seven days of its
decision
(sections 22 and
43(5)

Local authority or
Appropriate
Tribunal
may revoke or
vary EPO
(sections 43(5)
and
22(7)–(9))

Appeal to
Appropriate
Tribunal
within 28 days of
date of order
(section 45 and
schedule 2,
part 3)

Notes
1. For prohibition notices, see FC57.
2. The emergency prohibition order is to be registered as a local land charge (section 37).

The local authority may vary an emergency prohibition order either with the agreement of every person on whom a copy has been served or, for a suspended order, to alter the time or events to which the suspension is to end (section 25).

Appeals

There is an appeal to the Appropriate Tribunal against the making of an emergency prohibition order by those in receipt of a copy of it. This must be made within 28 days (the tribunal may agree a longer period) of the date on which the emergency prohibition order is made, but the order continues in effect until determination of the appeal. The tribunal may confirm, vary or revoke the emergency prohibition order. There is also an appeal against refusal to agree to alternative uses (section 45).

Enforcement

Failure to comply with an emergency prohibition order is an offence punishable by a fine of up to level 5 and daily fine of £20 (section 43(5)(b) and section 32).

Recovery of possession

Where it is necessary for the owner to take possession of the premises or part of it in order to comply with the order, this may be done by a notice to quit. The restrictions on the recovery of possession in the Rent Act 1977, the Rent (Agriculture) Act 1976 and part 1 of the Housing Act 1988 will not apply (section 43(5)(b) and section 33).

MEANS OF ESCAPE FROM CERTAIN HIGH BUILDINGS

Reference

Building Act 1984, section 72

Extent

These provisions apply to England and Wales.

Scope

The section applies to existing or proposed buildings which are, or are to be let as, flats or tenement dwellings and which exceed two storeys in height and in which the floor of any upper storey is more than 20 ft above the surface of the street or ground on any side of the building (section 72(1) and (6)). The local authority must consult the Fire and Rescue Authority before taking action.

The section is, however, now only a residual section and does not cover premises subject to the Fire Precautions Act 1971 or houses in multiple occupation, which are dealt with in the Housing Act 2004 (see FCs 77–83).

FC61 Means of escape from certain high buildings – section 72 Building Act 1984

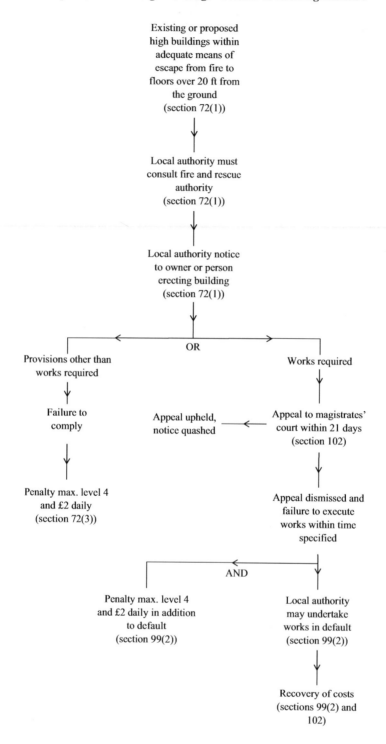

Existing or proposed
high buildings within
adequate means of
escape from fire to
floors over 20 ft from
the ground
(section 72(1))

Local authority must
consult fire and rescue
authority
(section 72(1))

Local authority notice
to owner or person
erecting building
(section 72(1))

OR

Provisions other than
works required

Works required

Failure to
comply

Appeal upheld,
notice quashed

Appeal to magistrates'
court within 21 days
(section 102)

Penalty max. level 4
and £2 daily
(section 72(3))

Appeal dismissed and
failure to execute
works within time
specified

AND

Penalty max. level 4
and £2 daily in addition
to default
(section 99(2))

Local authority
may undertake
works in default
(section 99(2))

Recovery of costs
(sections 99(2) and
102)

Note

1. For means of escape from fire in houses in multiple occupation, see FCs 77–83.

Notices

The local authority must serve notice requiring the owner of the building, or, as the case may be, the person proposing to erect the building, to execute work or make such other provision to provide the means of escape in the case of fire (see above).

The provisions of sections 99 and 102 apply to appeals against, and the enforcement of, notices under this procedure where works are required (p. 570).

If a notice requires a person to make provision otherwise than by the execution of works, he is, if he fails to comply with the notice, liable on summary conviction to a fine not exceeding level 4 on the standard scale and to a further fine not exceeding £2 for each day on which the offence continues after he is convicted (section 72(3)).

Appeals

No appeal is provided against notices which do not require works, but in any subsequent proceedings it is open to the defendant to question the reasonableness of the requirements (section 72(4)).

DEFECTIVE PREMISES

References

Building Act 1984, section 76
Building (Scotland) Act 2003
Pollution Control and Local Government (Northern Ireland) Order 1978

Extent

These provisions apply to England and Wales. Similar provisions apply in Scotland and Northern Ireland.

Scope

This procedure is available to deal with any premises which are in such a state as to be prejudicial to health (i.e. 'defective state'), but where unreasonable delay in dealing with the situation would occur if the normal provisions for dealing with statutory nuisances in section 80 of the Environmental Protection Act 1990 were used – see FC128 (section 76(1)).

Person responsible

Notices are to be served either:

(a) on the person responsible for the nuisance; or
(b) where the person in (a) cannot be found or the nuisance has not yet occurred, on the owner or occupier of the premises on which the nuisance arises; or
(c) where the nuisance arises from a structural defect, on the owner of the premises (section 27(1) and Environmental Protection Act 1990, section 80(2)).

FC62 Defective premises – section 76 Building Act 1984

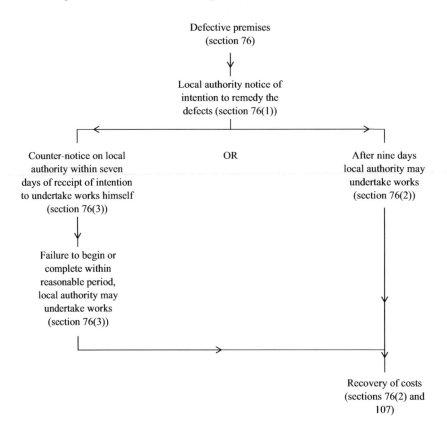

Defective premises
(section 76)

Local authority notice of
intention to remedy the
defects (section 76(1))

| Counter-notice on local authority within seven days of receipt of intention to undertake works himself (section 76(3)) | OR | After nine days local authority may undertake works (section 76(2)) |

Failure to begin or
complete within
reasonable period,
local authority may
undertake works
(section 76(3))

Recovery of costs
(sections 76(2) and
107)

Notes
1. For normal statutory nuisance procedure relating to defective premises, see FC128.
2. There is no provision for appeal against notices served under section 76 and there is no penalty for non-compliance with the notice.

Notices

Notices must be in writing and, in addition to giving the local authority's intention to remedy the defects, must specify the actual defects which it intends to remedy (section 76(1) and sections 92 and 99). If a counter-notice is served by the recipient of the local authority notice but works are not begun within a reasonable time or progress towards completion is not reasonable, the local authority may execute the works itself. 'Reasonable' is not defined and the period allowed by the local authority will therefore need to take account of all the relevant circumstances (section 76(2) and (3)).

Recovery of costs

In any proceedings by the local authority to recover costs of carrying out works in default, the court must have regard to:

(a) whether the local authority was right in concluding that the premises were in such a state as to be prejudicial to health or a nuisance;

(b) whether unreasonable delay would have occurred if the normal procedures for dealing with statutory nuisances had been followed;

(c) following a counter-notice, whether the defendant had been given a reasonable time to begin or complete the works; and

(d) whether any costs ought to be borne wholly or partly by someone other than the defendant (section 76(4)).

OVERCROWDING NOTICES FOR NON-HOUSES IN MULTIPLE OCCUPATIONS

References

Housing Act 1985, part 10
The Housing (Prescribed Forms) Regulations 1990 (as amended)
Housing (Prescribed Forms) (No. 2) Regulations 1990
Private Rented Housing (Scotland) Act 2011
Housing (Scotland) Act 1987

Extent

These provisions apply to England and Wales. Similar provision apply in Scotland.

Note

Section 216 of the Housing Act 2004 enables the Secretary of State to modify the operation of this procedure and the standards incorporated into it. No such order has yet been made, but a government response to consultation was published in December 2017 as part of the Houses in Multiple Occupation and residential property licensing reforms (Chapter 2).

Scope

This procedure applies to any dwelling, which is defined as 'premises used or suitable for use as a separate dwelling'. It is a standard base on sleeping capacity and is not related to the provision of facilities etc.

Statutory overcrowding

A dwelling-house is deemed to be overcrowded when either:

(a) the number of persons sleeping there is such that any two persons 10 years old or more of opposite sexes, not being man and wife, must sleep in the same room (i.e. a room normally used in the locality either as a bedroom or as a living room and is available as sleeping accommodation) (the room standard); or

(b) the permitted number as calculated below is exceeded (the space standard).

No account is taken of any child under one year and a child of an age between one year and under 10 years is counted as half a unit (sections 324, 325 and 326).

Permitted number – the space standard

This is the smaller of a number of persons calculated by the following two methods:

(a) where a house consists of:

one room – two persons
two rooms – three persons
three rooms – five persons
four rooms – seven and a half persons
five rooms or more – 10 persons, with additional two persons in respect of each room in excess of five (no regard is had to rooms of less than 50 sq. ft.); **or**

(b) where the aggregate for all rooms is determined by reference to the following:

110 sq. ft. or more – two persons
90–100 sq. ft. – one and a half persons
70–90 sq. ft. – one person
50–70 sq. ft. – half person
Less than 50 sq. ft. – nil (section 326).

Offences by occupiers

The occupier of a dwelling-house who causes or permits it to be overcrowded is guilty of an offence unless:

(a) the overcrowding is caused only by virtue of a child attaining the age of one or ten years and the occupier has applied to the local authority for suitable alternative accommodation (defined in section 342) provided:

(i) alternative accommodation is refused; or

(ii) it subsequently becomes reasonably practicable for a person living in the house, but not a member of the family, to move; or

(b) the overcrowding is caused only because a member of the family is sleeping in the house temporarily; or

(c) the local authority has issued a licence authorising temporary overcrowding (below) (section 327).

Offences by landlords

The landlord is guilty of an offence if he has either:

(a) received a notice of overcrowding from the local authority but fails to take steps reasonably open to him to abate the overcrowding; or

(b) when letting the house, he had reasonable cause to believe it would become overcrowded or failed to make enquiries of the proposed occupier; or

(c) failed to notify the local authority of overcrowding within seven days of becoming aware of it (sections 331 and 333).

Duty to inspect, report and prepare proposals

If it appears to the local housing authority a report on overcrowding in their district or part of it, is needed or if the Secretary of State directs, the authority must:

(a) inspection the premises,

(b) prepare and submit to the Secretary of State a report showing the result of the inspection and the number of new dwellings required in order to abate the overcrowding, and

(c) unless they are satisfied that the dwellings will be otherwise provided, prepare and submit to the Secretary of State proposals for providing the required number of new dwellings (section 334(1)).

Where the Secretary of State directs inspections, he may, after consultation with the local housing authority, fix dates for completion of the inspections (section 334(2)).

Power to require information about persons sleeping in dwelling

The local housing authority may serve notice on the occupier of a dwelling requiring him to give them within 14 days a written statement of the number, ages and sexes of the persons sleeping in the dwelling (section 335(1)).

The occupier commits a summary offence if he does not comply with the requirement, or he gives a statement which to his knowledge is false in a material particular, and is liable on conviction to a fine not exceeding level 1 on the standard scale (section 335(2)).

Power to require production of rent book

A duly authorised officer of the local housing authority may require an occupier of a dwelling to produce any rent book or similar document for inspection which is being used in relation to the dwelling and is in his custody or under his control (section 336(1)).

The occupier must produce any such book or document to the officer or at the offices of the authority within 7 days (section 336(2)).

An occupier who fails to do so commits a summary offence and is liable on conviction to a fine not exceeding level 1 on the standard scale (section 336(3)).

Notice to abate overcrowding

Where a dwelling is overcrowded and the occupier is guilty of an offence, the local housing authority may serve on the occupier notice in writing requiring him to abate the overcrowding within 14 days from the date of service of the notice (section 338(1)).

If at any time within three months from the end of that period the dwelling is in the occupation of the person on whom the notice was served or of a member of his family, and it is overcrowded and the occupier is guilty of an offence, the local housing authority may apply to the County Court which shall order vacant possession of the dwelling to be given to the landlord within such period, not less than 14 or more than 28 days, as the court may determine (section 338(2)).

Expenses incurred by the local housing authority in securing the giving of possession of a dwelling to the landlord may be recovered by them from him by action (section 338(3)).

Licence for temporary overcrowding

An occupier, or intending occupier, may apply to the local authority for a licence authorising the sleeping occupation of a house by a number of persons in excess of the permitted number.

Such a licence may be granted by the local authority where it thinks it to be expedient, having had regard to any exceptional circumstances, including any seasonal increase in the general population of this district.

Licences may not operate for in excess of 12 months and may be revoked by the giving of at least one month's notice. There is no appeal against a refusal by a local authority to grant a licence or against revocation (section 330).

Definitions

Agent in relation to the landlord of a dwelling:

(a) means a person who collects rent in respect of the dwelling on behalf of the landlord, or is authorised by him to do so; and

(b) in the case of a dwelling occupied under a contract of employment under which the provision of the dwelling for his occupation forms part of the occupier's remuneration, includes a person who pays remuneration on behalf of the employer, or is authorised by him to do so.

Landlord in relation to a dwelling:

(a) means the immediate landlord of an occupier of the dwelling; and

(b) in the case of a dwelling occupied under a contract of employment under which the provision of the dwelling for his occupation forms part of the occupier's remuneration, includes the occupier's employer.

FC63 Overcrowding notices for non-HMOs – sections 324 to 344 Housing Act 1985

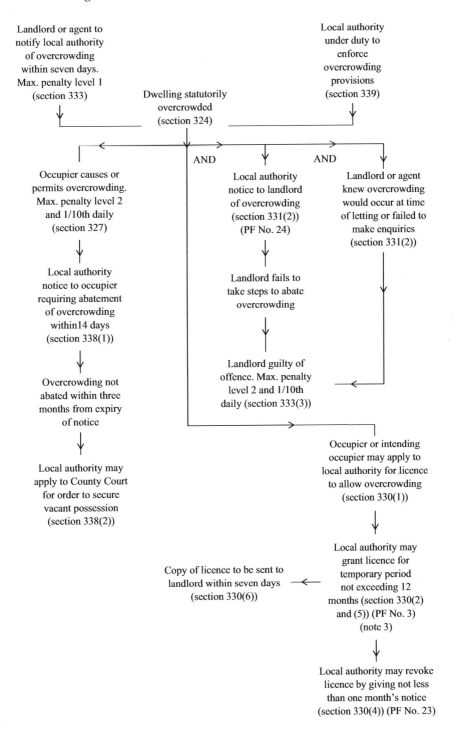

Landlord or agent to notify local authority of overcrowding within seven days. Max. penalty level 1 (section 333)

Local authority under duty to enforce overcrowding provisions (section 339)

Dwelling statutorily overcrowded (section 324)

AND

AND

Occupier causes or permits overcrowding. Max. penalty level 2 and 1/10th daily (section 327)

Local authority notice to landlord of overcrowding (section 331(2)) (PF No. 24)

Landlord or agent knew overcrowding would occur at time of letting or failed to make enquiries (section 331(2))

Local authority notice to occupier requiring abatement of overcrowding within 14 days (section 338(1))

Landlord fails to take steps to abate overcrowding

Overcrowding not abated within three months from expiry of notice

Landlord guilty of offence. Max. penalty level 2 and 1/10th daily (section 333(3))

Local authority may apply to County Court for order to secure vacant possession (section 338(2))

Occupier or intending occupier may apply to local authority for licence to allow overcrowding (section 330(1))

Local authority may grant licence for temporary period not exceeding 12 months (section 330(2) and (5)) (PF No. 3) (note 3)

Copy of licence to be sent to landlord within seven days (section 330(6))

Local authority may revoke licence by giving not less than one month's notice (section 330(4)) (PF No. 23)

Notes
1. For overcrowding in registered HMOs, see FC76.
2. There are no appeal provisions relating to overcrowding notices.
3. The prescribed forms in this procedure are specified in the Housing (Prescribed Forms) (No. 2) Regulations 1990, except for the licence for temporary occupation, which is contained with the Housing (Prescribed Forms) Regulations 1990.
4. In this flowchart, PF = prescribed form.

Owner in relation to premises:

(a) means a person (other than a mortgagee not in possession) who is for the time being entitled to dispose of the fee simple, whether in possession or in reversion; and

(b) includes also a person holding or entitled to the rents and profits of the premises under a lease of which the unexpired term exceeds three years (section 343).

Suitable alternative accommodation, in relation to the occupier of a dwelling, means a dwelling as to which the following conditions are satisfied –

(a) he and his family can live in it without causing it to be overcrowded;

(b) it is certified by the local housing authority to be suitable to his needs and those of his family as respects security of tenure, proximity to place of work and otherwise, and to be suitable in relation to his means;

(c) where the dwelling belongs to the local housing authority, it is certified by them to be suitable to his needs and those of his family as respects accommodation (section 342).

RESTORATION OR CONTINUATION OF SUPPLY OF WATER, GAS OR ELECTRICITY

Reference

Local Government (Miscellaneous Provisions) Act 1976, section 33

Extent

These provisions apply to England and Wales.

Scope

If any premises are occupied as a dwelling and the supply of water, gas or electricity to the premises:

(a) is cut off in consequence of the failure of the owner or former owner of the premises to pay the bill; or

(b) is in the opinion of the council likely to be cut off in consequence of such a failure,

the council may, at the request in writing of the occupier of the premises, make such arrangements as it thinks fit with the person who provided the supply for it to be restored to the premises or, as the case may be, for it to be continued to the premises (section 33 (1)).

Where a council makes a bill payment when the relevant supply to the premises was or became likely to be cut off or restored, the council can demand and recover those costs and interest (section 33(2)).

Such a demand must –

(a) be served on the recipient in writing; and

(b) give particulars of the payment to which the sum demanded relates; and

(c) in the case of a demand for a sum on which interest is payable, state the rate of the interest and that interest is payable (section 33(3)).

FC64 Restoration or continuation of supply of water, gas or electricity – section 33 Local Government (Miscellaneous Provisions) Act 1976

Water, gas or electricity supply cut off due to owner's failure to pay bill (section33 (1))

Occupier request Council in writing to restore supply (section 33(1))

Council contacts supplier to make arrangements to restore supply (section 33(1))

Council serves demand for recovery of costs and interest (section 33(2) and (3))

The costs involved may also be, until recovered, be a charge on the premises and if the owner is normally required to pay for a supply of the kind to which that sum relates; and the council has served a demand notice, the rent can be paid to the council (section 33(4)).

DISABLED FACILITIES GRANTS

References

Housing Grants, Construction and Regeneration Act 1996, sections 19 to 24
Housing (Scotland) Act 1987
The Housing (Northern Ireland) Order 2003

Extent

These provisions apply to England and Wales. Similar provisions are available in Northern Ireland.

Scope

This procedure involves Disabled Facilities Grants which are available from local authorities to pay for essential housing adaptations to help disabled people stay in their own homes.

Owner's and tenant's applications

When a grant application is made a local housing authority must be satisfied –

(a) that the applicant is or will be the owner of the land on which the relevant works are to be carried out, or
(b) that the applicant is a tenant (alone or jointly with others) of a dwelling or flat
(c) that the applicant is an occupier (alone or jointly with others) of a qualifying houseboat or a caravan (section 19(1)).

A grant application must be accompanied by an owner's or occupier's certificate in respect of the dwelling or flat to which the application relates or, in the case of a common parts application, in respect of each flat in the building occupied or proposed to be occupied by a disabled occupant.

An owner's certificate certifies that the applicant has or proposes to acquire a qualifying owner's interest. The disabled occupant intends to live in the dwelling or flat as his only or main residence throughout the grant condition period, or for such shorter period as his health and other relevant circumstances permit (sections 21 and 22).

Except where the authority consider it unreasonable in the circumstances to require such a certificate, the application must be accompanied by a consent certificate from each person who at the time of the application is entitled to possession of the premises at which the qualifying houseboat is moored or the land on which the caravan is stationed; or is entitled to dispose of the qualifying houseboat or the caravan (section 22A(3)).

Purposes for which grant must or may be given

The purposes for which an application for a grant must be approved are the following:

(a) facilitating access by the disabled occupant to and from –

   (i) the dwelling, qualifying houseboat or caravan, or
   (ii) the building in which the dwelling or, as the case may be, flat is situated;

(b) making –

   (i) the dwelling, qualifying houseboat or [caravan] 3, or
   (ii) the building,

   safe for the disabled occupant and other persons residing with him;

(c) facilitating access by the disabled occupant to a room used or usable as the principal family room;
(d) facilitating access by the disabled occupant to, or providing for the disabled occupant, a room used or usable for sleeping;
(e) facilitating access by the disabled occupant to, or providing for the disabled occupant, a room in which there is a lavatory, or facilitating the use by the disabled occupant of such a facility;
(f) facilitating access by the disabled occupant to, or providing for the disabled occupant, a room in which there is a bath or shower (or both), or facilitating the use by the disabled occupant of such a facility;
(g) facilitating access by the disabled occupant to, or providing for the disabled occupant, a room in which there is a washhand basin, or facilitating the use by the disabled occupant of such a facility;
(h) facilitating the preparation and cooking of food by the disabled occupant;
(i) improving any heating system in the dwelling, qualifying houseboat or caravan to meet the needs of the disabled occupant or, if there is no existing heating system there or any such system is unsuitable for use by the disabled occupant, providing a heating system suitable to meet his needs;
(j) facilitating the use by the disabled occupant of a source of power, light or heat by altering the position of one or more means of access to or control of that source or by providing additional means of control;
(k) facilitating access and movement by the disabled occupant around the dwelling, qualifying houseboat or caravan in order to enable him to care for a person who is normally resident there and is in need of such care;
(l) such other purposes as may be specified by order of the Secretary of State.

If in the opinion of the local housing authority the relevant works are more or less extensive than is necessary to achieve any of the purposes set out above, they may, with the consent of the applicant, treat the application as varied so that the relevant works are limited to or, as the case may be, include such works as seem to the authority to be necessary for that purpose (section 23).

Approval of application

The local housing authority must approve an application for a grant but where an authority entertain an owner's application for a grant made by a person who proposes to acquire a qualifying owner's interest, they must not approve the application until they are satisfied that he has done so (section 24(2)).

A local housing authority must not approve an application for a grant unless they are satisfied:

(a) that the relevant works are necessary and appropriate to meet the needs of the disabled occupant, and

(b) that it is reasonable and practicable to carry out the relevant works having regard to the age and condition of:

    (i) the dwelling, qualifying houseboat or [caravan] 4, or

    (ii) the building.

In considering the relevant works a local housing authority which is not itself a social services authority shall consult the social services authority (section 24(3)).

A local housing authority shall not approve a common parts application for a grant unless they are satisfied that the applicant has a power or is under a duty to carry out the relevant works (section 24(5)).

Definitions

Tenant includes –

(a) a secure tenant, introductory tenant or statutory tenant,

(b) a protected occupier under the Rent (Agriculture) Act 1976 or a person in occupation under an assured agricultural occupancy within the meaning of Part I of the Housing Act 1988,

(c) an employee (whether full-time or part-time) who occupies the dwelling or flat concerned for the better performance of his duties, and

(d) a person having a licence to occupy the dwelling or flat concerned which satisfies such conditions as may be specified by order of the Secretary of State.

Disabled occupant means the disabled person for whose benefit it is proposed to carry out any of the relevant works (section 20).

Practical step-by-step procedures

1. Formal recommendation. The Occupational Therapist makes a formal recommendation. The recommendation details the necessary adaptations that are needed to the property to make it suitable for use by the disabled person.

2. Initial contact. Completion of a grant application and consent forms. These are returned together with financial evidence to the Disabled Facilities Grant Officer.

    The officer completes the Formal Means Test and ascertains ownership. The applicant is advised in writing if eligible for financial assistance or there is a financial contribution to make towards the cost of the adaptations.

    Confirm that the person who is detailed to own the property is named on the deeds. The officer contacts the land registry to confirm these details.

    Complete the owner's certificate. This certificate states that it is the disabled person's intention to live in the property for period of five years after the completion of the works. In addition there may be a local land charge placed on the property for a period of 10 years from the date of completion.

FC65 Disabled Facilities Grants – Housing Grants, Construction and Regeneration Act 1996, sections 19 to 24

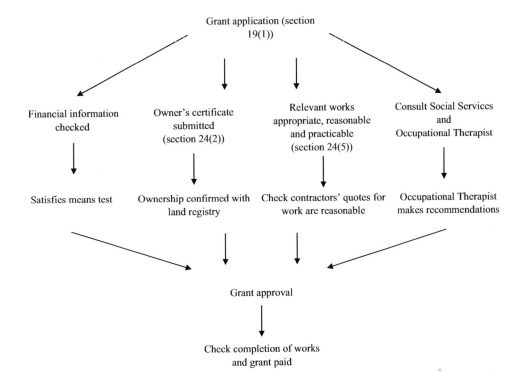

Grant application (section 19(1))

Financial information checked

Owner's certificate submitted (section 24(2))

Relevant works appropriate, reasonable and practicable (section 24(5))

Consult Social Services and Occupational Therapist

Satisfies means test

Ownership confirmed with land registry

Check contractors' quotes for work are reasonable

Occupational Therapist makes recommendations

Grant approval

Check completion of works and grant paid

3. Confirm formally the results of the grant means test, detailing any contribution that will need to be made.

   Prepare and obtain signature on consent form agreeing to pay contribution direct to the contractor, on completion of works.

4. Getting quotes. Applicant to obtain 2/3 quotes from contractors. Local Authorities usually can supply a list of contractors. The builder should provide a detailed itemised quote using the Occupational Therapists drawing and scheme. Quotes submitted to the local authority.

5. Check the quotes to determine a reasonable cost for the recommended works.

   The Council's assessed cost will be detailed on the grant approval.

   Once the cost checking exercise has been completed, the grant can be approved.

   No works included in a grant application should begin until the Council's written Notice of Approval has been sent to the applicant.

6. Approval. The approval document will detail the total cost deemed reasonable by the Council for works required, the formally assessed means test contribution and any costs the applicant will have to incur where costs exceed the maximum grant.

7. Commencement of works.

8. Completion of works. On completion of works, the final invoice is submitted to the Council. Check the completed works and carry out a post inspection. This post inspection will be for the purpose of ensuring that the works have been completed as per the Occupational Therapist Scheme and to ensure grant monies have been spent appropriately.

9. Payment will only be made following a satisfactory visit from the Council and on receipt of invoice and electrical certificate (if applicable) and on return of signed Completion of Installation of Adaptations form. Arrange payment to applicant.

10. Local land charge entry. The disabled facilities grant process allows the council to place a limited charge on adapted properties of owner-occupiers, if it is sold within 10 years. This is only applicable where the costs are between £5,000 and £10,000.

DANGEROUS BUILDINGS

References

Building Act 1984, sections 77 and 78
Building (Scotland) Act 2003
Pollution Control and Local Government (Northern Ireland) Order 1978

Extent

These provisions apply in England and Wales. Similar provisions apply in Scotland and Northern Ireland.

Scope

The procedure may be used where a building or structure, or part thereof, is:

(a) in such a state; or
(b) is used to carry such loads

as to be dangerous (sections 77(1) and 78(1)).

FC66 Dangerous buildings – sections 77 and 78 Building Act 1984

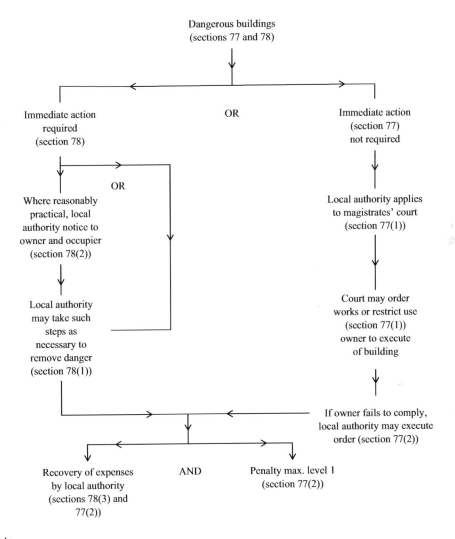

Note

1. This procedure is subject to the provisions of the Planning (Listed Buildings and Conservation Areas) Act 1990 in relation to listed buildings etc.

Court order

Where there is no call for immediate action, the local authority may apply to a magistrates' court and the court may order the owner to:

(a) execute necessary works to remove the danger; or
(b) if the owner elects, to demolish the building or structure and remove resulting rubble or rubbish; or
(c) where the danger arises from the overloading of the building or structure, restrict the use until the court withdraws or modifies that restriction following works having been carried out (section 77(1)).

Emergency measures

These may be taken where immediate action is required in the opinion of the local authority to remove the danger (section 78(1)).

The measures which may be taken are those necessary to remove the danger (section 78(1)). Where practicable, prior notice must be given to the owner and occupier (section 78(2)).

Recovery of expenses

The local authority's expenses in executing a court order are recoverable from the owner (section 77(2)). Where the emergency provisions have been used, the local authority may recover its costs, but in any dispute as to the recovery of these costs, the court may consider whether the local authority was justified in taking emergency action as opposed to an application to the court for an order under section 77 (section 78(5)).

Recovery of expenses connected with the fencing off of a building or structure or arranging for it to be watched are not recoverable after:

(a) the danger has been removed by emergency action by the local authority; or
(b) after works required under section 77 have been carried out (section 78(4)).

PROTECTION OF BUILDINGS

Reference

Local Government (Miscellaneous Provisions) Act 1982, sections 29–32 inclusive

Extent

This procedure applies in England and Wales.

Scope

This procedure applies to any building:

(a) which is unoccupied; or
(b) where the occupier is temporarily absent; (section 29(1)).

and which is:

(a) not effectively secured against unauthorised entry; or
(b) likely to become a danger to public health (section 29(2)).

Works

The local authority may undertake any works necessary to prevent unauthorised entry or to prevent the building becoming a danger to public health (section 29(2)). If either:

(a) works are immediately necessary to prevent it becoming a danger to public health; or
(b) the name and address of the owner, or the whereabouts of the occupier are not reasonably practical to obtain,

a notice of intention is not required. In all other cases, at least 48 hours' written notice must be given (section 29(6) and (9)).

Notices

(a) **Person.** The notice is to be served on either the owner or the occupier (section 29(6)) or on British Railways Board or any statutory undertaker in respect of their operational land (section 30(2)).
(b) **Content.** The notice must specify the works which the local authority intends to carry out (sections 29(7) and 30(3)).
(c) **Service.** This procedure is not incorporated in the provisions of either the Public Health Acts or the Housing Acts and therefore service should be by one of the methods specified in section 233 of the Local Government Act 1972 (p. 46).

Appeals

A person in receipt of the local authority notice of intention may appeal to the County Court within 21 days of service on any of the following grounds:

(a) the works specified are not authorised by the section 29 procedure;
(b) the works are unnecessary;
(c) it was unreasonable for the local authority to undertake the works (section 31(1), (2) and (3)).

In the event of an appeal, the local authority must stop any works which it has commenced and await the determination of the court. If the works had been completed, the court may still determine the position regarding recovering of costs (section 31(4)).

The court may confirm, vary or quash the notice and make provision in its order relating to recovery of costs by the local authority (section 31(5) and (6)).

Persons not in receipt of a local authority notice but from whom costs are being recovered may appeal to the County Court within 21 days of the request for payment and ask the court for a declaration that either:

(a) the works were unnecessary; or
(b) it was unreasonable for the local authority to undertake them (section 32).

FC67 Protection of buildings – sections 29 to 32 Local Government (Miscellaneous Provisions) Act 1982

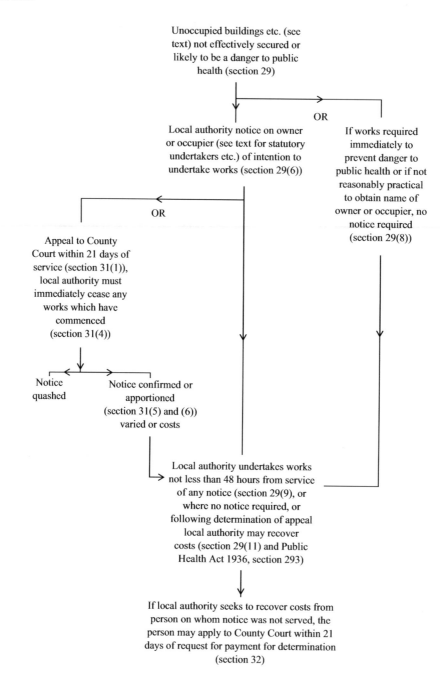

Unoccupied buildings etc. (see text) not effectively secured or likely to be a danger to public health (section 29)

OR

Local authority notice on owner or occupier (see text for statutory undertakers etc.) of intention to undertake works (section 29(6))

If works required immediately to prevent danger to public health or if not reasonably practical to obtain name of owner or occupier, no notice required (section 29(8))

OR

Appeal to County Court within 21 days of service (section 31(1)), local authority must immediately cease any works which have commenced (section 31(4))

Notice quashed

Notice confirmed or apportioned (section 31(5) and (6)) varied or costs

Local authority undertakes works not less than 48 hours from service of any notice (section 29(9), or where no notice required, or following determination of appeal local authority may recover costs (section 29(11) and Public Health Act 1936, section 293)

If local authority seeks to recover costs from person on whom notice was not served, the person may apply to County Court within 21 days of request for payment for determination (section 32)

Notes

1. No offences are committed under this procedure.
2. For operational land of British Railways Board or any statutory undertaker, in carrying out works the local authority must comply with any reasonable request which British Railways Board or the statutory undertakers may impose for the protection or safety of their undertaking (section 30).
3. The procedure can be applied to any building in the specified condition, including ones subject to orders under the Housing Acts.

Definitions

Building includes structure (section 29(3)).

EMPTY DWELLING MANAGEMENT ORDERS

References

Housing Act 2004, sections 132 to 138

Housing (Empty Dwelling Management Orders) (Prescribed Exceptions and Requirements) (England) Order 2006 (as amended)

Housing (Management Orders and Empty Dwelling Management Orders) (Supplemental Provisions) (England) Regulations 2006

Housing (Empty Dwelling Management Orders) (Prescribed Period of Time and Additional Prescribed Requirements) (England) Amendment Order 2012

Housing (Management Orders and Empty Dwelling Management Orders) (Supplemental Provisions) (Wales) Regulations 2006

Extent

These provisions apply to England only but there are similar provisions for Wales.

Scope

Where dwellings are empty for an extended period of time, local authorities can serve notice on the owner with the purpose of bringing the property back into use.

Interim empty dwelling management orders

A local authority can apply to the Appropriate Tribunal to issue an interim empty dwelling management order (EDMO) with the consent of the relevant proprietor to ensure the dwelling becomes and continues to be occupied (section 132 (2)). The local authority must give the relevant proprietor three months' notice before applying to the Appropriate Tribunal (article 4(1)(aa) of 2006 order).

The local authority must make reasonable efforts:

(a) to notify the relevant proprietor that they are considering making an interim EDMO in respect of the dwelling; and

(b) to ascertain what steps (if any) he is taking, or is intending to take, to secure that the dwelling is occupied (section 133(3)).

The local authority must also take into account the rights of the relevant proprietor of the dwelling and the interests of the wider community (section 133(4)).

The local authority may make an interim EDMO despite any pending appeal against the order of the tribunal (section 133(5)). Before making an interim EDMO, the local authority must:

(a) serve a copy of the proposed order on each relevant person; and

(b) consider any representations.

The notice must state that the local authority is proposing to make an interim EDMO and set out:

(a) the reasons for making the order;
(b) the main terms of the proposed order (including those of the management scheme to be contained in it); and
(c) the end of the consultation period (schedule 6).

The local authority must consider any representation made about the management order before they consider whether to apply to the Tribunal (schedule 6).

Following representations, the notice must set out:

(a) the proposed modifications;
(b) the reasons for them; and
(c) the end of the consultation period (schedule 6).

Details of the evidence that local housing authorities must provide for the Tribunal are contained in the Housing (Empty Dwelling Management Orders) (Prescribed Exceptions and Requirements) (England) Order 2006 as amended.

A Tribunal may authorise a local authority to make an interim EDMO if it is satisfied:

(a) that the dwelling has been wholly unoccupied for at least two years or such longer period as may be prescribed;
(b) that there is no reasonable prospect that the dwelling will become occupied in the near future;
(c) that, if an interim order is made, there is a reasonable prospect that the dwelling will become occupied;
(d) that the authority has made reasonable efforts to notify the relevant proprietor of the making of the order and what the proprietor intends to do with the property;
(e) that any prescribed requirements have been complied with; (section 134(2))
(f) that the relevant proprietor is not a public body and:
(g) that the case does not fall within one of the prescribed exceptions (section 133 (2)(b)).

In deciding whether to authorise a local authority to make an interim EDMO, the tribunal must take account of:

(a) the interests of the community; and
(b) the effect that the order will have on the rights of the relevant proprietor and may have on the rights of third parties (section 134).

The tribunal can confirm or vary the order.

Local authority's duties once interim EDMO in force

The local authority must take such steps as they consider appropriate for securing that the dwelling becomes and continues to be occupied and making proper management arrangements, including insurance, for the dwelling pending the making of a final EDMO or the revocation of the interim EDMO.

If the local authority concludes that there are no steps which they could appropriately take, the authority must either:

(a) make a final EDMO in respect of the dwelling; or
(b) revoke the order without taking any further action.

The owner would receive the rental minus the council's costs and benefit from any repairs when they retake control (section 135).

If none of these exemptions apply and the tribunal rules in favour of the council, it could take control of the property for seven years.

Exemptions

Many properties will be exempt – for example, those from where someone has gone temporarily abroad or is in hospital; where it is a holiday home; where the proprietor is a member of the armed forces and is serving abroad; or within six months of probate being granted where the home owner has died (Housing (Empty Dwelling Management Orders) (Prescribed Exceptions and Requirements) (England) Order 2006 as amended).

Final empty dwelling management orders

A final EDMO is an order made, in succession to an interim EDMO or a previous final EDMO, for the purpose of securing that a dwelling is occupied (section 132(3)).

A local authority may make a final EDMO to replace an interim EDMO if:

(a) they consider that, unless a final EDMO is made in respect of the dwelling, the dwelling is likely to become or remain unoccupied;
(b) where the dwelling is unoccupied, they have taken all such steps as it was appropriate for them to take under the interim EDMO with a view to securing the occupation of the dwelling (section 136(1)).

The local authority can make another final EDMO if the first one is not working (section 136(2)).

The same criteria and provisions for making an interim EDMO apply to a final EDMO.

Local housing authority's duties once final EDMO in force

The duties are the same as for interim EDMO but the authority must from time to time review:

(a) the operation of the order and in particular the management scheme contained in it;
(b) whether, if the dwelling is unoccupied, there are any steps which they could appropriately take under the order for the purpose of securing that the dwelling becomes occupied; and
(c) whether keeping the order in force in relation to the dwelling is necessary to secure that the dwelling becomes or remains occupied.

If, on a review, the local authority considers that any variations should be made, they must proceed to make those variations. The review can be done at the local authority's own volition or on application from a relevant proprietor.

If the dwelling is unoccupied and on a review the local authority concludes that either:

(a) there are no steps which they could appropriately take; or
(b) keeping the order in force is not necessary,

they must proceed to revoke the order (section 137).

Compensation payable to third parties

Before making an interim or final EDMO, the tribunal or local authority must consider whether compensation should be paid by them to any third party in respect of any interference in consequence of the order with the rights of the third party (section 134(4)).

A third party may apply to a Tribunal for an order requiring the local authority to pay to him compensation in respect of any interference in consequence of the order with his rights in respect of the dwelling.

The tribunal may make an order requiring the authority to pay to the third party an amount by way of compensation for such interference.

If a third party requests them to do so at any time, the local authority must consider whether an amount by way of compensation should be paid to him in respect of any interference in consequence of a final EDMO with his rights.

The authority must notify the third party of their decision as soon as practicable.

Where the local authority decides that compensation ought to be paid to a third party, they must vary the management scheme contained in the order to specify the amount of the compensation to be paid and to make provision for its payment (section 138).

Definitions

Dwelling means:

(i)   a building intended to be occupied as a separate dwelling; or
(ii) a part of a building intended to be occupied as a separate dwelling which may be entered otherwise than through any non-residential accommodation in the building.

Relevant proprietor, in relation to a dwelling, means:

(i)   if the dwelling is let under one or more leases with an unexpired term of seven years or more, the lessee under whichever of those leases has the shortest unexpired term; or
(ii) in any other case, the person who has the freehold estate in the dwelling.

Third party, in relation to a dwelling, means any person who has an estate or interest in the dwelling, other than the relevant proprietor and any person who is a tenant under a lease (section 132(4)).

Wholly unoccupied means that no part is occupied, whether lawfully or unlawfully (section 133)

Other powers in respect of empty dwellings

Local authorities have a range of powers it can use to deal with empty properties.

Section 77 of the Building Act 1984 deals with making the building safe.

FC68 Interim and final empty property management orders – sections 132 to 138 Housing Act 2004

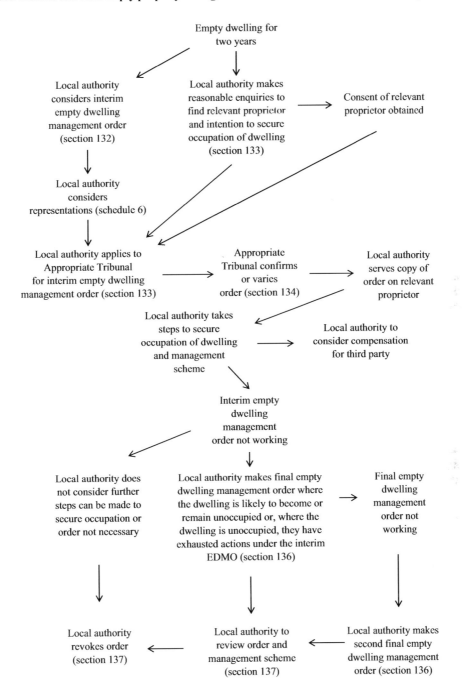

Section 79 of the Building Act 1984 enables councils to deal with ruinous and dilapidated buildings and neglected sites 'in the interest of amenity'.

Section 215 of the Town and County Planning Act 1990 allows councils to take action over unsightly land or the external appearance of a property.

Section 29 of the Local Government (Miscellaneous Provisions) Act 1982 allows councils to board up or make secure empty properties to stop people getting into the property or to stop the property being a danger to public health.

Additional powers that the council can use include those given by the Law of Property Act 1925, which enables councils to force the sale of empty properties where the local authority is owed money by the owner.

Empty dwelling management orders can be used on properties empty for two years or more. Here the council can take over the management of a property, essentially acting as the landlord, bringing the property back into use. These powers can only be used with the agreement of the owner.

The strongest power that the local authority can use is that of compulsory purchase where the local authority buys the property without the agreement of the owner. The local authority needs to apply to the Secretary of State in order to carry out such a purchase. They have to show that there is a public interest in compulsorily buying the property and that the other methods of bringing the property into use have not worked.

RUINOUS AND DILAPIDATED BUILDINGS

References

Building Act 1984, section 79
Building (Scotland) Act 2003
Pollution Control and Local Government (Northern Ireland) Order 1978

Extent

These provisions apply to England and Wales. Similar provisions apply in Scotland and Northern Ireland.

Scope

The procedure is applicable to any buildings or structures which are seriously detrimental to the amenities of the neighbourhood because of their ruinous or dilapidated condition (section 79(1)), and to rubbish resulting from the demolition of a building which renders the site seriously detrimental to the amenities of the neighbourhood (section 79(2)).

Notices

The local authority must serve notices in writing on the owner of the building or structure concerned and may require either:

(a) repair or restoration; or
(b) if the owner so elects, demolition of the building or structure or parts thereof and the removal of rubbish resulting from demolition (section 79(1)); or
(c) removal of the rubbish already existing (section 79(2)).

FC69 Ruinous and dilapidated buildings – section 79 Building Act 1984

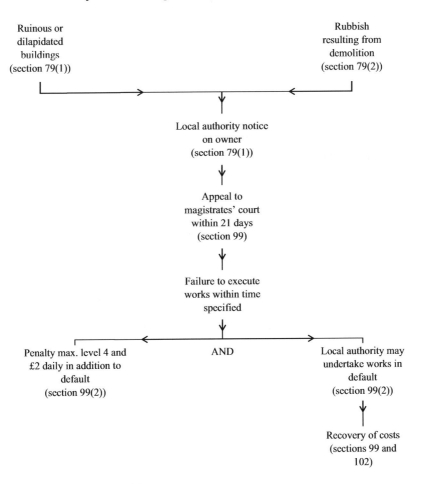

Notices must indicate both the nature of works of repair or restoration and the works of demolition and removal of rubbish or material (section 79(3)). The provisions of sections 99 and 102 apply with regard to appeals against and the enforcement of these notices (p. 570).

DEMOLITION ORDERS

References

Housing Act 1985, part 9, sections 265 to 275 (as amended by the Housing Act 2004, section 46 and schedule 15, paragraphs 13–21)
Housing (Scotland) Act 1987
Housing (Northern Ireland) Order 1981
Housing (Prescribed Forms) (No. 2) Regulations 1990
Enforcement Guidance – HHSRS, ODPM, February 2006, part 5, paragraphs 5.44–5.46

Extent

These provisions apply to England and Wales. Similar provisions apply in Scotland and Northern Ireland.

Scope

Where a category 1 hazard exists on any residential premises (dwelling, HMO or flat), the making of a demolition order is one of the courses of action available to a local authority in complying with its duty under section 5 to take the most appropriate course of action (section 265(1) and (2)).

In relation to a category 2 hazard, the local authority may use its discretion to make a demolition order as the most appropriate action but only where the circumstances are specified in an order made by the Secretary of State. No such orders have yet been made (section 265(3) and (4)).

Demolition orders are not to be made where the premises are subject to a management order or are listed buildings (sections 265(5) and 304).

In deciding whether to make a demolition order, the local authority should:

- take into account the availability of local accommodation for rehousing the occupants;
- take into account the demand for, and sustainability of, the accommodation if the hazard was remedied;
- consider the prospective use of the cleared site;
- consider the local environment, the sustainability of the area for continued residential use and the impact of a cleared site on the appearance and character of the neighbourhood (Enforcement Guidance, paragraph 5.44).

Content of demolition orders

Demolition orders must require:

- vacation within a period specified in the order but not less than 28 days from the date of operation of the order,

- the demolition of the premises within six weeks of the date of operation or the date of vacation, whichever is the longer. The local authority may agree a longer period (section 267(1)).

Service of demolition orders

Copies of the demolition order must be served within seven days of the date on which it was made on:

(a) owners and occupiers;
(b) persons authorised to permit occupation of it; and
(c) mortgagees (section 268(1)).

The service of these copies of the demolition order are subject to the provisions of section 246 of the Housing Act 2004 and not section 617 of the Housing Act 1985. Service on occupiers is deemed to have been properly made if the copy is posted on a conspicuous part of the premises (section 268(3)).

Service of copies of demolition orders is to be accompanied by a 'statement of reasons' for the local authority decision to make a demolition order (Housing Act 2004, section 8 – see p. 283).

Operation of a demolition order

A demolition order becomes operative 28 days from the date on which it was made or the date when the appeal is determined by a modification or confirmation (section 268(4)).

Appeals

Any person aggrieved by a demolition order, other than an occupier under a lease or agreement with an unexpired term of three years or less, may appeal to the Appropriate Tribunal within 28 days of the order being made. The Tribunal may vary, quash or confirm the order (section 269).

The grounds of appeal include that the demolition order was not the most appropriate course of action, which would have been an improvement notice, a hazard awareness notice, a prohibition order or declaring the area to be a clearance area (section 269A).

Cleansing before demolition

Before a demolition order becomes operative, the local authority may give notice to the owner that they intend to cleanse the property from vermin after operation of the order and before demolition takes place. Demolition may not proceed until this work is completed and the local authority has served a notice to this effect. However, the owner may, after vacation and by notice, request the local authority to carry out the treatment within 14 days from this notice and may demolish after the expiry of that period, whether or not the work has been done (section 273).

Enforcement

If the premises are not demolished by the owner within the time allowed, the local authority is required to do so, sell the materials and recover their expenses (sections 271 and 272).

FC70 Demolition orders (DOs) – sections 265 to 275 Housing Act 1985

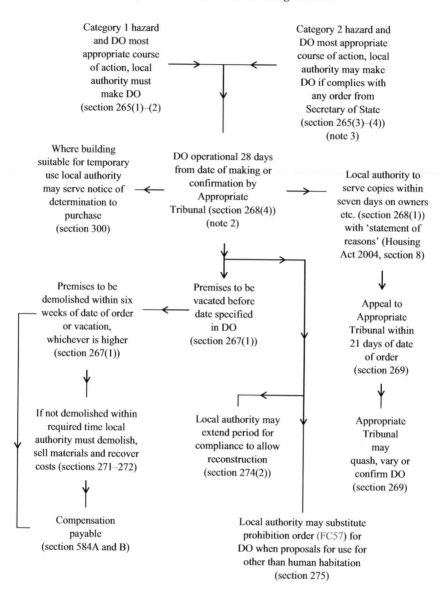

Notes

1. Section numbers relate to the Housing Act 1985 unless indicated otherwise.
2. Local authority may undertake cleansing of vermin – see text (section 273).
3. No order has yet been made by the Secretary of State.

Where a demolition order has become operative the owner of the premises, or any other person who in the opinion of the local housing authority is or will be in a position to put his proposals into effect, may submit proposals to the authority for the execution by him of works designed to secure the reconstruction, enlargement or improvement of the premises, or of buildings including the premises (section 274(1)).

If it is intended to carry out works that will result in the elimination of the category 1 hazard, the local authority may extend the period within which demolition is required to allow for the carrying out of the works (section 274(2)).

It is possible to substitute a demolition order with a prohibition order (see FC57), where proposals are made for the use of the premises other than for human habitation (section 275).

Compensation

Compensation is payable by the local authority to every owner of premises subject to a demolition order (section 584A) but is repayable where the demolition order is revoked to permit reconstruction under section 274 (section 584B).

Determinations to purchase

In circumstances where the local authority would be required to make a demolition order and it appears to them that the building can be rendered capable of providing accommodation which is adequate for the time being, the local authority may purchase the premises, by agreement or compulsory purchase order, instead of making a demolition order (section 300).

DEMOLITION OF BUILDINGS

References

Building Act 1984, sections 80–83 inclusive
Housing (Scotland) Act 1987
Housing (Northern Ireland) Order 1981

Extent

These provisions apply to England and Wales. Similar provisions apply in Scotland and Northern Ireland.

Scope

The notification procedure applies to the demolition of the whole or part of any building except:

(a) of an internal part of an occupied building which is to continue to be occupied;
(b) where the cubic content by external measurement does not exceed 1,750 cubic feet;
(c) of greenhouses, conservatories, sheds or prefabricated garages forming part of a larger building;

(d) agricultural buildings, unless it is contiguous to a non-agricultural building;
(e) demolition of a house subject to a demolition order under the Housing Act 1985 (section 80(1)).

The works that may be required are those which are consequential of the demolition i.e. demolition waste, nearby utilities and safety of adjacent buildings or land.

Notification to a local authority and others

The person undertaking the demolition of any building within the scope of this procedure must give prior notification to the local authority in writing indicating the building concerned and the works of demolition which are intended. Copies of this notice must be sent or given to:

(a) occupiers of adjacent buildings;
(b) the gas supplier(s);
(c) the electricity supplier(s).

Demolition must not commence until the local authority has served its notice under section 81 or the 'relevant period' (below) has expired (section 80(2) and (3)).

Local authority notices

In addition to those situations where notification must be made to the local authority, notices may also be served by the local authority in the situations identified in (i), (ii) and (iii) under (a) below.

(a) **Person.** These notices are to be served on either:

   (i) a person upon whom a demolition order under the Housing Act 1985 has been served; or
   (ii) a person who is not intending to comply with an order under section 77 of the Building Act 1984 (dangerous structure etc.) or a notice under section 79 of the Building Act 1984 (ruinous and dilapidated buildings, etc.); or
   (iii) a person who is intending, or has begun, a demolition to which section 80 of the Building Act 1984 applies ('Scope' above) (section 81(1)).

(b) **Relevant period.** This is the period within which the local authority is required to serve its notice and this is:

   (i) where a person has served a notification on the local authority under section 80, within six weeks from the giving of the notice to the local authority, or a longer period that may be agreed by the notifier in writing; and
   (ii) in relation to a demolition order under the Housing Act 1985, not less than seven days after service of the order, or a longer period as the person on whom the order was served may allow in writing (section 81(4)).

(c) **Contents.** The local authority notice may require all or any of the following:

    (i) shoring up of the adjacent buildings;

    (ii) weather-proofing of exposed surfaces of adjacent buildings and repairing or making good damage;

    (iii) removal of material from site;

    (iv) disconnection, sealing and removal of drains and sewers;

    (v) making good ground surface following disconnection or removal of drains and sewers;

    (vi) arrangements with the statutory undertakers (including water undertakers) for the disconnection of gas, electricity and water supplies;

    (vii) arrangements for the burning of structures and materials on site with the Fire and Rescue Authority;

    (viii) any other steps considered necessary for the protection of the public and the preservation of public amenity (section 82(1)).

(d) **Copies.** The local authority is required to send copies of its notices to:

    (i) owners *and* occupiers of adjacent buildings;

    (ii) if the notice requires disconnection of gas, electricity or water supplies, the statutory undertaker (including water undertakers) concerned;

    (iii) if the notice relates to the burning of structures or materials on site, the Fire and Rescue Authority (section 81(5) and (6)).

(e) **Enforcement of local authority notices.** These notices are subject to section 99 procedures (content of a notice for execution of works) (p. 570) and the local authority is, therefore, able to undertake works in default and recover costs as well as prosecuting for non-compliance (section 82(6)).

Appeals

Appeal against local authority notices served under section 81 may be made under the grounds set out in Building Act 1984, section 102 (p. 570) and in addition on the following grounds:

    (i) in relation to the shoring up of an adjacent building, that the owner is not entitled to support by the building to be demolished and he, therefore, should pay for, or contribute towards, the costs; and

    (ii) in relation to the weather-proofing of an adjacent building, that the owner should pay for, or contribute towards the costs (section 83(2)).

In both cases, the appellant must serve copies of his notices of appeal on the other persons concerned (section 83(3)).

On the hearing of the appeal, the court may make an order in respect of the payment of, or contribution towards, the cost of the works by any such person, or on how any expenses that may be recoverable by the local authority are to be borne between the appellant and any such person (section 83(3)(b)).

FC71 Demolition of buildings – sections 80 to 83 Building Act 1984

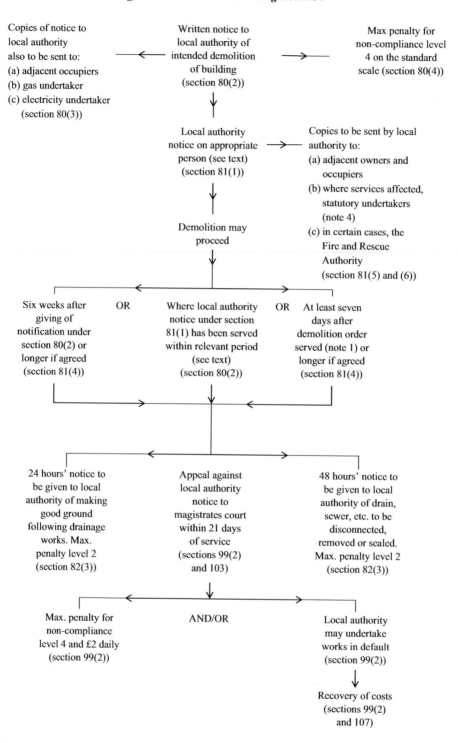

Notes
1. For demolition orders, see FC70.
2. For further procedure relating to rubbish following demolition, see FC69.
3. The section 99 procedure applies here for appeals against and the enforcement of local authority notices (p. 570).
4. The statutory undertakers include the water undertakers.

CLEARANCE AREAS (CAs)

References

Housing Act 1985, sections 289 to 298, as amended by the Housing Act 2004
Housing (Scotland) Act 2006, 2010 and 2014
Housing (Northern Ireland) Order 1981
Land Compensation Act 1973, section 39 (as amended)
Housing (Prescribed Forms) (No. 2) Regulations 1990 (as amended)
The Regulatory Reform (Housing Assistance) Order 2002
Enforcement Guidance, HHSRS, ODPM February 2006, part 5, paragraphs 5.47–5.49

Extent

These provisions apply to England and Wales. Similar provisions apply in Scotland and Northern Ireland.

Scope

A clearance area is an area which is to be cleared of all buildings in accordance with this procedure (section 289 (1)).

A local authority may declare a clearance area if either:

1. a category 1 hazard exists in each of the residential buildings in an area and if any other buildings in that area are dangerous or harmful to the health or safety of the inhabitants. This is one of the options available to local authorities in undertaking their duty to take the most appropriate action under section 5 of the Housing Act 2004; or
2. the local authority is satisfied that the residential buildings in the area are dangerous or harmful to the health or safety of the inhabitants because of their bad arrangement, or the narrowness or bad arrangement of the streets, and that any other buildings are dangerous or harmful to health. This is a discretionary power; or
3. a category 2 hazard exists in each of the residential premises and any other buildings to be included are dangerous or harmful to health and safety. This is also a discretionary power (section 289 as amended).

Before declaration of a clearance area, the local authority should consider the desirability of clearance in the context of the proposals for the wider neighbourhood of which the area forms part. In deciding to make a clearance area for an area in which hazardous dwellings are situated, the local authority should have regard to:

- the likely long-term demand for residential accommodation;
- the degree of concentration of dwellings containing serious and intractable hazards within the area;
- the density of the buildings and street pattern around them;
- the overall availability of housing accommodation in the wider neighbourhood in relation to housing needs and demands;
- the proportion of dwellings free of hazards and other, non-residential, premises in sound condition which would also need to be cleared to produce a suitable site;

Housing and health

FC72 Clearance areas (CAs) – sections 289 to 298 Housing Act 1985 (as amended)

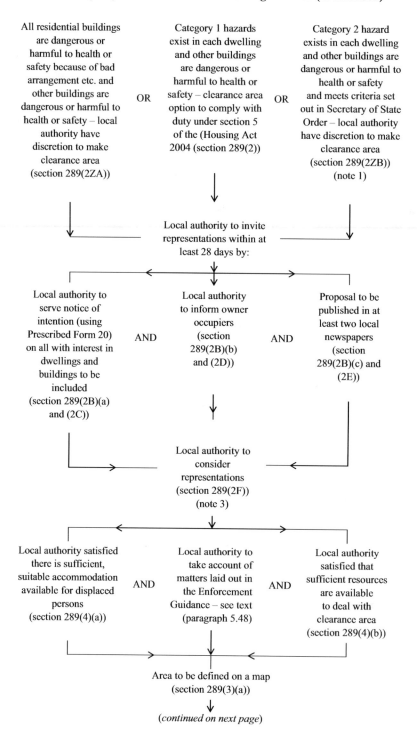

All residential buildings are dangerous or harmful to health or safety because of bad arrangement etc. and other buildings are dangerous or harmful to health or safety – local authority have discretion to make clearance area (section 289(2ZA))

OR

Category 1 hazards exist in each dwelling and other buildings are dangerous or harmful to health or safety – clearance area option to comply with duty under section 5 of the (Housing Act 2004 (section 289(2))

OR

Category 2 hazard exists in each dwelling and other buildings are dangerous or harmful to health or safety and meets criteria set out in Secretary of State Order – local authority have discretion to make clearance area (section 289(2ZB)) (note 1)

Local authority to invite representations within at least 28 days by:

Local authority to serve notice of intention (using Prescribed Form 20) on all with interest in dwellings and buildings to be included (section 289(2B)(a) and (2C))

AND

Local authority to inform owner occupiers (section 289(2B)(b) and (2D))

AND

Proposal to be published in at least two local newspapers (section 289(2B)(c) and (2E))

Local authority to consider representations (section 289(2F)) (note 3)

Local authority satisfied there is sufficient, suitable accommodation available for displaced persons (section 289(4)(a))

AND

Local authority to take account of matters laid out in the Enforcement Guidance – see text (paragraph 5.48)

AND

Local authority satisfied that sufficient resources are available to deal with clearance area (section 289(4)(b))

Area to be defined on a map (section 289(3)(a))

(*continued on next page*)

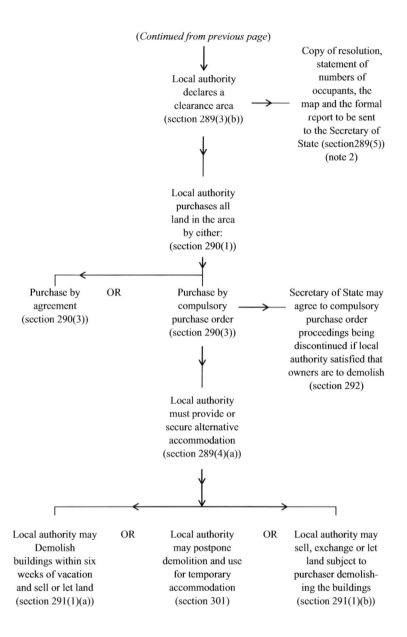

(*Continued from previous page*)

Local authority declares a clearance area (section 289(3)(b))

Copy of resolution, statement of numbers of occupants, the map and the formal report to be sent to the Secretary of State (section289(5)) (note 2)

Local authority purchases all land in the area by either: (section 290(1))

Purchase by agreement (section 290(3))   OR   Purchase by compulsory purchase order (section 290(3))

Secretary of State may agree to compulsory purchase order proceedings being discontinued if local authority satisfied that owners are to demolish (section 292)

Local authority must provide or secure alternative accommodation (section 289(4)(a))

Local authority may Demolish buildings within six weeks of vacation and sell or let land (section 291(1)(a))   OR   Local authority may postpone demolition and use for temporary accommodation (section 301)   OR   Local authority may sell, exchange or let land subject to purchaser demolishing the buildings (section 291(1)(b))

Notes

1. No order has yet been made.
2. The Secretary of State has no powers of intervention at this stage; these powers come with any subsequent compulsory purchase order.
3. After consideration of representations, the local authority may declare a clearance area as proposed, or with the exclusion of some residential buildings, or may decide not to proceed and take other action.

- whether it would be necessary to acquire land surrounding or adjoining the area and whether added land can be acquired by agreement;
- the existence of any listed buildings – listed buildings should only be included in a clearance area in exceptional circumstances and only when building consent has been given;
- the results of statutory consultation – see below;
- the arrangements necessary for the rehousing of displaced occupants and the extent to which occupants are satisfied with those arrangements;
- the impact of clearance on, and the scope for relocating, commercial premises;
- the suitability of the proposed after-use of the site, the needs of the wider neighbourhood and the socio-economic benefits which after-use would bring;
- the degree of support from local residents;
- the extent to which such after-use would create private investment in the area (Enforcement Guidance, paragraph 5.48).

Consultation

Before making a clearance area, the local authority must consult with those people who will be directly affected. This is to be done by service of a notice on every person who has an interest, including freeholders and sub-lease holders, and on all residential occupiers. A notice must also be placed in at least two local newspapers. In both cases, the local authority is to invite representations within a period of not less than 28 days (section 289).

Options after consultation

The local authority must consider all representations made during the stated period and must then take one of the following options:

1. Declare the area to be a clearance area.
2. Declare a clearance area with the exclusion of certain residential buildings.
3. Not declare a clearance area and consider other options (in relation to category 1 hazards, the local authority is still required to take action under section 5 of the Housing Act 2004) (section 289(2F)).

A map defining the proposed clearance area containing certain categories of buildings must be prepared before the resolution declaring the clearance area is passed (section 289(3)(a)).

Implementation

A copy of the resolution together with a statement of numbers of occupants, the map and the formal report to the council are to be sent to the Secretary of State, although there is no power for him to intervene at this stage.

The clearance area is implemented by the purchase of all land in the area by the local authority, either by agreement or by the use of a compulsory purchase order. The local authority may also acquire land that surrounds the area and is needed to secure an area of convenient shape and dimension and land necessary for satisfactory development.

Compulsory purchase orders are made under section 290 and are regulated by the Acquisition of Land Act 1981 and regulations made under it. The compulsory purchase order will require Secretary of State confirmation or modification before proceeding (sections 290–291).

Temporary use of residential buildings

Although purchased for demolition, if the local authority considers that any of the residential buildings can provide adequate temporary accommodation, with or without the carrying out of works, it may postpone demolition for such period as they determine. Other buildings in the area may have demolition postponed where:

- the demolition of the residential buildings has been postponed;
- they are required to support a residential building;
- there is some other special reason connected with the provision of temporary accommodation in the area (section 301).

Demolition

Having secured the land by agreement or compulsory purchase order, the local authority must arrange for the vacation and demolition of buildings within six weeks of vacation or such other period as the local authority determines to be reasonable. The land may subsequently be sold, let or exchanged. Alternatively, the local authority may sell or exchange the land with a condition that the purchaser undertakes the demolition (section 291).

Compensation

Open market value is payable for all dwellings purchased via a compulsory purchase order. Home loss and disturbance payments are also payable (Land Compensation Act 1973, sections 29–32 and 37–38). The Regulatory Reform (Housing Assistance) Order 2002 makes provision for a local authority to provide financial assistance where there has been demolition and contains power to purchase by agreement.

COMPULSORY PURCHASE

References

Housing and Planning Act 2016, sections 172 to 177
Planning and Compulsory Purchase Act 2004
Housing (Scotland) Act 1987
Housing (Northern Ireland) Order 1981

Extent

These provisions apply to England and Wales. Similar provisions apply in Scotland and Northern Ireland.

Scope

The procedure to compulsory purchase property is mainly dealt with by legal teams in local authorities. However, Environmental Health teams will be involved initially by using their powers to inspect property in preparation for compulsory purchase and redevelopment.

Right to enter and survey land

A person authorised in writing by an acquiring authority may enter and survey or value land in connection with a proposal to acquire an interest in or a right over land (section 172(1)).

The person may only enter and survey or value land at a reasonable time, and may not use force unless a Justice of the Peace has issued a warrant authorising the person to do so (section 172(2)).

The person must, if required when exercising or seeking to exercise the power, produce evidence of the authorisation, and a copy of any warrant issued (section 172(3)).

An authorisation may relate to the land which is the subject of the proposal or to other land (section 172(4)).

If the land is unoccupied or the occupier is absent from the land when the person enters it, the person must leave it as secure against trespassers as when the person entered it (section 172(5)).

Warrant authorising use of force to enter and survey land

A Justice of the Peace may issue a warrant authorising a person to use force to enter land if satisfied that another person has prevented or is likely to prevent the exercise of that power, and that it is reasonable to use force in the exercise of that power (section 173(1)).

The force that may be authorised by a warrant is limited to that which is reasonably necessary (section 173(2)).

A warrant authorising the person to use force must specify the number of occasions on which the authority can rely on the warrant when entering and surveying or valuing land (section 173(3)).

The number specified must be the number which the Justice of the Peace considers appropriate to achieve the purpose for which the entry and survey or valuation are required (section 173(4)).

Any evidence in proceedings for a warrant must be given on oath (section 173(5)).

Notice of survey and copy of warrant

The acquiring authority must give every owner or occupier of land at least 14 days' notice before the first day on which the authority intends to enter the land (section 174(1)).

Notice given must include a statement of the recipient's rights, and a copy of the warrant, if there is one (section 174(2)).

If the authority proposes to do any of the following, the notice must include details of what is proposed:

(a) searching, boring or excavating;
(b) leaving apparatus on the land;
(c) taking samples;
(d) an aerial survey;
(e) carrying out any other activities that may be required to facilitate compliance with the instruments mentioned in subsection (5) (section 174(3)).

These instruments are:

(a) Council Directive 85/337/EEC of 27 June 1985 on the assessment of the effects of certain public and private projects on the environment, as amended from time to time,

(b) Council Directive 92/43/EC of 21 May 1992 on the conservation of natural habitats and of wild fauna and flora, as amended from time to time, or

(c) any EU instrument from time to time replacing all or part of those Directives.

If the authority obtains a warrant after giving notice it must give a copy of the warrant to all those to whom it gave that notice (section 174(4)).

Enhanced authorisation procedures etc. for certain surveys

A written authorisation from the appropriate minister is required before a person enters and surveys or values land if:

(a) the land is held by a statutory undertaker,

(b) within the notice period, the statutory undertaker objects to the proposed entry and survey or valuation in writing to the acquiring authority, and

(c) the objection is that the proposed entry and survey or valuation would be seriously detrimental to the statutory undertaker carrying on its undertaking (section 175)

Right to compensation after entry on or survey of land

A person interested in land is entitled to compensation from the acquiring authority for damage as a result of the exercise of the power (section 176(1)).

Any disputes relating to compensation are to be determined by the Upper Tribunal (section 176(2)).

Offences in connection with powers to enter land

A person who without reasonable excuse obstructs another person in the exercise of this power commits an offence (section 177(1)).

A person who commits an offence is liable on summary conviction to a fine not exceeding level 3 on the standard scale (section 177(2)).

A person commits an offence if the person discloses confidential information, obtained in the exercise of this power, for purposes other than those for which the power was exercised (section 177(3)).

A person who commits an offence is liable on summary conviction to a fine, on conviction on indictment to imprisonment for a term not exceeding 2 years or to a fine, or both (section 177(4)).

Definitions

Confidential information means information –

(a) which constitutes a trade secret, or

(b) the disclosure of which would or would be likely to prejudice the commercial interests of any person (section 177(5)).

FC73 Compulsory purchase – Housing and Planning Act 2016, sections 172 to 177

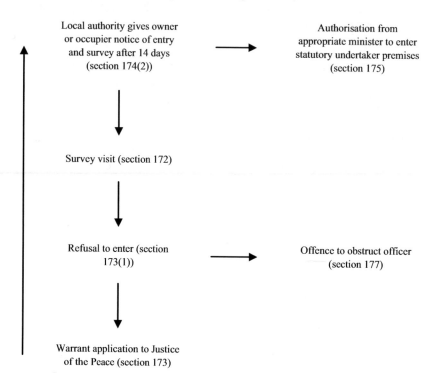

Local authority gives owner
or occupier notice of entry
and survey after 14 days
(section 174(2))

Authorisation from
appropriate minister to enter
statutory undertaker premises
(section 175)

Survey visit (section 172)

Refusal to enter (section
173(1))

Offence to obstruct officer
(section 177)

Warrant application to Justice
of the Peace (section 173)

Chapter 3

HOUSING LICENSING

SELECTIVE LICENSING OF RESIDENTIAL ACCOMMODATION

References

Housing Act 2004, sections 79 to 100
Selective licensing in the private rented sector – a guide for local authorities, March 2015

Extent

These provisions apply to England and Wales.

Scope

The Housing Act 2004 provides for the mandatory licensing of privately rented property including certain houses in multiple occupation (see p. 364), and the discretionary licensing of other houses in multiple occupation. Here the provisions relate to selective licensing of privately rented dwellings.

Designation of selective licensing areas

Local authorities can designate all or part of their district for selective licensing of rented dwellings (section 80(1)).
Before making this designation, it must consider:

(a) that the area is, or is likely to become, an area of low housing demand; and
(b) that making a designation will, when combined with other measures, contribute to the improvement of the social or economic conditions in the area (section 80(3)).

It must take into account:

(a) the value of residential premises in the area, in comparison to the value of similar premises in other areas which the authority consider to be comparable (whether in terms of types of housing, local amenities, availability of transport or otherwise);
(b) the turnover of occupiers of residential premises;
(c) the number of residential premises which are available to buy or rent and the length of time for which they remain unoccupied (section 80(4));

(d) whether the area is experiencing a significant and persistent problem caused by anti-social behaviour;

(e) whether some or all of the private sector landlords who have let premises in the area are failing to take action to combat the problem that it would be appropriate for them to take; and

(f) whether making a designation will, when combined with other measures taken in the area by the local authority, lead to a reduction in, or the elimination of, the problem (section 80(6)).

Before making a designation, the local authority must:

(a) take reasonable steps to consult persons who are likely to be affected by the designation; and

(b) consider any representations made in accordance with the consultation and not withdrawn (section 80(9)).

The local authority must also ensure that any exercise of the power is consistent with the authority's overall housing strategy (section 81(2)) and must also seek to adopt a coordinated approach in dealing with homelessness, empty properties and antisocial behaviour.

It must consider whether there are any other courses of action available to them that might provide an effective method of achieving the objective or objectives that the designation would be intended to achieve, and they consider that making the designation will significantly assist them to achieve the objective or objectives. The designation must be submitted to and confirmed by the Secretary of State or comply with prior approved schemes (section 82).

If confirmed, a designation comes into force on a date specified by the local authority, no earlier than three months after the date on which the designation is confirmed (section 82(3) and 82(4)).

Notification

As soon as the designation is confirmed or made, the local authority must publish a notice stating:

(a) that the designation has been made;

(b) whether the designation was required to be confirmed or that a general approval applied to it (giving details of the approval in question);

(c) the date on which the designation is to come into force; and

(d) any other prescribed information (section 83(2)).

Copies of the designation, and prescribed information must be made available (section 83(3)).

Duration, review and revocation of designations

A designation ceases to have effect at the time specified in the designation, no later than five years after the designation came into force. The local authority must from time to time review the operation of any designation made by them.

On review, the local authority may revoke the designation ceasing to have effect on the date specified. The local authority must publish notice of the revocation (section 84).

The designation applies to all privately rented dwellings except houses in multiple occupation in the designated area. The local authority must take all reasonable steps to secure that applications for licences are made to them for houses in their area which are required to be licensed (section 85).

Temporary exemption

A person having control of or managing a house which is required to be licensed may notify the local authority of his intention to take steps to ensure the house is no longer required to be licensed. If the local authority thinks fit, it may serve a temporary exemption notice in respect of the house.

A temporary exemption notice is in force for three months. If the local authority receives a further notification of intention to take steps to ensure licensing is not required, a second temporary exemption notice may be served for a further three months (section 86).

If the authority decides not to serve a temporary exemption notice, they must without delay serve on the person concerned a notice informing him of:

(a) the decision;
(b) the reasons for it and the date on which it was made;
(c) the right to appeal against the decision; and
(d) the period within which an appeal may be made.

The person concerned may appeal to the Appropriate Tribunal against the decision within 28 days of the decision. The tribunal may confirm or reverse the decision of the authority, and if it reverses the decision, must direct the local authority to issue a temporary exemption notice with effect from such date as the tribunal directs (section 86).

Applications for licences

An application for a licence must be made to the local authority accompanied by a fee fixed by the authority. In fixing the fee for application, the local authority should consider all costs incurred in carrying out their functions under the licensing scheme (section 87).

Grant or refusal of licence

On receiving an application for licensing, the authority must either grant or refuse a licence (section 88(1)).

If the local authority is satisfied:

(a) that the proposed licence holder is a fit and proper person and is the most appropriate person to be the licence holder;
(b) that the proposed manager of the house is either the person having control of the house, or a person who is an agent or employee of the person having control of the house;
(c) that the proposed manager of the house is a fit and proper person; and
(d) that the proposed management arrangements for the house are satisfactory,

they may grant a licence either to the applicant or to some other person, if both he and the applicant agree (section 88).

Tests for fitness etc. and satisfactory management arrangements

In deciding whether a person is a fit and proper person to be the licence holder or the manager of the house, the local authority must have regard to evidence that shows that the person has:

(a) committed offences involving fraud, other dishonesty, violence or drugs, or of a sexual nature;

(b) practised unlawful discrimination on grounds of sex, colour, race, ethnic or national origins or disability in, or in connection with, the carrying on of any business; or

(c) contravened any provision of the law relating to housing or of landlord and tenant law.

(d) that no banning order under section 16 of the Housing and Planning Act 2016 is in force against a person who owns or leases an estate or interest in the house or part of it (section 88).

They must also consider evidence that shows that any person associated or formerly associated with the person (whether on a personal, work or other basis) has done any of the things above and it appears to the authority that the evidence is relevant to the question of whether the person is a fit and proper person to be the licence holder or the manager of the house.

In deciding whether the proposed management arrangements for the house are satisfactory, the local authority must have regard to:

(a) whether any person proposed to be involved in the management of the house has a sufficient level of competence to be involved;

(b) whether any person proposed to be involved in the management of the house (other than the manager) is a fit and proper person to be involved; and

(c) whether any proposed management structures and funding arrangements are suitable (section 89).

Licences

The local authority may attach conditions to a licence and they may include:

(a) restrictions or prohibitions on the use or occupation of particular parts of the house by persons occupying it;

(b) requiring the taking of reasonable and practicable steps to prevent or reduce antisocial behaviour by persons occupying or visiting the house;

(c) requiring facilities and equipment to be made available in the house;

(d) requiring such facilities and equipment to be kept in repair and proper working order;

(e) requiring works needed for any such facilities or equipment to be made available (section 90(1) to (3)).

Local authorities must not use licensing to remove or reduce category 1 or category 2 hazards in the house but this does not prevent the authority from imposing licence conditions relating to the installation or maintenance of facilities or equipment (section 90).

Licences apply to only one house and come into force at the specified time and, unless previously terminated or revoked, continue in force for the specified period for at least five years. A licence cannot be transferred to another person. If the holder of the licence dies while the licence is in force, the licence ceases to be in force on his death. But for the three months after the licence holder's death, the house is to be treated as if on that date a temporary exemption notice had been served in respect of the house. If, at any time during that period, the personal representatives of the licence holder requests the local authority to do so, the authority may serve on them a notice which has the same effect as a temporary exemption notice (section 91).

Variation of licences

The local authority may vary a licence with the agreement of the licence holder, or if they consider that there has been a change of circumstances since the licence was granted, including any discovery of new information. A variation made with the agreement of the licence holder takes effect at the time when it is made. Otherwise, it comes into force 28 days later, unless there is an appeal against the variation.

The local authority can vary the licence on its own initiative or on application from the licence holder or a 'relevant person' (who has an estate or interest in the house concerned), or a person managing or having control of the house or a person who is subject to any restriction or obligation imposed by the licence (section 92).

Power to revoke licences

The local authority may revoke a licence:

(a) with the agreement of the licence holder;
(b) where the licence holder or any other person has committed a serious breach of a condition of the licence;
(c) where the licence holder is no longer considered a fit and proper person to be the licence holder; and
(d) where the management of the house is no longer being carried on by persons who are fit and proper persons to be involved in its management.

A revocation made with the agreement of the licence holder takes effect at the time when it is made. Otherwise, it takes effect 28 days from the notification of revocation, unless there is an appeal against the revocation to the Appropriate Tribunal.

The local authority can revoke a licence on its own initiative or on application made by the licence holder or a 'relevant person' (see above) (section 93).

The local housing authority must revoke a licence if a banning order is made under section 16 of the Housing and Planning Act 2016 against the licence holder (a person who owns or leases an estate or interest in the house or part of it). The notice served by the local housing authority under paragraph 24 of schedule 5 must specify when the revocation takes effect. The revocation must not take effect earlier than the end of the period of 7 days beginning with the day on which the notice is served (section 93A).

The procedural requirements and appeals against licence decisions and contained in schedule 5 of the Housing Act 2004, which deals with grant, refusal, variation or revocation of licences and appeals against licence decisions.

Offences

It is an offence not to license a house which should be licensed and to fail to comply with any condition of the licence.

FC74 Designation of areas for selective licensing of rented dwellings – sections 80 to 84 Housing Act 2004

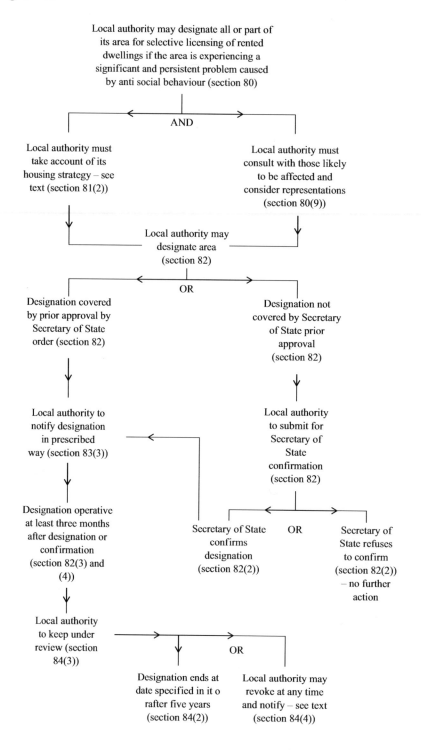

FC75 Selective licensing of rented dwellings – temporary exemption notices – section 86 Housing Act 2004

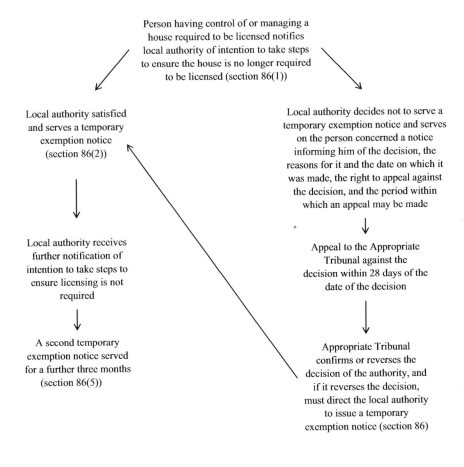

Person having control of or managing a house required to be licensed notifies local authority of intention to take steps to ensure the house is no longer required to be licensed (section 86(1))

Local authority satisfied and serves a temporary exemption notice (section 86(2))

Local authority receives further notification of intention to take steps to ensure licensing is not required

A second temporary exemption notice served for a further three months (section 86(5))

Local authority decides not to serve a temporary exemption notice and serves on the person concerned a notice informing him of the decision, the reasons for it and the date on which it was made, the right to appeal against the decision, and the period within which an appeal may be made

Appeal to the Appropriate Tribunal against the decision within 28 days of the date of the decision

Appropriate Tribunal confirms or reverses the decision of the authority, and if it reverses the decision, must direct the local authority to issue a temporary exemption notice (section 86)

FC76 Selective licensing of rented dwellings – sections 87 to 93 Housing Act 2004

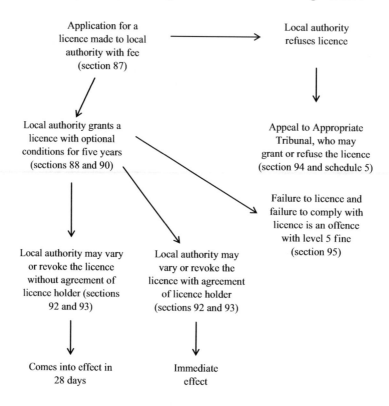

Notes

1. Local authorities must not use licensing to remove or reduce category 1 or category 2 hazards in the house (section 90).
2. The local authority can vary the licence on its own initiative or on application from the licence holder or a 'relevant person', or a person managing or having control of the house or a person who is subject to any restriction or obligation imposed by the licence (section 92).

It is a defence that a temporary exemption notice is in force or an application for a licence had been made. It is also a defence to have a reasonable excuse for having control of or managing the house or for failing to comply with the condition.

On summary conviction, offences are subject to a fine (section 95).

Definitions

Private sector landlord does not include a non-profit registered provider of social housing or a registered social landlord (section 80).

PRIVATE LANDLORDS' ACCREDITATION SCHEMES

A Landlord Accreditation Scheme aims to train potential and existing landlords in the various areas of managing property, as well as promoting good practice across the private rented sector. They will give tenants peace of mind that their chosen landlord has been vetted and trained.

A private landlords' accreditation scheme gives recognition to private landlords who offer good-quality, private rented accommodation. It usually covers all types of private rented accommodation – from purpose-built flats and houses in multiple occupation (HMOs), through to larger single houses and uses standards based on existing legislation, with additional facilities normally considered to constitute a satisfactory decent standard.

The objectives of a scheme are to recognise and encourage landlords who are prepared to provide good-quality accommodation at an appropriate rent and to improve and promote the public image of the private rented sector and private rented sector housing conditions. Schemes also aim to improve liaison and communication between landlords and local authorities and provide tenants with confidence in the quality and management of the accommodation they are renting.

Landlords in the scheme can gain a market advantage over other providers when attracting new tenants; can generate a better image with potential lenders; gain recognition that the landlord provides a high standard of accommodation, with fewer transient tenants, which means regular on-time rent payments; and have lower management costs and fewer void periods.

Tenants get clearer information on properties available to let and it is easier for them to identify good-quality accommodation. Schemes give confidence that landlords are professional and reputable, and that properties are safe, secure and free from hazards, as identified under the Housing Health and Safety Rating System, with improved energy efficiency measures.

Accreditation

Accreditation provides official recognition that landlord accommodation is of good quality, free from hazards and well managed. Accredited properties are those which have been inspected by an accreditation officer, who has certified that they meet the requirements stated in the scheme's Code of Standards, that the landlord displays good management standards and practices and that the relevant physical standards relating to the property have been verified.

An accreditation certificate is issued to the landlord confirming that the property meets the required standards. In addition, a property information folder is also presented to the landlord – this is to be kept at the property and contains useful information, contact details and copies of all relevant safety certificates.

Schemes are usually established by a local authority in partnership with landlords and others. Officers from the council oversee the promotion and day-to-day running of the scheme.

Schemes are usually voluntary, some with no cost involved in applying for accreditation. If all of the landlords' properties do not currently comply with the relevant standards, they can still participate in the scheme because it is property-specific. It is quite possible to have both accredited and non-accredited properties in a landlord's portfolio – though landlords will be actively encouraged to work towards accredited status for all of their properties.

Local authorities will inspect non-accredited properties, identify any work which is needed to bring them up to the accredited standard and, if necessary, prepare and issue schedules of work.

Standards can be chosen for each scheme and are usually drawn up by the local authority. Property standards are defined and detailed guidance is given.

Accreditation status will normally last for three years. At the end of this period, further checks and a re-inspection of the property will be undertaken before the accredited status is renewed.

Relevant certificates for each property will also be required annually and spot-check inspections of properties may be carried out within the three-year accreditation period.

Accredited status may be suspended where:

- the property has fallen below the required standard and the landlord does not undertake the necessary work;
- relevant certification is not submitted;
- the landlord contravenes the qualifying criteria for the scheme;
- any information provided supporting an application was found to be fraudulent or inaccurate.

Landlords of accredited properties are expected to:

- at all times comply with the management Code of Practice agreed with the accreditation officer;
- comply with the accreditation scheme's Code of Standards;
- once accreditation is achieved, maintain the property to those standards;
- ensure that any staff or persons acting on their behalf comply with the requirements of the scheme and any Codes of Practice.

All breaches of the requirements of the scheme in relation to the property standards will be reviewed and, if the landlord does not take satisfactory remedial action, then the property concerned will be removed from the accreditation scheme.

If the breach relates to the management standards and Code of Practice, then the incident will be referred to the accreditation officer, who will consider the nature of the breaches, the representations of the landlord and any other relevant parties, and decide on any appropriate sanctions.

Some schemes operate a complaint service. An accreditation officer will act as a mediator for complaints between landlords and tenants. The complainant may be either the landlord or the tenant.

If a complaint is from the tenant:

- a visit will be made to the tenant to determine the nature of the complaint and decide whether it is justified;
- discussions will take place with the landlord to inform them of the nature of the complaint;
- the Accreditation Officer will agree with the landlord any remedial works to be carried out and a timescale for those works.

If this fails, then the complaint will be passed to the relevant local authority department for statutory action.

If the complaint concerns management standards (and not the physical state of the property) then the complaint will be referred to the accreditation officer.

If a complaint is from the landlord, the complaint will be referred to the accreditation officer. The accreditation officer will discuss the nature of the problem with the landlord and provide the appropriate advice, support and assistance as necessary to attempt to resolve the problem.

Landlord accreditation or Registration schemes apply in all home countries in one form or another.

Chapter 4

HOUSES IN MULTIPLE OCCUPATION

This section deals exclusively with houses in multiple occupation, but the Housing Health and Safety Rating System and its enforcement procedures in the previous sections also apply to houses in multiple occupation, except in those cases where the existence of a management order removes it from that procedure. These situations are noted in the text relating to that procedure.

HOUSES IN MULTIPLE OCCUPATION: DEFINITIONS

Section 254 of the Housing Act 2004 defines three categories of privately rented buildings that are deemed to be houses in multiple occupation. These are as follows:

1. Converted blocks of flats as defined in section 257, which are buildings or parts of buildings which have been converted into and consist of self-contained flats and:

   (a) the works of conversion did not, and still do not, comply with appropriate building standards of the Building Regulations; and

   (b) less than two-thirds of the self-contained flats are owner-occupied.

   These are known as 'poorly converted buildings' or PCBs. Their inclusion as houses in multiple occupation will be made by regulations to be made by the Secretary of State (note – Houses in Multiple Occupation (Certain Blocks of Flats) (Modifications to the Housing Act 2004 and Transitional Provisions for section 257 HMOs) (England) Regulations 2007).

2. Buildings or parts thereof which are the subject of a house in multiple occupation declaration under section 255 – see FC77 below.

3. Buildings or parts thereof that meet one of the tests set out in section 254. These are as follows:

   (a) The 'standard test' where persons who do not form a single household (this term is defined in section 258), and occupy as their sole or main residence, share basic amenities in living accommodation that is not a self-contained flat or flats, or that the living accommodation lacks one or more of those amenities. In order to be a house in multiple occupation, the property must be used as the tenant's only or main residence and it should be used solely or mainly to house tenants. Properties let to students and migrant workers will be treated as their only or main residence and the same applies to properties that are used as domestic refuges (see section 259 and regulation 5 of the Licensing and Management of HMOs and other Houses (Miscellaneous Provisions) (England) Regulations 2006).

(b) The 'self-contained flat test' where, in part of a building, persons of different households share basic amenities or the flat lacks one or more of those amenities.

(c) The 'converted building test' where privately rented converted building unrelated occupiers share basic facilities, including flats where the basic amenities for the exclusive use of the occupant are situated outside of the main living accommodation.

The basic amenities referred to are:

- a toilet;
- personal washing facilities;
- cooking facilities (Housing Act 2004, section 254).

Persons do not form a single household unless they are members of the same family (defined in detail in section 258(3)–(5)) or they form a prescribed relationship under the Licensing and Management of Houses in Multiple Occupation and Other Houses (Miscellaneous Provisions) (England) Regulations 2006. These 'prescribed relationships' include persons carrying out domestic work, carers and fostering families (section 258).

In order to satisfy any of the three tests above, the property must be used as the tenant's only or main residence and it should be used solely or mainly to house tenants. (But see houses in multiple occupation declarations below.)

A reasonable precis of all of this is provided in 'Houses in multiple occupation (HMO): Licensing scope and schemes – May 2006' (DCLG).

A house in multiple occupation, subject to exemptions (see below) is either:

(a) an entire house which is let to three or more tenants who form two or more households and who share basic facilities;

(b) a house that has been converted entirely into bedsits or other non-self-contained accommodation and which is let to three or more tenants and who form two or more households and who share basic facilities;

(c) a converted house which contains one or more flats which are not wholly self-contained and which is occupied by three or more tenants who form two or more households;

(d) a building which is converted entirely into self-contained flats if the conversion did not meet the building regulation standards and more than one-third of the flats are let on short-term tenancies.

Exemptions

In relation to the licensing, management and overcrowding provisions of part 2, but not the enforcement provisions of part 1, the following properties are not to be regarded as houses in multiple occupation:

(a) those managed by local authorities, registered social landlords and other specified bodies (see the Houses in Multiple Occupation (Specified Educational Establishments) (England) Regulations 2012);

(b) those subject to other regulatory schemes specified in schedule 1 of the Licensing and Management of Houses in Multiple Occupation and other Houses (Miscellaneous Provisions) (England) Regulations 2006;

(c) student halls of residence;

(d) buildings occupied by religious communities;

(e) those occupied predominantly by owner-occupiers at a level to be specified in regulations;

(f) those occupied only by persons who form two households (section 254(5) and schedule 14).

Section 254(6) empowers the Secretary of State by regulations to alter any of these definitions of a house in multiple occupation if it is considered appropriate.

HOUSES IN MULTIPLE OCCUPATION DECLARATIONS

Reference

Housing Act 2004, sections 255–256

Extent

These provisions apply to England and Wales.

Scope

To satisfy any of the three tests which define a house in multiple occupation (see above), all occupants need to live there as their sole or main residence. Because of the rapidly changing occupancy of many such properties, this may lead to difficulties in the enforcement of standards. This procedure allows a local authority to designate a house in multiple occupation where the building, or part of it, is occupied by persons who do not form a single household, and occupancy as their sole or main residence forms only a 'significant use' of that accommodation. 'Significant use' is not defined. Such designation then removes doubt about its status, and the provisions of the housing authorities relating to houses in multiple occupation can then be applied (section 255(1)–(3)).

Service of house in multiple occupation declaration

The declaration is made by service of a notice on all 'relevant persons' (see below) within seven days of the decision which:

1. states the date of the decision to make the declaration;
2. states the day on which it will come into force if no appeal is made being not less than 28 days after the decision;
3. sets out the rights of appeal (section 255(4)).

Relevant persons

For the purposes of this procedure, these are:

(a) those having an estate or interest but not a tenant under a lease with three years or less unexpired;

(b) those managing or having control (section 255(12)).

Appeals

Any 'relevant person' may appeal to the Appropriate Tribunal within 28 days of the date of decision. And in this event the effect of the declaration is delayed until the appeal is

FC77 House in multiple occupation (HMO) declarations – sections 255 and 256 Housing Act 2004

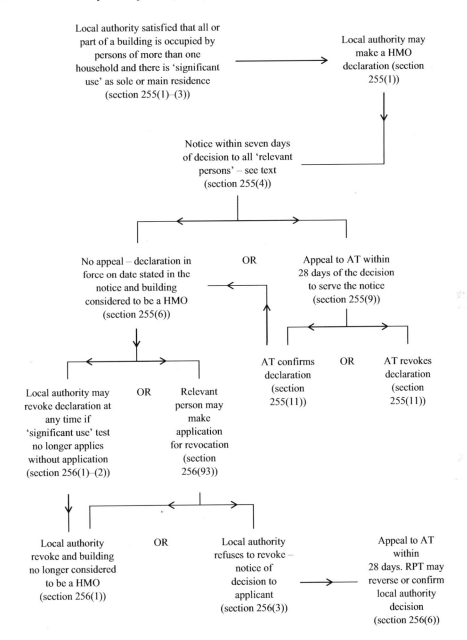

Local authority satisfied that all or part of a building is occupied by persons of more than one household and there is 'significant use' as sole or main residence (section 255(1)–(3))

Local authority may make a HMO declaration (section 255(1))

Notice within seven days of decision to all 'relevant persons' – see text (section 255(4))

No appeal – declaration in force on date stated in the notice and building considered to be a HMO (section 255(6))

OR

Appeal to AT within 28 days of the decision to serve the notice (section 255(9))

AT confirms declaration (section 255(11))

OR

AT revokes declaration (section 255(11))

Local authority may revoke declaration at any time if 'significant use' test no longer applies without application (section 256(1)–(2))

OR

Relevant person may make application for revocation (section 256(93))

Local authority revoke and building no longer considered to be a HMO (section 256(1))

OR

Local authority refuses to revoke – notice of decision to applicant (section 256(3))

Appeal to AT within 28 days. RPT may reverse or confirm local authority decision (section 256(6))

Note

1. AT in this flow chart = Appropriate Tribunal.

determined. The tribunal may confirm or revoke the declaration. If confirmed, the declaration comes into effect 28 days after the tribunal decision (section 255(7) and (9)).

Revocation

The local authority may revoke a declaration at any time, either on its own initiative or following representation from a 'relevant person', if satisfied that the 'significant use' test is no longer fulfilled. There is no requirement for notification of a decision to revoke. There is, however, a requirement for the local authority to serve notice on any person who made an application for revocation where that request is refused. In this case, a notice must be served giving:

(a) the decision and the date on which it was made;
(b) the reasons for that decision;
(c) the rights of appeal.

A person served with a refusal notice may, within 28 days, appeal to the Appropriate Tribunal, which may confirm or reverse the decision by revoking the declaration (section 256).

LICENSING OF HOUSES IN MULTIPLE OCCUPATION

References

Housing Act 2004, part 2
Management of Houses in Multiple Occupation (England) Regulations 2006
The Licensing of Houses in Multiple Occupation (Prescribed Descriptions) (England) Order 2018 (Prescribed Description Order)
The Licensing and Management of Houses in Multiple Occupation and Other Houses (Miscellaneous Provisions) (England) Regulations 2006 (Miscellaneous Provisions Regulations) (as amended)
The Housing Act 2004: Licensing of Houses in Multiple Occupation and Selective Licensing of Other Residential Accommodation (England) General Approval 2015
Enforcement Guidance, HHSRS, ODPM February 2006, part 6, paragraphs 6.1–6.5
Housing (Scotland) Act 2006
Houses in Multiple Occupation Act (Northern Ireland) 2016

Extent

These provisions apply to England only but there are similar provisions for Wales, Scotland and Northern Ireland.

Scope

Houses in multiple occupation with five or more tenants in two or more separate households meeting the standard tests must be licensed with the local authority, except those:

(i) where an exemption notice is in force (FC82);
(ii) subject to an interim or final management order (FC80 and FC81) (section 55(2), 61(1) and Prescribed Description Order 2018).

FC78 Licensing of houses in multiple occupation (HMOs) – sections 55 to 71 Housing Act 2004

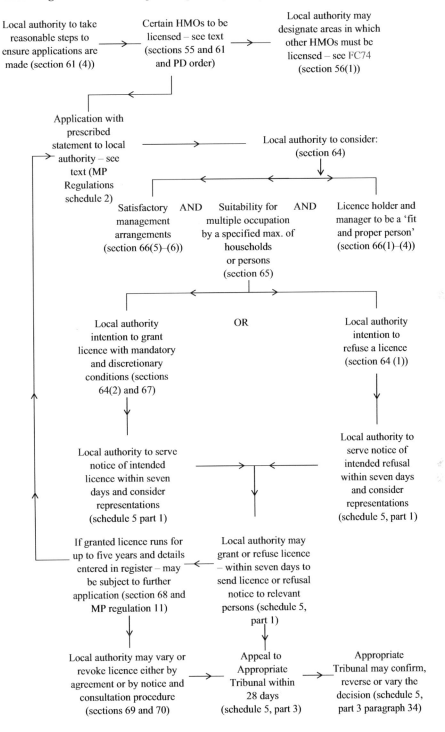

Notes
1. PD order = The Licensing of Houses in Multiple Occupation (Prescribed Descriptions) (England) Order 2018.
2. MP regulations = The Licensing and Management of Houses in Multiple Occupation and other Houses (Miscellaneous Provisions) (England) Regulations 2006.

The intention behind the licensing requirements for this group of houses in multiple occupation is that they present the highest risk to occupiers and therefore need to have additional controls to ensure adequate standards.

In addition, local authorities may designate areas of their district in which there is to be licensing for other types of houses in multiple occupation. The procedure for designation is at FC79 but the licensing procedure is the same as for other houses in multiple occupation (section 56(1)).

A licence for a house in multiple occupation is defined as 'authorising occupation of the house concerned by not more than a maximum number of households or persons specified in the licence' (section 61(2)).

Local authority duties connected with licensing

The local authority is required to:

1. promote the implementation of licensing – this includes a duty to take reasonable steps to ensure that applications are made where they appear to be required;
2. ensure that it deals with applications promptly;
3. satisfy itself that there are no part 1 enforcement functions remaining to be discharged in respect of a licensed house in multiple occupation as soon as possible but in any event not longer than five years from the first application (section 55(5) and 61(4)).

Applications for licences

The requirements are prescribed in section 63 and the Miscellaneous Provisions Regulations 2006 and these include:

- a prescribed statement from the local authority requiring the applicant to inform all persons with an interest etc. of the application and details relating to it (schedule 2, paragraph 1 of the 2006 regulations);
- details to be included in the application (regulation 7(2) and schedule 2);
- payment of a fee at the discretion of the local authority but subject to a maximum specified in the regulations (section 63(3)).

Grounds for consideration

A licence must be granted if the following criteria are satisfied:

(a) The house is suitable (see below) for occupation by a certain number of persons or households as specified in the application or by the local authority, or can be rendered suitable by the imposition of conditions.
(b) The proposed licence holder is a fit and proper person (see below) as well as being the most appropriate person to be granted a licence.
(c) The proposed manager is the person having control of the house or an agent or employee of that person and is also a fit and proper person (see below).
(d) The proposed management arrangements are satisfactory (see below).

If these criteria are not met, the application must be refused (section 64).

Suitability for multiple occupation

A house in multiple occupation is deemed reasonably suitable for occupation by a particular maximum of households or persons if it meets the standards contained in schedule 3 of the Miscellaneous Provisions Regulations 2006. These standards relate primarily to:

  (i) heating;
 (ii) washing facilities (bathrooms, toilets, washbasins and showers);
(iii) kitchens (areas for food storage, preparation and cooking);
(iv) units of living accommodation without shared basic facilities; and
 (v) fire precautions.

If it does not meet these standards, a licence cannot be granted. A local authority has the discretion to decide that the house is not suitable for occupation even if it complies with the prescribed standards and may require higher standards in relation to a maximum number of persons (section 65).

Fit and proper person

In determining whether a person is a fit and proper person to be a licence holder or manager, the local authority must have regard to whether that person, or a relevant associate, e.g. a spouse or business partner, has committed offences involving fraud, dishonesty, violence, drugs or sexual offences, unlawful discrimination in business, contravention of housing law or breach of an approved code of management, e.g. Housing (Codes of Management Practice) (Student Accommodation) (England) Order 2010 (section 66(1)–(4)).

Satisfactory management arrangements

In deciding if the proposed management arrangements are satisfactory, the local authority must have regard, amongst other matters, to:

(a) the competence of the manager;
(b) fitness of other persons involved in the management structure;
(c) management structure; and
(d) funding (section 66(5) and (6)).

Grant and refusal of licences

Before granting a licence, the local authority must:

(a) serve notice on the applicant and other relevant persons of its intention to grant the licence, together with a copy of it;
(b) consider any representations made within the period specified.

If, following this consultation, the local authority intends to modify the proposed licence, they must again serve notice and consider representations. Before refusing to grant a licence, the local authority must:

(a) serve notice on the applicant and other relevant persons of its intention; and

(b) state the reasons for refusal and indicate the length of the consultation period;

(c) consider representations made.

Following a grant or refusal, the local authority must serve notice within seven days on the applicant and other relevant persons setting out:

(a) the local authority's decision;

(b) the reasons for the decision and the date on which it was made; and

(c) rights of appeal and the appeal period (section 71 and schedule 5, part 1).

Licence conditions

Mandatory conditions

These are set out in schedule 4 and deal mainly with the safety of occupiers, including furniture and electrical, gas and fire safety appliances. These must be included in each licence (section 67(3) and schedule 4).

Discretionary conditions

The local authority may include any conditions relating to the management, use and occupation of the house and its contents and condition. These might include:

(a) restrictions on the use of particular parts;

(b) the reduction of antisocial behaviour;

(c) the installation and maintenance of facilities and equipment to meet the standards required under section 65 above;

(d) carrying out works to facilities and equipment within specified periods.

Local authorities are not to use licence conditions to deal with matters of health and safety but should use the enforcement procedures of part 1 (section 67).

General requirements and duration

A licence can only relate to a single house in multiple occupation and is valid for the period specified in it, which cannot exceed five years. They are not transferable but, on death of a licence holder, a three months' temporary exemption from the licensing requirement is given (section 68) (see p. 385). If the house is to continue in a licensable status after the expiration of the licence, there must be a further application for a new licence, which will be subject to the same procedure.

Variation

A local authority may vary a licence either by agreement with the owner or if there is a change of circumstances, e.g. the discovery of new information about issues present before the licence was granted, or there is a need to vary the appropriate number of households or persons.

Variation may be instigated by the local authority or on application from the owner or a relevant person. The procedure to be used in varying a licence is contained in part 2 of schedule 5 and provisions for appeal against refusals to vary in part 3.

Variations with the agreement of the owner come into effect immediately, otherwise when the period for appeal has elapsed or any appeal has disposed of (section 69).

Power to revoke

A local authority may revoke a licence:

- with the agreement of the licence holder;
- where the licence holder has committed a serious breach of a licence condition or there are repeated breaches of conditions;
- where the local authority considers that the holder is no longer a fit and proper person; and
- where the local authority believes that the property no longer meets the standards required for a licence.

The procedure to be adopted is set out in part 2 of schedule 5, with appeals dealt with in part 3 of schedule 5. Revocations with the agreement of the owner come into effect immediately, otherwise when the period for appeal has elapsed or any appeal has disposed of (section 70).

Appeals to the Appropriate Tribunal

Appeals against licensing decisions are dealt with in part 3 of schedule 5 and deal with appeals:

(a) against the refusal or grant of a licence, including licence conditions; and
(b) decisions to vary or revoke, or refusals to do so.

Generally appeals are to be made within 28 days of the date specified in the local authority notice.

Register of licences

A local authority must keep a register of the licences granted containing the details laid out in regulation 11 of the Miscellaneous Provisions Regulations. The register must be available to the public at the head office of the local authority and copies of it, or extracts from it, must be provided by the local authority on payment of a fee (section 232).

Offences

It is an offence, punishable by a fine, to own or manage a house in multiple occupation without a licence or knowingly allow occupation by more persons or households than permitted. Breaching the licence conditions carries a fine of up to level 5. There is a defence of 'reasonable excuse' in each case (section 72).

There is also a provision for the making of rent repayment orders (see FC85) by the Appropriate Tribunal in relation to unlicensed houses in multiple occupation, whereby there is a penalty equivalent to the rent received during unlicensed operation of up to 12 months (sections 73–74).

Management regulations

The Management of Houses in Multiple Occupation (England) Regulations 2006, as amended, apply to all houses in multiple occupation, except those converted blocks of flats under section 257. They impose duties on managers concerning:

- provision of information to occupiers;
- safety measures, including fire safety;
- maintenance of water, drainage, gas and electricity;
- maintaining common parts and living accommodation; and
- providing waste disposal facilities.

They also place a duty on occupiers to ensure that the manager can effectively carry out these duties.

There are no provisions for the service of notices by a local authority to secure compliance but a person in breach of these requirements is liable to a fine of up to level 5.

It should be noted that there are duties placed on the manager of a house in multiple occupation through the Licensing and Management of Houses in Multiple Occupation (Additional Provisions) (England) Regulations 2007.

DESIGNATION OF AREAS BY LOCAL AUTHORITIES FOR THE ADDITIONAL LICENSING OF HMOs

References

Housing Act 2004, sections 55–60
Housing (Scotland) Act 2006
Houses in Multiple Occupation Act (Northern Ireland) 2016
The Licensing and Management of Houses in Multiple Occupation and other Houses (Miscellaneous Provisions) (England) Regulations 2006

Extent

These provisions apply to England and Wales. Similar provisions apply in Scotland and Northern Ireland.

Scope

This procedure allows a local authority to designate the whole or part of its district as an area in which all or specified types of houses in multiple occupation will require to be registered in addition to those already designated by the Licensing of Houses in Multiple Occupation (Prescribed Descriptions) (England) Order 2018 (section 56(1)). Following designation, the procedure for their licensing is the same as in FC78 above.

Before making a designation, the local authority must be satisfied that a significant proportion of the houses in multiple occupation to be included are being managed sufficiently ineffectively as to give rise, or be likely to give rise, to one or more particular problems, either for their occupants or for members of the public, e.g. antisocial behaviour.

FC79 Designation of areas for additional licensing of houses in multiple occupation – sections 55 to 60 Housing Act 2004

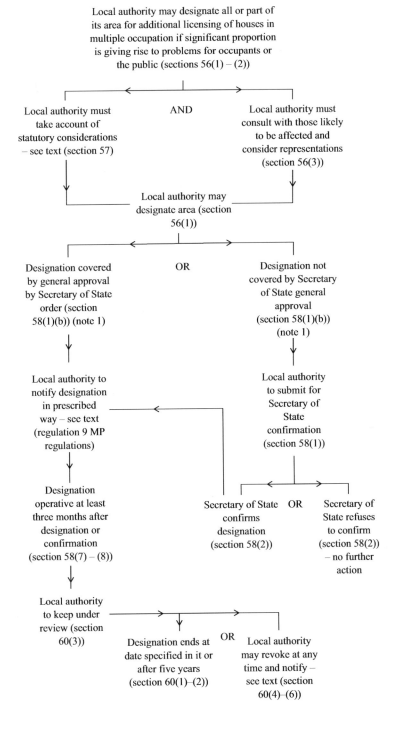

Local authority may designate all or part of its area for additional licensing of houses in multiple occupation if significant proportion is giving rise to problems for occupants or the public (sections 56(1) – (2))

Local authority must take account of statutory considerations – see text (section 57)

AND

Local authority must consult with those likely to be affected and consider representations (section 56(3))

Local authority may designate area (section 56(1))

Designation covered by general approval by Secretary of State order (section 58(1)(b)) (note 1)

OR

Designation not covered by Secretary of State general approval (section 58(1)(b)) (note 1)

Local authority to notify designation in prescribed way – see text (regulation 9 MP regulations)

Local authority to submit for Secretary of State confirmation (section 58(1))

Designation operative at least three months after designation or confirmation (section 58(7) – (8))

Secretary of State confirms designation (section 58(2))

OR

Secretary of State refuses to confirm (section 58(2)) – no further action

Local authority to keep under review (section 60(3))

Designation ends at date specified in it or after five years (section 60(1)–(2))

OR

Local authority may revoke at any time and notify – see text (section 60(4)–(6))

Notes

1. No order has yet been made.
2. MP regulations in this flow chart – The Licensing and Management of Houses in Multiple Occupation and other Houses (Miscellaneous Provisions) (England) Regulations 2006.

In forming this opinion, the local authority must have regard to any Codes of Practice issued under section 233 (section 56(2)–(6)).

Designation

Before making a designation, the local authority must consult with those most likely to be affected and take account of any representations made (section 56(3)).

There are further matters for the local authority to consider:

- The designation must be consistent with the local authority's overall housing strategy.
- There must be a coordinated approach to dealing with homelessness, empty properties and antisocial behaviour in the private rented sector, which considers both combining additional licensing with other possible courses of action and measures to be taken by others.
- The local authority must also consider alternative actions.
- The local authority must be satisfied that designation will significantly assist in dealing with the problems of the area (section 57).

Confirmation by Secretary of State and operation

Unless covered by a general approval (none have yet been issued), the designation requires confirmation by the Secretary of State (section 58(1)). There is no specified procedure for this part of the process. The Secretary of State may confirm or refuse to confirm a designation. The designation comes into effect no earlier than three months after it has been confirmed or, in the case of general approvals, three months after the designation was made (section 58).

Notification of designation

As soon as designation is confirmed or made, the local authority must publicise it by:

1. within seven days:

    (a) placing notices on municipal buildings in or closest to the area;
    (b) publishing the notice on the internet site;
    (c) publishing it in the next edition of two local newspapers and five times more at intervals of between two and three weeks;

2. within two weeks, sending a copy of the notice to:

    (a) those who responded to the consultation;
    (b) organisations representing landlords, tenants, agents;
    (c) organisations providing advice on landlord and tenant matters, e.g. housing advice centres.

The notice must contain:

    (a) the date of designation;
    (b) details of whether the designation requires confirmation;

(c) a brief description of the area;

(d) contact details for the local authority, the place where the designation may be inspected and where applications for licences may be obtained;

(e) a statement advising persons affected to seek advice from the local authority;

(f) a warning of the consequences of not applying for a licence (section 59(2) and the Licensing and Management of Houses in Multiple Occupation and other Houses (Miscellaneous Provisions) (England) Regulations 2006, regulation 9).

Duration

The designation must be reviewed from time to time and may be revoked at any time at the discretion of the local authority but ends no later than five years after it has been made (section 60(1)–(3)). Revocation does not require a confirmation by the Secretary of State.

Revocation

When a scheme is revoked, the local authority must within seven days:

- place notices on municipal buildings in or closest to the area;
- publish the notice on the internet site;
- publish it in the next edition of two local newspapers.

The notice must give:

- a brief description of the area;
- a summary of the reasons for revocation;
- the date revocation takes effect;
- contact details for the local authority (section 60(6) and the Licensing and Management of Houses in Multiple Occupation and other Houses (Miscellaneous Provisions) (England) Regulations 2006, regulation 10).

INTERIM MANAGEMENT ORDERS

References

Housing Act 2004, sections 101 to 112

Housing (Interim Management Orders) (Prescribed Circumstances) (England) Order 2006

Housing (Management Orders and Empty Dwelling Management Orders) (Supplemental Provisions) (England) Regulations 2006

Housing (Scotland) Act 1987

Housing (Northern Ireland) Order 1981

Extent

These provisions apply to England only. There are similar provisions for Wales, Scotland and Northern Ireland.

Scope

An interim management order is an order made by a local authority in relation to a house (see below) which cannot be licensed or where there is a management problem that requires intervention by the local authority. It runs for up to 12 months and is intended to secure that:

- immediate steps are taken to protect the health, safety or welfare of occupants or persons having an estate or interest in the vicinity; and
- appropriate steps are taken to secure the proper management of the house pending the grant of a house in multiple occupation licence, or one under part 3, or a final management order (see FC81) (section 101(3)).

It is therefore an interim measure designed to provide time for a longer-term management solution to be found. Through an interim management order, the local authority takes over most of the rights and duties of the landlord and right to the possession of the property – see below.

In this procedure, a 'house' is defined as a house in multiple occupation or a part 3 house (selective licensing of other residential accommodation – see p. 349) (section 101(5)).

That part of a house occupied by a resident landlord may be excluded from the terms of the order (section 102(8)).

Making of an interim management order

An interim management order *must* be made when:

(a) a house in multiple occupation or part 3 house ought to be licensed but is not and either there is no reasonable prospect of the house becoming licensed in the near future or the health and safety condition is met (see below); or

(b) the local authority intends to revoke an existing licence and either there is no prospect of a new licence being issued in the near future or the health and safety condition will be satisfied (section 102(1)–(3)).

The local authority *may* make an interim management order for:

(a) a house in multiple occupation not licensable under part 2 where the Appropriate Tribunal considers that the health and safety condition is satisfied (see below) and takes account of any relevant, approved Code of Practice (section 101(4)–(6)); and

(b) a house which satisfies the conditions of section 103 relating to special management orders and the Appropriate Tribunal authorises the local authority to make the interim management order (section 102(7)).

These discretionary interim management orders therefore require the submission to the Appropriate Tribunal, who may authorise them, either on the terms contained in the draft order or on such other terms as they consider appropriate (section 102 (4(b)) and 7(b)).

The procedure following the making of an interim management order is contained in schedule 6, part 1, paragraph 7. As soon as possible for occupiers and within seven days for relevant persons, the local authority should serve notices with copies of order stating:

(a) reasons for making the order;
(b) its general effect;

(c) date of operation;

(d) details of appeals to the Residential Property Tribunal to be made within 28 days.

The health and safety condition

It is necessary to make an interim management order for the purpose of protecting the health, safety or welfare of the occupants of the house or persons occupying or owning property in its vicinity. This condition can be satisfied if there is a threat to evict occupiers in order to avoid licensing under part 2 but cannot be met where the threat relates to a category 1 or category 2 hazard and the most appropriate course of enforcement action can be taken (section 104).

Operation and duration of interim management order

The interim management order comes into effect when it is made, except when it is made to follow a revocation of a licence, in which case it takes effect upon that revocation being made. The interim management order ceases to have effect after 12 months, unless an earlier date had been set or is continued in force pending the disposal of an appeal against the making of a final management order (section 105).

Local authority duties under an interim management order

These are to:

(a) take immediate steps to ensure the health, safety or welfare of occupants and those involved with property in the vicinity;

(b) put in place the long-term management arrangements;

(c) where the house is licensable, grant a licence or make a final management order;

(d) where the house is not licensable, consider if it should make a final management order or revoke the interim management order and take no further action (section 106).

General effect of an interim management order

Whilst the interim management order is operative, the local authority takes over most of the rights and responsibilities of the landlord, including the right to possession. With the written consent of the landlord, the local authority may grant occupation rights (section 107). Tenancies and leases granted by the local authority in this period are to be treated as if they were legal grants by the landlord (section 108). There are special provisions in the Housing (Management Orders and Empty Dwelling Management Orders) (Supplemental Provisions) (England) Regulations 2006 which apply where the local authority is to be treated as the lessee.

Effect of interim management order on landlords and mortgagees

The landlord is no longer able to receive rents and cannot exercise any management functions or grant tenancies. His right to sell the house remains, as do the rights of any mortgagee, except where they are being used to grant tenancies or licences under section 107 (section 109).

FC80　Interim management orders (IMOs) – sections 102 to 112 Housing Act 2004

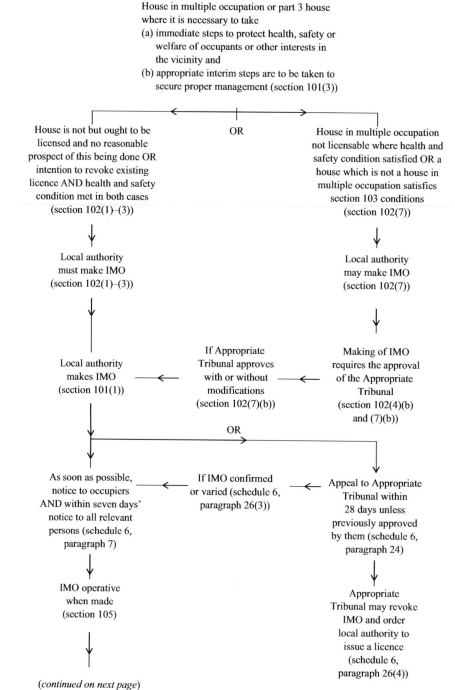

House in multiple occupation or part 3 house
where it is necessary to take
(a) immediate steps to protect health, safety or
　　welfare of occupants or other interests in
　　the vicinity and
(b) appropriate interim steps are to be taken to
　　secure proper management (section 101(3))

House is not but ought to be
licensed and no reasonable
prospect of this being done OR
intention to revoke existing
licence AND health and safety
condition met in both cases
(section 102(1)–(3))

OR

House in multiple occupation
not licensable where health and
safety condition satisfied OR a
house which is not a house in
multiple occupation satisfies
section 103 conditions
(section 102(7))

Local authority
must make IMO
(section 102(1)–(3))

Local authority
may make IMO
(section 102(7))

Local authority
makes IMO
(section 101(1))

If Appropriate
Tribunal approves
with or without
modifications
(section 102(7)(b))

Making of IMO
requires the approval
of the Appropriate
Tribunal
(section 102(4)(b)
and (7)(b))

OR

As soon as possible,
notice to occupiers
AND within seven days'
notice to all relevant
persons (schedule 6,
paragraph 7)

If IMO confirmed
or varied (schedule 6,
paragraph 26(3))

Appeal to Appropriate
Tribunal within
28 days unless
previously approved
by them (schedule 6,
paragraph 24)

IMO operative
when made
(section 105)

Appropriate
Tribunal may revoke
IMO and order
local authority to
issue a licence
(schedule 6,
paragraph 26(4))

(*continued on next page*)

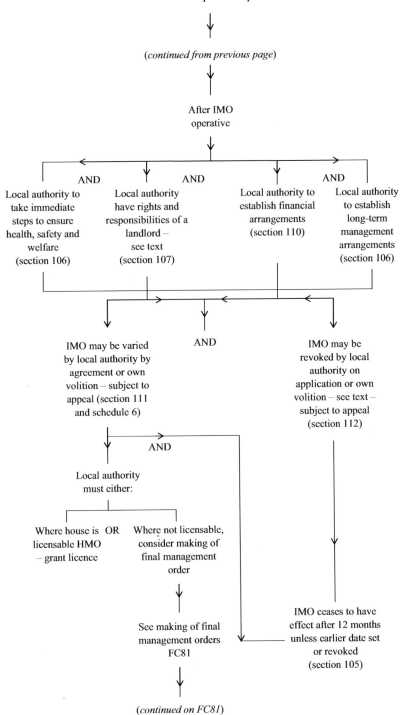

(*continued from previous page*)

After IMO
operative

AND
Local authority to take immediate steps to ensure health, safety and welfare (section 106)

AND
Local authority have rights and responsibilities of a landlord – see text (section 107)

AND
Local authority to establish financial arrangements (section 110)

AND
Local authority to establish long-term management arrangements (section 106)

IMO may be varied by local authority by agreement or own volition – subject to appeal (section 111 and schedule 6)

AND

IMO may be revoked by local authority on application or own volition – see text – subject to appeal (section 112)

AND

Local authority must either:

Where house is licensable HMO – grant licence

OR

Where not licensable, consider making of final management order

See making of final management orders FC81

IMO ceases to have effect after 12 months unless earlier date set or revoked (section 105)

(*continued on FC81*)

Compensation

Compensation is payable to third parties (a person with an estate or interest in the house other than the immediate landlord and any tenants under leases granted by the local authority) for any interference or loss with their rights due to the interim management order. Compensation disputes are dealt with by the Appropriate Tribunal (section 128).

Financial arrangements

The local authority is allowed to spend the rent receipts on 'relevant expenditure', which is that reasonably incurred in performing their duties under section 106(1)–(3) (see above), with any balance going to the landlord. The local authority must keep accounts and make them available for inspection. Persons entitled to review the accounts may seek a declaration from the Appropriate Tribunal that the amount of expenditure by the local authority is unreasonable and seek an order making financial adjustments (section 110).

At the end of an interim management order, any surpluses must be paid to the landlord unless a following final management order disallows this. Deficits on balance may be recovered from the landlord (section 129).

Variation of interim management order

Interim management orders may be varied at the request of a relevant person or by the local authority without application. The procedure for dealing with this is set out in schedule 6, part 2 and includes:

- notices to relevant persons;
- consideration of representations;
- notice of decision within seven days;
- details of appeals to the Appropriate Tribunal to be within 28 days (section 111 and schedule 6).

Revocation of interim management order

The local authority may revoke an interim management order where:

(a) the house ceases to need a licence;
(b) the local authority grants a licence;
(c) the local authority makes a final management order; or
(d) other appropriate circumstances apply.

The revocation comes into effect after 28 days of being made or the resolution of any appeal.

An interim management order cannot be revoked if the immediate landlord is subject to a banning order under section 16 of the Housing and Planning Act 2016; there is an agreement in force which, under section 108, has effect as a lease or licence granted by the authority, and revoking the interim management order would cause the immediate landlord to breach the banning order (section 112(2A)).

The procedure for dealing with revocations is the same as for variations above (section 112).

Appeals

A relevant person may appeal to the Appropriate Tribunal against:

  (i) the making of an interim management order;
 (ii) the terms of it;
(iii) decision or refusal to vary it;
(iv) decision or refusal to revoke it;
 (v) compensation payable.

In the case of (i) and (ii), there is no appeal if the interim management order had previously been authorised by the Appropriate Tribunal. Generally the appeal must be made within 28 days of the making or decision involved and the tribunal has powers to confirm, vary or revoke the local authority's order or other decisions (schedule 6, part 3).

Special interim management orders

These may be applied to properties that are not houses in multiple occupation but which could be licensed under part 3 (see p. 349) and have circumstances as defined in the Housing (Interim Management Orders) (Prescribed Circumstances) (England) Order 2006. They must be confirmed by the Appropriate Tribunal. This procedure is not covered here (section 103).

FINAL MANAGEMENT ORDERS

References

Housing Act 2004, sections 113 to 131
Housing (Management Orders and Empty Dwelling Management Orders) (Supplemental
    Provisions) (England) Regulations 2006
Housing (Management Orders and Empty Dwelling Management Orders) (Supplemental
    Provisions) (Wales) Regulations 2006

Extent

These provisions apply in England only but there are similar provisions in Wales.

Scope

The making of a final management order is one of the options a local authority can take after the ending of an interim management order; the others being revocation or the granting of a licence (section 106(4)).

A final management order is an order made by a local authority in relation to a house (see below) which cannot be licensed or at which there is some management problem which requires intervention by the local authority. A final management order runs for up to five years and is intended to secure the proper management of the house on a long-term basis in accordance with a management scheme contained in the order (section 101(4)).

In this procedure, a 'house' is defined as a house in management order or a part 3 house (for selective licensing of other residential accommodation, see p. 349) (section 101(5)).

FC81 Final management orders (FMOs) – sections 113 to 122 Housing Act 2004

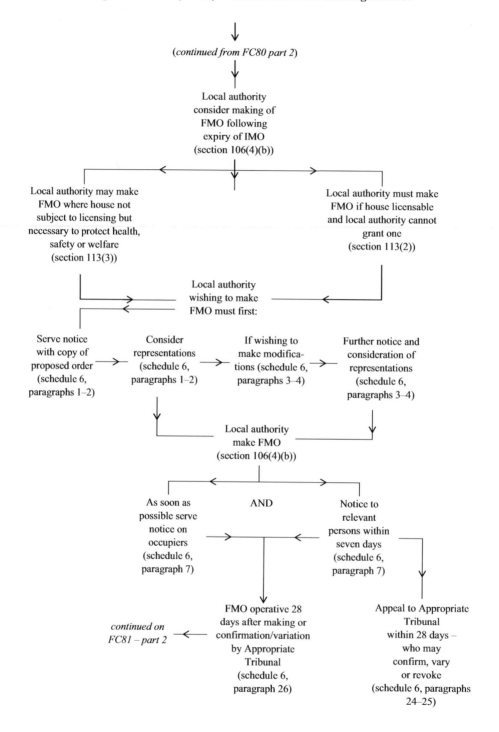

(continued from FC80 part 2)

Local authority
consider making of
FMO following
expiry of IMO
(section 106(4)(b))

Local authority may make
FMO where house not
subject to licensing but
necessary to protect health,
safety or welfare
(section 113(3))

Local authority must make
FMO if house licensable
and local authority cannot
grant one
(section 113(2))

Local authority
wishing to make
FMO must first:

Serve notice
with copy of
proposed order
(schedule 6,
paragraphs 1–2)

Consider
representations
(schedule 6,
paragraphs 1–2)

If wishing to
make modifica-
tions (schedule 6,
paragraphs 3–4)

Further notice and
consideration of
representations
(schedule 6,
paragraphs 3–4)

Local authority
make FMO
(section 106(4)(b))

As soon as
possible serve
notice on
occupiers
(schedule 6,
paragraph 7)

AND

Notice to
relevant
persons within
seven days
(schedule 6,
paragraph 7)

*continued on
FC81 – part 2*

FMO operative 28
days after making or
confirmation/variation
by Appropriate
Tribunal
(schedule 6,
paragraph 26)

Appeal to Appropriate
Tribunal
within 28 days –
who may
confirm, vary
or revoke
(schedule 6, paragraphs
24–25)

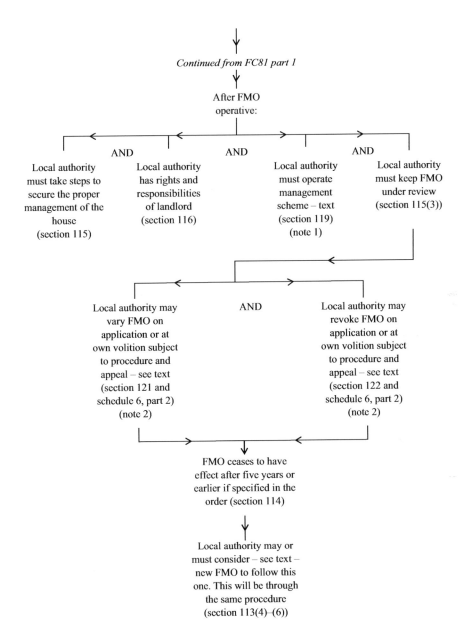

Continued from FC81 part 1

After FMO operative:

AND — **AND** — **AND**

Local authority must take steps to secure the proper management of the house (section 115)

Local authority has rights and responsibilities of landlord (section 116)

Local authority must operate management scheme – text (section 119) (note 1)

Local authority must keep FMO under review (section 115(3))

Local authority may vary FMO on application or at own volition subject to procedure and appeal – see text (section 121 and schedule 6, part 2) (note 2)

AND

Local authority may revoke FMO on application or at own volition subject to procedure and appeal – see text (section 122 and schedule 6, part 2) (note 2)

FMO ceases to have effect after five years or earlier if specified in the order (section 114)

Local authority may or must consider – see text – new FMO to follow this one. This will be through the same procedure (section 113(4)–(6))

Notes
1. Relevant landlord or third party may appeal to Appropriate Tribunal if management by local authority is not in accordance with scheme (section 120).
2. There is an appeal where the local authority refuses an application to vary or revoke (schedule 6, paragraph 28).
3. In this procedure, IMO = interim management order.
4. In this procedure FMO = final management order.

The final management order may be made to exclude part of a property occupied by a person who is the owner or long leaseholder of the entire house (section 113(7)).

Making of a final management order

The local authority:

(a) *must* make a final management order to replace an interim management order that is expiring if the house is licensable under part 2 or 3 and the local authority cannot grant a licence;

(b) *may* make a final management order for a non-licensable house where an interim management order is ending and the local authority considers it necessary to protect on a long-term basis the health, safety or welfare of the occupants or of those in the vicinity (section 113(1)–(5)).

The procedural requirements for the making of a final management order are as follows.

Before making the order

(a) Serve notice containing details of and copy of order on relevant persons, together with end date of consultation period.

(b) Consider any representations.

Before making the final management order with modifications

(a) Serve notice of proposals and reasons and indicate length of consultation period.

(b) Consider representations.

After making final management order

Local authority must serve notices with copies of order:

(a) within seven days on all relevant persons, and

(b) as soon as possible on occupiers, stating:

    (i) reasons for making the order;

    (ii) its general effect;

    (iii) a description of the way the house is to be managed in accordance with the management scheme;

    (iv) date of operation;

    (v) details of appeals to the Appropriate Tribunal to be made within 28 days (schedule 6 part 1).

Operation of final management order

A final management order will come into force 28 days after being made or, on appeal, 28 days after confirmation by the Appropriate Tribunal. If there is a further appeal to the Upper Tribunal, it will become operative on the date of confirmation by the Upper Tribunal.

It ceases to have effect after five years or on an earlier date if specified in the order. If its cessation is to be followed by a new final management order (see below), which is the subject of an appeal, the original final management order will continue until the appeal has been determined (section 114).

New final management orders

On the expiry of a final management order, the local authority must or may, depending on the particular situation (see 'Making of a final management order' above), make a further final management order if the circumstances causing the existing final management order to be made are still present (section 113(4)–(6)). The procedure to be followed will be the same as for the making of the existing final management order.

Local authority duties when final management order becomes operative

The local authority must take steps to secure the proper management of the house through the management scheme (see below). They must review the order and the scheme from time to time and consider whether the final management order should remain in force (section 115).

General effect of a final management order

The local authority takes over the duties and responsibilities of the landlord and, unlike the position with an interim management order, does not need the consent of the landlord before granting occupation rights, unless that right is for a fixed term to expire after the final management order is to end. Tenancies and leases granted by the local authority in this period are to be treated as if they were legal grants by the landlord (section 116). There are special provisions in the Housing (Management Orders and Empty Dwelling Management Orders) (Supplemental Provisions) (England) Regulations 2006 which apply where the local authority is to be treated as the lessee.

Sections 126–127 apply detailed provisions to deal with the status of the local authority as landlord in legal proceedings and the supply of furniture.

The landlord is no longer able to receive rents and cannot exercise any management functions or grant tenancies. His right to sell the house remains, as do the rights of any mortgagee, except where they are being used to grant tenancies or licences (sections 116–118). The occupiers retain the same legal status as before the final management order was made and nothing prevents the local authority from granting a private sector-type tenancy, although the tenant does not become a secure tenant (section 124).

Management schemes and accounts

The final management order must contain a management scheme which only becomes operative after the landlord has had the opportunity to agree its terms or after an appeal to the Appropriate Tribunal is settled. The scheme is to be divided into two parts. The first will deal with financial matters, including:

- carrying out of works;
- estimates of expenditure to be incurred;
- amounts of rents to be sought;

- amounts and provision of compensation;
- any payments to the landlord;
- financial arrangements with the landlord when the final management order ends.
- The second part of the scheme will cover those issues that caused the local authority to make the order, including:
- steps to be taken to require occupiers to comply with their obligations under lease or general law;
- description of any repairs required (section 119).

Any relevant landlord or third party may apply to the Appropriate Tribunal if it is considered that management of the house by the local authority is not in accordance with the management scheme. The tribunal may make an order requiring the local authority to act in accordance with the scheme and award damages. It may also revoke the final management order (section 120).

At the end of a final management order, any surpluses must be paid to the landlord unless any following final management order disallows this. Deficits on balance may be recovered from the landlord (section 129).

Variation of final management order

Final management orders may be varied at the request of a relevant person or by the local authority without application (section 121). The procedure for dealing with this includes:

- notices to relevant persons with reasons for the decision and indicating the length of the consultation period;
- consideration of representations;
- notice of decision to vary within seven days;
- appeals to the Appropriate Tribunal within 28 days.

Before refusing to vary a final management order, the local authority must:

(a) give notice to each relevant person with reasons and length of consultation period;
(b) consider representations;
(c) if still intending to revoke, give notice to same persons within seven days giving the decision, the reasons for it and the rights of appeal to be made within 28 days (schedule 6, part 2).

Revocation of final management order

The local authority is able to revoke a final management order either on its own volition or following a request to do so:

(a) if the house ceases to be licensable;
(b) if the local authority grants a licence;
(c) if the local authority makes a further final management order; or
(d) in other appropriate circumstances (section 122).

The procedure for dealing with the revocation of a final management order is contained in schedule 6, part 2 and mirrors that for the variation of them above.

Compensation

Compensation is payable to third parties (a person with an estate or interest in the house other than the immediate landlord and any tenants under leases granted by the local authority) for any interference or loss with their rights due to the final management order. Compensation disputes are dealt with by the Appropriate Tribunal (section 128).

Appeals

A relevant person may appeal to the Appropriate Tribunal against:

- the making of a final management order;
- the terms of it and of the management scheme;
- decision or refusal to vary it;
- decision or refusal to revoke it;
- compensation payable.

Generally the appeal must be made within 28 days of the making or decision involved and the Appropriate Tribunal has powers to confirm, vary or revoke the local authority's order or other decisions (schedule 6, part 3).

TEMPORARY EXEMPTION NOTICES FOR HMOs

References

Housing Act 2004, section 62
Housing (Scotland) Act 2006
Houses in Multiple Occupation Act (Northern Ireland) 2016

Extent

These provisions apply in England and Wales. Similar provisions apply in Scotland and Northern Ireland.

Scope

This procedure allows a local authority to grant a temporary exemption from the need to licence a house in multiple occupation in circumstances where the person having control or management notifies the local authority of his intention to take steps to create a situation where the house will no longer require a licence, e.g. by reducing occupants or households (section 62(1)).

Temporary exemption notices (TENs)

Having notified the local authority of his intent, the person does not commit the offence of using a house as a house in multiple occupation without a licence until a temporary exemption notice expires. This applies both to licensing requirements under the Licensing of Houses in Multiple Occupation (Prescribed Descriptions) (England) Order 2018 and those houses in multiple occupation included in designated areas (section 72(4)(a)).

FC82 Temporary exemption notices (TENs) – section 62 Housing Act 2004

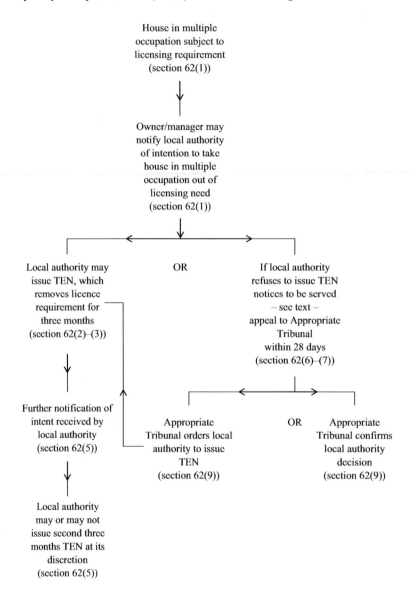

The local authority upon receipt of such a notification may, at its discretion, serve a temporary exemption notice, which will be in force for three months from the date of service (section 62(4)(a)).

If the local authority receives a further notification, and considers that there are exceptional circumstances which justify it, they may serve a second temporary exemption notice, which will operate for three months from the date of expiration of the first notice. No further temporary exemption notices may be issued for that property (section 62(5)).

Appeals

If the local authority refuses to serve a temporary exemption notice following a first notification, they must, without delay, serve notice on the person informing him of:

(a) the decision;
(b) the reasons for it and the date on which it was made;
(c) the rights of appeal; and
(d) the period in which an appeal can be made (section 62(6)).

The person concerned may appeal to the Appropriate Tribunal within 28 days of the date of the decision and the tribunal may confirm or reverse the local authority's decision. If the latter, the tribunal will direct the local authority to issue a temporary exemption notice, which will come into force on a date which the tribunal directs (section 62(7)–(9)).

Temporary exemption notices where there is a death of a holder of a house in multiple occupation licence

Upon the death of a house in multiple occupation licence holder, the licence ceases to have effect but, for a period of three months from the date of death, the house is treated as if a temporary exemption notice has been served. This period may be extended for a further three months by the service of a temporary exemption notice at the request of the person's representatives made during that initial period. There is an appeal against a local authority decision not to serve a temporary exemption notice for that additional period and the same notice procedure as for temporary exemption notices generally, above, is then applicable (section 68(7)–(10)).

OVERCROWDING NOTICES FOR HMOs

References

Housing Act 2004, sections 139–144
Civil penalties under the Housing and Planning Act 2016 – Guidance for Local Housing Authorities
Housing (Scotland) Act 1987
Housing (Northern Ireland) Order 1992

Extent

These provisions apply to England and Wales. Similar provisions apply in Scotland and Northern Ireland.

Scope

This procedure deals with the control of overcrowding in houses in multiple occupation (HMOs) that are not required to be licensed under part 2 (see FC78) and which are not the subject of operative interim management orders or final management orders (see FC80 and FC81) (section 139(1)). The notices relate to situations where, having regard to the rooms available, the local authority considers that an excessive number of persons is being, or is likely to be, accommodated in the house in multiple occupation (section 139(2)).

The level of occupation in licensed houses in multiple occupation is controlled through the licence in that it will only permit occupation by a specified number.

Service of overcrowding notices

At least seven days before any overcrowding notice is served, the local authority must inform the following persons of its intention:

(a) persons having an estate or interest in the house;
(b) the person managing or in control of it; and
(c) as far as is reasonably possible, all occupiers.

In the case of (a) and (b), the notification must be in writing.

Each of these must be given the opportunity to make representations for the local authority to consider. The overcrowding notice is then served on one or more of the relevant persons ((a) and (b)) above (section 139(1)–(4)).

Contents of overcrowding notice

The notice must state in relation to each room either:

(i) the maximum number of persons who may occupy it as sleeping accommodation at any one time; or
(ii) that the room is regarded as unsuitable for sleeping accommodation.

Special maxima may be applied relating to the age of occupants.

The notice must also *either*:

(a) require that the terms of the notice are not breached;
(b) disallow numbers of occupants of the house such as results in occupation of a sleeping room by opposite sexes who are not living together as man and wife (ignoring children under the age of ten); *or*

apply (a) and (b) but only in relation to new occupants not resident when the notice was served (sections 140–142).

Operation of an overcrowding notice

The notice becomes operative 21 days after service. However, where an appeal is lodged, it becomes effective after confirmation by the Appropriate Tribunal or subsequently by the Upper Tribunal (sections 139(5) and 143(4)).

FC83 Overcrowding notices (ONs) for houses in multiple occupation

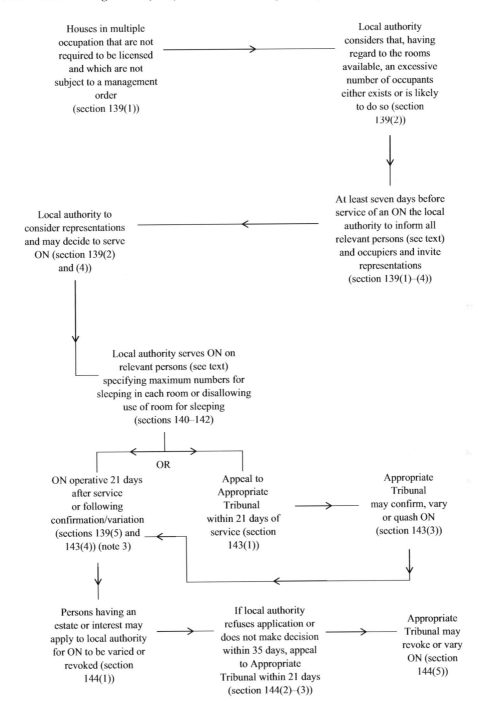

Houses in multiple occupation that are not required to be licensed and which are not subject to a management order (section 139(1))

Local authority considers that, having regard to the rooms available, an excessive number of occupants either exists or is likely to do so (section 139(2))

At least seven days before service of an ON the local authority to inform all relevant persons (see text) and occupiers and invite representations (section 139(1)–(4))

Local authority to consider representations and may decide to serve ON (section 139(2) and (4))

Local authority serves ON on relevant persons (see text) specifying maximum numbers for sleeping in each room or disallowing use of room for sleeping (sections 140–142)

OR

ON operative 21 days after service or following confirmation/variation (sections 139(5) and 143(4)) (note 3)

Appeal to Appropriate Tribunal within 21 days of service (section 143(1))

Appropriate Tribunal may confirm, vary or quash ON (section 143(3))

Persons having an estate or interest may apply to local authority for ON to be varied or revoked (section 144(1))

If local authority refuses application or does not make decision within 35 days, appeal to Appropriate Tribunal within 21 days (section 144(2)–(3))

Appropriate Tribunal may revoke or vary ON (section 144(5))

Notes

1. For the control of occupancy in licensed houses in multiple occupation, see FC78.
2. For the control of overcrowding in houses other than houses in multiple occupation, see FC63.
3. Offence to breach ON with penalty up to level 4 on the standard scale (section 139 (7)–(8)).

Appeals

An appeal against the service of the overcrowding notice may be made to the Appropriate Tribunal within 21 days of service, although that period may be extended by the tribunal if it feels that there is good reason. The tribunal may confirm, vary or quash the notice with a further appeal to the Upper Tribunal (section 143).

Revocation and variation

On the application of a person having an estate or interest or a person managing or having control, the local authority may, at any time:

- revoke the overcrowding notice; or
- vary it to allow higher occupancy.

The local authority must make a decision and give written notice of it within 35 days. There is an appeal to the Appropriate Tribunal against a refusal or where the matter has not been dealt with within the 35 days. Appeals are to be made within 21 days but the tribunal has power to extend that period if there is good reason. The tribunal may revoke the notice or confirm it (section 144).

Offences

Any person contravening an overcrowding notice is liable, on conviction, to a fine up to level 4 on the standard scale. There is a defence of 'reasonable excuse' (section 139(7)–(8)).

Chapter 5

ROGUE LANDLORDS

BANNING ORDERS

Reference

Housing and Planning Act 2016, sections 15 to 23

Extent

These provisions apply to England and Wales.

Scope

These provisions allow a banning order to be made where a landlord or property agent has been convicted of a banning order offence; requires a database of rogue landlords and property agents to be established and allows a rent repayment order to be made against a landlord who has committed an offence.

The objectives are threefold. It introduces new financial sanctions against rogues who break the law, by extending the rent repayment order provisions. It also enables local authorities to identify rogues operating in the private sector in their area and place them on a database, which other local authorities in England will have access to. Finally, it provides a regime for removing the worst offenders from the sector through banning orders (Hansard).

Application for banning order and notice of intended proceedings

A local housing authority may apply for a banning order against a person who has been convicted of a banning order offence (section 15(1)).

If an authority applies for a banning order against a body corporate that has been convicted of a banning order offence, it must also apply for a banning order against any officer who has been convicted of the same offence in respect of the same conduct (section 15(2)).

Before applying for a banning order, the authority must give the person a notice of intended proceedings:

(a) informing the person that the authority is proposing to apply for a banning order and explaining why,
(b) stating the length of each proposed ban, and
(c) inviting the person to make representations within a period specified in the notice of not less than 28 days (section 15(3)).

The authority must consider any representations made during that notice period (section 15(4)). The authority must wait until the notice period has ended before applying for a banning order (section 15(5)). A notice of intended proceedings cannot be made after 6 months of the person being convicted of the offence (section 15(6)).

Making a banning order

The First-Tier Tribunal may make a banning order against a person who has been convicted of a banning order offence, and was a residential landlord or a property agent at the time the offence was committed (section 16(1)).

A banning order may only be made on an application by a local housing authority (section 16(2)).

Where an application against an officer of a body corporate is made, the First-Tier Tribunal may make a banning order against the officer even if the officer was not the landlord, or property agent, at the time of the offence (section 16(3)).

In deciding whether to make a banning order against a person, and in deciding what order to make, the Tribunal must consider –

(a) the seriousness of the offence of which the person has been convicted,
(b) any previous convictions that the person has for a banning order offence,
(c) whether the person is or has at any time been included in the database of rogue landlords and property agents, and
(d) the likely effect of the banning order on the person and anyone else who may be affected by the order (section 16(4)).

Duration and effect of banning order

A banning order must specify the length of each ban imposed by the order and it must last at least 12 months (section 17(1) and (2)).

A banning order may contain exceptions to a ban for some or all of the period to which the ban relates and the exceptions may be subject to conditions (section 17(3) and (4)).

Power to require information

A local housing authority may require a person to provide specified information for the purpose of enabling the authority to decide whether to apply for a banning order against the person (section 19(1).

It is an offence for the person to fail to comply with a requirement, unless the person has a reasonable excuse (section 19(2)).

It is an offence for the person to provide information that is false or misleading if the person knows that the information is false or misleading or is reckless as to whether it is false or misleading (section 19(3)).

Revocation or variation of banning orders

A person against whom a banning order is made may apply to the First-Tier Tribunal for an order revoking or varying the order (section 20(1)).

The power to vary a banning order may be used to add new exceptions to a ban or to vary the banned activities, the length of a ban, or existing exceptions to a ban (section 20(5)).

Offence of breach of banning order

A person who breaches a banning order commits an offence and is liable on summary conviction to imprisonment for a period not exceeding 51 weeks or to a fine or to both (section 21(1) and (2)).

If a financial penalty under section 23 has been imposed in respect of the breach, the person may not be convicted of an offence (section 21(3)).

Where a person is convicted of breaching a banning order and the breach continues after conviction, the person commits a further offence and is liable on summary conviction to a fine not exceeding one-tenth of level 2 on the standard scale for each day or part of a day on which the breach continues (section 21(4)).

In proceedings for an offence it is a defence to show that the person had a reasonable excuse for the continued breach (section 21(5)).

Offences by bodies corporate

Where an offence committed by a body corporate is proved to have been committed with the consent or connivance of, or to be attributable to any neglect on the part of, an officer of a body corporate, the officer as well as the body corporate commits the offence (section 22(1)).

Where the affairs of a body corporate are managed by its members, the member is regarded as an officer of the body corporate (section 22(2)).

Financial penalty for breach of banning order

The local housing authority may impose a financial penalty on a person if satisfied, beyond reasonable doubt, that the person's conduct amounts to an offence (section 23(1)).

Only one financial penalty may be imposed in respect of the same conduct unless a breach continues for more than six months, where a financial penalty may be imposed for each additional six-month period for the whole or part of which the breach continues (section 23 (3) and (4)).

The amount of a financial penalty imposed is to be determined by the authority imposing it, but must not be more than £30,000 (section 23(5)).

The authority may not impose a financial penalty if the person has been convicted of an offence relating to conduct, or criminal proceedings for the offence have been instituted against the person and the proceedings have not been concluded (section 23(6)).

A local housing authority must have regard to any guidance given by the Secretary of State (section 23(10)).

Definitions

Banning order means an order, made by the First-Tier Tribunal, banning a person from:

- (a) letting housing,
- (b) engaging in letting agency work,
- (c) engaging in property management work, or
- (d) doing two or more of those things (section 14(1)).

FC84 Rogue landlord banning orders – Housing and Planning Act 2016, sections 15 to 23

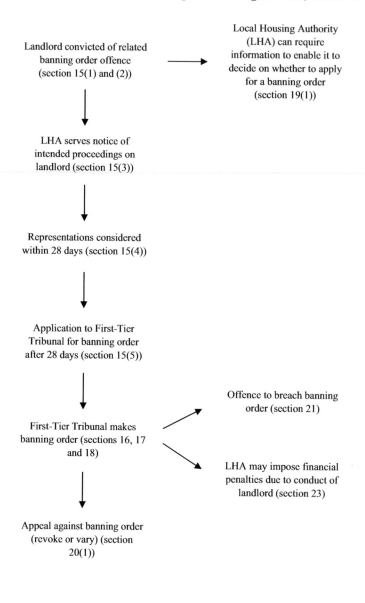

Landlord convicted of related banning order offence (section 15(1) and (2))

Local Housing Authority (LHA) can require information to enable it to decide on whether to apply for a banning order (section 19(1))

LHA serves notice of intended proceedings on landlord (section 15(3))

Representations considered within 28 days (section 15(4))

Application to First-Tier Tribunal for banning order after 28 days (section 15(5))

First-Tier Tribunal makes banning order (sections 16, 17 and 18)

Offence to breach banning order (section 21)

LHA may impose financial penalties due to conduct of landlord (section 23)

Appeal against banning order (revoke or vary) (section 20(1))

DATABASE OF ROGUE LANDLORDS AND PROPERTY AGENTS

References

Housing and Planning Act 2016 – sections 28–39
Database of rogue landlords and property agents under the Housing and Planning Act 2016:
  statutory guidance

Extent

These provisions apply to England only.

Scope

The Secretary of State must establish and operate a database of rogue landlords and property agents (section 28(1)). Local housing authorities in England are responsible for maintaining the content of the database (section 28(1) and (2)).

The Secretary of State must ensure that local housing authorities are able to edit the database for the purpose carrying out their functions and updating the database (section 28(3)).

Duty to include person with banning order

A local housing authority must make an entry in the database of a person subject to a banning order (section 29(1)). An entry must be maintained for the period of the banning order and must then be removed (section 29(2)).

Power to include person convicted of banning order offence

A local housing authority may make an entry if the person has been convicted of a banning order offence, and the offence was committed when the person was a residential landlord or a property agent (section 30(1)).

The authority can make an entry if a person who has, at least twice within a period of 12 months, received a financial penalty for a banning order offence committed when the person was a residential landlord or a property agent (section 30(2)).

A financial penalty is to be taken into account only if the period for appealing the penalty has expired and any appeal has been finally determined or withdrawn (section 30(3)).

Procedure for inclusion under section 30

If a local housing authority decides to make an entry in the database it must give the person a decision notice before the entry is made (section 31(1)).

The decision notice must explain that the authority has decided to make the entry 21 days after the notice is given, and specify the period for which the person's entry will be maintained, which must be at least 2 years from the entry being made (section 31(2)).

The decision notice must also summarise the person's appeal rights (section 31(3)).

The authority must wait until the notice period has ended before making the entry in the database (section 31(4)).

If a person appeals within the notice period, the local housing authority cannot make the entry in the database until the appeal has been determined or withdrawn, and there is no possibility of further appeal (section 31(5)).

A decision notice cannot be given more than 6 months after the person was convicted of the banning order offence, or received the second of the financial penalties to which the notice relates (section 31(6)).

Appeals

A person who has been given a decision notice may appeal to the First-Tier Tribunal against the decision to make the entry in the database, or the period for which the person's entry is to be maintained (section 32(1)).

An appeal must be made before the end of the notice period specified in the decision notice (section 32(2)).

The Tribunal may allow an appeal to be made to it after the end of the notice period if satisfied that there is a good reason for the person's failure to appeal within the period (section 32(3)).

On an appeal the tribunal may confirm, vary or cancel the decision notice (section 32(4)).

Information to be included in the database

Regulations can be made about the information that must be included in a person's entry in the database (section 33(1)).

The regulations may require a person's entry to include:

(a) the person's address or other contact details,
(b) the period for which the entry is to be maintained;
(c) details of properties owned, let or managed by the person;
(d) details of any banning order offences of which the person has been convicted;
(e) details of any banning orders made against the person, whether or not still in force;
(f) details of financial penalties that the person has received (section 33(2)).

Where a body corporate is entered in the database, the regulations may also require information to be included about its officers (section 33(3)).

Updating

A local housing authority must take reasonable steps to keep information in the database up to date (section 34).

Power to require information

A local housing authority may require a person to provide specified information for the purpose of enabling the authority to decide whether to make an entry in the database (section 35(1)).

The authority that makes an entry in the database, or that is proposing to make an entry in the database, may require the person to provide any information needed to complete the entry or keep it up to date (section 35(2)).

It is an offence for the person to fail to comply with a requirement, unless the person has a reasonable excuse for the failure (section 35(3)).

It is an offence to provide information that is false or misleading if the person knows that the information is false or misleading or is reckless as to whether it is false or misleading (section 35(4)). A person who commits an offence is liable on summary conviction to a fine (section 35(5)).

Removal or variation of entries

An entry made in the database may be removed or varied (section 36(1)). If the entry was made on the basis of one or more convictions all of which are overturned on appeal, the responsible local housing authority must remove the entry (section 36(2)).

If the entry was made on the basis of more than one conviction and some of them have been overturned on appeal, or spent, the responsible local housing authority may remove the entry, or reduce the period for which the entry must be maintained (section 36(3) and (4)).

If the entry was made on the basis that the person has received two or more financial penalties and at least one year has elapsed since the entry was made, the local housing authority may remove the entry, or reduce the period for which the entry must be maintained (section 36(5)).

These powers may even be used to remove or reduce the period of an entry before the end of the two-year period (section 36(6)).

If a local housing authority removes an entry in the database, or reduces the period for which it must be maintained, it must notify the person to whom the entry relates (section 36(7)).

Requests for exercise of powers under section 36 and appeals

A person whose entry is made in the database may request, in writing, the local housing authority to remove the entry, or reduce the period for which the entry must be maintained (section 37(1) and (2)).

Where a request is made, the authority must decide whether to comply with the request, and give the person notice of its decision (section 37(3)).

If the authority decides not to comply with the request, the notice must include reasons for that decision, and a summary of the appeal rights (section 37(4)).

Where a person is given notice that the authority has decided not to comply with the request the person may appeal to the First-Tier Tribunal (section 37(5)).

An appeal to the First-Tier Tribunal must be made within 21 days of the notice being given (section 37(6)).

On an appeal the tribunal may order the local housing authority to remove the entry, or reduce the period for which the entry must be maintained (section 37(8)).

Access to database and use of its information

The Secretary of State must give every local housing authority access to information in the database (section 38).

A local housing authority may only use information obtained from the database for the following purposes:

(a) in connection with its functions under the Housing Act 2004,
(b) a criminal investigation or proceedings relating to a banning order offence,
(c) an investigation or proceedings relating to a contravention of the law relating to housing or landlord and tenant,
(d) promoting compliance with the law relating to housing or landlord and tenant by any person in the database, or
(e) statistical or research purposes (section 39(4)).

RENT REPAYMENT ORDERS

References

Housing Act 2004, sections 96 to 100 (H)
Housing and Planning Act 2016, sections 41–53 (HP)
Rent Repayment Orders (Supplementary Provisions) (England) Regulations 2007
Rent Repayment Orders and Financial Penalties (Amounts Recovered) (England) Regulations 2017
Rent repayment orders under the Housing and Planning Act 2016 – Guidance for Local Housing Authorities, April 2017

Extent

These provisions apply to England but there are similar provisions in Wales.

Scope

Rent repayment orders can be used for tackling unlicensed houses. They are a means by which a tenant or local authority can seek to have up to 12 months of rent, Housing Benefit, or universal credit repaid, usually in addition to other fines.

The Housing Act 2004 introduced rent repayment orders to cover situations where the landlord of a property had failed to obtain a licence for a property that was required to be licensed, specifically offences in relation to licensing of Houses in Multiple Occupation (section 72(1)) and offences in relation to licensing of houses under Part 3 of the Act (section 95(1)).

Rent repayment orders have now been extended through the Housing and Planning Act 2016 to cover a much wider range of offences.

Housing Act 2004

An Appropriate Tribunal can make a rent repayment order on application of the local authority or an occupier of the whole or part of the house. The tribunal must be satisfied:

(a) that within that previous 12 months, the appropriate person has committed an offence in relation to the house (whether or not he has been charged or convicted);

(b) that housing benefit or universal credit has been paid rent during any period during that the offence was being committed;

(c) that the authority has served on the appropriate person a 'notice of intended proceedings':

    (i) informing him that the authority are proposing to make an application;

    (ii) setting out the reasons why;

    (iii) stating the amount that they will seek to recover and how that amount is calculated; and

    (iv) inviting him to make representations to them within a period specified in the notice of not less than 28 days;

(d) that such period must have expired; and

(e) that the authority has considered any representations made to them within that period by the appropriate person.

Where a local authority serves a notice of intended proceedings, they must ensure that a copy of the notice is received by the department of the authority responsible for administering the housing benefit and that that department is subsequently kept informed of any matters relating to the proceedings that are likely to be of interest to it in connection with the administration of housing benefit (section 96 H).

If the tribunal is satisfied that a person has been convicted of an offence; and that housing benefit or universal credit was paid in respect of rent, the tribunal must make a rent repayment order requiring the appropriate person to pay to the authority an amount equal to the total amount of housing benefit paid or an amount reasonable in the circumstances.

If the total of rent is less than the total amount of housing benefit paid, the amount required to be paid through a rent repayment order is limited to the rent total.

Any amount payable to a local authority under a rent repayment order does not constitute an amount of housing benefit recovered by them, and is, until recovered by them, a legal charge on the house.

If the authority subsequently grants a licence in respect of the house to the appropriate person or any person acting on his behalf, the conditions contained in the licence may include a condition requiring the licence holder to pay (possibly by instalments) to the authority any amount payable to them under the rent repayment order not recovered by them (section 97).

A local authority that has made an application for a rent repayment order can ask the Tribunal to amend its application where it believes that there has been an overpayment of housing benefit so that the application is in respect of the amount of housing benefit that the local authority believes is properly payable (regulation 2).

A local housing authority may apply any amount recovered under a rent repayment order to meet the costs and expenses incurred in, or associated with, carrying out any of its enforcement functions in relation to the private rented sector (regulations 3 and 4 of 2017 regulations).

Housing and Planning Act 2016

Duty to consider applying for rent repayment orders

If a local housing authority becomes aware that a person has been convicted of an offence in relation to housing in its area, the authority must consider applying for a rent repayment order (section 48 HP).

FC85 Rent repayment orders – sections 96 to 98 Housing Act 2004 and sections 41 to 53 Housing and Planning Act 2016.

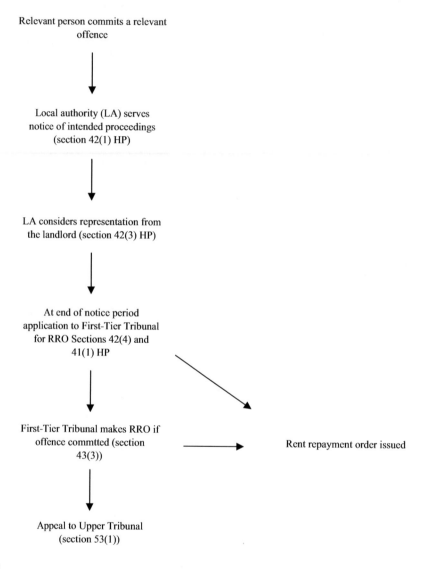

Relevant person commits a relevant offence

Local authority (LA) serves notice of intended proceedings (section 42(1) HP)

LA considers representation from the landlord (section 42(3) HP)

At end of notice period application to First-Tier Tribunal for RRO Sections 42(4) and 41(1) HP

First-Tier Tribunal makes RRO if offence commtted (section 43(3))

Rent repayment order issued

Appeal to Upper Tribunal (section 53(1))

Notes

1. HP = Housing and Planning Act 2016.

Application for rent repayment order

A tenant or a local housing authority can apply to the First-Tier Tribunal for a rent repayment order against a person who has committed an offence below (section 41(1) HP).

A local housing authority may apply for a rent repayment order if the offence relates to housing in the authority's area, and the authority has given notice of proceedings below (section 41(3)HP).

In deciding whether to apply for a rent repayment order a local housing authority must have regard to any guidance given by the Secretary of State (section 41(4) HP).

Notice of intended proceedings

Before applying for a rent repayment order a local housing authority must give the landlord a notice of intended proceedings (section 42(1) HP).

A notice of intended proceedings must:

(a) inform the landlord that the authority is proposing to apply for a rent repayment order and explain why,

(b) state the amount that the authority seeks to recover, and

(c) invite the landlord to make representations within a period specified in the notice of not less than 28 days (section 42(2) HP).

The authority must consider any representations made during the notice period (section 42(3)). The authority must wait until the notice period has ended before applying for a rent repayment order (section 42(4) HP).

A notice of intended proceedings cannot be given 12 months after the landlord committed the offence (section 42(5) HP).

Making of rent repayment order

The First-Tier Tribunal may make a rent repayment order if satisfied that a landlord has committed an offence (section 43(1) HP).

The amount of a rent repayment order is to be determined in accordance with:

(a) section 44 (where the application is made by a tenant);

(b) section 45 (where the application is made by a local housing authority);

(c) section 46 (in certain cases where the landlord has been convicted etc.) (section 43(3) HP).

Amount of order: local housing authorities

Where the First-Tier Tribunal decides to make rent repayment order in favour of a local housing authority the amount must relate to universal credit paid 12 months before the date of the offence and not exceeding 12 months depending on the offence.

The amount that the landlord may be required to repay in respect of a period must not exceed the amount of universal credit that the landlord received in respect of rent for that period (section 45(3) HP).

In determining the amount the tribunal must, in particular, take into account of the conduct of the landlord, the financial circumstances of the landlord, and whether the landlord has at any time been convicted of an offence (section 45(4) HP).

Amount of order following conviction

Where the First-Tier Tribunal decides to make a rent repayment order and both of the following conditions are met, the amount above is to be the maximum that the tribunal has power to order (section 46(1) HP).

Condition 1 is that the order:

(a) is made against a landlord who has been convicted of the offence, or
(b) is made against a landlord who has received a financial penalty in respect of the offence and is made at a time when there is no prospect of appeal against that penalty (section 46(2) HP).

Condition 2 is that the order is made:

(a) in favour of a tenant on the ground that the landlord has committed certain section 40 offences or,
(b) in favour of a local housing authority (section 46(3) HP).

Enforcement of rent repayment orders

An amount payable to a tenant or local housing authority under a rent repayment order is recoverable as a debt (section 47(1) HP).

Appeals from the First-Tier Tribunal

A person aggrieved by a decision of the First-Tier Tribunal may appeal to the Upper Tribunal (both must give permission) (section 53(1) and (4) HP).

Offences

- Section 6(1) Criminal Law Act 1977 – violence for securing entry
- Section 1(2), (3) Or (3A) Protection from Eviction Act 1977 – eviction or harassment of occupiers
- Section 30(1) Housing Act 2004 – failure to comply with improvement notice
- Section 32(1) Housing Act 2004 – Failure to comply with prohibition order
- Section 72(1) Housing Act 2004 – control or management of unlicensed HMO
- Section 95(1) Housing Act 2004 – control or management of unlicensed HMO
- Section 21 Housing and Planning Act 2016 – breach of banning order

PREVENTING RETALIATORY EVICTION

References

Deregulation Act 2015, section 33
Retaliatory Eviction and the Deregulation Act 2015 – A guidance note on the changes coming into force on 1 October 2015
Rent (Scotland) Act 1984
The Private Tenancies (Northern Ireland) Order 2006
Rent (Northern Ireland) Order 1978

Extent

These provisions apply to England and Wales. Similar provisions apply in Scotland and Northern Ireland.

Scope

Tenants should report any disrepair or poor conditions that may arise to the landlord as soon as possible. They should put their complaint in writing.

If, after 14 days from the tenant making a complaint, the landlord does not reply, that reply is inadequate, or they respond by issuing a section 21 eviction notice, the tenant should approach their local authority and ask them to step in and carry out an inspection to verify the need for a repair. The local authority should then arrange to inspect the property using the Housing Health and Safety Rating System.

If the inspection verifies the tenant's complaint, the inspector will take appropriate action, including Improvement Notices and Notices of Emergency Remedial Action. If the local authority serves an Improvement Notice or Notice of Emergency Remedial Action, the landlord cannot evict the tenant for six months using the no-fault eviction procedure.

Preventing retaliatory eviction

Where an improvement notice relating to category 1 or 2 hazards, or an emergency remedial action notice is served, a possession on termination of shorthold tenancy notice cannot be given:

(a) within six months beginning with the day of service of the relevant notice, or
(b) where the operation of such a notice has been suspended, within six months beginning with the day on which the suspension ends (section 33(1) and (13)).

This does not apply where the possession notice is given after:

(a) the relevant notice has been wholly revoked having been served in error,
(b) the relevant notice has been quashed,
(c) a decision of the local authority to refuse to revoke the relevant notice has been reversed, or
(d) a decision of the local authority to take the action has been reversed (section 33(8)).

A possession on termination of shorthold tenancy notice given in relation to an assured shorthold tenancy is invalid where:

(a) before the notice was given, the tenant made a complaint in writing to the landlord regarding the condition of the dwelling-house at the time of the complaint,
(b) the landlord –

(i) did not provide a response to the complaint within 14 days beginning with the day on which the complaint was given,
(ii) provided a response to the complaint that was not an adequate response, or
(iii) gave such a notice in relation to the dwelling-house following the complaint,

FC86 Preventing retaliatory eviction – Deregulation Act 2015, section 33

Tenants report disrepair or
poor conditions in writing (or
not) to landlord (section
33 (4))

If landlord does not reply
after 14 days, reply
inadequate, or section 21
eviction notice issued (section 33)

If tenant approach local
authority to carry out an
inspection to verify need for a
repair (section 33)

Local authority inspects
property using Housing
Health and Safety Rating
System (section 33)

If inspection verifies tenant's
complaint, inspector will take
appropriate action (section 33)

If local authority serves
Improvement Notice or
Notice of Emergency
Remedial Action, landlord
cannot evict tenant for six
months using the no-fault
eviction procedure (section
33(1) and (13))

(c) the tenant then made a complaint to the relevant local housing authority about the same subject matter as the complaint to the landlord,

(d) the relevant local housing authority served a relevant notice in relation to the dwelling-house in response to the complaint, and

(e) if the possession notice was not given before the tenant's complaint to the local housing authority, it was given before the service of the relevant notice (section 33(2)).

This does not apply where the operation of the relevant notice has been suspended (section 33(9)).

An adequate response by the landlord is to a response in writing which:

(a) provides a description of the action that the landlord proposes to take to address the complaint, and

(b) sets out a reasonable timescale within which that action will be taken (section 33(3)).

The procedure applies despite the complaint not being in writing where the tenant does not know the landlord's postal or email address (section 33(4)).

It also applies despite the tenant having made reasonable efforts to contact the landlord to complain about the condition of the dwelling-house but was unable to do so (section 33(5)).

The court must strike out proceedings for an order for possession if, before the order is made, the section 21 notice that would otherwise require the court to make an order for possession, has become invalid (section 33(6)).

An order for possession must not be set aside on the ground that a relevant notice was served in relation to the dwelling-house after the order for possession was made (section 33(7)).

Definitions

Relevant notice means a notice served under section 11 of the Housing Act 2004 (improvement notices relating to category 1 hazards), a notice served under section 12 of that Act (improvement notices relating to category 2 hazards), or a notice served under section 40(7) of that Act (emergency remedial action) (section 33(13)).

Section 21 notice means a notice given under section 21(1)(b) or (4)(a) of the Housing Act 1988 (recovery of possession on termination of shorthold tenancy) (section 33(13)).

PROTECTION FROM EVICTION

References

Protection from Eviction Act 1977
Rent (Scotland) Act 1984
The Private Tenancies (Northern Ireland) Order 2006
Rent (Northern Ireland) Order 1978

Extent

These provisions apply to England and Wales. Similar provisions apply in Scotland and Northern Ireland.

Scope

This act describes offences in relation to eviction of a tenant from his or her home, and for the correct service of notices of eviction.

Offences

It is an offence to:

- do acts likely to interfere with the peace or comfort of a tenant or anyone living with him or her; or
- persistently withdraw or withhold services for which the tenant has a reasonable need to live in the premises as a home.

It is also an offence to do any of the things described above intending, knowing, or having reasonable cause to believe, that they would cause the tenant to leave their home, or stop using part of it, or stop doing the things a tenant should normally expect to be able to do. It is also an offence to take someone's home away from him or her unlawfully.

A person who is convicted by magistrates of an offence under the Protection from Eviction Act may have to pay a maximum fine of £5,000, or be sent to prison for six months, or both. If the case goes to the Crown Court, the punishment can be prison for up to two years, or a fine, or both.

Legal proceedings

Local authorities have the power to start legal proceedings for offences of harassment and illegal eviction. If the evidence justifies it, they can carry out an investigation and prosecute if they believe an offence has been committed. In extreme cases of harassment, and where the property is in poor condition, a local authority may take over the management of a house in multiple occupation, by making it subject to a control order.

The Housing Act 1988 now makes it a general requirement for a licensor to obtain a court order before he or she can evict a licensee. However, certain licences and tenancies are excluded from this requirement.

Although it is not necessary to get a court order to evict someone in the excluded categories, there is a common law requirement for a landlord to serve a periodic tenant with notice equivalent to the period of the tenancy.

This means, for example, that if the tenancy was from month to month, the landlord must give a month's notice. (In the case of yearly tenancies, he or she must give six months' notice.) At common law, a licensee must be given notice which is reasonable in all the circumstances.

A landlord who evicts the tenant without going through the proper legal processes because he or she consider that bad behaviour on the tenant's part has provoked him or her, may say so in defence in court. The court may reduce the damages described here if:

- it considers that the tenant's behaviour or the behaviour of anybody living with the tenant, justifies awarding him or her less than the full damages; or
- it considers that, if the landlord did offer to let the tenant back into his or her home before the court proceedings began, it would have been reasonable for the tenant to accept that offer.

Chapter 6

TEMPORARY ACCOMMODATION

CONTROL OF MOVEABLE DWELLINGS
(TENTS, VANS AND SHEDS)

Reference

Public Health Act 1936, section 269 (as amended by Caravan Sites and Control of Development Act 1960)

Extent

These provisions apply to England and Wales.

Scope

The procedure applies to all moveable dwellings (other than caravans) and licences are required for:

(a) the use of land for camping purposes on more than 42 consecutive days or more than 60 days in any 12 consecutive months; or

(b) the keeping of a moveable dwelling on any one site, or two or more sites in succession if any of those sites is within 100 yards of another of them, for more than the same periods in (a).

In respect of (a), land which is in the occupation of the same person, and within 100 yards of a site on which a moveable dwelling is stationed during any part of a day, is regarded as being used for camping on that day (section 269(1)–(3)). The removal of a moveable dwelling for not more than 48 hours does not constitute an interruption of the 42 days' period (section 269(8)).

Exemptions

Apart from the use of land or moveable dwellings for periods less than those in (a) and (b) above, the following situations do not require licensing by the local authority:

(a) moveable dwellings kept by the owner on land occupied by him in connection with his dwelling if used only by him or members of his household;

(b) moveable dwellings kept by the owner on agricultural land occupied by him and used for habitation only at certain times of the year and only by persons employed in farming operations on that land;

FC87 Licensing of camping sites – section 269 Public Health Act 1936

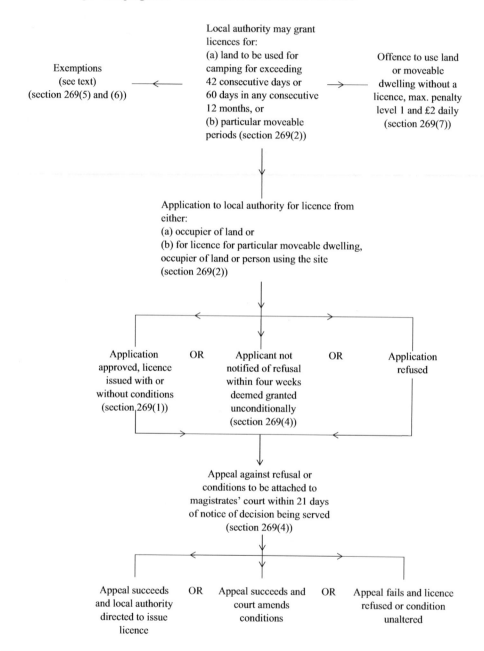

Exemptions
(see text)
(section 269(5) and (6))

Local authority may grant licences for:
(a) land to be used for camping for exceeding 42 consecutive days or 60 days in any consecutive 12 months, or
(b) particular moveable periods (section 269(2))

Offence to use land or moveable dwelling without a licence, max. penalty level 1 and £2 daily (section 269(7))

Application to local authority for licence from either:
(a) occupier of land or
(b) for licence for particular moveable dwelling, occupier of land or person using the site (section 269(2))

Application approved, licence issued with or without conditions (section 269(1))

OR

Applicant not notified of refusal within four weeks deemed granted unconditionally (section 269(4))

OR

Application refused

Appeal against refusal or conditions to be attached to magistrates' court within 21 days of notice of decision being served (section 269(4))

Appeal succeeds and local authority directed to issue licence

OR

Appeal succeeds and court amends conditions

OR

Appeal fails and licence refused or condition unaltered

Notes
1. This procedure does not apply to caravans. For licensing of caravan sites, see FC88.
2. For removal of unauthorised campers under the Criminal Justice and Public Order Act 1994, see FC112.

(c) moveable dwellings not in use for human habitation kept on premises by an occupier who does not permit moveable dwellings to be kept there for habitation;

(d) organisations granted a certificate of exemption by the minister, being satisfied that:

    (i) the camping sites belonging to, provided by or used by the organisation are properly managed and kept in good sanitary condition; and

    (ii) the moveable dwellings are used so as not to give rise to any nuisance (section 269(5) and (6)).

Applicants

Applications for licences may be made by:

(a) occupiers in respect of either the use of land or the stationing of moveable dwellings on that land; or

(b) a person intending to station a moveable dwelling on land (section 269(2) and (3)).

Conditions

Local authorities may attach to licences such conditions as they think fit with respect to:

(a) for licences authorising the use of land:

    (i) number and classes of moveable dwellings;

    (ii) the space between them;

    (iii) water supply; and

    (iv) for securing sanitary conditions;

(b) for licences authorising the use of a moveable dwelling for:

    (i) the use of that dwelling, including space to be kept free between dwellings;

    (ii) securing its removal at the end of a specified period; and

    (iii) for securing sanitary conditions (section 269(1)).

Definitions

Caravan – see p. 417.

Moveable dwelling includes any tent, any van or other conveyance (other than a caravan) whether on wheels or not, and . . . any shed or similar structure, being a tent, conveyance or structure which is used either regularly or at seasons only, or intermittently, for human habitation but does not include a structure to which the building regulations apply (section 269(8)).

MOBILE HOMES AND CARAVAN SITES

References

Caravan Sites and Control of Development Act 1960 Part I (as amended)

Mobile Homes (Site Licensing) (England) Regulations 2014

Advice to local authorities on the new regime for applications for the grant or transfer of a site licence, March 2015

Model Standards 2008 for Caravan Sites in England: Caravan Sites and Control of
    Development Act 1960 – section 5, April 2008
Caravans Act (Northern Ireland) 1963

Extent

The procedure applies in England, Wales and with modifications in Scotland (section 50).
Similar provisions apply in Northern Ireland.

Scope

The occupier of any land must not cause or permit the land to be used as a caravan site unless
he holds a site licence (section 1(1)).

Exemptions

The long list of exemptions from licensing requirements is set out in schedule 1 of the Act
in the following categories:

(a) incidental use within the curtilage of a dwelling-house;
(b) a single caravan for not more than two nights and 28 days in 12 months;
(c) holdings of five acres or more if not more than 28 days in 12 months and maximum
    three caravans at a time;
(d) sites occupied and supervised by organisations exempted by the minister;
(e) sites approved by exempted organisations for up to five caravans;
(f) meetings organised by exempted organisations;
(g) agriculture and forestry workers;
(h) building and engineering sites;
(i) travelling showmen;
(j) sites occupied by a local authority; and
(k) gypsy sites occupied by county councils or regional councils (section 2 and schedule 1).

Applications

The occupier of the land may apply to the local authority in writing and must specify the land
in question and, either at the time of making the application or later, give the local authority
such information as it may reasonably require (section 3). A fee may be payable.

Issue of licence

The local authority cannot issue the licence if:

(a) there has not been a formal grant of planning permission (section 3(3)); or
(b) the applicant has had a site licence revoked within the previous three years (section
    3(6)).

Licences must be issued within two months of giving the local authority of any information which it requires, or a longer period which may be agreed in writing between them

(section 3(4)). If it does not occur, no offence is committed in relation to not having a site licence (section 6).

Licences may not be time limited unless this is the case with the planning permission, in which case the licence must expire on the same date (section 4).

Conditions

The local authority may attach to the licence any conditions which it sees necessary or desirable to impose in the interest of persons living in the caravans, or of other classes of persons, or of the public at large. In forming conditions, the local authority must have regard to model standards issued by the minister from time to time.

In respect of conditions which require the carrying out of works, the works may be required within a specified period and caravans may be prohibited or restricted until the works have been completed. The local authority has power to undertake these works in default at the end of the period allowed and recover costs (other than with protected sites) (sections 5 and 9(3)).

The local authority must consult with the Fire and Rescue Authority before granting a site licence or making or varying conditions relating to fire precautions (section 5(3A) and (3B)).

Where the Regulatory Reform (Fire Safety) Order 2005 applies to the land concerned, no condition may be attached to the site licence in relation to any requirements or prohibitions which are or could be imposed under that Order (section 5 (2A)).

Appeal to magistrates' court against conditions attached to site licence

Any person aggrieved by any condition in a site licence may, within 28 days of the date the licence was issued, appeal to a magistrates' court or, in a case relating to land in England, to the Appropriate Tribunal. The court or tribunal, if satisfied that the condition is unduly burdensome, may vary or cancel the condition (section 7(1)).

In a case where the tribunal varies or cancels a condition, it may also attach a new condition to the licence (section 7(1A)).

If the effect of a condition is to require the carrying out of any works, the condition will not have effect during the appeal period against the condition or while an appeal is pending (section 7(2)).

Power of local authority to alter conditions attached to site licences

The conditions attached to a site licence may be altered at any time by the local authority, but before exercising their powers the local authority shall give the holder of the licence an opportunity of making representations (section 8(1)).

A local authority in England may require a fee for an application for the alteration of the conditions on the licence (section 8(1B)).

Where the holder of a site licence is aggrieved by any alteration or refusal of the conditions in an application, he may, within 28 days of the date on which written notification of the alteration or refusal is received by him, appeal to a magistrates' court or, in a case relating to land in England, to the Appropriate Tribunal. The court or tribunal may, if they allow the appeal, give the local authority directions (section 8(2)).

Such alterations do not have effect during the appeal against the alteration or while an appeal against the alteration is pending (section 8(3)).

The local authority must consult the Fire and Rescue Authority before exercising the powers in related conditions attached to a site licence (section 8(5)).

Breach of condition: land other than relevant protected sites in England

If an occupier of land, other than on a relevant protected site, fails to comply with any condition attached to a site licence, he is guilty of an offence and liable on summary conviction, in the case of the first offence to a fine not exceeding level 4 of the standard scale (section 9(1)).

Where a person convicted for failing to comply with a condition attached to a site licence has on two or more previous occasions been convicted for failing to comply with a condition attached to that licence, the court may, if the local authority applies, make an order for the revocation of the site licence to come into force on a date the court specify.

The person convicted or the local authority who issued the site licence may apply to the magistrates' court which has made the order revoking a site licence for an order extending the period at the end of which the revocation is to come into force, and the magistrates' court may, if satisfied that adequate notice of the application has been given to the local authority or, the person convicted, make an order extending that period (section 9(2)).

Where an occupier of land, other than a relevant protected site, fails within the time specified in a condition attached to a site licence to complete to the satisfaction of the local authority any works required by the condition, the local authority may carry out those works, and recover as a simple contract debt from that person any costs (section 9(3)).

Breach of condition: relevant protected sites in England – compliance notice

If it appears to a local authority who have issued a site licence in respect of a relevant protected site in their area that the occupier of the land concerned is failing or has failed to comply with a condition attached to the site licence, they may serve a compliance notice on the occupier (section 9A(1)).

A compliance notice is a notice which:

(a) sets out the condition in question and details of the failure to comply with it,
(b) requires the occupier of the land to take such steps as the local authority consider appropriate and specified in the notice in order to ensure that the condition is complied with,
(c) specifies the period within which those steps must be taken, and
(d) explains the right of appeal (section 9A(2)).

An occupier of land served with a compliance notice may appeal to the Appropriate Tribunal against that notice (section 9A(3)).

A local authority may revoke or vary a compliance notice by extending the period specified in the notice (section 9A(4)).

The power to revoke or vary a compliance notice is exercisable by the local authority on an application made by the occupier of land on whom the notice was served, or on the authority's own initiative (section 9A(5)).

Where a local authority revoke or vary a compliance notice, they must notify the occupier of the land of the decision as soon as is reasonably practicable (section 9A(6)).

Where a compliance notice is revoked, the revocation comes into force at the time it is made (section 9A(7)).

Where a compliance notice is varied –

(a) if the notice has not become operative when the variation is made, the variation comes into force at such time as the notice becomes operative in accordance with section 9H;
(b) if the notice has become operative when the variation is made, the variation comes into force at same the time (section 9A(8)).

Compliance notice: offence and multiple convictions

An occupier of land who has been served with a compliance notice which has become operative commits an offence if the occupier fails to take the steps specified in the notice within the period specified and is liable on summary conviction to a fine not exceeding level 5 on the standard scale (section 9B(1) and (2)).

In proceedings against an occupier of land for an offence, it is a defence that the occupier had a reasonable excuse for failing to take the steps within the period (section 9B(3)).

Where an occupier of land is convicted of an offence and has been convicted on two or more previous occasions of an offence in relation to the site licence, on an application by the local authority who served the compliance notice, the court may make an order revoking the site licence on the date specified in the order (section 9B(4) and (5)).

An order must not specify a date which is before the end of the period within which notice of appeal may be given against the conviction (section 9B(6)).

Where an appeal against the conviction is made by the occupier of the land before the date specified in an order, the order does not take effect until the appeal is finally determined, or the appeal is withdrawn (section 9B(7)).

On an application by the occupier of the land or by the local authority, the court which made the order may make an order specifying a date on which the revocation of the site licence takes effect later than the date specified in the order (section 9B(8)).

But the court must not make an order unless it is satisfied that adequate notice of the application has been given to the occupier of the land or to the local authority (section 9B(9)).

Compliance notice: power to demand expenses

When serving a compliance notice on an occupier of land, a local authority may impose a charge on the occupier as a means of recovering expenses incurred by them:

(a) in deciding whether to serve the notice, and
(b) in preparing and serving the notice or a demand (section 9C(1)).

The expenses include in particular the costs of obtaining expert advice (including legal advice) (section 9C(2)).

The power is exercisable by serving the compliance notice together with a demand which sets out:

(a) a detailed breakdown of the total expenses the local authority seek to recover,
(b) where the local authority propose to charge interest, the rate at which the relevant expenses carry interest (section 9C(3)).

Where a tribunal allows an appeal against the compliance notice with which a demand was served, it may make such order confirming, reducing or quashing any charge made in respect of the notice, and varying the demand as appropriate (section 9C(4)).

Power to take action following conviction of occupier

Where an occupier of land is convicted of an offence of failure to take steps required by a compliance notice, the local authority who issued the compliance notice may:

(a) take any steps required by the compliance notice to be taken by the occupier, but which have not been so taken; and
(b) take further action as the authority consider appropriate for ensuring that the condition specified in the compliance notice is complied with (section 9D(1)).

Where a local authority propose to take such action they must serve on the occupier of the land a notice which:

(a) identifies the land and the compliance notice,
(b) states that the authority intend to enter onto the land,
(c) describes the action the authority intend to take on the land,
(d) if the person whom the authority propose to authorise to take the action on their behalf is not an officer of theirs, states the name of that person, and
(e) sets out the dates and times on which it is intended that the action will be taken (start and completion) (section 9D(2)).

Power to take emergency action

A local authority in England who have issued a site licence of a relevant protected site may take action in relation to the land concerned if it appears to the authority that:

(a) the occupier of the land is failing or has failed to comply with a condition for the time being attached to the site licence, and
(b) as a result of that failure there is an imminent risk of serious harm to the health or safety of any person who is or may be on the land (section 9E(1)).

Such action a local authority must remove the imminent risk of serious harm (section 9E(2)).

Where a local authority propose to take emergency action, the authority must serve on the occupier of the land a notice similar to that of a compliance notice under section 9D(2) (section 9E(3)).

A notice may state that, if entry onto the land were to be refused, the authority would propose to apply for a warrant (section 9E(4)).

A notice must be served sufficiently in advance of when the local authority intend to enter onto the land as to give the occupier of the land reasonable notice of the intended entry (section 9E(5)).

Within seven days of the authority starting to take the emergency action, the authority must serve on the occupier of the land a notice which:

(a) describes the imminent risk of serious harm to the health or safety of persons who are or may be on the land,

(b) describes the emergency action which has been, and any emergency action which is to be, taken by the authority on the land,

(c) sets out when the authority started taking the emergency action and when the authority expect it to be completed,

(d) if the person whom the authority have authorised to take the action on their behalf is not an officer of theirs, states the name of that person, and

(e) explains the right of appeal (section 9E(8)).

An occupier of land in respect of which a local authority has taken or is taking emergency action may appeal to the Appropriate Tribunal against the taking of the action by the authority (section 9E(9)).

The grounds on which the appeal may be brought are:

(a) that there was no imminent risk of serious harm;

(b) that the action the authority has taken was not necessary to remove the imminent risk of serious harm (section 9E(10)).

Action under section 9D or 9E: power to demand expenses

Similar powers to works in default of a compliance notice apply to emergency (section 9F(1) and (2)).

In the case of emergency action, no charge may be imposed while appeal process is in operation (section 9F (3) and (4)).

The power to demand expenses is exercisable by serving on the occupier of the land a demand for the expenses similar to that of a compliance notice works in default (section 9F(6)).

An occupier of land who is served with a demand may appeal to the Appropriate Tribunal against the demand (section 9F(7)). A demand must be served with two months regarding section 9D of the charge being imposed or the action is completed (section 9F(8)).

Appeals under section 9A, 9E or 9F

An appeal under section 9A, 9E or 9F must be made within 21 days of the relevant document compliance notice, notice under subsection (8) or demand being served (section 9G(1) and (2)).

The tribunal may allow an appeal after the end of the appeal period if it is satisfied that there is a good reason for the failure to appeal in time (section 9G(3)).

An appeal is by re-hearing, but may be determined having regard to matters of which the local authority who made the decision were unaware (section 9G(4)).

The tribunal may by order, confirm, vary or quash the compliance notice, the decision of the local authority, or the demand (section 9G(5)).

Section 9H details when notices and demands become operative.

Recovery of expenses and interest

When a demand becomes operative, the relevant expenses set out in the demand will carry interest at such a rate the local authority may fix until recovery of all sums due under the demand; and the expenses and any interest are recoverable by them as a debt (section 9I(1)).

The expenses and any interest are, until recovery, are a charge on the land to which the compliance notice or emergency action (section 9I(2) and (3)).

Transfer of site licences, and transmission on death, etc.

When the holder of a site licence ceases to be the occupier of the land, he may, with the consent of the local authority, transfer the licence to the person who then becomes the occupier of the land with a fee (section 10(1) and (1A)).

Where a local authority give their consent to the transfer of a site licence, they shall endorse on the licence the name of the person to whom it is transferred and the date agreed (section 10(2)).

If an application is made for consent to the transfer of a site licence, other than one for a relevant protected site, to a person who is to become the occupier of the land, that person may apply for a site licence, and if the local authority at any time before issuing a site licence in compliance with that application give their consent to the transfer they need not proceed with the application for the site licence (section 10(3)).

Where any person becomes entitled to an estate or interest in land in respect of which a site licence is in force and is to become the occupier of the land, he shall be treated as having become the holder of the licence on the day on which he became the occupier of the land, and the local authority must, if an application in that behalf is made to them, endorse his name and the date on the licence (section 10(4)).

Duty of licence holder to surrender licence for alteration

A local authority may at any time require the licence holder to provide it to enable them to enter in it any alteration of the conditions or other terms of the licence (section 11(1)).

If the holder of a site licence fails without reasonable excuse to comply with a requirement, he is liable on summary conviction to a fine not exceeding level 1 on the standard scale (section 11(2)).

Registers of site licences

Every local authority must keep a register of site licences, and it must be open for inspection by the public at all reasonable times (section 25(1)).

Where a local authority endorse on a site licence the name of any person they must record his name, and the date entered in the licence, in the register (section 25(2)).

The Mobile Homes (Site Licensing) (England) Regulations 2014 prescribe the detail of how local authorities exercise their discretion when deciding whether or not to issue or consent to the transfer of a site licence in respect of a relevant protected site.

Prescribed matters for issuing or transferring relevant protected site licences

Regulation 3 details the matters for issuing and transferring site licences where the site is a relevant protected site.

Information to be provided to the local authority

A local authority may specify information or documents it requires in relation to an application for consent to transfer a site licence in respect of a relevant protected site –

(a) to accompany an application; or
(b) following receipt of an application, in which case, the authority may specify a date by which the information or documents must be submitted (regulation 4).

Right of appeal

The applicant may appeal to the Appropriate Tribunal against a local authority's decision not to issue, or consent to the transfer of, a site licence in respect of a relevant protected site within 28 days of notification of the decision by the local authority (regulation 4(1)).

The appeal shall be a re-hearing of the local authority's decision and determined having regard to:

(a) any undertaking given to the tribunal in relation to one or more of the matters set out in regulation 3(4); and

(b) any other matters that the tribunal thinks are relevant (regulation 4(2)).

On determining an appeal, the tribunal may confirm, reverse the local authority's decision by ordering that the local authority issues a site licence, or consents to the transfer of a site licence (regulation 4(3)).

Definitions

Caravan means any structure designed or adapted for human habitation which is capable of being moved from one place to another (whether by being towed, or by being transported on a motor vehicle or trailer) and any motor vehicle so designed or adapted, but does not include:

(a) any railway rolling stock which is for the time being on rails forming part of a railway system; or

(b) any tent (section 29(1)),

but a structure designed or adapted for human habitation which –

(a) is composed of not more than two sections separately constructed and designed to be assembled on a site by means of bolts, clamps or other devices; and

(b) is, when assembled, physically capable of being moved by road from one place to another (whether by being towed, or by being transported on a motor vehicle or trailer),

shall not be treated as not being (or as not having been) a caravan within the meaning of Part I of the Caravan Sites and Control of Development Act 1960 by reason only that it cannot lawfully be so moved on a highway when assembled.

In addition, the expression 'caravan' shall not include a structure designed or adapted for human habitation which falls within paragraphs (a) and (b) above if its dimensions when assembled exceed any of the following limits, namely –

(a) length (exclusive of any drawbar): 65.616 feet (20 metres);

(b) width: 22.309 feet (6.8 metres);

(c) overall height of living accommodation (measured internally from the floor at the lowest level to the ceiling at the highest level): 10.006 feet (3.05 metres) (Caravan Sites Act 1968, section 13).

Caravan site means land upon which a caravan is stationed for the purposes of human habitation and land which is used in conjunction with land on which a caravan is so stationed (section 1(4)).

Relevant Protected Site means a site with planning permission or a site licence or both that permit the site to be used for residential use all year round. If the site benefits from a permission that allows a mixture of holiday and residential use then the site will still qualify as a Relevant Protected Site, unless the residential units on a mixed site are solely occupied by the owner and/or person(s) employed by the occupier but who do not occupy the caravan under an agreement to which the Mobile Homes Act 1983 applies (section 29(2)).

FC88 Licensing of caravan sites – Caravan Sites and Control of Development Act 1960

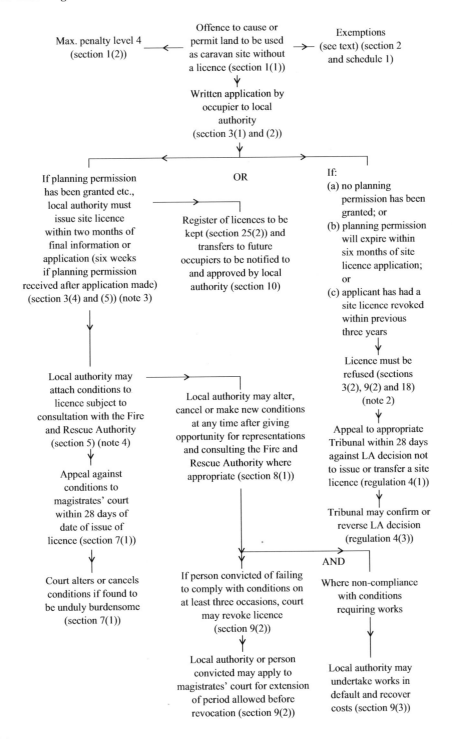

Notes

1. For licensing of camping sites (other than for caravans), see FC87.
2. There is no power for the local authority to refuse the licence for reasons other than these.
3. The licence is not to be issued for a limited period unless this is the case with the planning consent, in which case the two must be brought together (section 4).
4. Maximum penalty for failure to comply with conditions – level 4 (section 9(1)).

**FC89 Compliance notices and emergency action for relevant protected caravan sites –
Caravan Sites and Control of Development Act 1960 Part I (as amended) and Mobile Homes
(Site Licensing) (England) Regulations 2014**

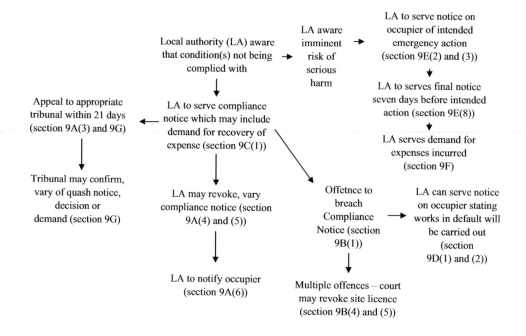

Local authority (LA) aware that condition(s) not being complied with

LA aware imminent risk of serious harm

LA to serve notice on occupier of intended emergency action (section 9E(2) and (3))

LA to serves final notice seven days before intended action (section 9E(8))

LA serves demand for expenses incurred (section 9F)

Appeal to appropriate tribunal within 21 days (section 9A(3) and 9G)

LA to serve compliance notice which may include demand for recovery of expense (section 9C(1))

Tribunal may confirm, vary of quash notice, decision or demand (section 9G)

LA may revoke, vary compliance notice (section 9A(4) and (5))

Offetnce to breach Compliance Notice (section 9B(1))

LA can serve notice on occupier stating works in default will be carried out (section 9D(1) and (2))

LA to notify occupier (section 9A(6))

Multiple offences – court may revoke site licence (section 9B(4) and (5))

PART 6

ANTISOCIAL BEHAVIOUR

ANTISOCIAL BEHAVIOUR: LANDLORDS' POLICIES AND POSSESSION PROCEDURES

References

Housing Act 1985 sections 82A to 86
Housing Act 1988 sections 6A to 9A, 20B and 20C
Housing Act 1996 section 218A
The Antisocial Behaviour (Northern Ireland) Order 2004

Extent

These provisions apply to England, Wales and Scotland. Similar provisions apply in Northern Ireland.

Scope

These provisions cover the requirement for local housing authorities to prepare a policy about dealing with antisocial behaviour of its tenants, and obtaining possession of the house as a result of antisocial behaviour of the tenant.

Demotion orders reduce the rights of the tenants and can lead to tenants being more easily evicted due to antisocial behaviour. There are separate provisions for secured tenancies and assured tenancies.

Antisocial behaviour policies

A local housing authority, housing action trust or registered social landlord must prepare a policy in relation to antisocial behaviour and procedures for dealing with occurrences of antisocial behaviour (section 218A(1) and (2)).

The landlord must publish a statement of the policy and procedures and from time to time keep the policy and procedures under review and, when it thinks appropriate, publish a revised statement (section 218A(3)and (4)).

A copy of a statement must be available for inspection at all reasonable hours at the landlord's principal office and must be provided on payment of a reasonable fee to any person who requests it (section 218A(5)).

The landlord must also prepare a summary of its current policy and procedures and provide a free copy of the summary to any person who requests it (section 218A(6)).

In preparing and reviewing the policy and procedures the landlord must have regard to issued guidance (section 218A(7)).

Demotion orders due to antisocial behaviour – secure tenancy (sections 82A to 86 Housing Act 1985)

A local housing authority, housing action trust or registered social landlord may apply to a county court for a demotion order (section 82A(1) and (2)).

The effect of a demotion order is that it can terminate a secure tenancy with effect from the date specified in the order and if the tenant remains in occupation of the house after that date, a demoted tenancy is created with effect from that date. It is a term of the demoted tenancy that any arrears of rent payable at the termination of the secure tenancy become payable under the demoted tenancy and any rent paid in advance or overpaid at the termination of the secure tenancy is credited to the tenant's liability to pay rent under the demoted tenancy (section 82A(3)).

The court must be satisfied that the tenant or a person residing in or visiting the dwelling-house has engaged or has threatened to engage in antisocial behaviour or use of premises for unlawful purposes and that it is reasonable to make the order (section 82A(4)).

A demoted tenancy has the same effect in relation to the secure tenancy with regard to the parties to the tenancy, the period of the tenancy, the amount of the rent and the dates on which the rent is payable (section 82A(5)). This does not apply if the secure tenancy was for a fixed term and in such a case the demoted tenancy is a weekly periodic tenancy (section 82A(6)).

If the landlord of the demoted tenancy serves on the tenant a statement of any other express terms of the secure tenancy which are to apply to the demoted tenancy such terms are also terms of the demoted tenancy (section 82A(7)).

Before court proceedings, the landlord must serve a prescribed notice on the tenant specifying the ground and its particulars on which the court will be asked to make the order (section 83(1) and (2)). The grounds are in Schedule 2 and depending on the ground certain dates need also to be included.

Demotion orders – assured shorthold tenancies (section 6A to 9A Housing Act 1988)

If a tenant has an assured tenancy, a registered social landlord may apply to a county court for a demotion order (section 6A(1) and (2)). If granted a demotion order the assured tenancy is terminated with effect from the date specified in the order and if the tenant remains in occupation of the dwelling-house after that date a demoted tenancy is created with effect from that date. It is a term of the demoted tenancy that any arrears of rent payable at the termination of the assured tenancy become payable under the demoted tenancy and that any rent paid in advance or overpaid at the termination of the assured tenancy is credited to the tenant's liability to pay rent under the demoted tenancy (section 6A(3)).

The court must be satisfied that the tenant or a person residing in or visiting the dwelling-house has engaged or has threatened to engage in antisocial behaviour or use of premises for unlawful purposes, and that it is reasonable to make the order (section 6A(4)).

The court must not consider a demotion order unless the landlord has served on the tenant a notice below, or it thinks it is just and equitable to dispense with the requirement of the notice (section 6A(5)).

The notice must give particulars of the conduct in respect of which the order is sought; state that the proceedings will not begin before the date specified in the notice and that the proceedings will not begin until twelve months and 2 weeks after the date of service of the notice (section 6A(6)and (7)).

A demoted tenancy has the same terms as the assured tenancy (section 6A(8)). This does not apply if the assured tenancy was for a fixed term and in such a case the demoted tenancy is a weekly periodic tenancy.

If the landlord of the demoted tenancy serves on the tenant a statement of any other express terms of the assured tenancy which are to apply to the demoted tenancy such terms are also terms of the demoted tenancy (section 6A(10)).

An assured tenancy is an assured shorthold tenancy (a demoted assured shorthold tenancy) if the tenancy is created by court order under section 82A of the Housing Act 1985 or section 6A of the Housing Act 1988 (a demotion order), and the landlord is a registered social landlord (section 20B(1)).

One year after the demotion order takes effect a demoted assured shorthold tenancy ceases to be an assured shorthold tenancy (section 20B(2)), however, if before the end of that period the landlord gives notice of proceedings for possession of the dwelling-house the tenancy continues to be a demoted assured shorthold tenancy until the notice of proceedings for possession is withdrawn; the proceedings are determined in favour of the tenant or the period of six months beginning with the date on which the notice is given ends and no proceedings for possession have been brought (section 20B(3) and (4)).

Absolute grounds for possession for antisocial behaviour

If the court is satisfied that any of the conditions in schedule 2 of the Housing Act 1985 is met, it must make an order for the possession of a dwelling-house let under a secure tenancy (section 84A(1)).

This only applies only where the landlord has complied with any obligations under a review of decision to seek possession (section 84A(2)).

The conditions under which proceedings can be initiated are:

Condition 1 – the tenant has been convicted of a serious offence in or near the house.
Condition 2 – the tenant has breached an injunction.
Condition 3 – the tenant has been convicted of an offence under section 30 of the Antisocial Behaviour, Crime and Policing Act 2014.
Condition 4 – the house has been subject to a relevant closure order.
Condition 5 – the tenant has been convicted of an offence statutory nuisance legislation relating to noise.

Condition 1, 2, 3, 4 or 5 is not met if –

(a) there is an appeal against the conviction, finding or order concerned which has not been finally determined, abandoned or withdrawn, or
(b) the final determination of the appeal results in the conviction, finding or order being overturned (section 84A(8)).

Notice requirements

In proceedings for possession of a dwelling-house for *absolute ground for possession* for antisocial behaviour the court must not entertain the proceedings unless the landlord has served a notice on the tenant which must:

(a) state that the court will be asked to make an order for the possession of the dwelling-house,
(b) set out the reasons for the landlord's decision to apply for the order, and
(c) inform the tenant of any right that the tenant to request a review of the landlord's decision and of the time within which the request must be made.

In a case where possession is also sought on one or more of the grounds set out in schedule 2 of the Housing Act 1985, the notice must also –

(a) specify the ground on which the court will be asked to make the order, and
(b) give particulars of that ground.

A notice which states which conditions that the landlord proposes to rely:

(a) must also state the conviction on which the landlord proposes to rely, and
(b) must be served on the tenant within 12 months of the conviction, or

if there is an appeal against the conviction, within 12 months of the appeal being finally determined, abandoned or withdrawn.

A notice which states that the landlord proposes to rely upon condition 4:

(a) must also state the closure order concerned, and
(b) must be served on the tenant within three months of the closure order, or if there is an appeal against the order, within three months of the appeal being finally determined, abandoned or withdrawn.

Notices must also inform the tenant that, if the tenant needs help or advice about the notice and what to do about it, the tenant should take it immediately to a Citizens' Advice Bureau, a housing aid centre, a law centre or a solicitor.

They must also specify the date after which proceedings for the possession of the dwelling-house may be begun (not earlier than a notice to quit date or one month after the notice is served), and ceases to be in force 12 months after the date specified (section 83ZA).

Review of decision to seek possession on absolute ground for antisocial behaviour

A tenant may request a review of a landlord's decision to seek an order for possession of a dwelling-house if the landlord is a local housing authority, or a housing action trust (section 85ZA(1)).

Such a request must be made in writing within 7 days of the notice being served and the landlord must review its decision (section 85ZA(2) and (3)).

The landlord must notify the tenant in writing of the decision on the review and if the decision is to confirm the original decision, the landlord must also notify the tenant of the reasons for the decision (section 85ZA(4) and (5)).

The review must be carried out, and the tenant notified, before the date on which proceedings for the possession of the dwelling-house may start (section 85ZA(6)).

Where a court is considering whether it is reasonable to make an **order for possession** due to the conduct of tenant or other person it must consider, in particular, the effect that the nuisance or annoyance has had on persons other than the person against whom the order is sought; any continuing effect the nuisance or annoyance is likely to have on such persons and the effect that the nuisance or annoyance would be likely to have on such persons if the conduct is repeated (section 85A(1) and(2) and section (9A).

Definitions

Relevant proceedings means proceedings for contempt of court or proceedings under Schedule 2 to the Antisocial Behaviour, Crime and Policing Act 2014.

Serious offence means an offence which –

(a) was committed on or after the day on which subsection (3) comes into force,

(b) is specified, or falls within a description specified, in Schedule 2A at the time the offence was committed and at the time the court is considering the matter, and

(c) is not an offence that is triable only summarily by virtue of section 22 of the Magistrates' Courts Act 1980 (either-way offences where value involved is small) (section 84A(9)).

CLOSURE OF PREMISES ASSOCIATED WITH NUISANCE OR DISORDER ETC.

References

Antisocial Behaviour, Crime and Policing Act 2014 sections 76 to 93
Antisocial Behaviour etc. (Scotland) Act 2004
The Antisocial Behaviour (Northern Ireland) Order 2004
Anti-social Behaviour, Crime and Policing Act 2014: Anti-social behaviour powers statutory guidance for frontline professionals, Updated December 2017

Extent

These provisions apply to England and Wales. Similar provisions apply to Scotland and Northern Ireland.

Scope

These powers should be used appropriately to provide a proportionate response to the specific behaviour that is causing harm or nuisance without impacting adversely on behaviour that is neither unlawful nor antisocial. Those who work within and for local communities are best placed to understand what is driving the behaviour in question, the impact that it is having, and to determine the most appropriate response.

The guidance focuses on putting victims at the heart of the response to antisocial behaviour. It also focuses on the use of the powers provided by the 2014 Act which are designed to be flexible to ensure that local agencies have the tools they need to respond to different forms of antisocial behaviour.

Closure notices

The closure of premises associated with nuisance or disorder has two stages – the closure notice and the closure order.

A notice is issued out of court, can be issued for a maximum of 24 hours, and cannot prohibit access by the owner of the premises or people who habitually live on the premises. The notice can be designed to prohibit access to particular people at particular times. For example, where a property is closed in anticipation of a party publicised through social media, the family who lived there would not be prohibited, and additional people could also be exempted (such as other family members) where appropriate.

A police officer of at least the rank of inspector, or the local authority, may issue a closure notice if satisfied on reasonable grounds that the use of particular premises has resulted, or (if the notice is not issued) is likely soon to result, in nuisance to members of the public, or that there has been, or (if the notice is not issued) is likely soon to be, disorder near those premises associated with the use of those premises, and that the notice is necessary to prevent the nuisance or disorder from continuing, recurring or occurring (section 76(1)).

A closure notice is a notice prohibiting access to the premises for a period specified in the notice (section 76(2)). The maximum period is 24 hours (section 77).

A closure notice may prohibit access by all persons except those specified, or by all persons except those of a specified description; at all times, or at all times except those specified; or in all circumstances, or in all circumstances except those specified. (section 76(3)).

A closure notice cannot prohibit access by people who habitually live on the premises, or the owner of the premises (section 76(4)).

A closure notice must

(a) identify the premises;
(b) explain the effect of the notice;
(c) state that failure to comply with the notice is an offence;
(d) state that an application will be made for a closure order;
(e) specify when and where the application will be heard;
(f) explain the effect of a closure order;
(g) give information about the names of, and means of contacting, persons and organisations in the area that provide advice about housing and legal matters. (section 76(5)).

A closure notice may be issued only if reasonable efforts have been made to inform people who live on the premises (whether habitually or not), and any person who has control of or responsibility for the premises or who has an interest in them, that the notice is going to be issued (section 76(6)).

Before issuing a closure notice the police officer or local authority must ensure that any body or individual the officer or authority thinks appropriate has been consulted (section 76(7)).

Examples of use: closing a nightclub where the police have intelligence to suggest that disorder is likely in the immediate vicinity on a specific Friday night. closing a property where loud music is consistently being played at unsociable hours in a residential area, where negotiation had failed to resolve the issue.

It could be a premises used for drug dealing that is associated with serious antisocial behaviour in the immediate vicinity, or premises where the persistent behaviour of the residents – for example, visitors coming and going at all hours; frequent loud parties; and harassment and intimidation of neighbours – is associated with serious antisocial behaviour in the immediate vicinity (Hansard).

Duration of closure notices

The maximum period that may be specified in a closure notice is 24 hours (section 77(1)). However, the maximum period is 48 hours if, in the case of a notice issued by a local authority, the notice is signed by the chief executive officer or a person designated by him or her (section 77(2)).

The 24-hour period specified in a closure notice may be extended by up to 24 hours if the authority issues an extension notice signed by the chief executive officer of the authority or a person designated by the chief executive officer (section 77(4)). An extension notice is a notice which identifies the closure notice, and specifies the period of the extension (section 77(5)).

Cancellation or variation of closure notices

If the circumstances leading to the issue of a closure order no longer apply, it should be cancelled or varied, it must be signed off by the person who originally issued the notice or, if they are not available, the chief executive of the local authority or a person delegated by him. A cancellation notice is a notice cancelling the closure notice. A variation notice is a notice varying the closure notice so that it does not apply to the part of the premises (section 78 (1) to (5)).

Service of notices

A closure notice, an extension notice, a cancellation notice or a variation notice must be served by a constable, or a representative of the authority that issued the notice by fixing it and giving it to the relevant premises, person or to occupiers (section 79(1), (2) and (3)).

The constable or local authority representative may enter any premises, using reasonable force if necessary, to fix the notice on a prominent place on the premises (section 79(4)).

Closure orders

Whenever a closure notice is issued an application must be made to a magistrates' court for a closure order unless it has been cancelled (section 80(1)).

An application for a closure order must be made by a constable or by the authority that issued the closure notice (section 80(2)).

The application must be heard by the magistrates' court not later than 48 hours after service of the closure notice (section 80(3)).

The court may make a closure order if it is satisfied –

(a) that a person has engaged, or (if the order is not made) is likely to engage, in disorderly, offensive or criminal behaviour on the premises, or
(b) that the use of the premises has resulted, or (if the order is not made) is likely to result, in serious nuisance to members of the public, or
(c) that there has been, or (if the order is not made) is likely to be, disorder near those premises associated with the use of those premises,

and that the order is necessary to prevent the behaviour, nuisance or disorder from continuing, recurring or occurring (section 80(5)).

A closure order is an order prohibiting access to the premises (or part of it) for a period specified in the order which may not exceed three months (section 80(6)).

A closure order may prohibit access by all persons, or by all persons except those specified, or by all persons except those of a specified description; at all times, or at all times except those specified; or in all circumstances, or in all circumstances except those specified (section 80(7)).

The court must notify the relevant licensing authority if it makes a closure order in relation to premises with a premises licence (section 80(9)).

A person or court deciding whether to issue a closure notice or order may take into account things that happened before the commencement day, and would have given rise to the power to issue one of the notices or to make an order (section 93(4) and (5)).

Temporary orders

Where an application has been made to a magistrates' court for a closure order, if the court does not make a closure order it may nevertheless order that the closure notice continues in force for a specified further period of not more than 48 hours, if satisfied:

(a) that the use of particular premises has resulted, or (if the notice is not continued) is likely soon to result, in nuisance to members of the public, or
(b) that there has been, or (if the notice is not continued) is likely soon to be, disorder near those premises associated with the use of those premises,

and that the continuation of the notice is necessary to prevent the nuisance or disorder from continuing, recurring or occurring (section 81(1) and (2)).

The court may adjourn the hearing of the application for not more than 14 days to enable the occupier of the premises, the person with control of or responsibility for the premises, or any other person with an interest in the premises, to show why a closure order should not be made (section 81(3)).

If the court adjourns the hearing it may order that the closure notice continues in force until the end of the period of the adjournment (section 81(4)).

Extension of closure orders

At any time before the expiry of a closure order, an application may be made to a justice of the peace for an extension of the period for which the order is in force (section 82(1)).

The application can be made only on reasonable grounds that it is necessary for the period of the order to be extended to prevent the occurrence, recurrence or continuance of:

(a) disorderly, offensive or criminal behaviour on the premises,
(b) serious nuisance to members of the public resulting from the use of the premises, or
(c) disorder near the premises associated with the use of the premises,

and also satisfied that the local authority or police has been consulted about the intention to make the application (section 82(3) and (4)).

Where an application is made, the justice of the peace may issue a summons directed to any person on whom the closure notice was served, or any other person with an interest in the premises, requiring the person to appear before the magistrates' court to respond to the application (section 82(5)).

If such a summons is issued a notice stating the date, time and place of the hearing of the application must be served on the persons to whom the summons is directed (section 82(6)).

If the magistrates' court is satisfied the order should be extended it may make an order extending (or further extending) the period of the closure order by a period not exceeding 3 months (section 82(7)).

The period of a closure order cannot be extended so that the order lasts for more than 6 months (section 82(8)).

Discharge of closure orders

At any time before the expiry of a closure order, an application may be made to a justice of the peace for the order to be discharged (section 83(1)).

Application can be made by the constable or the authority that applied for the closure order or a person on whom the closure notice was served or anyone else who has an interest in the premises (section 83(2)).

Where a person other than a constable or authority makes an application for the discharge of an order that was made on the application of a constable or authority, the justice may issue a summons to the constable or authority considered appropriate by the justice requiring him or her to appear before the magistrates' court to respond to the application (section 83(3) and (5)). If such a summons is issued, a notice stating the date, time and place of the hearing of the application must be served on that person (section 83(4) and (6)).

The magistrates' court may not make an order discharging the closure order unless satisfied that the closure order is no longer necessary to prevent the occurrence, recurrence or continuance of disorderly, offensive or criminal behaviour on the premises, serious nuisance to members of the public resulting from the use of the premises, or disorder near the premises associated with the use of the premises (section 83(7)).

Appeals

An appeal against a decision to make or extend a closure order may be made by a person on whom the closure notice was served or anyone else who has an interest in the premises but on whom the closure notice was not served (section 84(1)).

A constable or local authority may appeal against –

(a) a decision not to make a closure order applied for by a constable or local authority;
(b) a decision not to extend a closure order made on the application of a constable or local authority;
(c) a decision not to order the continuation in force of a closure notice issued by a constable or local authority (section 84(2) and (3)).

An appeal is to the Crown Court (section 84(4)). An appeal must be made within 21 days of the decision (section 84(5)). On appeal, the Crown Court may make whatever order it thinks appropriate (section 84(6)).

The Crown Court must notify the relevant licensing authority if it makes a closure order (section 84(7)).

Enforcement of closure orders

An authorised person may enter premises using reasonable force in respect of which a closure order is in force and do anything necessary to secure the premises against entry (section 85(1) and (3)). A person seeking to enter premises must, if required to do so by or on behalf of the owner, occupier or other person in charge of the premises, produce evidence of his or her identity and authority before entering the premises (section 85(4)). An authorised person may also enter premises in respect of which a closure order is in force to carry out essential maintenance or repairs to the premises (section 85(5)).

Offences

A person who without reasonable excuse remains on or enters premises in contravention of a closure notice (including a temporary closure notice) commits an offence and is liable on

FC90 Closure of licensed premises (Licensing Act 2003)

Closure order made under
ASB, C and P Act 2014 or
Immigration Act 2016 (note 1)

↓

Licensing authority to review
premise licence with 28 days
(section 167)

↓

LA sends notice of review
and closure to licence holder
and responsible authorities
(section 167)

↓

LA advertises review and
invites representatives
(section 167)

↓

Hearing to consider order
(section 167)

LA holds representations
frivolous etc. and notifies
persons make representations
with reasons (section 167)

LA may decide to revoke the
licence (section 168)   →   Offence to open premises
(section 168)

summary conviction to imprisonment for a period not exceeding 3 months, or to a fine, or to both (section 86(1) and (4)).

A person who without reasonable excuse remains on or enters premises in contravention of a closure order commits an offence and is liable on summary conviction to imprisonment for a period not exceeding 51 weeks, or to a fine, or to both. (section 86(2) and (5)).

Reimbursement of costs

A local policing body or a local authority that incurs expenditure for the purpose of clearing, securing or maintaining premises in respect of which a closure order is in force may apply to the court that made the order for an order to reimburse their costs (section 88(1)).

On an application the court may make whatever order it thinks appropriate for the reimbursement (in full or in part) by the owner or occupier of the premises of the expenditure (section 88(2)).

An application for an order must be made within three months of the date the closure order ceases to have effect (section 88(3)).

An order may be made only against a person who has been served with the application for the order (section 88(4)).

An application must also be served on the local policing body for the area in which the premises are situated, if the application is made by a local authority, the local authority, if the application is made by a local policing body (section 88(5)).

Example – If the owner of the premises has rented out the property, they may be wholly unaware of the problems being caused by the occupiers. Alternatively, they may have worked actively with either the police or the local authority to try to stop the behaviour. In such circumstances, it would, on the face of it, be unreasonable for an order for the reimbursement of costs to be made against the owner rather than the occupier.

However, it is not necessarily always the case that where the premises have been rented out or leased, the liability should fall on the tenant or lessee. There may be instances when the landlord refuses to engage with the police or local authority to help tackle an issue relating to their premises. In such cases, it may be that liability to meet the costs incurred by the police or local authority in clearing, securing or maintaining the premises should fall on the owner rather than the occupier (Hansard).

Compensation

A person who claims to have incurred financial loss in consequence of a closure notice or a closure order may apply to the appropriate court for compensation (section 90(1)).

The appropriate court is the magistrates' court that considered the application for a closure order and the Crown Court, in the case of a closure order that was made or extended by an order of that Court on an appeal (section 90(2)).

An application may not be heard unless it is made within three months of the closure notice being cancelled; a closure order being refused or the closure order ceasing to have effect (section 90(3)).

On an application the court may order the payment of compensation out of central funds if it is satisfied that the applicant:

- is not associated with the use of or behaviour on the premises, on the basis of which the closure notice was issued or the closure order made;

- took reasonable steps to prevent that use or behaviour, has incurred financial loss in consequence of the notice or order, and
- that having regard to all the circumstances it is appropriate to order payment of compensation in respect of that loss (section 90(5)).

Definitions

An authorised person is:

(a) in relation to a closure order made on the application of a constable, means a constable or a person authorised by the chief officer of police for the area in which the premises are situated;

(b) in relation to a closure order made on the application of a local authority, means a person authorised by that authority (section 85(2)).

Offensive behaviour means behaviour by a person that causes or is likely to cause harassment, alarm or distress to one or more other persons not of the same household as that person (section 92).

COMMUNITY PROTECTION NOTICES

References

Antisocial Behaviour, Crime and Policing Act 2014, sections 43 to 57
Antisocial Behaviour etc. (Scotland) Act 2004
The Antisocial Behaviour (Northern Ireland) Order 2004

Extent

These provisions apply to England and Wales. Similar provisions apply to Scotland and Northern Ireland.

Scope

The community protection notice (CPN) is intended to deal with unreasonable, ongoing problems or nuisances which negatively affect the community's quality of life by targeting the person responsible.

For instance, where a dog was repeatedly escaping from its owner's back garden due to a broken fence, the owner could be issued with a notice requiring that they fix the fence to avoid further escapes and also, if appropriate, ensure that the owner and dog attended training sessions to improve behaviour (if this was also an issue).

It is not meant to replace the statutory nuisance regime, although there is no legal bar to it being used where behaviour is such as to amount to a statutory nuisance. For example, a local authority could issue a CPN to address antisocial behaviour while investigating whether it constitutes statutory nuisance.

The notice should be issued to someone who can be held responsible for the antisocial behaviour. For instance, if a small shop were allowing litter to be deposited outside the property and not dealing with the issue, a notice could be issued to the business owner, whereas if a large national supermarket were to cause a similar issue, the company itself or the store manager could be issued with a notice.

Before issuing a notice, an authorised person is required to inform whatever agencies or persons he or she considered appropriate (for example the landlord of the person in question, or the local authority), partly in order to avoid duplication. The authorised person would also have to have issued a written warning in advance and allowed an appropriate amount of time to pass. This is to ensure that the perpetrator is aware of their behaviour and allows them time to rectify the situation. It will be for the person issuing the written warning to decide how long is appropriate before serving a notice. In the example above where a dog owner's fence needs to be fixed, this could be days or weeks, in order to allow the individual to address the problem. However, it could be minutes or hours in a case where, for example, skateboarders were causing nuisance to a local community.

Remedial works or works in default can be added to the notice immediately or once the individual, business or organisation has had sufficient time to comply with any requirements. For instance, if the behaviour related to a front garden full of rubbish, the individual could be given a period of seven days to clear the waste. The issuing officer could also make clear on the face of the notice that if this were not complied with, they would authorise the works in default on a given date and at a given cost. Consent would only be required when that work necessitated entry to the perpetrator's property – those issuing a notice would be able to carry out remedial works in default in areas 'open to the air', for instance clearing rubbish from a front garden.

CPNs have the following effects:

a. They cover a wider range of behaviour (all behaviour that is detrimental to the local community's quality of life) rather than specifically stating the behaviour covered (for example, litter or graffiti);

b. Noise disturbance could be tackled, particularly if it is demonstrated to be occurring in conjunction with other antisocial behaviour;

c. Notices can be issued by a wider range of agencies: the police, local authorities and private registered providers of social housing (if approved by local authorities), thereby enabling the most appropriate agency to deal with the situation;

d. Notices can apply to businesses and individuals (which is the same as for some of the notices they will replace but not all); and

e. It would be a criminal offence if a person did not comply, with a sanction of a fine (or fixed penalty notice) for non-compliance (Hansard).

Power to issue notices

An authorised person may issue a community protection notice to an individual aged 16 or over, or a body, if satisfied on reasonable grounds that the conduct of the individual or body is having a detrimental effect, of a persistent or continuing nature, on the quality of life of those in the locality, and the conduct is unreasonable (section 43(1)).

A CPN is a notice that imposes any of the following requirements on the individual or body issued with it –

- a requirement to stop doing specified things;
- a requirement to do specified things or
- a requirement to take reasonable steps to achieve specified results (section 43(3)).

The only requirements that may be imposed are ones that are reasonable to impose in order to prevent the detrimental effect from continuing or recurring, or to reduce that detrimental effect or to reduce the risk of its continuance or recurrence (section 43(4)).

A person (A) may issue a CPN to an individual or body (B) only if –

(a) B has been given a written warning that the notice will be issued unless B's conduct ceases to have the detrimental effect, and
(b) A is satisfied that, despite B having had enough time to deal with the matter, B's conduct is still having that effect (section 43(5)).

Before an authorised person issues a community protection notice he/she must inform any body or individual the person thinks appropriate (section 43(6)).

A CPN must identify the relevant conduct and explain the effect of the notice and its related issues and may specify periods within which, or times by which, requirements are to be complied with (section 43(7) and (8)).

Occupiers of premises etc.

Conduct on, or affecting, premises that a particular person owns, leases, occupies, controls, operates, or maintains, is treated as conduct of that person (section 44(1)).

Conduct on, or affecting, premises occupied for the purposes of a government department is treated as conduct of the Minister in charge of that department (section 44(2)).

An individual's conduct is not regarded as that of another person if that person cannot reasonably be expected to control or affect it (section 44(3)).

Occupier or owner unascertainable

Where the authorised person has made reasonable enquiries to find out the name or proper address of the occupier of the premises (or, if the premises are unoccupied, the owner) but without success (section 45(1)), he/she may post the community protection notice on the premises and enter the premises, or other premises, to the extent reasonably necessary for that purpose (section 45(2)).

The CPN is treated as having been issued to the occupier of the premises (or, if the premises are unoccupied, the owner) at the time the notice is posted (section 45(3)).

Appeals against notices

A person issued with a CPN may appeal to a magistrates' court against the notice on any of the following grounds.

1. That the conduct specified in the CPN did not take place; has not had a detrimental effect on the quality of life of those in the locality; has not been of a persistent or continuing nature, is not unreasonable, or is conduct that the person cannot reasonably be expected to control or affect.
2. That any of the requirements in the notice, or any of the periods within which or times by which they are to be complied with, are unreasonable.
3. That there is a material defect or error in, or in connection with, the notice.
4. That the notice was issued to the wrong person (section 46(1)).

An appeal must be made within 21 days beginning with the day on which the person is issued with the notice (section 46(2)).

While an appeal against a CPN is in progress a requirement imposed by the notice to stop doing specified things remains in effect, unless the court orders otherwise, but any other requirement imposed by the notice has no effect.

An appeal is 'in progress' until it is finally determined or is withdrawn (section 46(3)).

For example, where rubbish has accumulated in someone's front garden and a notice issued to the owner, a requirement to stop adding to the rubbish would continue in effect but a requirement to clear the garden would not (Hansard).

A magistrates' court hearing an appeal against a community protection notice must quash the notice, modify the notice (for example by extending a period specified in it), or dismiss the appeal (section 46(4)).

Remedial action by local authority

Where a person issued with a CPN ('the defaulter') fails to comply with a requirement of the notice, the relevant local authority may take action (section 47(1)). The relevant local authority may have work carried out to ensure that the failure is remedied, but only on land that is open to the air (section 47(2)).

As regards premises other than land open to the air, if the relevant local authority issues the defaulter with a notice specifying work it intends to have carried out to ensure that the failure is remedied; specifying the estimated cost of the work, and inviting the defaulter to consent to the work being carried out, the authority may have the work carried out if the necessary consent of the defaulter and the owner of the premises is given. This does not apply where the authority has made reasonable efforts to contact the owner of the premises but without success (section 47(3) and (4)).

A person authorised by a local authority to carry out work may enter any premises to the extent reasonably necessary or, where relevant, enter land that is open to the air (section 47(5)).

If work is carried out and the relevant local authority issues a notice to the defaulter giving details of the work that was carried out, and specifying an amount that is no more than the cost to the authority of having the work carried out, the defaulter is liable to the authority for that amount (subject to the outcome of any appeal (section 47(6)). A person issued with such a notice may appeal to a magistrates' court, within 21 days beginning with the day on which the notice was issued, on the ground that the amount specified is excessive (section 47(7)). A magistrates' court hearing an appeal must confirm the amount, or substitute a lower amount (section 47(8)).

Offence of failing to comply with notice

A person issued with a CPN who fails to comply with it, commits an offence (section 48(1)). A person guilty of an offence is liable on summary conviction to a fine not exceeding level 4 on the standard scale, in the case of an individual; to a fine, in the case of a body (section 48(2)). A person does not commit an offence if the person took all reasonable steps to comply with the notice, or there is some other reasonable excuse for the failure to comply with it (section 48(3)).

Remedial orders

A court before which a person is convicted of an offence in respect of a CPN may make whatever order the court thinks appropriate for ensuring that what the notice requires to be done is done (section 49(1)).

An order may in particular require the defendant to carry out specified work, or to allow specified work to be carried out by or on behalf of a specified local authority (section 49(2)). A requirement imposed does not authorise the person carrying out the work to enter the defendant's home without the defendant's consent. But this does not prevent a defendant who fails to give that consent from being in breach of the court's order (section 49(4)).

If work is carried out and the local authority specified issues a notice to the defaulter giving details of the work that was carried out, and specifying an amount that is no more than the cost to the authority of having the work carried out, the defaulter is liable to the authority for that amount (subject to the outcome of any appeal (section 49(6)).

A person issued with a notice may appeal to a magistrates' court, within 21 days beginning with the day on which the notice was issued, on the ground that the amount specified is excessive (section 49(7)).

A magistrates' court hearing an appeal must confirm the amount, or substitute a lower amount (section 49(8)).

Forfeiture of item used in commission of offence

A court before which a person is convicted of an offence may order the forfeiture of any item that was used in the commission of the offence (section 50(1)).

An order may require a person in possession of the item to hand it over as soon as reasonably practicable to a constable, or to a person employed by a local authority or designated by a local authority (section 50(2)).

An order may require the item to be destroyed, or to be disposed of in whatever way the order specifies (section 50(3)).

Where an item ordered to be forfeited is kept by or handed over to a constable or local authority person, the police force or local authority who employs them must ensure that arrangements are made for its destruction or disposal, either in accordance with the order, or if no arrangements are specified in the order, in whatever way seems appropriate (section 50(4) and (5)).

Seizure of item used in commission of offence

If a justice of the peace is satisfied on information on oath that there are reasonable grounds for suspecting that an offence under section 48 has been committed, and that there is an item used in the commission of the offence on premises specified in the information, the justice may issue a warrant authorising any constable or designated person to enter the premises within 14 days from the date of issue of the warrant to seize the item (section 51(1)).

A constable or designated person may use reasonable force, if necessary, in executing a warrant (section 51(3)).

A constable or designated person who has seized an item under a warrant may retain the item until any relevant criminal proceedings have been finally determined, if such proceedings are started within 28 days following the day on which the item was seized, otherwise, must before the end of that period return the item to the person from whom it was seized (section 51(4)).

Fixed penalty notices

An authorised person may issue a fixed penalty notice to anyone who that person has reason to believe has committed an offence (section 52(1)).

A fixed penalty notice is a notice offering the person to whom it is issued the opportunity of discharging any liability to conviction for the offence by payment of a fixed penalty to a local authority specified in the notice (section 52(3)).

The local authority specified must be the local authority that issued the CPN to which the fixed penalty notice relates, or if the CPN was not issued by a local authority, the local authority (or, as the case may be, one of the local authorities) that could have issued it (section 52(4)).

Where a person is issued with a notice in respect of an offence, no proceedings may be taken for the offence before 14 days following the date of the notice and the person may not be convicted of the offence if the person pays the fixed penalty before the end of that period (section 52(5)).

A fixed penalty notice must:

(a) give reasonably detailed particulars of the circumstances alleged to constitute the offence;
(b) state the period during which proceedings will not be taken for the offence;
(c) specify the amount of the fixed penalty, an amount specified must not be more than £100;
(d) state the name and address of the person to whom the fixed penalty may be paid;
(e) specify permissible methods of payment (section 52(6) and (7)).

A fixed penalty notice may specify two amounts and that, if the lower of those amounts is paid within a specified period (of less than 14 days), that is the amount of the fixed penalty (section 52(8)).

Whatever other method may be specified payment of a fixed penalty may be made by pre-paying and posting to the person whose name is stated, at the stated address, a letter containing the amount of the penalty (in cash or otherwise) (section 52(9)).

Where a letter is sent mentioning payment is regarded as having been made at the time at which that letter would be delivered in the ordinary course of post (section 52(10)).

In any proceedings, a certificate that purports to be signed by or on behalf of the chief finance officer of the local authority concerned, and states that payment of a fixed penalty was, or was not, received by the dated specified in the certificate, is evidence of the facts stated (section 52(11)).

Authorised persons

A CPN or a fixed penalty notice may be issued by a constable, the relevant local authority or a person designated by the relevant local authority (section 53(1)).

Exemption from liability

A local authority exercising or purporting to exercise these powers is not liable to an occupier or owner of land for damages or otherwise (whether at common law or otherwise) arising out of anything done or omitted to be done in the exercise of that power (section 54(1)).

A person carrying out work, or a person by or on whose behalf work is carried out, is not liable to an occupier or owner of land for damages or otherwise (whether at common

FC91 Community protection notices – Antisocial Behaviour, Crime and Policing Act 2014

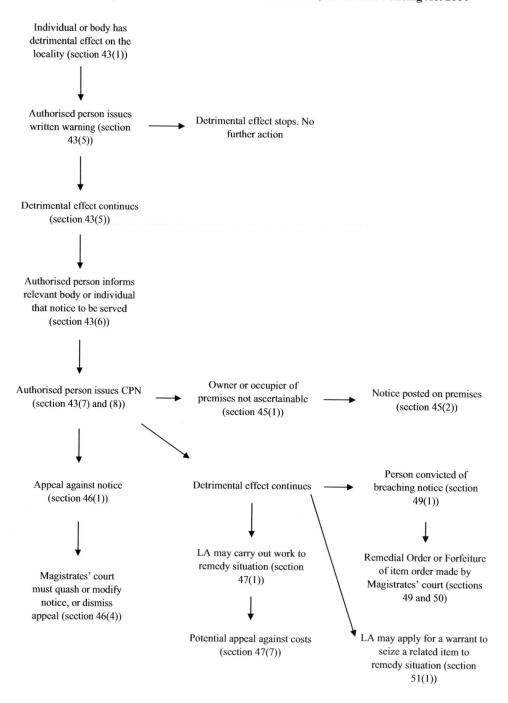

law or otherwise) arising out of anything done or omitted to be done in carrying out that work (section 54(2)).

The above exemptions do not apply to an act or omission shown to have been in bad faith, or to liability arising out of a failure to exercise due care and attention (section 54(3)). They also do not apply so as to prevent an award of damages made in respect of an act or omission on the ground that the act or omission was unlawful by of the Human Rights Act 1998 (section 54(4)).

Guidance

The Secretary of State may issue or revise guidance, and must publish it, to chief officers of police or to local authorities about the exercise of their functions in relation to community protection notices.

Definitions

Authorised person means a person on whom section 53 (or an enactment amended by that section) confers power to issue community protection notices (section 43(2)) or a person on whom section 53 power to issue fixed penalty notices is conferred (section 52(2)).

Conduct includes a failure to act.

Defendant's home means the house, flat, vehicle or other accommodation where the defendant usually lives, or is living at the time when the work is or would be carried out (section 49(5))).

Relevant local authority

- for a CPN, means the local authority (or, as the case may be, any of the local authorities) within whose area the conduct specified in the notice has, according to the notice, been taking place. (section 53(2)).
- for a fixed penalty notice, means the local authority (or, as the case may be, any of the local authorities) within whose area the offence in question is alleged to have taken place (section 53(3)).

PUBLIC SPACES PROTECTION ORDERS

References

Antisocial Behaviour, Crime and Policing Act 2014, sections 59 to 74
Antisocial Behaviour etc. (Scotland) Act 2004
The Antisocial Behaviour (Northern Ireland) Order 2004
Antisocial Behaviour, Crime and Policing Act 2014 (Publication of Public Spaces Protection Orders) Regulations 2014
Public Spaces Protection Orders Guidance for councils – February 2018
Relevant case – Summers v Richmond Upon Thames LBC [2018] EWHC 782 (Admin).

Extent

These provisions apply to England and Wales. Similar provisions apply in Scotland and Northern Ireland.

Scope

A public spaces protection order is intended to deal with a particular nuisance or problem in a particular area that is detrimental to the local community's qualify of life, by imposing conditions on the use of that area. The order can also be used to deal with likely future problems. Examples of where a new order could be used include prohibiting the consumption of alcohol in public parks or ensuring dogs are kept on a leash in children's play areas. It could also prohibit spitting in certain areas (if the problem were persistent and unreasonable).

Only a local authority can issue the order, and before doing so, they must consult with the Chief Officer of Police, the Police and Crime Commissioner and any representatives of the local community they consider appropriate – for example, a local residents group or a community group that regularly uses the public place.

The order can prohibit certain things (for example, drinking alcohol), require specific things to be done (for example, keeping dogs on leashes), or both.

Failure to comply with the order can result in a fine or a fixed penalty notice. When deciding whether an order should be issued the local authority must consider the behaviour is have a detrimental effect and whether the effect is of such a persistent and serious nature as to justify a notice.

A PSPO can be made to apply to all people or be limited to certain people. It can be also restricted to certain times. It can last no longer than three years but can be renewed if necessary.

Dog Control Orders in operation when these powers came into force automatically became a PSPO. All orders not confirmed under PSPO legislation automatically cease on 19 October 2020.

Power to make orders

A local authority may make a public spaces protection order if satisfied on reasonable grounds that two conditions are met.

The first condition is that activities carried on in a public place within the authority's area have had a detrimental effect on the quality of life of those in the locality, or it is likely that activities will be carried on in a public place within that area and that they will have such an effect (section 59 (1) and (2)).

The second condition is that the effect, or likely effect, of the activities is, or is likely to be, of a persistent or continuing nature; is, or is likely to be, such as to make the activities unreasonable, and justifies the restrictions imposed by the notice (section 59(3)).

The order identifies the public place 'the restricted area' and prohibits specified things being done in the restricted area; requires specified things to be done by persons carrying on specified activities in that area, or does both of those things (section 59(4)).

The only prohibitions or requirements that may be imposed are ones that are reasonable to impose in order to prevent the detrimental effect from continuing, occurring or recurring, or to reduce that detrimental effect or to reduce the risk of its continuance, occurrence or recurrence (section 59(5)).

A prohibition or requirement may be framed:

- to apply to all persons, at all times, or in all circumstances; or
- only to persons in specified categories, at specified times, or specified circumstances; or
- to all persons except those in specified categories; at all times and all circumstances except those specified (section 59(6)).

A public spaces protection order must:

(a) identify the activities referred to in subsection (2);
(b) explain the effect of sections 63 and 67;
(c) specify the period for which the order has effect (section 59(7)).

A public spaces protection order must be published in accordance with regulations (section 59(8)).

Duration of orders

A public spaces protection order lasts for up to three years, unless extended (section 60(1)).

Before the time when a public spaces protection order is due to expire, the local authority can extend the period if satisfied on reasonable grounds that doing so is necessary to prevent occurrence or recurrence after that time of the activities identified in the order, or an increase in the frequency or seriousness of those activities after that time (section 60(2)).

An extension cannot be more than 3 years and must be published in accordance with regulations (section 60(3)). An order may be extended more than once (section 60(4)).

Variation and discharge of orders

The local authority may vary an order by increasing or reducing the restricted area; by altering or removing a prohibition or requirement included in the order, or adding a new one (section 61(1)).

The local authority may make a variation that results in the order applying to an area to which it did not previously apply only if the conditions in section 59(2) and (3) (section 61(2)).

The local authority may make a variation that makes a prohibition or requirement more extensive, or adds a new one, only if the prohibitions and requirements imposed by the order, as varied, are ones that section 59(5) allows to be imposed (section 61(3)).

A public spaces protection order may be discharged by the local authority that made it (section 61(4)).

Where an order is varied, it must be published in accordance with the relevant regulations made by the Secretary of State (section 61(5)).

Where an order is discharged, a notice identifying the order and stating the date when it ceases to have effect must be published in accordance with the relevant regulations (section 61(6)).

Premises etc. to which alcohol prohibition does not apply

A prohibition in a public spaces protection order on consuming alcohol does not apply to:

(a) premises including its curtilage authorised by a premises licence to be used for the supply of alcohol;
(b) premises authorised by a club premises including its curtilage certificate to be used by the club for the supply of alcohol;
(c) premises which by the Licensing Act 2003 used for the supply of alcohol or which, could have been so used within the 30 minutes before that time;
(d) a place where facilities or activities relating to the sale or consumption of alcohol permitted under highway-related uses (section 62(1), (2) and (4)).

Consumption of alcohol in breach of prohibition in order

Where a constable or an authorised person reasonably believes that a person (P) is or has been consuming alcohol in breach of a prohibition in a public spaces protection order, or intends to consume alcohol in circumstances in which doing so would be a breach of the prohibition, he or she may require P not to consume, alcohol or anything which he or she reasonably believes to be alcohol or to surrender anything in P's possession (section 63 (1) and (2)).

A constable or an authorised person who imposes such a requirement must tell P that failing without reasonable excuse to comply with the requirement is an offence (section 63(3)).

A requirement imposed by an authorised person is not valid he or she is asked by P to show evidence of his or her authorisation, and fails to do so (section 63(4)).

A constable or an authorised person may dispose of anything surrendered in whatever way he or she thinks appropriate (section 63(5)).

A person who fails without reasonable excuse to comply with a requirement imposed on him or her commits an offence and is liable on summary conviction to a fine not exceeding level 2 on the standard scale (section 63(6)).

Note: Where there is no threat of antisocial behaviour, they need not challenge the individuals, for example a family picnic with a bottle of wine (Hansard).

Orders restricting public right of way over highway

A local authority cannot make a public spaces protection order that restricts the public right of way over a highway without considering the likely effect of making the order on the occupiers of premises adjoining or adjacent to the highway; the likely effect of making the order on other persons in the locality, and in a case where the highway constitutes a through route, the availability of a reasonably convenient alternative route (section 64(1)).

Before making such an order a local authority must:

(a) notify potentially affected persons of the proposed order,
(b) inform those persons how they can see a copy of the proposed order,
(c) notify those persons of the period within which they may make representations about the proposed order, and
(d) consider any representations made (section 64(2)).

If an order would restrict the public right of way over a highway that is also within the area of another local authority, the other authority must be consulted if appropriate to do so (section 64(3)).

An order cannot restrict the public right of way over a highway for the occupiers of premises adjoining or adjacent to the highway or the only or principal means of access to a dwelling (section 64(4) and (1)).

In relation to a highway that is the only or principal means of access to premises used for business or recreational purposes, an order cannot restrict the public right of way over the highway during periods when the premises are normally used (section 64(6)).

An order that restricts the public right of way over a highway can authorise the installation, operation and maintenance of a barrier or barriers for enforcing the restriction (section 64(7)). A local authority can install, operate and maintain such barriers (section 64(8)).

A highway over which the public right of way is restricted by a public spaces protection order does not cease to be regarded as a highway by reason of the restriction (or by reason of any barrier (section 64(9)).

Categories of highway over which public right of way may not be restricted

A public spaces protection order cannot restrict the public right of way over a highway that is a special road; a trunk road; a classified or principal road; a strategic road; a highway in England or Wales of a description prescribed by regulations (section 65(1)).

Challenging the validity of orders

An interested person may apply to the High Court to question the validity of a public spaces protection order, or its variation.

The grounds on which an application may be made are that the local authority did not have power to make the order or variation, or to include particular prohibitions or requirements imposed by the order (or by the order as varied); and that a requirement was not complied (section 66(2)).

An application must be made within the period of 6 weeks beginning with the date on which the order or variation is made (section 66(3)).

On an application, the High Court may by order suspend the operation of the order or variation, or any of the prohibitions or requirements imposed by the order (or variation), until the final determination of the proceedings (section 66(4)).

If on an application, the High Court is satisfied that the local authority did not have power to make the order or variation, or to include particular prohibitions or requirements imposed by the order (or varied); or the interests of the applicant have been substantially prejudiced by a failure to comply with a requirement, the Court may quash the order or variation, or any of the prohibitions or requirements imposed by the order (or varied) (section 66(5)).

An order, or any of the prohibitions or requirements imposed by the order (or varied), may be suspended or quashed generally, or so far as necessary for the protection of the interests of the applicant (section 66(6)).

An interested person can only challenge the validity of an order, or variation, in any legal proceedings (either before or after it is made), or where the interested person is charged with an offence under section 67(3) (section 66(7)).

Offence of failing to comply with order

It is an offence for a person without reasonable excuse to do anything that the person is prohibited from doing by a public spaces protection order; or to fail to comply with a requirement to which the person is subject under an order (section 67(1)).

A person guilty of an offence is liable on summary conviction to a fine not exceeding level 3 on the standard scale (section 67(2)).

A person does not commit an offence by failing to comply with a prohibition or requirement that the local authority did not have power to include in the order (section 67(3)).

Consuming alcohol in breach of a public spaces protection order is not an offence under this section (but note section 63) (section 67(4)).

Fixed penalty notices

A constable or an authorised person may issue a fixed penalty notice to anyone he or she has reason to believe has committed an offence in relation to a public spaces protection order (section 68(1)). Where a person is issued with a notice in respect of an offence, no proceedings may be taken for the offence before the end of the period of 14 days following the date of the notice; and the person cannot be convicted of the offence if the person pays the fixed penalty before the end of that period (section 68(4)).

A fixed penalty notice must:

(a) give reasonably detailed particulars of the circumstances alleged to constitute the offence;
(b) state the period during which proceedings will not be taken for the offence;
(c) specify the amount of the fixed penalty;
(d) state the name and address of the person to whom the fixed penalty may be paid;
(e) specify permissible methods of payment (section 68(5)).

The penalty must not be more than £100 (section 68(6)).

A fixed penalty notice may specify two amounts and specify that, if the lower of those amounts is paid within a specified period (of less than 14 days), that is the amount of the fixed penalty (section 68(7)).

Whatever other method is specified, payment of a fixed penalty may be made by pre-paying and posting to the person whose name is stated, at the stated address, a letter containing the amount of the penalty (in cash or otherwise) (section 68(8)).

Where a letter is sent as mentioned, payment is regarded as having been made at the time at which that letter would be delivered in the ordinary course of post (section 68(9)).

In any proceedings, a certificate that purports to be signed by or on behalf of the chief finance officer of the local authority concerned, and states that payment of a fixed penalty was, or was not, received by the dated specified in the certificate, is evidence of the facts stated (section 68(10)).

Byelaws

A byelaw that prohibits, by the creation of an offence, an activity regulated by a public spaces protection order is of no effect in relation to the restricted area during the time of the order (section 70).

Convention rights, consultation, publicity and notification

A local authority, in deciding:

(a) whether to make a public spaces protection order, and if so, what it should include,
(b) whether to extend the period for which the order has effect and if so for how long,
(c) whether to vary an order and if so how, or
(d) whether to discharge an order,

must have particular regard to the rights of freedom of expression and freedom of assembly set out in articles 10 and 11 of the Human Rights Convention (section 72(1) and (2)).

A local authority must carry out the necessary consultation, publicity and notification (if any), before making an order; extending the period for which an order has effect, or varying or discharging an order (section 72(3)).

The requirement to consult with the owner or occupier of land within the restricted area does not apply to land that is owned and occupied by the local authority; and applies only if, or to the extent that, it is reasonably practicable to consult the owner or occupier of the land (section 72(5)).

Guidance

The Secretary of State may issue, revise and publish guidance to local authorities and chief officers of police about the exercise of their functions and those of persons authorised (section 73).

Definitions

Community representative means in relation to a public spaces protection order that a local authority proposes to make or has made, means any individual or body appearing to the authority to represent the views of people who live in, work in or visit the restricted area (section 74).

Dwelling means a building or part of a building occupied, or intended to be occupied, as a separate dwelling; (section 64(10)).

Interested person means an individual who lives in the restricted area or who regularly works in or visits that area (section 66(1)).

Necessary consultation means consulting with the chief officer of police, and the local policing body, for the police area that includes the restricted area; whatever community representatives the local authority thinks it appropriate to consult; and the owner or occupier of land within the restricted area (section 72(4)).

Necessary publicity means in the case of a proposed order or variation, publishing the text of it; and in the case of a proposed extension or discharge, publicising the proposal (section 72(4)).

Necessary notification means notifying the following authorities of the proposed order, extension, variation or discharge the parish council or community council (if any) for the area that includes the restricted area; and in the case of a public spaces protection order made or to be made by a district council in England, the county council (if any) for the area that includes the restricted area (section 72(4)).

Potentially affected persons means occupiers of premises adjacent to or adjoining the highway, and any other persons in the locality who are likely to be affected by the proposed order (section 64(2)).

Public place means any place to which the public or any section of the public has access, on payment or otherwise, as of right or by virtue of express or implied permission (section 74).

FC92 Public spaces protection orders – Antisocial Behaviour, Crime and Policing Act 2014

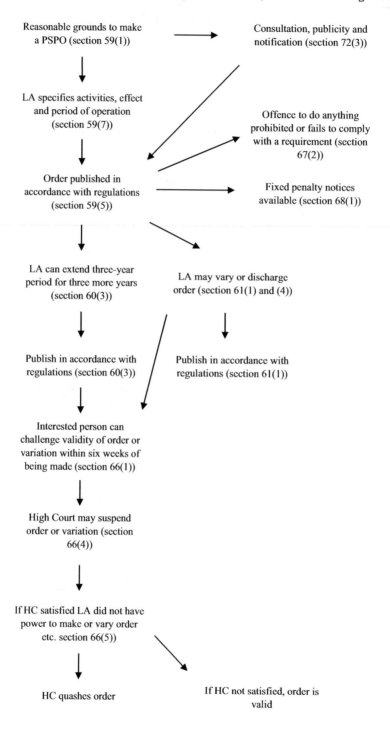

PART 7

PUBLIC HEALTH

Chapter 1

ANIMAL WELFARE

ANIMAL WELFARE LICENSING

References

Animal Welfare (Licensing of Activities Involving Animals) (England) Regulations 2018
Guidance for Animal Activity Licensing 2018
Guidance notes for conditions for Breeding Dogs 2018
Guidance for Boarding Cats 2018
Guidance for Boarding Dogs in Kennels 2018
Guidance for Dog Day Care 2018
Guidance for Exhibition of Animals 2018
Guidance for Hiring Out Horses
Guidance for Home Boarding of Dogs 2018
Guidance for Selling Animals as Pets 2018
The Animal Welfare (Breeding of Dogs) (Wales) Regulations 2014
Animal Health and Welfare (Scotland) Act 2006
Welfare of Animals Act (Northern Ireland) 2011

Extent

These provisions apply to England only. Similar provisions apply in Wales, Scotland and Northern Ireland.

Scope

These regulations replace a number of existing licensing by providing for the licensing of persons involved in selling animals as pets, providing or arranging for the provision of boarding for cats or dogs, hiring out horses, breeding dogs and keeping or training animals for exhibition.

A local authority is the licensing authority for any licensable activity carried on at premises in its area (regulation 2).

Transitional provisions

Any unexpired or provisional licences granted in accordance with the provisions of:

Pet Animals Act 1951,
Animal Boarding Establishments Act 1963,
Riding Establishments Act 1964 and 1970,

Breeding of Dogs Act 1973,
Breeding of Dogs (Licensing Records) Regulations 1999,
Breeding and Sale of Dogs (Welfare) Act 1999 and
Sale of Dogs (Identification Tag) Regulations 1999
continues in force for the remainder of its term subject to the provisions of that Act
(regulation 27).

Conditions of grant or renewal of a licence

Where a local authority has received from an operator an application in writing for the grant or
renewal of a licence to carry on a licensable activity on premises in the local authority's area,
and the application contains information the local authority has required, the authority must:

(a) appoint one or more suitably qualified inspectors to inspect premises on which the
licensable activity is being or is to be carried on, and
(b) following that inspection, grant a licence to the operator, or renew the operator's
licence, if it is satisfied that:

(i) the licence conditions will be met,
(ii) any appropriate fee has been paid and
(iii) the grant or renewal is appropriate having taken into account the required report
(regulation 4(1) and (2)).

A local authority must attach the general conditions and relevant specific conditions to the
licence (regulation 4(3)).

For applications for the grant or renewal of a licence for hiring out horses or breeding
of dogs, if no inspector appointed is a listed veterinarian, the local authority must appoint
a listed veterinarian to inspect the premises with the inspector (regulation 4(4) and (5)).

In considering whether the licence conditions will be met, a local authority must take
account of the applicant's conduct as the operator of the licensable activity, whether the
applicant is a fit and proper person to be the operator of that activity and any other relevant
circumstances (regulation 4(7)).

All licences granted or renewed for the licensable activities are subject to the licence
conditions (regulation 4(9)).

Period of licence

A local authority may grant or renew a licence:

(a) for a period of one, two or three years for selling pets, boarding cats and dogs, hiring
horses, breeding dogs if it is satisfied that the period is appropriate on the basis of its
assessment, having regard to guidance issued by the Secretary of State, of:

(i) the risk of an operator breaching any licence conditions;
(ii) the impact on animal welfare of any such breaches; and
(iii) whether the operator is already meeting higher standards of animal welfare than
are required by the licence conditions;

(b) for three years for keeping or training animals for exhibition (regulation 5).

This means the regulations incorporate 'earned recognition' into the licensing system. Local authorities will be able to issue licences of one, two or three years, with longer licences going to high-performing, low-risk businesses.

Power to take samples from animals

An inspector may, for the purposes of ensuring the licence conditions are being complied with, take samples for laboratory testing from any animals on premises occupied by an operator (regulation 6).

An operator must comply with any reasonable request of an inspector to facilitate the identification and examination of an animal and the taking of samples and, in particular, must arrange the suitable restraint of an animal where requested by an inspector (regulation 7).

Hiring out horses

Where such a licence is in force the local authority must appoint a listed veterinarian to inspect the premises on which the activity is being carried on and inspect the premises annually (regulation 8(1) and (2)).

Variation of a licence on the application, or with the consent, of a licence holder

A local authority may at any time vary a licence on the application in writing of the licence holder, or on its own initiative, with the consent in writing of the licence holder (regulation 9).

Inspector's report

Where a local authority arranges an inspection, a report by the inspector must be submitted to it (regulation 10(1)).

The inspector's report must contain information about the operator, any relevant premises, any relevant records, the condition of any animals and any other relevant matter, and state whether or not the inspector considers that the licence conditions will be met (regulation 10(2)).

Persons who may not apply for a licence

The following persons cannot apply for a licence for any licensable activity:

(a) a disqualified person where the time limit for any appeal against that disqualification has expired or where, if an appeal was made, that appeal was refused;

(b) a person having held a licence which was revoked where the time limit for any appeal against that revocation has expired or where, if an appeal was made, that appeal was refused (regulation 11(1)).

Any licence granted or renewed, or held by, a person mentioned above is automatically revoked (regulation 11(2)).

Death of a licence holder

In the event of the death of a licence holder, the licence is deemed to have been granted to, or renewed in respect of, the personal representatives of that former licence holder (regulation 12(1)).

The licence remains in force for three months beginning after the death of the licence holder or for as long as it was due to remain in force but for the death (whichever period is shorter) (regulation 12(2)).

The personal representatives must notify in writing the local authority which granted or renewed the licence that they are now the licence holders within 28 days of the death of the former licence holder (regulation 12(3)).

If the personal representatives fail to notify the local authority within that period, the licence ceases to have effect on the expiry of that period (regulation 12(4)).

The local authority which granted or renewed the licence may, on the application of the personal representatives, extend this period for up to three months if it is satisfied that the extension is necessary for the purpose of winding up the estate of the former licence holder and is appropriate in all the circumstances (regulation 12(5)).

Fees

A local authority may charge fees for an application for the grant, renewal or variation of a licence including any inspection, providing information, ensuring compliance with the Regulations and any relevant enforcement (regulation 13(1) and 29).

The fee must not exceed the reasonable costs incurred (regulation 13(2)).

Guidance

A local authority must have regard in the carrying out of its functions to guidance issued by the Secretary of State (regulation 14).

Grounds for suspension, variation without consent or revocation of a licence

A local authority may, without any requirement for the licence holder's consent, decide to suspend, vary or revoke a licence at any time on being satisfied that:

(a) the licence conditions are not being complied with,
(b) there has been a breach of the Regulations,
(c) information supplied by the licence holder is false or misleading or
(d) it is necessary to protect the welfare of an animal (regulation 15).

Procedure for suspension or variation without consent

The suspension or variation of a licence comes into effect seven working days after the notice of the decision is issued to the licence holder (regulation 16(1)).

If it is necessary to protect the welfare of an animal, the local authority may specify in the notice of its decision that the suspension or variation has immediate effect (regulation 16(2)).

A decision to suspend or vary a licence must:

(a) be notified to the licence holder in writing,
(b) state the local authority's grounds for suspension or variation,
(c) state when it comes into effect,
(d) specify measures that the local authority considers are necessary in order to remedy the grounds and
(e) explain the right of the licence holder to make written representations and give details of the person to whom such representations may be made and the date by the end of which they must be received (regulation 16(3)).

The licence holder may make written representations which must be received by the local authority within seven working days of the date of the notice of the decision to suspend or vary the licence (regulation 16(4)).

Where a licence holder makes written representations, the suspension or variation is not to have effect unless the local authority, after considering the representations, suspends or varies the licence below (regulation 16(5)).

Within seven working days of any representations, the local authority must, after considering the representations:

(a) suspend or vary the licence,
(b) cancel its decision to suspend or vary the licence,
(c) confirm the suspension or variation of the licence or
(d) reinstate the licence if it has been suspended, or cancel its variation if it has been varied (regulation 16(6)).

The local authority must issue to the licence holder written notice of its decision and the reasons for it within seven working days of receipt of any representations (regulation 16(7)). The local authority's decision has effect on service of its notice above (regulation 16(8)).

If the local authority fails to do this, after seven working days of receipt of any representations:

(a) a licence suspended is to be deemed to be reinstated;
(b) a licence varied is to be deemed to have effect as if it had not been so varied (regulation 16(9) and (10)).

Once a licence has been suspended for 28 days, the local authority must on the next working day:

(a) reinstate it without varying it,
(b) vary and reinstate it as varied or
(c) revoke it (regulation 16(11)).

If the local authority fails to do this, the licence is to be deemed to have been reinstated without variation with immediate effect (regulation 16(12)).

Reinstatement of a suspended licence by a local authority

A local authority must reinstate a suspended licence by written notice once it is satisfied that the grounds specified in the notice of suspension have been or will be remedied (regulation 17(1)).

Where a local authority reinstates a licence, it may reduce the period for which it is reinstated (regulation 17(2)).

Notice of revocation

A revocation decision must:

(a) be notified in writing to the licence holder,
(b) state the local authority's grounds for revocation and
(c) give notice of the licence holder's right of appeal to the First-Tier Tribunal and the appeal period (regulation 18(1)).

The decision has effect on service of the notice (regulation 18(2)).

Obstruction of inspectors

A person must not intentionally obstruct an inspector in the exercise of any powers under the Act (regulation 19).

Offences

It is an offence for a person, without lawful authority or excuse to breach a licence condition or to fail to comply with regulation 7 or 19 (regulation 20(1)).

A person who commits an offence is liable on summary conviction to a fine (regulation 20(2)).

Appeals

Any operator who is aggrieved by a decision by a local authority to refuse to grant or renew a licence, or to revoke or vary a licence, may appeal to the First-Tier Tribunal within 28 days of the decision (regulation 24(1) and (2)).

The First-Tier Tribunal may on application and until the appeal is determined or withdrawn in the case of a decision to refuse to renew a licence, permit a licence holder to continue to carry on a licensable activity or any part of it subject to the licence conditions, or suspend a revocation or variation (regulation 24(3)).

On appeal, the First-Tier Tribunal may overturn or confirm the local authority's decision, with or without modification (regulation 24(4)).

Provision of information to the Secretary of State

Each local authority must provide the following information to the Secretary of State in writing for the purpose of assisting a review of the operation of the regulations:

(a) the number of licences in force for each licensable activity in its area on each reference date, and
(b) the average level of fees it has charged for licences it has granted or renewed for each licensable activity in each reference period (regulation 29(1)).

Each local authority must provide the information to the Secretary of State in electronic form, or secure that it is accessible in electronic form, and no later than the next 31 May each year.

Definitions

Disqualified person or persons who may not apply for a licence:

A person who:

- has at any time held a licence which was revoked under regulation 15 of these Regulations
- has at any time held a licence which was revoked under regulation 17 of the Animal Welfare (Breeding of Dogs) (Wales) Regulations 2014
- has at any time held a licence which was revoked under regulation 13 of the Welfare of Wild Animals in Travelling Circuses (England) Regulations 2012
- is disqualified under section 33 of the Welfare of Animals Act (Northern Ireland) 2011
- has at any time held a licence which was revoked under regulation 12 of the Welfare of Racing Greyhounds Regulations 2010
- is disqualified under section 34 of the Act
- is disqualified under section 40(1) and (2) of the Animal Health and Welfare (Scotland) Act 2006
- is disqualified under section 4(1) of the Dangerous Dogs Act 1991
- is disqualified under Article 33A of the Dogs (Northern Ireland) Order 1983
- is disqualified under section 6(2) of the Dangerous Wild Animals Act 1976 from keeping a dangerous wild animal
- is disqualified under section 3(3) of the Breeding of Dogs Act 1973 from keeping a breeding establishment for dogs
- is disqualified under section 4(3) of the Riding Establishments Act 1964 from keeping a riding establishment
- is disqualified under section 3(3) of the Animal Boarding Establishments Act 1963 from keeping a boarding establishment for animals
- is disqualified under section 5(3) of the Pet Animals Act 1951 from keeping a pet shop
- is disqualified under section 1(1) of the Protection of Animals (Amendment) Act 1954 from having custody of an animal
- is disqualified under section 4(2) of the Performing Animals (Regulation) Act 1925
- is disqualified under section 3 of the Protection of Animals Act 1911 from the ownership of an animal (paragraphs 1 to 17 of schedule 8).

Licensable activity means selling animals as pets, providing or arranging for the provision of boarding for cats or dogs, hiring out horses, breeding dogs or keeping or training animals for exhibition (paragraph 2, 4, 6, 8 or 10 of Schedule 1).

FC93 Animal Welfare Licensing – Animal Welfare (Licensing of Activities Involving Animals) (England) Regulations 2018

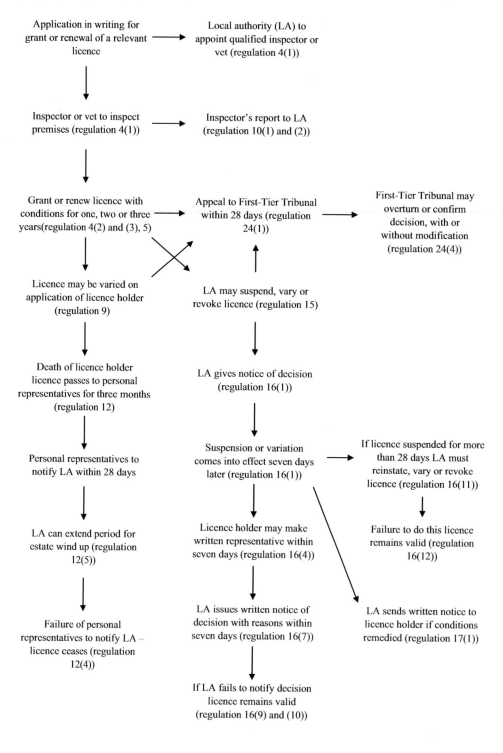

ANIMAL BOARDING ESTABLISHMENTS

References

Animal Boarding Establishments Act 1963 as amended by the Animal Welfare Act 2006
DEFRA Guidance Notes for Conditions for Providing Boarding for Cats (formerly the CIEH
   Model Licence Conditions and Guidance for Cat Boarding Establishments 2013)
CIEH Model Licence Conditions and Guidance for Dog Breeding Establishments
Welfare of Animals Act (Northern Ireland) 2011.
Welfare of Animals Act (Northern Ireland) 1972 (as amended)
Animal Boarding Establishments Regulations (Northern Ireland) 1974

Extent

This licensing procedure applies in Wales and Scotland.
   Similar provisions apply in Northern Ireland.
   In England, these provisions are replaced by the Animal Welfare (Licensing of Activities
Involving Animals) (England) Regulations 2018 (see p. 451).

Scope

A person keeping an Animal Boarding Establishment for cats and/or dogs is required to hold
a licence issued by the local authorities (section 1(1)).
   The Act defines the keeping of an Animal Boarding Establishment as:

'to the carrying on by him at premises of any nature (including a private dwelling) of a
business of providing accommodation for other people's animals:

Provided that:

(a) a person shall not be deemed to keep a boarding establishment for animals by reason
only of his providing accommodation for other people's animals in connection with a
business of which the provision of such accommodation is not the main activity; and
(b) nothing in this Act shall apply to the keeping of an animal at any premises in pursuance of a requirement imposed under, or having effect by virtue of, the Animal Health
Act 1981 (section 5(1)).'

Applications

Applications made to the local authority must specify the premises concerned and a fee
determined by the local authority must be paid before any licence is granted (section 1(2)).
The factors which may be taken into account in deciding whether or not to issue a licence are
listed under 'Conditions' below but it is at the local authority's discretion to refuse a licence
on any other grounds (section 1(3)).

Licences

A licence normally remains in force until the end of the year in which it is granted; however,
the applicant determines whether it comes into force on the date it is granted or the first day
of the following calendar year (section 1(5) and (6)).

In the event of the death of a licence holder, the licence passes to his personal representative and operates for a period of 3 months from the date of death before expiring. The local authority may, on application, agree to extension of that period (section 1(7)).

Conditions

The local authority is required to specify conditions in the licence as appears to it necessary or expedient for securing all or any of the following objectives:

(a) that animals will at all times be kept in accommodation suitable as respects construction, size of quarters, number of occupants, exercising facilities, temperature, lighting, ventilation and cleanliness;

(b) that animals will be adequately supplied with suitable food, drink and bedding material, adequately exercised, and (so far as necessary) visited at suitable intervals;

(c) that all reasonable precautions will be taken to prevent and control the spread among animals of infectious or contagious diseases, including the provision of adequate isolation facilities;

(d) that appropriate steps will be taken for the protection of the animals in case of fire or other emergency;

(e) that a register be kept containing a description of any animals received into the establishment, date of arrival and departure, and the name and address of the owner, such register to be available for inspection at all times by an officer of the local authority, veterinary surgeon or veterinary practitioner authorised under section 2(1) (section 1(3)).

Guidance on conditions relating to these licences is given in two booklets published by the Chartered Institute of Environmental Health in November 1995, Guide to Dog Boarding Establishments and the DEFRA Guide to Cat Boarding Establishments.

Local authorities should discuss with establishment owners appropriate standards and timescales based on the documents.

Power of entry

Local authority officers, veterinary surgeons or practitioners authorised in writing by the local authority for this purpose, may, upon producing their authority, if required, inspect a licensed Animal Boarding Establishment and any animals found there at all reasonable times. Persons wilfully obstructing or delaying authorised officers are subject on conviction to a maximum penalty at level 2 on the standard scale (sections 2 and 3(2)).

There is no power of entry to unlicensed premises.

Disqualifications and cancellations

Where a person is convicted under the Animal Boarding Establishment Act 1963, the Protection of Animals Act 1911, the Protection of Animals (Scotland) Act 1912, the Pet Animals Act 1951 or the Animal Welfare Act 2006, the court may cancel any Animal Boarding Establishment licence held by the person and may disqualify him from holding such a licence, whether or not he currently holds one, for any specified period. The cancellation or disqualification may be suspended by the court pending an appeal (sections 3(3) and (4)).

FC94 Animal boarding establishments – Animal Boarding Establishments Act 1963

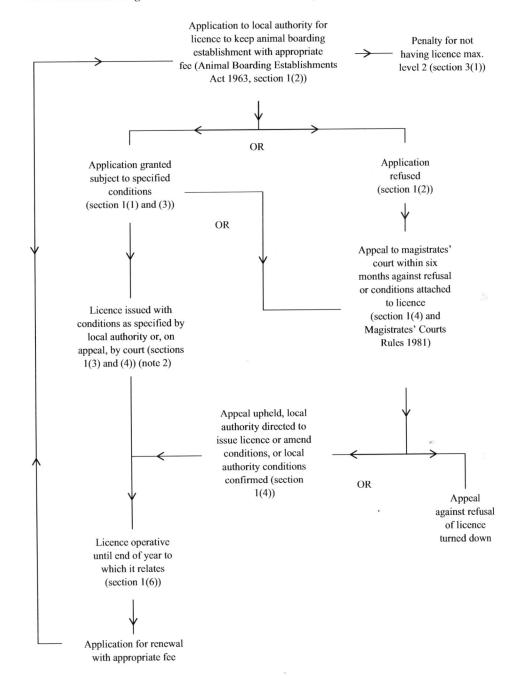

Application to local authority for licence to keep animal boarding establishment with appropriate fee (Animal Boarding Establishments Act 1963, section 1(2))

Penalty for not having licence max. level 2 (section 3(1))

OR

Application granted subject to specified conditions (section 1(1) and (3))

Application refused (section 1(2))

OR

Appeal to magistrates' court within six months against refusal or conditions attached to licence (section 1(4) and Magistrates' Courts Rules 1981)

Licence issued with conditions as specified by local authority or, on appeal, by court (sections 1(3) and (4)) (note 2)

Appeal upheld, local authority directed to issue licence or amend conditions, or local authority conditions confirmed (section 1(4))

OR

Appeal against refusal of licence turned down

Licence operative until end of year to which it relates (section 1(6))

Application for renewal with appropriate fee

Notes
1. This procedure applies in Wales and Scotland but not in Northern Ireland.
2. Penalty for contravening licence conditions is max. level 2 fine and up to three months' imprisonment, or both (section 3(1)).
3. Conditions may only be varied on the renewal of licence.

Local authorities must refuse applications for Animal Boarding Establishments licences from persons disqualified under:

(a) the Animal Boarding Establishment Act 1963;
(b) the Pet Animals Act 1951 from keeping a pet shop;
(c) the Protection of Animals (Amendment) Act 1954 from having the custody of animals; or
(d) the Animal Welfare Act 2006 (section 1(2)).

Definition

Animal means any cat or dog (section 5(2)).

DANGEROUS WILD ANIMALS

References

Dangerous Wild Animals Act 1976 as amended by the Animal Welfare Act 2006
Dangerous Wild Animals Act 1976 (Modification) (No. 2) Order 2007
DEFRA Circular 1/2002 The Keeping of Wild Animals: An Introductory Guide to: The Protection of Animals Act 1911, The Performing Animals (Regulation) Act 1925, The Pet Animals Act 1951, The Dangerous Wild Animals Act 1976, The Zoo Licensing Act 1981 (Archived)
The Dangerous Wild Animals (Northern Ireland) Order 2004

Extent

This licensing procedure applies in England, Wales and Scotland (section 10(3)). Similar provisions apply in Northern Ireland.

Scope

Any person keeping any dangerous wild animal is required to hold a licence from a local authority (section 1(1)).

The animals covered by this provision are listed in the schedule to the Act and this has been amended from time to time. The latest modification was in 2008 (Scotland) and the animals requiring their keeper to be licensed are given in paragraph 1 of schedule 1 of the Act and paragraph 1 of the schedule in the Dangerous Wild Animals Act 1976 (Modification) (No. 2) Order 2007.

This licensing procedure does not apply to animals kept in:

(a) a zoo within the meaning of the Zoo Licensing Act 1981;
(b) a circus;
(c) premises in England on which the activity described in paragraph 2 of Schedule 1 to the Animal Welfare (Licensing of Activities Involving Animals) (England) Regulations 2018 is carried on under a licence under those Regulations;
(c) pet shops in Wales; and
(d) places which are designed establishments under the Animals (Scientific Procedures) Act 1986 (section 5).

A person is held to be the keeper of the animal if he has it in his possession and the assumption of possession continues even if the animal escapes or it is being transported, etc. This removes the need for carriers or veterinary surgeons to be licensed (section 7(1)).

Applications for licence

Any application made to a local authority for a licence must be made (unless in exceptional circumstances) by the person who proposes to own and possess the animal and must:

(a) specify the species and number of animals to be kept;
(b) specify the premises where the animals will normally be kept;
(c) be made to the local authority for those premises;
(d) be made by a person 18 years of age or over and not disqualified from holding a licence under the Act; and
(e) be accompanied by a fee stipulated by the local authority at a level sufficient to meet the direct and indirect costs involved.

Applications not complying with these requirements may not be granted (section 1(2) and (4)).

Reports

Before granting any licence, the local authority is required to consider a report of an inspection of the premises by a veterinary surgeon or practitioner authorised by it (section 1(5)).

Matters for consideration

The local authority may not grant a licence unless:

(a) it will not be contrary to the public interest on grounds of safety, nuisance or otherwise to issue a licence;
(b) the applicant is suitable;
(c) animals will:

    (i) be held in secure accommodation suitable in size for the animals kept and which is suitable as regards construction, temperature, lighting, ventilation, drainage and cleanliness; and

    (ii) have adequate and suitable food, drink and bedding and be visited at regular intervals;

(d) be appropriately protected in case of fire or other emergency;
(e) be subject to precautions to control infectious diseases;
(f) be provided with adequate exercise facilities (section 1(3)).

Licences

Licences last for two years. According to the wishes of the applicant, a licence either comes into force on the day on which it is granted, in which case it expires on 31 December of that same year, or it comes into force on 1 January of the next year, in which case it expires on 31 December of that next year. If an application for renewal is made before the date of expiration, the licence continues until the application is determined.

On the death of a licence holder, the licence continues in the name of the personal representative for 28 days only and then expires unless application is made for a new licence within that time, in which case it continues until the new application is determined (sections 2(2), (3) and (4)).

Conditions

The local authority is required to specify conditions which:

(a) require the animals to be kept only by persons specified in the licence;
(b) require the animals to be normally held at the premises specified in the licence;
(c) require the animals not to be moved from those premises unless in circumstances allowed for in the licence;
(d) require the licence holder and person keeping the animals to be insured against liability for damage caused by the animals to the satisfaction of the local authority;
(e) restrict the species and numbers of animals;
(f) require a copy of the licence to be made available by the licence holder to persons entitled to keep the animals; and
(g) are necessary or desirable to secure the objectives specified in paragraphs (c)–(f) listed under 'Matters for consideration' above (section 1(6) and (7)).

The local authority may attach any other conditions which it thinks fit but if it is to permit the animal to be taken into another local authority area for more than 72 hours, it must consult that local authority (section 1(8)).

Conditions not required by the Act to be attached to the licence may be revoked or modified by the local authority or new conditions may be added. These variations come into effect immediately if they were requested by the licence holder but otherwise the local authority must notify him and allow a reasonable time for compliance (section 1(10)).

Disqualifications and cancellations

Where a person is convicted of an offence under the Dangerous Wild Animals Act 1976 or under the:

(a) Protection of Animals Act 1911;
(b) Protection of Animals (Scotland) Act 1912–64;
(c) Pet Animals Act 1951;
(d) Performing Animals (Regulation) Act 1925;
(e) Animals (Cruel Poisons) Act 1962;
(f) Animal Boarding Establishments Act 1963;
(g) Riding Establishments Act 1964 and 1970;
(h) Breeding of Dogs Act 1973;
(i) Animals Welfare Act 2006, sections 4–9 and 11,

the court may cancel any licence he may hold to keep a dangerous wild animal and disqualify him, whether or not he is a current holder, from holding such a licence for such period as the court thinks fit. The cancellation or disqualification may be suspended by the court in the event of an appeal (section 6(2) and (3)).

FC95 Dangerous wild animals – Dangerous Wild Animals Act 1976

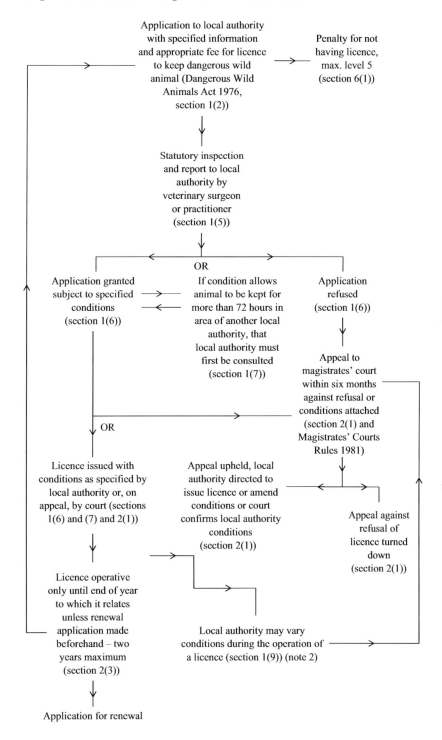

Application to local authority with specified information and appropriate fee for licence to keep dangerous wild animal (Dangerous Wild Animals Act 1976, section 1(2))

Penalty for not having licence, max. level 5 (section 6(1))

Statutory inspection and report to local authority by veterinary surgeon or practitioner (section 1(5))

OR

Application granted subject to specified conditions (section 1(6))

If condition allows animal to be kept for more than 72 hours in area of another local authority, that local authority must first be consulted (section 1(7))

Application refused (section 1(6))

Appeal to magistrates' court within six months against refusal or conditions attached (section 2(1) and Magistrates' Courts Rules 1981)

OR

Licence issued with conditions as specified by local authority or, on appeal, by court (sections 1(6) and (7) and 2(1))

Appeal upheld, local authority directed to issue licence or amend conditions or court confirms local authority conditions (section 2(1))

Appeal against refusal of licence turned down (section 2(1))

Licence operative only until end of year to which it relates unless renewal application made beforehand – two years maximum (section 2(3))

Local authority may vary conditions during the operation of a licence (section 1(9)) (note 2)

Application for renewal

Notes

1. Penalty for contravening licence condition is maximum level 5 (section 6(1)).
2. Unless the variation was requested by the licence holder, the local authority must notify him of the variation and allow a reasonable time for compliance.

Power of entry

Local authorities may authorise competent persons to enter premises either licensed under the Act or specified in an application for a licence, at all reasonable times, producing, if required, their authority, and the authorised officers may inspect these premises and an animal in them. The local authority may charge the person making the application, for the inspection.

The penalty for wilfully obstructing or delaying an authorised officer is a maximum fine at level 5 (sections 3 and 6(1)).

Seizure of animals

If a dangerous wild animal is being kept without the authority of a licence or in contravention of a licence condition, the local authority may seize the animal and retain it, destroy it or otherwise dispose of it. The local authority is not liable to compensation and may recover costs from the keeper of the animal at the time of this seizure (section 4).

PET SHOPS

References

Pet Animals Act 1951 (as amended)
DEFRA Circular 1/2002
Welfare of Animals Act (Northern Ireland) 2011
Welfare of Animals Act (Northern Ireland) 1972 (as amended)
Petshops Regulations (Northern Ireland) 2000

Extent

This licensing procedure applies in Wales and Scotland. Similar provisions apply in Northern Ireland.

In England these provisions are replaced by the Animal Welfare (Licensing of Activities Involving Animals) (England) Regulations 2018 (see p. 451).

Scope

A person keeping a pet shop requires a licence from the local authority (section 1(1)). The Act defines 'the keeping of a pet shop' as:

> the carrying on at premises of any nature (including a private dwelling) of a business of selling animals as pets, and as including references to the keeping of animals in any such premises . . . with a view to their being sold in the course of such a business, whether by the keeper thereof or by any other person.
> Provided that:
>
> (a) a person shall not be deemed to keep a pet shop by reason only of his keeping or selling pedigree animals bred by him, or the offspring of an animal kept by him as a pet;
> (b) where a person carries on a business of selling animals as pets in conjunction with a business of breeding pedigree animals, and the local authority are satisfied

that the animals so sold by him (in so far as they are not pedigree animals bred by him) are animals which were acquired by him with a view to being used, if suitable, for breeding or show purposes but have subsequently been found by him not to be suitable or required for such use, the local authority may if they think fit direct that the said person shall not be deemed to keep a pet shop by reason only of his carrying on the first-mentioned business.

References in this Act to the selling or keeping of animals as pets shall be construed in accordance with the following provisions:

(a) as respects cats and dogs, such references shall be construed as including references to selling or keeping, as the case may be, wholly or mainly for domestic purposes; and

(b) as respects any animal, such references shall be construed as including references to selling or keeping, as the case may be, for ornamental purposes.

(section 7(1) and (2))

Applications

Applications made to the local authority must specify the premises concerned and a fee as determined by the local authority must be paid before a licence is granted (section 1(2)).

The factors which may be taken into account in deciding whether or not to issue a licence are listed under 'Conditions' below but the local authority has discretion to refuse a licence on any other grounds (section 1(3)).

Licences

Licences remain in force until the end of the year to which they relate and the latter is determined by the applicant as being either the year in which it is granted or the following year. In the first case, i.e. the year in which it is granted, the licence comes into force on the day it is granted and expires on 31 December of that year; in the second case, i.e. the year following that in which it is granted, it comes into force on 1 January of that year and expires on 31 December of that year (section 1(5) and (6)).

Conditions

The local authority must attach any conditions which it considers to be necessary or expedient for securing all or any of the following objectives:

(a) that animals will at all times be kept in accommodation suitable as respects size, temperature, lighting, ventilation and cleanliness;

(b) that animals will be adequately supplied with suitable food and drink and (so far as necessary) visited at suitable intervals;

(c) that animals, being mammals, will not be sold at too early an age;

(d) that all reasonable precautions will be taken to prevent the spread among animals of infectious diseases;

(e) that appropriate steps will be taken in case of fire or other emergency (section 1(3)).

No conditions may be specified which relate to a matter dealt with by the Regulatory Reform (Fire Safety) Order 2005 (section 1(3A)).

FC96 Pet shops – Pet Animals Act 1951

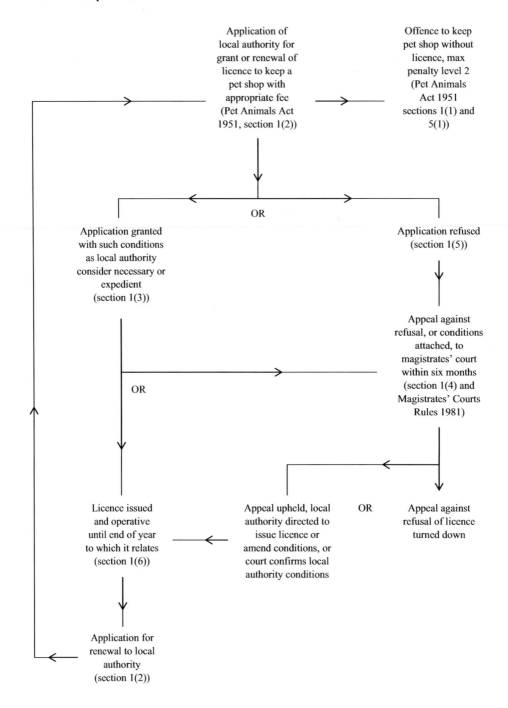

Application of local authority for grant or renewal of licence to keep a pet shop with appropriate fee (Pet Animals Act 1951, section 1(2))

Offence to keep pet shop without licence, max penalty level 2 (Pet Animals Act 1951 sections 1(1) and 5(1))

OR

Application granted with such conditions as local authority consider necessary or expedient (section 1(3))

Application refused (section 1(5))

Appeal against refusal, or conditions attached, to magistrates' court within six months (section 1(4) and Magistrates' Courts Rules 1981)

OR

Licence issued and operative until end of year to which it relates (section 1(6))

Appeal upheld, local authority directed to issue licence or amend conditions, or court confirms local authority conditions

OR

Appeal against refusal of licence turned down

Application for renewal to local authority (section 1(2))

Notes
1. Penalty for contravening licence conditions, maximum level 2 (section 5(1)).
2. Conditions may only be varied on the renewal of a licence.

Power of entry

Local authority officers, veterinary surgeons or practitioners authorised in writing by the local authority for this purpose, may, upon producing their authority if required, inspect a licensed pet shop and any animals there at all reasonable times. Persons wilfully obstructing authorised officers are subject on conviction to a maximum fine of level 2 on the standard scale (sections 4 and 5(1)).

There does not appear to be a power of entry for premises which do not hold a pet shop licence, e.g. which may be suspected of operating an unlicensed pet shop or in respect of which an application has been made. In the latter case, refusal of the licence application would be the obvious remedy.

A local authority in Wales can prosecute proceedings for any offence under this Act committed in the area of the authority but a local authority in England can only prosecute proceedings for an offence under section 2 (any person carrying on a business of selling animals as pets in any part of a street road or public place, or at a stall or barrow in a market).

Disqualifications and cancellations

Where a person is convicted under the Pet Animals Act 1951, the Protection of Animals Act 1911, the Protection of Animals (Scotland) Act 1912 or the Animal Welfare Act 2006, sections 4–9 and 11, the court may cancel any pet shop licence held by the person and may disqualify him from holding such a licence, whether or not he currently holds one, for any specified period. The cancellation or disqualification may be suspended by the court pending an appeal (section 5(3) and (4)).

Where proceedings for cruelty or neglect are brought under the Protection of Animals Act 1911, the Protection of Animals (Amendment) Act 2000 enables courts to make orders regarding the care, disposal or slaughter of animals kept for sale in pet shops.

Local authorities must refuse licence applications from persons currently disqualified by a court from holding a pet shop licence (section 1(2)). This provision differs from that relating to animal boarding establishments, where disqualifications under other Acts are also relevant.

Definitions

Animal includes any description of vertebrate.

Pedigree animal means an animal of any description which is, by its breeding, eligible for registration with a recognised club or society keeping a register of animals of that description (section 7(3)).

There is no definition of the word 'premises', this having been removed by the Amendment Act of 1983, but the sale of animals as pets as a business is prohibited in any part of a street or public place or at a stall or barrow in a market (section 2 as amended).

RIDING ESTABLISHMENTS

References

Riding Establishments Acts 1964 and 1970 as amended by the Animal Welfare Act 2006
Welfare of Animals Act (Northern Ireland) 2011

Welfare of Animals Act (Northern Ireland) 1972 (as amended)
Riding Establishments Regulations (Northern Ireland) 1980

Extent

This licensing procedure applies in Wales and Scotland (section 9(2)). Similar provisions apply in Northern Ireland.

In England some of these provisions remain enacted at the time of writing but in effect are replaced by the Animal Welfare (Licensing of Activities Involving Animals) (England) Regulations 2018.

Scope

A person keeping a riding establishment is required to hold a licence issued by the local authority (section 1(1)).

The 1964 Act defines the keeping of a riding establishment as:

> the carrying on of a business of keeping horses for either or both of the following purposes, that is to say, the purpose of their being let out on hire for riding or the purpose of their being used in providing, in return for payment, instruction in riding, but as not including a reference to the carrying on of such a business:
>
> (a) in a case where the premises where the horses employed for the purposes of the business are kept are occupied by or under the management of the Secretary of Defence; or
> (b) solely for police purposes; or
> (c) by the Zoological Society of London; or
> (d) by the Royal Zoological Society of Scotland.
>
> (section 6(1))

Horses kept by a university providing veterinary courses are also exempt and the place at which a riding establishment is run is to be taken as the place at which the horses are kept (section 6(2) and (3)).

Applications

Applications made to the local authority must specify the premises concerned and a fee as determined by the local authority must be paid before any licence is granted (section 1(2)).

Reports

The local authority is required to receive a report by a listed veterinary surgeon or practitioner (see 'Power of entry' below) before making a decision, and the report must be based on an inspection made not more than 12 months before the application was received (section 1(3)).

Matters for consideration

In deciding whether or not to grant a licence, or provisional licence, the local authority must, without prejudice to its right to refuse a licence on other grounds, have regard to the following matters:

(a) whether that person appears to them to be suitable and qualified, either by experience in the management of horses or by being the holder of an approved certificate or by employing in the management of the riding establishment a person so qualified, to be the holder of such a licence; and

(b) the need for securing:

    (i) that paramount consideration will be given to the condition of the horses and that they will be maintained in good health, and in all respects physically fit and that, in the case of a horse kept for the purpose of its being let out on hire for riding or a horse kept for the purpose of its being used in providing instruction in riding, the horse will be suitable for the purpose for which it is kept;

    (ii) that the feet of all animals are properly trimmed and that, if shod, their shoes are properly fitted and in good condition;

    (iii) that there will be available at all times, accommodation for horses suitable as respects construction, size, number of occupants, lighting, ventilation, drainage and cleanliness and that these requirements be complied with not only in the case of new buildings but also in the case of buildings converted for use as stabling;

    (iv) that in the case of horses maintained at grass there will be available for them at all times during which they are so maintained adequate pasture and shelter and water and that supplementary feeds will be provided as and when required;

    (v) that horses will be adequately supplied with suitable food, drink and (except in the case of horses maintained at grass, so long as they are so maintained) bedding material, and will be adequately exercised, groomed and rested and visited at suitable intervals;

    (vi) that all reasonable precautions will be taken to prevent and control the spread among horses of infectious or contagious disease and that veterinary first-aid equipment and medicines shall be provided and maintained in the premises;

    (vii) that appropriate steps will be taken for the protection and extrication of horses in case of fire and, in particular, that the name, address and telephone number of the licence holder or some other responsible person will be kept displayed in a prominent position on the outside of the premises and that instructions as to action to be taken in the event of fire, with particular regard to the extrication of horses, will be kept displayed in a prominent position on the outside of the premises;

    (viii) that adequate accommodation will be provided for forage, bedding, stable equipment and saddlery (section 1(4)).

Persons under 18 years old or persons or bodies corporate disqualified under the following provisions may not be given a licence:

(a) from keeping a riding establishment under the Riding Establishment Act 1964;

(b) from keeping a pet shop under the Pet Animals Act 1951;

(c) from having custody of animals under the Protection of Animals (Amendment) Act 1954;

(d) from keeping an Animal Boarding Establishment under the Animal Boarding Establishment Act 1963;

(e) under section 30(2), (3) or (4) of the Animal Welfare Act 2006;

(f) under subsection (1) of section 40 of the Animal Health and Welfare (Scotland) Act 2006 (asp 11) (section 1(2)).

Licences

Full licences continue for one year beginning on the day on which they came into force and then expire. The date of operation, depending upon the wishes of the applicant, is either the day on which it is granted or 1 January next (section 1(6) and (7)).

Provisional licences operate for three months from the day on which they are granted and are used where the local authority is satisfied that it would not be justified in issuing a full licence. The three months' period of operation may, on application before the expiration of the three months, be extended for a further period of not exceeding three months so long as this would not exceed a six-month period in one year (Riding Establishments Act 1970, section 1).

On the death of a licence holder, licences pass to his personal representative for a period of three months and then expire. The three-month period may be extended at the local authority's discretion (section 1(8)).

Any person guilty of an offence under any provision of this Act other than section 2(4) (level 2 on the standard scale) is liable to a fine not exceeding level 3 on the standard scale or three months' imprisonment, or both (section 4(1)).

Conditions

On granting a licence, the local authority is required to specify conditions as appear necessary or expedient to achieve all the objectives set out in (b)(i)–(viii) in the paragraph headed 'Matters for consideration' above. In addition, the following conditions are required by the Act, whether specified in the licence or not:

(a) a horse found on inspection of the premises by an authorised officer to be in need of veterinary attention shall not be returned to work until the holder of the licence has obtained at his own expense and has lodged with the local authority a veterinary certificate that the horse is fit for work;

(b) no horse will be let out on hire for riding or used for providing instruction in riding without supervision by a responsible person of the age of 16 years or over unless (in the case of a horse let out for hire for riding) the holder of the licence is satisfied that the hirer of the horse is competent to ride without supervision;

(c) the carrying on of the business of a riding establishment shall at no time be left in the charge of any person under 16 years of age;

(d) the licence holder shall hold a current insurance policy which insures him against liability for any injury sustained by those who hire a horse from him for riding and those who use a horse in the course of receiving from him in return for payment, instruction in riding and arising out of the hire or use of a horse and which also insures such persons in respect of any liability which may be incurred by them in respect of injury to any person caused by, or arising out of, the hire or use of a horse;

(e) a register shall be kept by the licence holder of all horses in his possession aged three years and under and usually kept on the premises, which shall be available for inspection by an authorised officer at all reasonable times (section 1(4A)).

Disqualifications and cancellations

In making a conviction under this Act, under the Protection of Animals Act 1911, the Protection of Animals (Scotland) Act 1912, the Pet Animals Act 1951, the Animal Boarding

FC97 Riding establishments – Riding Establishments Act 1964 as amended

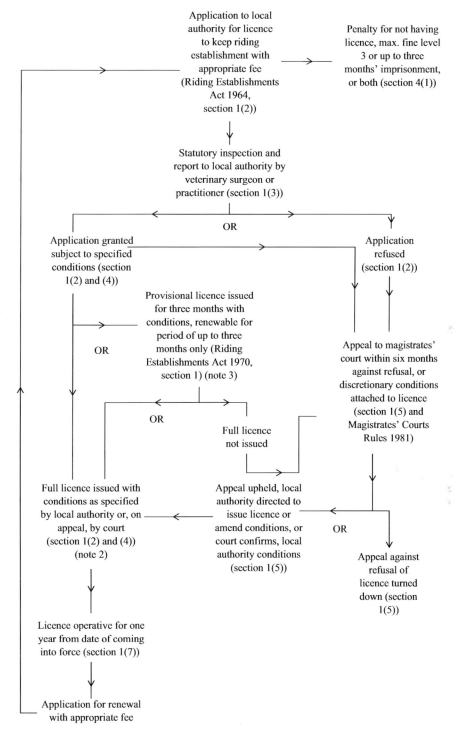

Application to local authority for licence to keep riding establishment with appropriate fee (Riding Establishments Act 1964, section 1(2))

Penalty for not having licence, max. fine level 3 or up to three months' imprisonment, or both (section 4(1))

Statutory inspection and report to local authority by veterinary surgeon or practitioner (section 1(3))

OR

Application granted subject to specified conditions (section 1(2) and (4))

Application refused (section 1(2))

Provisional licence issued for three months with conditions, renewable for period of up to three months only (Riding Establishments Act 1970, section 1) (note 3)

OR

OR

Full licence not issued

Appeal to magistrates' court within six months against refusal, or discretionary conditions attached to licence (section 1(5) and Magistrates' Courts Rules 1981)

Full licence issued with conditions as specified by local authority or, on appeal, by court (section 1(2) and (4)) (note 2)

Appeal upheld, local authority directed to issue licence or amend conditions, or court confirms, local authority conditions (section 1(5))

OR

Appeal against refusal of licence turned down (section 1(5))

Licence operative for one year from date of coming into force (section 1(7))

Application for renewal with appropriate fee

Notes
1. This procedure applies in Wales and Scotland but not in Northern Ireland (section 9(2)).
2. Penalty for contravening licence condition is maximum level 3 and up to three months' imprisonment, or both (section 4(1) as amended).
3. There is no appeal against the decision to issue a provisional licence only or against the conditions attached to it.
4. Conditions may only be varied at the renewal of the licence.

Establishment Act 1963 or sections 4–9 and 11 of the Animal Welfare Act 2006, the court may cancel any riding establishment licence held by the convicted person and may disqualify him from holding such a licence, whether or not he is a current holder, for such period as the court thinks fit. The cancellation or disqualification may be suspended by the court pending an appeal (section 4(3) and (4)).

Power of entry

Local authority officers and veterinary surgeons and practitioners, authorised in writing by the local authority, may, upon producing their authority if required, enter and inspect the following premises at all reasonable times:

(a) licensed riding establishments;
(b) premises subject to an application for a licence to run a riding establishment; and
(c) unlicensed premises suspected of being used as a riding establishment.

This power extends to the inspection of any horses found on the premises (section 2(1) to (3)).

Persons wilfully obstructing or delaying authorised officers are subject on conviction to a maximum fine not exceeding level 2 (sections 2(4) and 4(1)).

Definitions

Approved certificate means:

(a) any one of the following certificates issued by the British Horse Society, namely, Assistant Instructor's Certificate, Instructor's Certificate and Fellowship;
(b) Fellowship of the Institute of Horse; or
(c) any other Certificate for the time being prescribed by order of the Secretary of State.

Horse includes any mare, gelding, pony, foal, colt, filly or stallion and also any ass, mule or jennet.

Premises includes land (section 6(4)).

Chapter 2

DISEASE CONTROL

INFECTIOUS DISEASES CONTROL

References

Public Health (Control of Disease) Act 1984 as amended by the Health and Social Care Act 2008
Health protection legislation guidance 2010 – Department of Health
Health Protection Regulations 2010 Toolkit – Chartered Institute of Environmental Health, Department of Health and Lewes District Council
Public Health etc. (Scotland) Act 2008
Public Health Act (Northern Ireland) 1967

Extent

These provisions apply to England and Wales (section 79(3)). Similar provisions apply in Scotland and Northern Ireland.

Power of entry

A proper officer of a council, producing if required a duly authenticated document showing his authority, has a right to enter any premises at all reasonable hours for the following purposes:

(a) ascertaining if there has been a contravention of the Act or bye-laws made under it;
(b) seeing if circumstances exist which would authorise or require the council to take any action or execute work;
(c) to take any action or execute any work authorised or required;
(d) generally for the council's performance of its functions.

For premises other than a factory, workshop or workplace, at least 24 hours' notice must be given before entry may be demanded.
  Where there is reasonable ground for entry under (a)–(d) above and:

(a) admission has been refused; or
(b) refusal is apprehended; or
(c) the premises are unoccupied; or
(d) the occupier is temporarily absent; or

(e)   the case is one of urgency; or

(f)   application for admission would defeat the object of entry,

the local authority may make application to a Justice of the Peace by way of sworn information in writing, and the Justice of the Peace may authorise entry by an authorised officer, if necessary by force. Unless the case falls as one of (c)–(f) above, notice to apply for the warrant must be given to the occupier (section 61). The authorised officer may take with him any other persons as may be necessary but must leave any unoccupied premises as secure as he found them.

The officer may for the purpose for which entry is authorised:

(a)   search the premises;

(b)   carry out measurements and tests of the premises or of anything found on them;

(c)   take and retain samples of the premises or of anything found on them;

(d)   inspect and take copies or extracts of any documents or records found on the premises;

(e)   require information stored in an electronic form and accessible from the premises to be produced in a form in which it can be taken away and in which it is visible and legible or from which it can readily be produced in a visible and legible form; and

(f)   seize and detain or remove anything which the officer reasonably believes to be evidence of any contravention relevant to the purpose for which entry is authorised (section 62).

The penalty for obstructing any person acting in the execution of these Acts, regulations and orders is a maximum of level 1 on the standard scale and in any other case a fine (section 63).

Key to powers available for the control of disease

The following powers are available to local authorities for the control of communicable disease in the Health Protection (Local Authority Powers) Regulations 2010:

- power to order health measures in relation to persons;
- power to order health measures in relation to things;
- power to order health measures in relation to premises;
- orders in respect of groups;
- procedure for making, varying and revoking Part 2A orders;
- enforcement of Part 2A orders.

Note

Special arrangements exist for the control of infectious diseases in ships and aircraft in the Public Health (Aircraft) Regulations 1979 and the Public Health (Ships) Regulations 1979 and, in respect of the Channel Tunnel, the Public Health (International Trains) Regulations 1994 (see 'Port health').

Proper officer

All of the powers and procedures for the notification and control of communicative diseases identified in this chapter are enforced by the local authority (for definition, see p. 40). Local authorities are, however, advised that the majority of such powers should be delegated to the

Consultant in Communicative Disease Control (CCDC) appointed by each Health Protection Agency (Annex B to NHS Management Executive Guidance HSG (93)56). This will require a formal resolution of the local authority detailing each of the powers being delegated and is to be done in the context of a local communicable disease plan which identifies the role and responsibilities of the various agencies and individuals.

Local authorities need to provide appropriate professional and other staff support to the Consultant in Communicative Disease Control and ensure that proper arrangements have been made for the receipt of notifications (FC98).

NOTIFICATION OF DISEASES

References

Health Protection (Notification) Regulations 2010
The Health Protection (Notification) (Wales) Regulations 2010
The Public Health (Notification of Infectious Diseases) (Scotland) Regulations 1988
Public Health Act (Northern Ireland) 1967

Extent

These regulations referred to below apply to England only but there are similar provisions in Wales, Scotland and Northern Ireland.

Duty to notify

A registered medical practitioner who becomes aware or suspects that a patient whom he is attending:

- has a notifiable disease; or
- has an infection or is contaminated in a manner which presents or could present significant harm to human health

must notify the proper officer of the local authority within three days of the patient being attended to by the practitioner.

The duty does not apply where the registered medical practitioner believes the proper officer has already been notified (regulation 2).

Diseases to be notified

The diseases (confirmed or suspected) which are required to be notified by this procedure are:

Acute encephalitis
Acute infectious hepatitis
Acute meningitis
Acute poliomyelitis
Anthrax
Botulism
Brucellosis

Cholera
Diphtheria
Enteric fever (typhoid or paratyphoid fever)
Food poisoning
Haemolytic uraemic syndrome (HUS)
Infectious bloody diarrhoea
Invasive group A streptococcal disease and scarlet fever
Legionnaires' disease
Leprosy
Malaria
Measles
Meningococcal septicaemia
Mumps
Plague
Rabies
Rubella
SARS
Smallpox
Tetanus
Tuberculosis
Typhus
Viral haemorrhagic fever (VHF)
Whooping cough
Yellow fever

(Schedule 1 Health Protection (Notification) Regulations 2010)
   Note – for the notification of occupation-related diseases, see FC50.

Details to be notified

Notification must include the following details of the patient:

(a) name, date of birth and sex;
(b) home address, including postcode;
(c) current residence (if not home address);
(d) telephone number;
(e) NHS number;
(f) occupation (if considered relevant);
(g) the name, address and postcode of his/her place of work or education (if considered relevant);
(h) relevant overseas travel history;
(i) ethnicity;
(j) contact details for a parent of the patient (where a child) (where the child is living);
(k) the disease or infection which the patient has or is suspected of having, or the nature of the patient's contamination or suspected contamination;
(l) the date of onset of symptoms;
(m) the date of diagnosis; and
(n) name, address and telephone number of the registered medical practitioner (regulation 2(2))

Notification etc. by a proper officer

The proper officer of the local authority must disclose the notification and its contents to:

(a) the Health Protection Agency;
(b) the proper officer of the local authority in whose area the patient usually resides (if different); and
(c) the proper officer of the port health authority or local authority in whose district or area a ship, hovercraft, aircraft or international train is or was situated from which the patient has disembarked (if known to the disclosing proper officer and if that officer considers disclosure appropriate).

The disclosure must be made in writing within three days, beginning with the day that the proper officer receives the notification.

If the proper officer considers that the case is urgent, disclosure must be made orally as soon as reasonably practicable.

In determining whether a case is urgent, the proper officer must have regard to:

(a) the nature of the disease, infection or contamination or the suspected disease, infection or contamination notified;
(b) the ease of spread of the disease, infection or contamination;
(c) the ways in which the spread of the disease, infection or contamination can be prevented or controlled; and
(d) where known, the patient's circumstances (including age, sex and occupation) (regulation 6).

In addition, proper officers should inform the Department of Health of any case or outbreak of infectious disease which may have wider than local significance – this is to allow consideration of action at national level, e.g. withdrawal of a food for sale. Proper officers also have a responsibility to consult as appropriate with the Communicable Disease Surveillance Centre, specifically with cases of outbreaks which present unusual features. Detailed guidance is found in Communicable Disease Outbreak Management Operational guidance 2014 PHE.

The proper officer must make weekly and quarterly returns of notifications to the Registrar General and these are collated by the Communicable Disease Surveillance Centre for weekly publication in the Communicable Disease Report.

Notifications, information, disclosures, lists and reports, which are required to be in writing, may be communicated electronically if:

(a) the recipient has consented in writing to receiving the notification, information, disclosure list or report (as the case may be) by an electronic communication; and
(b) the communication is sent to the number or address specified by the recipient when giving that consent (regulation 7).

Duty to provide information to Public Health England

Public Health England may request that the person who solicited the laboratory test which identified the causative agent to which the notification relates provides to it the information included in the notification in writing within three days, beginning with the day on which the request is made.

If Public Health England considers the case to be urgent, the information must be provided orally as soon as reasonably practicable.

In determining whether the case is urgent, the Public Health England must have regard to:

(a) the nature of the causative agent to which the notification relates;
(b) the nature of the disease which the causative agent causes;
(c) the ease of spread of the causative agent;
(d) the ways in which the spread of the causative agent can be prevented or controlled; and
(e) where known, the circumstances of the person from whom the sample was taken (including age, sex and occupation) (regulation 5).

Causative agents

The operator of a diagnostic laboratory must notify the Public Health England in accordance with this regulation where the diagnostic laboratory identifies a causative agent in a human sample.

The notification must include the following information insofar as it is known to the operator of the diagnostic laboratory:

(a) name and address of the diagnostic laboratory;
(b) details of the causative agent identified;
(c) date of the sample;
(d) nature of the sample;
(e) name of person from whom the sample was taken;
(f) the person's date of birth and sex, current home address, including postcode, current residence (if not home address), ethnicity and NHS number;
(g) and the name, address and organisation of the person who solicited the test which identified the causative agent.

The notification must be provided in writing within seven days, beginning with the day on which the causative agent is identified.

If the operator of the diagnostic laboratory considers that the case is urgent, the notification must be provided orally as soon as reasonably practicable.

In determining whether the case is urgent, the operator of the diagnostic laboratory must have regard to:

(a) the nature of the causative agent;
(b) the nature of the disease which the causative agent causes;
(c) the ease of spread of the causative agent;
(d) the ways in which the spread of the causative agent can be prevented or controlled; and
(e) where known, the person's circumstances (including age, sex and occupation).

This regulation does not apply where the operator of the diagnostic laboratory reasonably believes that Public Health England has already been notified in accordance with this regulation by the operator of another diagnostic laboratory in relation to the same causative agent being found in a sample from the same person.

It is an offence for the operator of a diagnostic laboratory to fail without reasonable excuse to comply with this regulation.

FC98 Notification of diseases – Health Protection (Notification) Regulations 2010

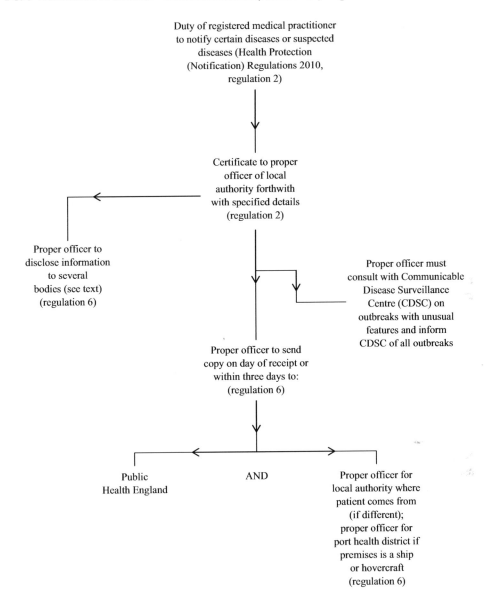

Duty of registered medical practitioner to notify certain diseases or suspected diseases (Health Protection (Notification) Regulations 2010, regulation 2)

Certificate to proper officer of local authority forthwith with specified details (regulation 2)

Proper officer to disclose information to several bodies (see text) (regulation 6)

Proper officer must consult with Communicable Disease Surveillance Centre (CDSC) on outbreaks with unusual features and inform CDSC of all outbreaks

Proper officer to send copy on day of receipt or within three days to: (regulation 6)

Public Health England

AND

Proper officer for local authority where patient comes from (if different); proper officer for port health district if premises is a ship or hovercraft (regulation 6)

Notes
1. This procedure does not apply to Scotland or Northern Ireland.
2. For special procedures relating to the notification of infectious diseases on board ships or aircraft, see the Public Health (Ships) Regulations 1979 and the Public Health (Aircraft) Regulations 1979.

Any person who commits an offence under this regulation is liable on summary conviction to a fine not exceeding level 5 on the standard scale (regulation 4).

List of causative agents

Bacillus anthracis
Bacillus cereus (only if associated with food poisoning)
Bordetella pertussis
Borrelia spp
Brucella spp
Burkholderia mallei
Burkholderia pseudomallei
Campylobacter spp
Chikungunya virus
Chlamydophila psittaci
Clostridium botulinum
Clostridium perfringens (only if associated with food poisoning)
Clostridium tetani
Corynebacterium diphtheriae
Corynebacterium ulcerans
Coxiella burnetii
Crimean-Congo haemorrhagic fever virus
Cryptosporidium spp
Dengue virus
Ebola virus
Entamoeba histolytica
Francisella tularensis
Giardia lamblia
Guanarito virus
Haemophilus influenzae (invasive)
Hanta virus
Hepatitis A, B, C, delta, and E viruses
Influenza virus
Junin virus
Kyasanur Forest disease virus
Lassa virus
Legionella spp
Leptospira interrogans
Listeria monocytogenes
Machupo virus
Marburg virus
Measles virus
Mumps virus
Mycobacterium tuberculosis complex
Neisseria meningitidis
Omsk haemorrhagic fever virus
Plasmodium falciparum, vivax, ovale, malariae, knowlesi
Polio virus (wild or vaccine types)

Rabies virus (classical rabies and rabies-related lyssa viruses)
Rickettsia spp
Rift Valley fever virus
Rubella virus
Sabia virus
Salmonella spp
SARS coronavirus
Shigella spp
Streptococcus pneumoniae (invasive)
Streptococcus pyogenes (invasive)
Varicella zoster virus
Variola virus
Verocytotoxigenic Escherichia coli (including E.coli O157)
Vibrio cholerae
West Nile virus
Yellow fever virus
Yersinia pestis

(Schedule 2 Health Protection (Notification) Regulations 2010).

Definitions

Causative agent means:

(a) a causative agent listed in schedule 2; or
(b) evidence of an infection caused by such an agent.

Child means a person under the age of 18 years.
Parent has the meaning given to it by section 576 of the Education Act 1996.

LOCAL AUTHORITY POWERS TO PREVENT SPREAD OF INFECTION

References

Public Health (Control of Disease) Act 1984
Health Protection (Local Authority Powers) Regulations 2010
Health Protection (Local Authority Powers) (Wales) Regulations 2010
Public Health etc. (Scotland) Act 2008
Public Health Act (Northern Ireland) 1967

Extent

These regulations apply to England only. There are similar powers in Wales, Scotland and Northern Ireland.

Scope

The following procedures are available to local authorities as the regulations confer discretionary powers on local authorities (including powers to impose restrictions and requirements) for the purposes of preventing, protecting against, controlling or providing a public health response to the incidence or spread of infection or contamination which presents or could present significant harm to human health.

Requirement to keep a child away from school

Where a local authority is satisfied that:

(a) a child is or may be infected or contaminated;
(b) the infection or contamination is one which presents or could present significant harm to human health;
(c) there is a risk that the child might infect or contaminate others;
(d) it is necessary to keep the child away from school in order to remove or reduce that risk; and
(e) keeping the child away from school is a proportionate response to the risk to others presented by the child,

the local authority may, by serving notice on the child's parent, require that he or she keep the child away from school.

The notice must include the following information:

(a) the date from which the requirement commences;
(b) the duration of the requirement (up to a maximum of 28 days);
(c) why the requirement is believed to be a necessary and proportionate measure;
(d) the penalty for failing to comply with the notice; and
(e) contact details for an officer of the local authority who is able to discuss the notice.

The local authority must, as soon as reasonably practicable after serving, inform the headteacher of the child's school that it has served such a notice in relation to the child and of the contents of that notice.

The parent may request that the local authority review the notice at any time before the requirement lapses.

The local authority must review the notice within five working days, beginning with the day on which the request is made where the parent is requesting a review of that notice for the first time or may review the notice in the case of all other requests.

The local authority must inform the parent and the headteacher of the child's school of the outcome of any review it conducts as soon as reasonably practicable after the review is concluded.

The local authority may vary or revoke a notice and must, as soon as reasonably practicable, inform the parent and the headteacher of the child's school that the notice has been varied or revoked and, if varied, the nature of the variation.

The local authority may serve consecutive notices and must inform the parent and the headteacher of the child's school as soon as reasonably practicable where a notice has expired and no further notice is to be served.

It is an offence for the parent to fail without reasonable excuse to comply with a notice.

Any person who commits an offence under this regulation is liable on summary conviction to one or both of:

(a) a fine not exceeding level 2 on the standard scale;

(b) a further fine not exceeding an amount equal to 50 per cent of level 1 on the standard scale for each day on which the default continues after conviction (regulation 2).

Requirement to provide details of children attending school

A local authority may serve notice to require a headteacher of a school in its area to provide it with a list of the names, addresses and contact telephone numbers for all the pupils of that school, or such group of pupils attending that school as it may specify, where:

(a) a person who is or has recently been on the school's premises is or may be infected or contaminated;

(b) the infection or contamination is one which presents or could present significant harm to human health;

(c) there is a risk that the person may have infected or contaminated pupils at the school;

(d) it is necessary for the local authority to have the list in order to contact those pupils with a view to ascertaining whether they are or may be infected or contaminated; and

(e) requiring the list (and contacting those pupils who may be infected or contaminated) is a proportionate response to the risk presented by the person.

The notice must:

(a) specify a time limit for meeting the requirement;

(b) specify an address where the list is to be sent;

(c) provide contact details for an officer of the local authority who is able to discuss the notice.

It is an offence for a headteacher to fail without reasonable excuse to comply with a notice.

Any person who commits an offence under this regulation shall be liable on summary conviction to a fine not exceeding level 1 on the standard scale (regulation 3).

Requests for cooperation for health protection purpose

A local authority may serve notice on any person or group of persons requesting that the person or group of persons do, or refrain from doing, anything for the purpose of preventing, protecting against, controlling or providing a public health response to the incidence or spread of infection or contamination which presents or could present significant harm to human health. The notice must provide contact details for an officer of the local authority who is able to discuss the notice.

The local authority may offer compensation or expenses in connection with its request (regulation 8).

Restriction of contact with and access to dead bodies

Where a local authority is satisfied that:

(a) a dead body is or may be infected or contaminated;
(b) the infection or contamination is one which presents or could present significant harm to human health;
(c) there is a risk that the dead body might infect or contaminate people;
(d) it is necessary to restrict contact with or access to the dead body in order to remove or reduce that risk; and
(e) prohibiting any person from having contact with or access to the dead body is a proportionate response to the risk presented by that dead body,

the local authority may serve on the person having charge or control of the premises in which the dead body is located a notice prohibiting any person from having contact with the dead body or entering the room where it is located.

On receipt of a notice served on the person having charge or control of the premises, he/she must arrange for a copy of the notice to be conspicuously displayed at each of the entry points to the room and/or near the dead body without delay. The notice must include:

(a) a statement to the effect that entry into the room and contact with the body near which the notice has been displayed is prohibited;
(b) a statement to the effect that breach of the prohibition is a criminal offence;
(c) contact details for an officer of the local authority who is able to discuss the notice; and
(d) the legal authority for the prohibition.

It is an offence for:

(a) any person having charge or control of the premises to fail to arrange for a copy of the notice to be displayed in the appropriate places;
(b) any person to remove or deface a notice displayed; or
(c) any person to fail to comply with a notice displayed,

but no offence is committed if

(a) the person has the local authority's consent to enter the room and/or have contact with the dead body; or
(b) the person is exercising the functions of a coroner or is acting under the authority of a coroner (regulations 9 and 10).

Relocation of dead bodies

Where a local authority is satisfied that:

(a) a dead body is or may be infected or contaminated;
(b) the infection or contamination is one which presents or could present significant harm to human health;
(c) there is a risk that the dead body might infect or contaminate people;

and it is necessary to relocate the dead body in order to remove or reduce that risk and relocating the body is a proportionate response to the risk to people presented by the dead body in its current location, it can relocate, or cause to be relocated, the dead body to a place where it considers that the risk of the dead body infecting or contaminating people is reduced or removed.

The local authority cannot relocate, or cause to be relocated, the dead body if:

(a) a coroner has jurisdiction over the dead body; or
(b) it has failed to take reasonable steps to inform the person with charge or control of the premises where the dead body is located of its intention to relocate the body.

Any person having charge or control of the premises in which a dead body is located must co-operate with a local authority that intends to take action and it is an offence for any person to fail without reasonable excuse to comply (regulation 11).

Any person who commits an offence under regulations 9, 10 and 11 is liable on summary conviction to a fine not exceeding level 3 on the standard scale.

Definitions

Child means a person under 18 years of age.
Headteacher means the headteacher or, if the headteacher is absent, the person deputising for the headteacher.
Parent has the meaning given to it by section 576 of the Education Act 1996.
School has the meaning given to it by section 4 of the Education Act 1996.
Working day means any day which is not Saturday, Sunday, Christmas Day, Good Friday or a day which is a bank holiday in England under the Banking and Financial Dealings Act 1971.

POWER TO ORDER HEALTH MEASURES IN RELATION TO PERSONS, THINGS AND PREMISES (PART 2A ORDERS)

References

Public Health (Control of Disease) Act 1984
Health Protection (Part 2A orders) Regulations 2010
Health Protection (Part 2A orders) (Wales) Regulations 2010
(Note – where section is mentioned, it refers to the above Act and, where regulation is mentioned, it refers to the above regulations.)
Public Health etc. (Scotland) Act 2008
Public Health Act (Northern Ireland) 1967

Extent

The Public Health (Control of Disease) Act 1984 applies in England and Wales. Health Protection (Part 2A orders) Regulations 2010 apply in England only, but there are similar provisions in Wales. Similar provisions apply in Scotland and Northern Ireland.

FC99 Requirement to keep a child away from a school – Regulation 2 Health Protection (Local Authority Powers) Regulations 2010

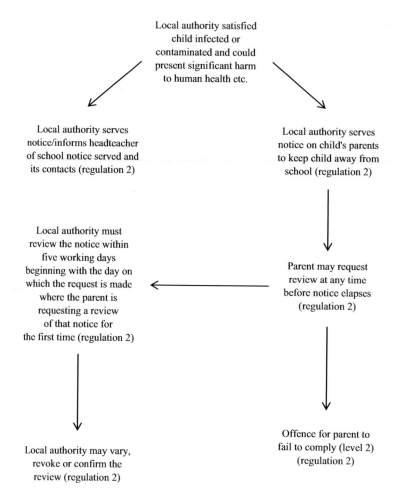

Note

1. There seems to be no appeal against this notice.

FC100 Request for cooperation for health protection purpose – Regulation 8 Health Protection (Local Authority Powers) Regulations 2010

Local authority responding
to an incidence or spread of
infection or contamination
which presents or could
present significant harm to
human health

↓

Local authority may
serve notice on person
or group requesting to
do something or refrain
from doing something
(regulation 8)

Local authority officer
to provide contact
details of officer
(regulation 8)

Local authority can offer
compensation or
expenses in connection
with request
(regulation 8)

Note
1. There seems to be no appeal against this notice.

FC101 Restriction of contact with and access to dead bodies – Regulations 9 and 10 Health Protection (Local Authority Powers) Regulations 2010

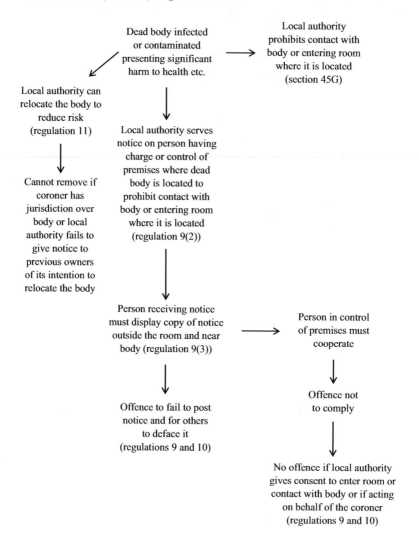

Part 2A orders and their content

A local authority can apply to a Justice of the Peace to make an order for health measures in relation to persons, things and premises (sections 45G, 45H and 45I). If the Justice is satisfied that:

(a) a person, thing or premises is or may be infected or contaminated;
(b) the infection or contamination is one which presents or could present significant harm to human health;
(c) there is a risk that the person, thing or premises might infect or contaminate others; and
(d) it is necessary to make the order in order to remove or reduce that risk,

he may make an order to impose one or more of the following restrictions or requirements – that the *person*:

(a) submit to medical examination;
(b) be removed to a hospital or other suitable establishment;
(c) be detained in a hospital or other suitable establishment;
(d) be kept in isolation or quarantine;
(e) be disinfected or decontaminated;
(f) wear protective clothing;
(g) provide information or answer questions about the person's health or other circumstances (including, in particular, information or questions about the identity of a related party);
(h) be monitored in their health and the results reported;
(i) attend training or advice sessions on how to reduce the risk of infecting or contaminating others;
(j) be subject to restrictions on where the person goes or with whom the person has contact;
(k) abstain from working or trading.

The Justice may also make an order to require that the person provide information or answer questions about the person's health or other circumstances (including, in particular, information or questions about the identity of a related party). This also applies to a person with parental responsibility (within the meaning of the Children Act 1989) for the person to secure that the person submits to or complies with the restrictions or requirements imposed by the order (section 45G).

In relation to *things*, the order may impose one or more of the following restrictions or requirements – that

(a) the thing be seized or retained;
(b) the thing be kept in isolation or quarantine;
(c) the thing be disinfected or decontaminated;
(d) in the case of a dead body, the body be buried or cremated;
(e) in any other case, the thing be destroyed or disposed of.

The Justice may also make an order to require that the owner or person having control of the thing provide information or answer questions about the thing (including, in particular, information or questions about where the thing has been or about the identity of any related person or the whereabouts of any related thing) (section 45H).

In relation to *premises*, the order may impose one or more of the following restrictions or requirements – that

    (a) the premises be closed;

    (b) in the case of a conveyance or movable structure, the conveyance or structure be detained;

    (c) the premises be disinfected or decontaminated;

    (d) in the case of a building, conveyance or structure, the premises be destroyed.

The Justice may also make an order to require the owner or any occupier of the premises to provide information or answer questions about the premises (including, in particular, information about the identity of any related person or the whereabouts of any related thing) (section 45I).

The Justice has the power to make an order in relation to a group of persons, things or premises (section 45J).

Orders for health measures in relation to persons, things and premises may include such other restrictions or requirements as the Justice considers necessary for the purpose of reducing or removing the risk in question, including conditions and directions. It may also order the payment of compensation or expenses in connection with the taking of measures related to the order and authorise those persons to whom it is addressed to do such things as may be necessary to give effect to it (section 45K).

Making, varying and revoking Part 2A orders

Local authorities may apply to a Justice of the Peace to make a Part 2A order and they must co-operate with each other in deciding which of them should apply for a Part 2A order in any particular case.

A Part 2A order may be varied or revoked by a Justice of the Peace on the application of:

    (a) an affected person;

    (b) a local authority; or

    (c) any other authority with the function of executing or enforcing the order in question.

For an order involving a person, the following persons are affected persons:

    (a) the person;

    (b) a person with parental responsibility (within the meaning of the Children Act 1989) for the person;

    (c) the person's husband, wife or civil partner;

    (d) a person living with the person as the person's husband, wife or civil partner; and

    (e) such other persons as may be prescribed by regulations.

For an order involving a thing, the following persons are affected persons:

    (a) the owner of the thing;

    (b) any person with custody or control of the thing; and

    (c) such other persons as may be prescribed by regulations.

For an order involving premises, the following persons are affected persons:

(a) the owner of the premises;
(b) any occupier of the premises; and
(c) such other persons as may be prescribed by regulations.

Variation or revocation of a Part 2A order does not invalidate anything done under the order prior to the variation or revocation (section 45M).

Notice of Part 2A applications

A local authority must make reasonable enquiries as to the existence and location of persons subject to a Part 2A application.

Having made reasonable enquiries, the local authority must give notice of the application to the persons specified in the application, but if a Justice of the Peace considers it necessary to do so, the Justice may make a Part 2A order without a person having been given such notice.

The persons specified are:

(a) the person subject to the application;
(b) a person with parental responsibility for that person, if a child; and
(c) that person's decision-maker (if any).

If the subject of the application is a thing, the persons specified are:

(a) the owner of the thing; and
(b) the person with custody or control of the thing.

Where a dead body or human remains is concerned, the person specified is the deceased's next of kin.

Where premises are the subject of an application, the persons specified are:

(a) the owner of the premises; and
(b) the occupier of the premises, if any.

If it is likely the person will abscond or otherwise take steps to undermine the order applied for, or where exceptional circumstances exist which mean that notifying such a person would not be in the person's best interests, the local authority is not required to give notice (regulation 3).

Evidence required for a Part 2A application in relation to persons

Before an order can be granted, the local authority should provide the Justice of the Peace with the following evidence:

(a) a report which gives known and relevant details, or reasons for omission of details, of:

   (i) the signs and symptoms of the infection or contamination in the person who is the subject of the application;
   (ii) the person's diagnosis;

    (iii) the outcome of clinical or laboratory tests; and

    (iv) the person's recent contacts with, or proximity to, a source or sources of infection or contamination;

(b) a summary of the characteristics and effects of the infection or contamination which the person has or may have, which includes an explanation of:

    (i) the mechanism by which the infection or contamination spreads;

    (ii) how easily the infection or contamination spreads amongst humans; and

    (iii) the impact of the infection or contamination on human health (by reference to pain, disability and the likelihood of death);

(c) where relevant an assessment of the risk to human health that the person or related party presents, including a description of any acts or omissions, or anticipated acts or omissions, of the person which affect that risk, and an assessment of the options available to deal with the risk that the person or related party presents.

The evidence must be given by persons who are suitably qualified to give the evidence and may be given orally or in writing (regulation 4).

Period for which Part 2A order may be in force

The period for which the restriction or requirement imposed by or under the order may be in force must not exceed 28 days beginning with the day on which the order was made.

The period of any extension or variation of a restriction or requirement must not exceed 28 days (section 45L and regulation 5).

Charges

Where a local authority has incurred costs taking measures under a Part 2A order in relation to things and premises, the local authority may impose a charge on the owner or person with custody or control of the thing or the owner or occupier of the premises.

The amount of the charge imposed must not exceed the actual costs (including staff costs) incurred by the local authority in taking measures in relation to the thing or premises and must be reasonable in the circumstances.

A local authority cannot impose a charge in connection with orders which relate to a dead body or human remains (regulation 7).

Understanding of the person subject to the order

The local authority must take all reasonable steps to ensure that the person who is the subject of the order understands:

(a) the effect of the order, the reason it has been made, the power under which it has been made and the person's right to apply for a variation or revocation of the order; and

(b) the relevant support services available to the person (and how to access them).

Where the person is under the age of 18 years, the person with parental responsibility for the person must understand the relevant matters (regulation 8).

Welfare of dependants

Where a person is detained in a hospital or other suitable establishment or is kept in isolation or quarantine under a Part 2A order, the local authority must have regard to the impact of the order on the welfare of the person and his/her dependants, if any, for the duration of the order (regulation 9).

Report to Public Health England

A local authority must provide a written report to the Chief Executive of Public Health England each time it makes a Part 2A application and where it is varied or evoked.

The report must include:

(a) the name of the local authority;
(b) contact details for the officer of the local authority responsible for the report;
(c) a copy of the Part 2A application (with information that would enable the identification of the person who is the subject of the application removed);
(d) if an order is made, a copy of that order (with information that would enable the identification of the person who is the subject of the order removed); and
(e) if a Part 2A order is not made, the reason for it not being made.

The report must be provided as soon as practicable after the application is determined and no later than ten days beginning with the day on which the application is determined (regulations 10 and 11).

Offences

A person commits an offence if the person:

(a) fails without reasonable excuse to comply with a restriction or requirement imposed by or under a Part 2A order; or
(b) wilfully obstructs anyone acting in the execution of a Part 2A order.

A person guilty of an offence is liable on summary conviction to a fine.

If a person is convicted of an offence above, and the court is satisfied that the failure or wilful obstruction constituting the offence has caused premises or things to become infected or contaminated or otherwise damaged them in a material way, it may order the person to take or pay for such remedial action as may be specified in the order.

If a Part 2A order imposes a requirement that a person be detained or kept in isolation or quarantine in a place, and the person leaves that place contrary to the requirement, a constable may take the person into custody and return the person to that place.

However, a person may not be taken into custody after expiry of the period for which the requirement is in force (section 45O).

The provisions of this part have effect in relation to the territorial sea adjacent to England or Wales (section 45S).

FC102 Part 2 orders – Public Health (Control of Disease) Act 1984

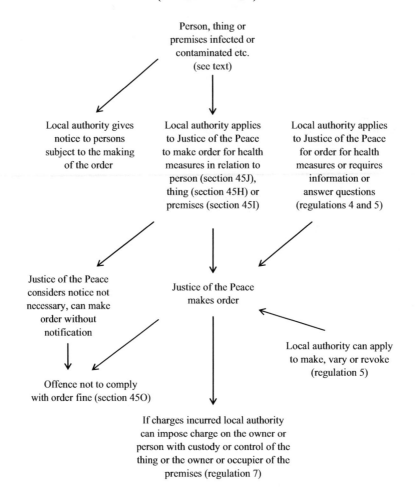

Note

1. For groups of persons, things or premises, order can be made by Justice of the Peace (section 45J).

Definitions

Affected persons – in section 45G, the person's decision-maker (if any). In section 45H(2), in relation to a dead body or human remains, the deceased's next of kin.

Medical examination includes microbiological, radiological and toxicological tests (section 45T).

Next of kin means the person accorded highest rank in the following list (but if two or more persons are accorded equal ranking, then each of those persons is to be treated as next of kin):

(a) a person with parental responsibility for the deceased person (P);
(b) P's husband, wife or civil partner;
(c) a person who had been living with P up to the time of P's death as P's husband, wife or civil partner;
(d) P's child where aged 18 years or over;
(e) P's parent;
(f) P's brother or sister where aged 18 years or over (regulation 3).

Person's decision-maker means the person's donee of enduring power of attorney or lasting power of attorney under the Mental Capacity Act 2005 or a deputy appointed by the Court of Protection in relation to the person, where decisions in connection with Part 2A applications or orders are within the scope of that person's authority (regulation 3).

Related party means:

(a) a person who has or may have infected or contaminated the person; or
(b) a person whom the person has or may have infected or contaminated (section 45G).

Thing includes:

(a) human tissue;
(b) a dead body or human remains;
(c) animals; and
(d) plant material (section 45T).

DISINFECTION OR DECONTAMINATION OF THINGS AND PREMISES

References

Health Protection (Local Authority Powers) Regulations 2010, regulations 4 to 7
The Health Protection (Local Authority Powers) (Wales) Regulations 2010
Public Health etc. (Scotland) Act 2008
Public Health Act (Northern Ireland) 1967

Extent

These provisions apply to England only. Similar provisions apply in Wales, Scotland and Northern Ireland.

**FC103 Disinfection or decontamination of premises and things on request of owner –
regulations 4 to 7 Health Protection (Local Authority Powers) Regulations 2010**

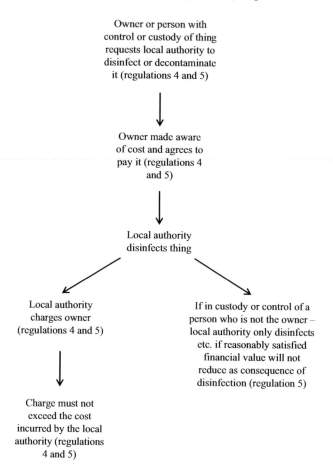

Notes
1. This procedure applies to things. The same procedure applies to premises by substituting 'premises' for 'things' and 'regulations 4 and 5' with 'regulations 6 and 7'.
2. Also substitute 'tenant' for 'person with control or custody of thing'.

Scope

These powers cover disinfection or decontamination of things and premises on request of the owner or person with custody or control of a premises or a thing.

A local authority may disinfect or decontaminate, or cause to be disinfected or decontaminated, a thing where requested to do so by the owner of the thing or the person with custody or control, and may charge the owner or person with custody or control for the disinfection or decontamination of the thing if the owner is made aware of the charge prior to disinfection or decontamination being carried out and agrees to pay it.

The local authority's charge must not exceed the cost incurred by the local authority in carrying out the disinfection or decontamination (regulations 4 and 5).

Where the thing is in the custody or control of a person, the local authority should only disinfect or decontaminate the thing if it is reasonably satisfied that the financial value of the thing will not be reduced as a consequence of the disinfection or decontamination.

This procedure also applies to premises. For premises with a tenant, he/she can request the action (regulations 6 and 7). See FC102 and FC103.

CLEANSING OF FILTHY OR VERMINOUS PREMISES, ARTICLES AND PERSONS

References

Public Health Act 1936, sections 83 to 85 (as amended)
Public Health Act 1961, sections 36 and 37
Public Health etc. (Scotland) Act 2008

Extent

These provisions apply to England and Wales only. Similar provisions apply in Scotland and Northern Ireland

Scope

These provisions apply to any premises except those forming part of a mine or quarry but including ships, boats, tents, vans and sheds (sections 83(4), 267(4) and 268(1)), and cover situations which are either:

(a) filthy or unwholesome so as to be prejudicial to health; or
(b) verminous.

Section 84 Public Health Act 1936 deals with articles that are filthy and verminous.
Section 85 Public Health Act 1936 deals with persons that are verminous.
Section 42 Care Act 2014 may be relevant in cases where the person is hoarding and leads to the premises or themselves being filthy and/or verminous. It is suggested that where an EH is aware of a filthy and/or verminous person or premises, the powers under the Care Act be considered (p. 520).

FC104 Cleansing of filthy or verminous premises – section 83 Public Health Act 1936

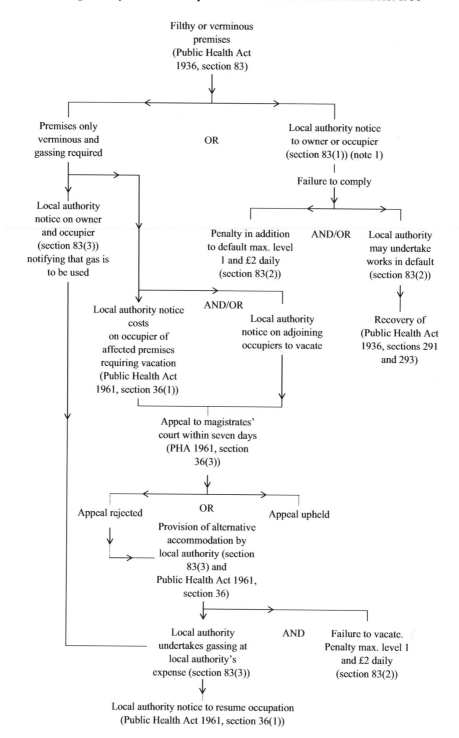

Notes

1. There seems to be no appeal against this notice.
2. For cleansing of verminous articles, see Public Health Act 1936, section 84, verminous persons and clothing, section 85, and sale of verminous articles, Public Health Act 1961, section 37.

FC105 Cleansing of verminous persons and their clothing – section 85 Public Health Act 1936

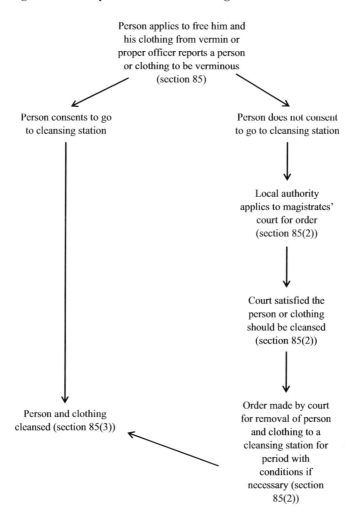

Person applies to free him and his clothing from vermin or proper officer reports a person or clothing to be verminous (section 85)

Person consents to go to cleansing station

Person does not consent to go to cleansing station

Local authority applies to magistrates' court for order (section 85(2))

Court satisfied the person or clothing should be cleansed (section 85(2))

Order made by court for removal of person and clothing to a cleansing station for period with conditions if necessary (section 85(2))

Person and clothing cleansed (section 85(3))

Notes
1. Cleansing of females done by a registered medical practitioner or woman authorised by the proper officer (section 85(4)).
2. No charge for cleansing of person or clothing (section 85(6)).

Cleansing verminous premises and articles

For *premises*, the steps which are required to be taken must be specified in the notice and may include:

(a) removal of wallpaper or other wall covering;
(b) destruction or removal of vermin;
(c) interior surfaces of houses, shops or offices to be papered, painted or distempered and in all other cases painted, distempered or whitewashed (section 83(1)).

A special procedure is set out for situations where the local authority wishes to use gas to destroy vermin, and this includes the power to require vacation of both the affected premises and ones adjoining at a date specified in the notice, provided that temporary accommodation has been provided, whether it be dwelling accommodation or otherwise. The period of vacation must be specified in the notice (Public Health Act 1961, section 36).

Where a certificate from the proper officer of the authority that any *article* in any premises:

(a) is in so filthy a condition as to render its cleansing, purification or destruction necessary in order to prevent injury, or danger of injury, to the health of any person in the premises; or
(b) is verminous, or likely to be verminous,

the local authority must cleanse, purify, disinfect or destroy at their expense and, if necessary, remove it from the premises.

If a person fails to comply with the requirements of the notice, the authority may themselves carry out the requirements and recover from him the expenses reasonably incurred, and he is liable to a fine not exceeding level 1 on the standard scale and to a further fine not exceeding £2 for each day on which the offence continues after conviction (section 83).

Notices

Notices must be in writing and the time allowed for compliance must be reasonable. In the case of notices indicating the local authority's intention to gas, the period of notice should be not less than the appeal period, i.e. seven days.

Appeals

There seems to be no appeal against the service of notices under section 83 (except those relating to gassing), but in any subsequent proceedings by the local authority for recovery of costs following work in default, but the defendant can question the reasonableness of the requirements and the fact that the notice was served on him and not the owner or occupier as the case may be (section 83(2)).

Cleansing of verminous persons and their clothing

Where a verminous person applies to a county council or a local authority, they may take measures necessary to free him and his clothing from vermin. However, where the proper officer reports that any person, or his clothing, is verminous, and the person consents, he can be taken to a cleansing station. Yet if he does not consent, they may apply to the magistrates' court, and if satisfied that it is necessary that he or his clothing should be cleansed, the court may make an order for his removal to a cleansing station to detain him for a period with conditions specified in the order. The local authority must take such measures necessary to free him and his clothing from vermin.

No charge can be made for cleansing of a person or his clothing or removal to, or maintenance in, a cleansing station (section 85).

Definition

Vermin, in its application to insects and parasites, includes their eggs, larvae and pupae, and the expression 'verminous' is constructed accordingly (section 90(1)).

SKIN PIERCING AND OTHER SPECIAL TREATMENTS

References

Local Government (Miscellaneous Provisions) Act 1982, sections 13–17 as amended by the Local Government Act 2003 (section 120)

Body Art – Skin piercing, cosmetic therapies and other special treatments – CIEH Good Practice Guidelines 2001

Local Government Act 2003: Regulation of cosmetic piercing and skin colouring businesses – guidance on section 120 and schedule 6, Department of Health, 26 February 2004

In London, similar powers also exist through the London Local Authorities Act 1991.

Public Health (Wales) Act 2017

The Civic Government (Scotland) Act 1982 (Licensing of Skin Piercing and Tattooing) Order 2006

The Local Government (Miscellaneous Provisions) (Northern Ireland) Order 1985

Extent

The provisions apply only in England. Similar provisions apply in Wales, Scotland and Northern Ireland.

Scope

The activities which may be controlled by registration and subsequent application of bye-laws are:

(a) acupuncture;
(b) tattooing;
(c) semi-permanent skin colouring;
(d) cosmetic piercing;
(e) electrolysis (sections 14–15).

Businesses carried on under the supervision of a registered medical practitioner do not require registration under these provisions (sections 14(8) and 15(8)) and they do not apply to skin piercing on animals (section 16(12)).

The only one of these activities which is defined in the Act is 'semi-permanent skin colouring', which is defined as 'the insertion of semi-permanent colouring into a person's skin' (section 15(9)).

Adoption

These provisions do not operate unless they are adopted by a local authority in respect of its own area by the procedure shown in FC100. The adoption may relate to all or any of the activities mentioned below and different dates of operation may be applied to different activities (section 13).

Registration

Both the practitioner and the premises at which the activity is to be carried out are required to be registered (sections 14(1) and (2) and 15(1) and (2)). Where a registered practitioner sometimes visits people to give treatment, the premises where the treatment takes place is not required to be registered (sections 14(2) and 15(2)).

Applications are to be accompanied by particulars the local authority reasonably require and these include details of:

(a) the premises where the applicant desires to practise; and
(b) any convictions under section 16 of this Act,

but the local authority cannot require information about people to whom treatment services have been given (sections 14(5) and 15(5)).

Registration is required by the local authority unless the applicant has had a previous registration cancelled by a court, in which case the consent of that court is required (sections 14(3), 15(3) and 16(8)(b)).

Any person who fails to register themselves and the premises they operate from is guilty of an offence and liable on summary conviction to a fine not exceeding level 3 and if they contravene bye-laws the court can suspend or cancel their registration (section 16).

Fees

The local authority may charge reasonable fees for registration at their discretion (sections 14(6) and 15(6)).

Bye-laws

The local authority operating these procedures may (but does not need to) make bye-laws which may cover:

(a) cleanliness of premises and fittings;
(b) cleanliness of persons;
(c) cleansing and sterilisation of instruments, materials and equipment (sections 14(7) and 15(7)).

FC106 Skin piercing and other special treatments – sections 13 to 17 Local Government (Miscellaneous Provisions) Act 1982

Resolution of local authority to adopt all or
some of these provisions (Local Government
(Miscellaneous Provisions) Act 1982,
section 13(2) and (3))

Local authority to publish notice for two
consecutive weeks in local newspaper, the
first at least 28 days before operative date
(section 13(6) and (7))

Resolution operative on date(s) specified not
less than one month of date of resolution
(section 13(2))

| Offence to practise any/all activities in area covered by resolution without registration of person and premises (sections 14(1) and 15(1). Max. penalty level 3 (section 16(1)) | AND | Local authority may make bye-laws (sections 14(7) and 15(7)) | AND | Application for registration of person and premises to local authority accompanied by particulars specified by local authority and fee (sections 14(3) to (6) and 15(3) to (6)) |

Local authority
must register
(sections 14(3) and
15(3)) and send
certificate of
registration
(note 1)

(continued on next page)

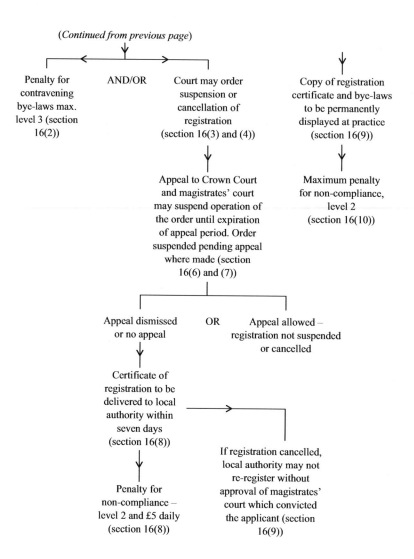

(*Continued from previous page*)

Penalty for contravening bye-laws max. level 3 (section 16(2))

AND/OR

Court may order suspension or cancellation of registration (section 16(3) and (4))

Copy of registration certificate and bye-laws to be permanently displayed at practice (section 16(9))

Appeal to Crown Court and magistrates' court may suspend operation of the order until expiration of appeal period. Order suspended pending appeal where made (section 16(6) and (7))

Maximum penalty for non-compliance, level 2 (section 16(10))

Appeal dismissed or no appeal

OR

Appeal allowed – registration not suspended or cancelled

Certificate of registration to be delivered to local authority within seven days (section 16(8))

If registration cancelled, local authority may not re-register without approval of magistrates' court which convicted the applicant (section 16(9))

Penalty for non-compliance – level 2 and £5 daily (section 16(8))

Note

1. Where a previous registration has been cancelled by that local authority, future registrations require the consent of the magistrates' court which convicted the applicant (section 16(8)(b)).

Power of entry

This is only available through a warrant by a Justice of the Peace and may be granted if the Justice of the Peace is satisfied that there is reasonable ground for entry on suspicion that an offence is being committed and that:

(a) admission has been refused; or
(b) admission is apprehended; or
(c) the case is one of urgency; or
(d) application for admission would defeat the object of entry.

Unless the situation falls under (c) or (d) above, notice of intention to apply for the warrant must be given to the occupier.

The warrant is operative for seven days or until entry is secured, whichever is the shorter period. There is no mention in these provisions of entry by force and authorised officers effecting entry may be required to show their authority.

The maximum penalty for refusing to permit entry to an authorised officer acting under warrant is a fine not exceeding level 3 (section 17).

Chapter 3

DOG AND HORSE CONTROL

CONTROL OF HORSES

References

Animals Act 1971 (all references to section numbers concern this Act)
Control of Horses Act 2015
Control of Horses (Wales) Act 2014

Extent

These provisions apply to England and Wales.

Scope

Where horses are unlawfully present on land, whether they have strayed there or been placed there deliberately, local authorities can deal with those present unlawfully in public places. The provisions also enable freeholders and occupiers to deal with horses which are present unlawfully on their land.

These changes were regarded as essential to follow the example set by the Welsh Assembly to ensure that fly-grazing is tackled consistently and with the necessary resolve to protect horses from neglect and to safeguard the public from danger. By conferring powers on landowners and occupiers alongside local authorities, the legislation provides a fairer and more equitable remedy against fly-grazing and horse abandonment (Hansard).

Power of local authorities to detain horses

A local authority may detain a horse which is in any public place in its area, if the following conditions are met (section 7A(1)).

The local authority may detain the horse where it has reasonable grounds for believing that the horse is there without lawful authority, and if the land is lawfully occupied by a person and:

(i) that person consents to the detention of the horse, or
(ii) the local authority has reasonable grounds for believing that that person would consent to the detention of the horse (but this does not require the authority to seek consent) (section 7A(2)).

FC107 Control of horses – Animals Act 1971 and Control of Horses Act 2015

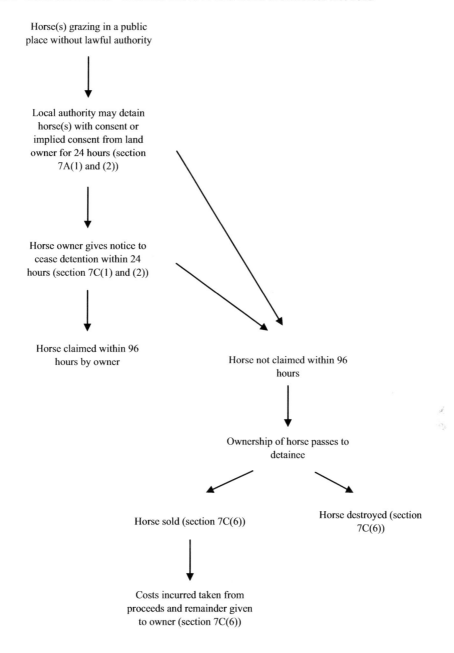

Horse(s) grazing in a public place without lawful authority

Local authority may detain horse(s) with consent or implied consent from land owner for 24 hours (section 7A(1) and (2))

Horse owner gives notice to cease detention within 24 hours (section 7C(1) and (2))

Horse claimed within 96 hours by owner

Horse not claimed within 96 hours

Ownership of horse passes to detainee

Horse sold (section 7C(6))

Horse destroyed (section 7C(6))

Costs incurred taken from proceeds and remainder given to owner (section 7C(6))

Powers of freeholders and occupiers to detain horses

Where a horse is on any land without lawful authority, the horse may be detained in any case, by the occupier of the land, and if the freeholder is not the occupier, by the freeholder with the occupier's consent (section 7B(1) and (2)).

Detention of horses under sections 7A and 7B

A horse can be detained for 24 hours beginning with the time when it is first detained unless, within that period, the person detaining the horse gives notice of the detention to cease to the officer in charge of a police station, and if the person detaining the horse knows to whom the horse belongs, that person (section 7C(1) and (2)).

Where notice is given, the right to detain the horse ceases if, within the period of 96 hours beginning with the time when it is first detained, the person entitled to possession of the horse claims it, and complies with the condition that the person tenders to each person with a claim under section 4A (below) in respect of the horse an amount sufficient to satisfy the claim (section 7C(3) and (4)).

If by the end of the 96-hour period the right to detain the horse has not ceased, ownership of the horse passes to the person detaining the horse, and the person detaining the horse may dispose of it by selling it, arranging for it to be destroyed or in any other way (section 7C(5)).

Where a horse is sold and the proceeds of sale, less the costs of the sale and any costs incurred in connection with it, exceed the amount of any claims under section 4A in respect of the horse, the excess is recoverable from the person detaining the horse by the person who would have been entitled to possession of the horse but for this section. (section 7C(6)).

A person detaining a horse is liable for any damage caused to it by a failure to treat it with reasonable care and supply it with adequate food and water while it is detained (section 7C(7)).

References to a claim under section 4A do not include a claim for damage done by or expenses incurred before the horse was on the land without lawful authority (section 7C(8)).

Liability for damage and expenses due to horses on land without lawful authority

Where a horse is on any land without lawful authority, the person to whom the horse belongs is liable for any damage done by the horse to the land, or any property on it which is in the ownership or possession of the freeholder or occupier of the land, and any expenses which are reasonably incurred by a person detaining the horse

(i)   in keeping the horse while it cannot be restored to the person to whom it belongs or while it is detained, or

(ii)   in ascertaining to whom it belongs (section 4A(1)).

A horse belongs to the person in whose possession it is (section 4A(3)).

SEIZURE OF STRAY DOGS

References

Environmental Protection Act 1990, sections 149 and 150 as amended by the Clean Neighbourhoods and Environment Act 2005
The Environmental Protection (Stray Dogs) Regulations 1992
The Control of Dogs Order 1992

Dogs Act 1906
The Dogs (Northern Ireland) Order 1983

Extent

This procedure applies in England, Wales and Scotland but not in Northern Ireland, where there is separate legislation (section 164(4)).

Scope

Local authorities must appoint an officer to deal with stray dogs (section 149(1)).

There are no particular powers of entry provided for any authorised local authority officers.

Where the authorised officer finds a stray dog in a public place or on any other land or premises, he must seize it and detain it provided, in the case of other than a public place, the owner or occupier agrees (section 149(3)). Dogs not wearing collars with the name and address of the owner may be seized and treated as a stray dog under this procedure (Control of Dogs Order 1992).

A member of the public may take possession of a stray dog and must then take it to the local authority. Unless the finder wishes to keep the dog, the local authority is required to deal with it by the procedure in FC109.

Notification to owner

Where the authorised officer knows the dog's owner or this is shown on its collar, the authorised officer must give that person notice in writing that the dog has been seized, inform them of where it is being kept and that he will dispose of the dog unless it is claimed within seven days and the local authority's expenses are paid (section 149(4)).

Identification of dog owners

The Control of Dogs Order 1992, which is enforced by local authorities, requires dogs, with certain exceptions, e.g. pack hounds, whilst on a highway or in a public place, to wear a collar with the name and address of the owner inscribed on it or on a place attached to it. If not so worn, the owner commits an offence under the Animal Health Act 1981. In these circumstances, local authorities have been given the power to prosecute under the Environmental Protection Act (Environmental Protection Act 1990, section 151).

Detention of dogs

The local authority must provide or arrange for the detention of dogs and for them to be properly fed and confined. Costs involved in the kennelling of dogs are recoverable from owners who require return of the animal (section 149(9)) and additionally the owner is required to pay a sum prescribed by regulations (regulation 2).

Register of seized dogs

The authorised officer must keep a register of seized dogs, which is kept by the officer, including:

   (a) a brief description of each dog, including its breed (if known), and any distinctive physical characteristics or markings, tattoos or scars;

(b) any information which is recorded on a tag or collar worn by, or which is otherwise carried by, the dog;

(c) the date, time and place of the seizure;

(d) where a notice has been served under section 149(4), the date of service of the notice, and the name and address of the person on whom it has been served;

(e) where the officer disposes of the dog under section 149(6):

    (i) the date of disposal;

    (ii) whether disposal was by destruction, gift or sale, and if by sale, the price obtained;

    (iii) the name and address of the purchaser, donee or person effecting the destruction; and

(f) where the dog was returned to a person claiming to be its owner, the name and address of that person, and the date of return.

The register shall be available, at all reasonable times, for inspection by the public free of charge (section 149(8) and Environmental Protection (Stray Dogs) Regulations 1992, regulation 3).

Dog to be kept by finder

The finder of a stray dog may indicate to the authorised officer of the local authority a wish to keep the dog. The authorised officer must record specific details and contact the dog's owner where this proves to be possible. The owner may elect to collect it. However, where the owner is not known or does not wish to collect the dog, the authorised officer can allow the finder to keep the dog if he is satisfied that the person is fit and proper and is able to feed and care for it. The authorised officer must inform the finder both verbally and in writing that he must keep the dog for at least one month and that failure to do so is a criminal offence (section 150(2) and Environmental Protection (Stray Dogs) Regulations 1992, regulation 4).

Disposal of dogs

The authorised officer may dispose of the dog by either:

(a) giving it or selling it to:

    (i) any person who will care for it;

    (ii) an establishment for the reception of stray dogs; or

(b) destroying it so as to cause as little pain as possible.

No dog may be sold or given away for vivisection (section 149(6)).

Definitions

Public place means:

(a) as respects England and Wales, any highway and any other place to which the public are entitled or permitted to have access;

(b) as respects Scotland, any road (within the meaning of the Roads (Scotland) Act 1984) and any other place to which the public are entitled or permitted to have access (section 149(11)).

FC108 Seizure of stray dogs by the public – section 150 Environmental Protection Act 1990

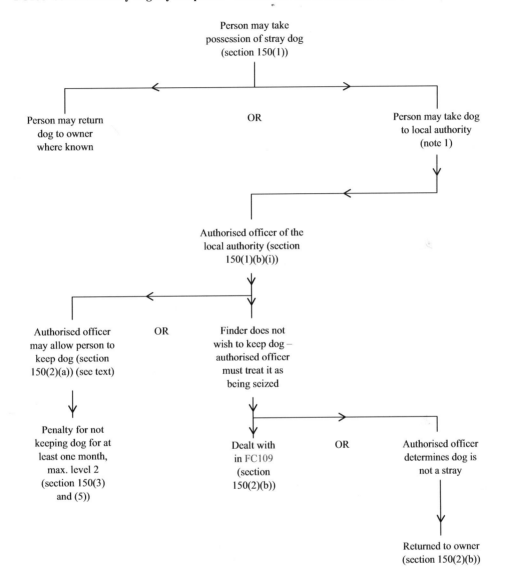

Notes

1. Having taken possession of a stray dog, the member of the public must bring it either to the owner or to the local authority – maximum penalty for not doing so, level 2 (section 150(1) and (5)).

2. In Scotland, a person who keeps a dog by virtue of this section for a period of two months without its being claimed by the person who has right to it shall at the end of that period become the owner of the dog (section 150(4)).

FC109 Seizure of stray dogs by local authorities

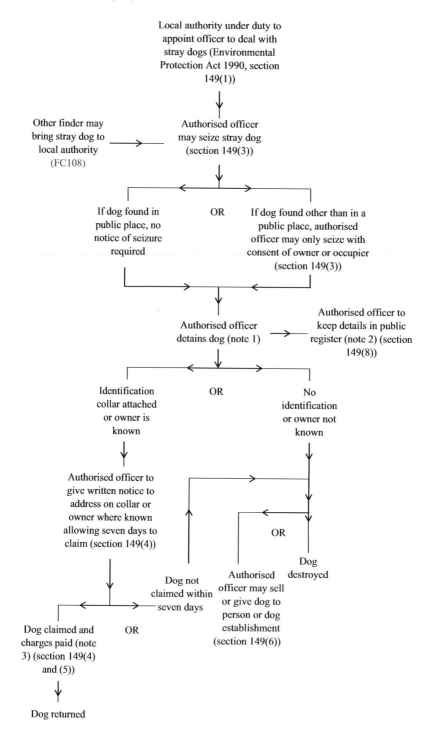

Local authority under duty to appoint officer to deal with stray dogs (Environmental Protection Act 1990, section 149(1))

Other finder may bring stray dog to local authority (FC108)

Authorised officer may seize stray dog (section 149(3))

If dog found in public place, no notice of seizure required

OR

If dog found other than in a public place, authorised officer may only seize with consent of owner or occupier (section 149(3))

Authorised officer detains dog (note 1)

Authorised officer to keep details in public register (note 2) (section 149(8))

Identification collar attached or owner is known

OR

No identification or owner not known

Authorised officer to give written notice to address on collar or owner where known allowing seven days to claim (section 149(4))

Dog not claimed within seven days

Authorised officer may sell or give dog to person or dog establishment (section 149(6))

Dog destroyed

OR

Dog claimed and charges paid (note 3) (section 149(4) and (5))

OR

Dog returned

Notes

1. The authorised officer may arrange for the dog to be destroyed immediately if this is necessary to avoid suffering (section 149(10)).

2. The details to be kept in the register are prescribed in the Environmental Protection (Stray Dogs) Regulations 1992.

3. The charges to be made are controlled by the Environmental Protection (Stray Dogs) Regulations 1992.

Chapter 4

EXHUMATIONS

EXHUMATIONS

Reference

Burial Act 1857, section 25

Extent

These provisions apply in England and Wales. There is no specific legislation or precise guidance for exhumation in Scotland or Northern Ireland.

Scope

Bodies may need to be exhumed for many reasons: lack of burial space; the redevelopment of old cemeteries or crypts; archaeological reasons; individual requests for reburial, repatriation or cremation; criminal investigations; or even to correct errors made at the time – for instance, the person was buried in the wrong plot.

A body cannot be removed from a consecrated place without a licence to do so. Licences are available from the Ministry of Justice and the licence will include conditions or precautions to be taken. It is an offence to remove a body or body remains without a licence or to neglect to observe the precautions prescribed under the conditions of the licence. A person guilty of such an offence is liable, on summary conviction, to a level 1 fine (section 25).

Practice

There are three categories of legitimate exhumation – under Ministry of Justice licence; ecclesiastical faculty, where remains are exhumed from consecrated ground; or a coroner's order.

Of these, environmental health practitioners are only required for those exhumations sanctioned under the Ministry of Justice licence. The Ministry of Justice licence will stipulate that exhumation should be carried out early in the morning. In practice, the licence is sent to the local authority, who may devise and attach other conditions.

It is the responsibility of the licence holder to ensure the conditions are adhered to. This may be a relative of the deceased or a company if a licence has been applied for on behalf of an individual.

Consents

Any person who wishes to exhume buried remains has to apply to the Ministry of Justice for a licence to permit it. There are a number of factors which the Ministry of Justice considers before issuing a licence to exhume human remains. The Ministry of Justice will normally grant exhumation licences to the next of kin, subject to any other necessary consents, where the application is made for private family reasons. The consents of all the next of kin of the deceased are required. The priority given is in accordance with that set out in the Administration of Estates Act 1925 or the Civil Partnerships Act 2004.

If contact has been lost with any of the surviving relatives, it is the applicant's responsibility to undertake a search for that relative. Licences are unlikely to be issued without all of the required consents. Applications will also be considered from any person, but in such cases it will be important to explain why the application is not being made by a relative.

Where a non-family member, e.g. the coroner, applies for a licence, they will need the consent of the next of kin/relatives.

Consent will also be needed from the owner of the exclusive rights of burial relating to the grave. This is because a Ministry of Justice licence is permissive only: it does not require the remains to be removed, and the grave owner must therefore be prepared to grant access. If the grave is a public or common one, which could mean that unrelated remains may be buried there, and if so may have to be disturbed, then the permission of any surviving relatives would be required.

If the remains are buried within consecrated ground, then consent from the Churches of England, Scotland and Wales and the Roman Catholic Church will also be required. If the present burial plot is within consecrated ground, you may need to apply for faculty consent. It is the applicant's responsibility to obtain these permissions.

A licence from a straightforward application would be given a three-month timescale from the date of issue unless there are other extraneous circumstances.

Objections

It is the applicant's responsibility, even if they are the nearest surviving relative, to state whether there could be objections to the exhumation from other relatives, as it may be necessary to take such objections into account. However, this does not necessarily mean that a licence would be refused.

If the consent of the grave owner is not given, then the Ministry of Justice would not normally be prepared to issue a licence.

The Ministry of Justice is unable to become involved in any family dispute and will not normally issue a licence until objections or disagreements between next of kin with the same degree of kinship have been resolved.

Definitions

Consecrated means a burial ground that has been consecrated in accordance with the rites of the Church of England and not merely blessed by the vicar during the funeral service, e.g. consecrated by the Roman Catholic Church or the Church in Wales.

Exhumation means 'any disturbance' of buried human remains.

Faculty means consent from the ecclesiastical courts of the Church of England to exhume.

Note

The responsibility for protecting the health and ensuring the safety of the EHP in attendance lies with the company carrying out the exhumation. See 'Controlling the risks of infection at work from human remains: A guide for those involved in funeral services (including embalmers) and those involved in exhumation', Health and Safety Executive.

FC110 Exhumations – Burial Act 1857

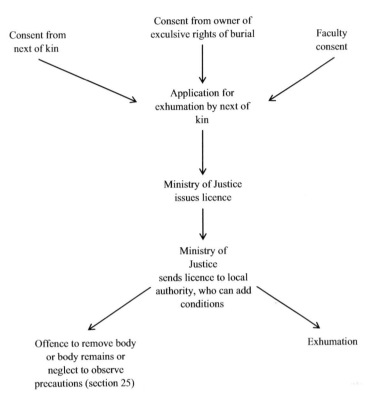

Chapter 5

HOARDING

HOARDING

References

Care Act 2014, section 42
Professional Practice Note: Hoarding and how to approach it – Guidance for Environmental
   Health Officers and others – CIEH

Extent

These provisions apply to England and Wales.

Scope and background information

Hoarding is acquiring or saving lots of things regardless of their value.

Many people have some belongings they consider special and it's common to save some things because they could come in useful in the future. Hoarding is when your need to keep things causes you distress or interferes with your day-to-day life.

It is increasingly being recognised that hoarding can be a condition by itself, as well as sometimes being a symptom of other mental health problems. People used to think hoarding was a form of obsessive-compulsive disorder (OCD), but research suggests they are not the same.

For these reasons hoarding disorder has been listed as a distinct mental health problem in the DSM-5 and ICD-11 (manuals that doctors use to categorise and diagnose mental health problems). Hoarding can also be caused by some other conditions (for example, dementia or brain injury) which are generally diagnosed and treated differently to mental health problems.

Hoarding disorder is a fairly new term. Doctors or healthcare professionals might also call this compulsive hoarding. People hoard if they:

- find it really difficult to discard or part with possessions, regardless of their value
- feel distressed at the thought of discarding things, and that they need to save them
- are unable to use parts of their home because they are very cluttered
- are experiencing distress due to hoarding, or it's affecting other areas of their life
- aren't hoarding because of another mental health problem, or other health condition.

If a person keeps more pets than they can adequately look after and doesn't provide basic care for them – including food, shelter, toilet facilities and vet care – this is sometimes viewed as a type of hoarding behaviour.

Hoarding is likely to be a combination of things:

- Difficult feelings
- Perfectionism and worrying
- Childhood experiences
- Trauma and loss
- Family history or habits
- Other mental health problems

If hoarding causes distress to a person, they might want to consider seeking treatment. A growing number of professionals are aware of hoarding, including the need to help the person take things at their own pace and not pressure them to make changes faster than they want to. These include their GP, therapists, social workers and EHPs.

Research suggests it can help if a therapist visits the home, so they can understand more about the situation and help work out how to make changes. Some people also seem to find it helpful to have treatment in a familiar environment.

There aren't any specific medications for hoarding disorder, but some people find medication helps with other problems they are experiencing alongside hoarding, for example antidepressants.

In dealing with people who hoard:

- **Respect their decisions.** Most people have some attachment to things they own. You might not understand why they keep particular things, but try to remember that the items they hoard feel important to them (even if they don't seem valuable to you). For example, try to avoid describing them as junk or rubbish.
- **Don't take over.** It's understandable to want to help them improve things. But if you try to take charge, they might shut you out and not accept any help at all. For example, don't touch or move things without their permission.
- **Be gentle – you can't force someone to change their behaviour.** Trying hard to persuade, trick or force someone into clearing up or throwing things away is unlikely to help them change in the long term and could make them withdraw from you.
- **Help them to seek treatment and support.** For example, you could encourage them to talk to their doctor.
- **Don't pressure them to let you into their home.** They might feel really anxious about having visitors, so it's important not to take it personally if they don't want you to come in. It might help to consider other places you could meet instead.
- **Try to be patient.** Once someone seeks help with hoarding, it can still take a long time before they are ready to make changes.
- **Help them celebrate successes**, such as clearing a small area. They might feel very anxious about what's left to do, so it could help if you encourage them to notice their achievements. You could also remind them to take things one step at a time.
- **Try peer support.** Some people find it really helpful to connect with others who are also supporting someone with hoarding. To find peer support, contact Mind's Infoline or local Mind to see what support there is in the area, or try online peer support, such as Elefriends and OCD Action's support for family members.

Forced clear-ups

If someone is supporting someone who is hoarding, it's understandable to want to help them clear up and to believe it might be doing them a favour if it is arranged to clean and tidy things for them. **But this is very unlikely to help in the long term and it could make things worse.**

Family members and carers sometimes believe they might be helping if they turn up without advance warning and without permission, or pay someone to tidy or declutter behind the person's back. However, **professionals who understand hoarding should never agree to make surprise visits**, and should know that it's unhelpful to tidy up against someone's wishes.

Legislation

When investigating case of hoarding it is common to find that the dwelling is infested with pests and the occupant may not be disposing of faeces and urine in the facilities provided for that purpose. Following the course of persuasion and the help of other agencies it may be that the only course is to serve notice(s) under the Public Health Act 1936 to remove rubbish and clean the house (i.e. see p. 499). It may also be the case that notice(s) under the Housing Act 2004 may need to be served to reinstate facilities such as WC, bath, sink and hot and cold water. See p. 302. It may be that if you have built a good relationship with the owner/occupier, this can be done with their agreement and arrangements made for payment for these works through the powers available under the relevant legislation. The person(s) involved in the hoarding are often vulnerable people.

The Care Act 2014 is relevant in dealing with them in a sensitive and coordinated manner.

Safeguarding adults at risk of abuse or neglect

Where a local authority has reasonable cause to suspect that an adult in its area (whether or not ordinarily resident there):

(a) has needs for care and support (whether or not the authority is meeting any of those needs),
(b) is experiencing, or is at risk of, abuse or neglect, and
(c) as a result of those needs is unable to protect himself or herself against the abuse or neglect or the risk of it,

the local authority must make whatever enquiries it thinks necessary to enable it to decide whether any action should be taken in the adult's case and, if so, what and by whom (Care Act 2014 section 42(1) and (2)).

Safeguarding Adults Boards

Each local authority must establish a Safeguarding Adults Board (an 'SAB') for its area (section 43(1).

The objective of an SAB is to help and protect adults in its area in cases of the kind described (section 43(2).

The way in which an SAB must seek to achieve its objective is by co-ordinating and ensuring the effectiveness of what each of its members does (section 43(3).

An SAB may do anything which appears to it to be necessary or desirable for the purpose of achieving its objective (section 43(4).

Two or more local authorities can exercise their respective duties by establishing an SAB for their combined areas (section 43(6)).

Safeguarding adult reviews

An SAB must arrange for there to be a review of a case involving an adult in its area with needs for care and support (whether or not the local authority has been meeting any of those needs) if –

    (a) there is reasonable cause for concern about how the SAB, members of it or other persons with relevant functions worked together to safeguard the adult, and

    (b) condition 1 or 2 is met (section 44(1)).

Condition 1 is met if –

    (a) the adult has died, and

    (b) the SAB knows or suspects that the death resulted from abuse or neglect (whether or not it knew about or suspected the abuse or neglect before the adult died) (section 44(2)).

Condition 2 is met if –

    (a) the adult is still alive, and

    (b) the SAB knows or suspects that the adult has experienced serious abuse or neglect (section 44(3)).

An SAB may arrange for there to be a review of any other case involving an adult in its area with needs for care and support (whether or not the local authority has been meeting any of those needs) (section 44(4)).

Each member of the SAB must co-operate in and contribute to the carrying out of a review with a view to:

    (a) identifying the lessons to be learnt from the adult's case, and

    (b) applying those lessons to future cases (section 44(5)).

Supply of information

If an SAB requests a person to supply information to it, or to some other person specified in the request, the person to whom the request is made must comply with the request if:

    (a) conditions 1 and 2 are met, and

    (b) condition 3 or 4 is met (section 45(1)).

Condition 1 is that the request is made for the purpose of enabling or assisting the SAB to exercise its functions (section 45(2)).

Condition 2 is that the request is made to a person whose functions or activities the SAB considers to be such that the person is likely to have information relevant to the exercise of a function by the SAB (section 45(3)).

Condition 3 is that the information relates to –

    (a) the person to whom the request is made,

    (b) a function or activity of that person, or

    (c) a person in respect of whom that person exercises a function or engages in an activity (section 45(4)).

Condition 4 is that the information –

    (a) is information requested by the SAB from a person to whom information was supplied in compliance with another request under this section, and

    (b) is the same as, or is derived from, information supplied (section 45(5)).

Information may be used by the SAB, or other person to whom it is supplied, only for the purpose of enabling or assisting the SAB to exercise its functions (section 45(6)).

Protecting property of adults being cared for away from home

Where an adult is having needs for care and support met above, in a way that involves the provision of accommodation, or is admitted to hospital (or both), and it appears to a local authority that there is a danger of loss or damage to movable property of the adult's in the authority's area because the adult is unable (whether permanently or temporarily) to protect or deal with the property, and no suitable arrangements have been or are being made, the local authority must take reasonable steps to prevent or mitigate the loss or damage (section 47(1) and(2)).

The local authority may at all reasonable times and on reasonable notice enter any premises which the adult was living in immediately before being provided with accommodation or admitted to hospital, and may deal with any of the adult's movable property in any way which is reasonably necessary for preventing or mitigating loss or damage (section 47(3)).

A local authority cannot exercise this power unless it has obtained the consent of the adult concerned or, where the adult lacks capacity to give consent, the consent of a person authorised under the Mental Capacity Act 2005 to give it on the adult's behalf, or if there is no person so authorised, the local authority is satisfied that exercising the power would be in the adult's best interests (section 47(4)).

Where a local authority is proposing to exercise the power under subsection (3)(a), the officer it authorises to do so must, if required, produce valid documentation setting out the authorisation to do so (section 47(5)).

A person who, without reasonable excuse, obstructs the exercise of this power commits an offence, and is liable on summary conviction to a fine not exceeding level 4 on the standard scale (section 47(6)).

A local authority may recover from an adult whatever reasonable expenses the authority incurs in the adult's case (section 47(7)).

Definitions

Abuse includes financial abuse which includes:

    (a) having money or other property stolen,

    (b) being defrauded,

    (c) being put under pressure in relation to money or other property, and

    (d) having money or other property misused (section 42(3)).

FC111 Hoarding – Care Act 2014, section 42

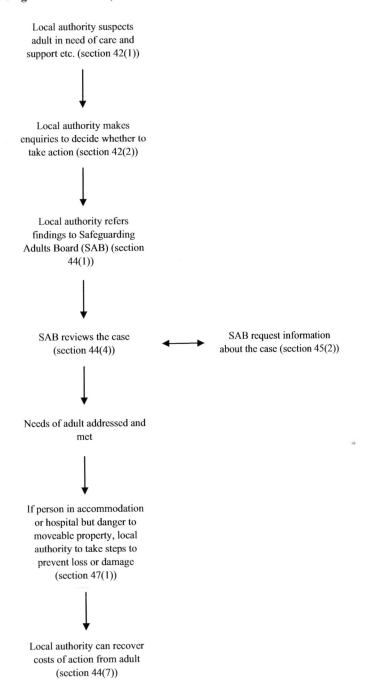

Local authority suspects adult in need of care and support etc. (section 42(1))

↓

Local authority makes enquiries to decide whether to take action (section 42(2))

↓

Local authority refers findings to Safeguarding Adults Board (SAB) (section 44(1))

↓

SAB reviews the case (section 44(4)) ←→ SAB request information about the case (section 45(2))

↓

Needs of adult addressed and met

↓

If person in accommodation or hospital but danger to moveable property, local authority to take steps to prevent loss or damage (section 47(1))

↓

Local authority can recover costs of action from adult (section 44(7))

Chapter 6

REMOVAL OF UNAUTHORISED CAMPERS

REMOVAL OF UNAUTHORISED CAMPERS

References

Criminal Justice and Public Order Act 1994, sections 61 to 62 and 77 to 80

Department of the Environment Circular 18/94 – Gypsy Sites Policy and Unauthorised Camping (amended 2000)

Guidance on Managing Unauthorised Camping, Office of the Deputy Prime Minister, February 2004

Supplement to 'Managing Unauthorised Camping: A Good Practice Guide', Office of the Deputy Prime Minister, March 2005

Guide to effective use of enforcement powers, Part 1: Unauthorised encampments 2006

Extent

This procedure is applicable in England and Wales but not in Scotland or Northern Ireland.

Scope

Local authorities are able to direct persons residing unlawfully in vehicles:

(a) on highway land;
(b) on unoccupied land; or
(c) on occupied land without the consent of the occupier

to leave the land and remove their vehicles and any other property they have with them (section 77(1)).

The procedure may be applied to all persons residing in vehicles (as defined below); however, in Department of the Environment Circular 18/94, local authorities are asked not to use these powers needlessly in relation to gypsy encampments, particularly where they are causing no nuisance (paragraphs 6 to 9). Complementary provisions in section 61 give powers to the police to deal with collective trespass or nuisance on land in circumstances where the trespassers have refused requests from the occupier to leave and either:

(a) they have caused damage to land/property; or
(b) threatening, abusive or insulting behaviour has been used; or
(c) there are at least six vehicles (section 61).

The operation of both provisions therefore requires close liaison between the police and the local authorities.

In implementing these procedures, local authorities are asked to take account of the government's objective of ensuring that the settled communities and gypsies and travellers live together peacefully. Stress is also placed on the importance of choosing the most appropriate power to deal with any situation and that this should be done within the umbrella of a protocol between a local authority, the police and other involved agencies.

Under part 7 of the Anti-Social Behaviour Act 2003, the police have powers to move on unauthorised camping provided that there is adequate site provision elsewhere.

Vehicles

Both provisions deal with the presence of vehicles, defined as including:

(a) any vehicle, whether or not it is in a fit state for use on roads, and includes any body, with or without wheels, appearing to have formed part of such a vehicle, and any load carried by, and anything attached to, such a vehicle; and

(b) a caravan as defined in section 29(1) of the Caravan Sites and Control of Development Act 1960;

and a person may be regarded for the purposes of this section as residing on any land even if he has a home elsewhere (section 61(9) and 77(6)).

Gypsies

In Circular 18/94, local authorities are asked in enforcing this procedure to take account of the definition of 'gypsies' given by the Court of Appeal in R v South Hams DC, *ex parte* Gibb (*The Times*, 8 June 1994; i.e. 'gypsies' meant any persons who wandered or travelled for the purpose of making or seeking their livelihood, and did not include persons who moved from place to place without any connection between their movement and their means of livelihood).

Directions

Directions to be given by local authorities must be served on the persons to whom the direction is to be applied but where there is more than one person the direction can merely specify the land and be addressed to all occupants of the vehicles without naming them (section 77(2)).

Where it is impracticable to serve on the persons named in the direction, a copy may be fixed prominently on the vehicle concerned and, if directed to, unnamed persons may be served in the same manner on every vehicle (section 79(2)).

Copies of the direction must also be displayed on the land (other than on a vehicle) in such a way that it can easily be seen by persons camping on the land and copies sent to the owner and any occupier of the land, unless after reasonable enquiry they cannot be found (section 79(3) and (4)).

The direction requires the persons concerned to leave the land and remove the vehicles etc., as soon as practicable. It is an offence:

(a) not to comply; and

(b) having removed the vehicles etc., to again enter the land with a vehicle within three months of the day on which the direction was given (section 77(3)).

On summary conviction, level 3 penalties are available to the court.

Defences

In any proceedings taken by the local authority for non-compliance with a direction, it will be a defence to show that non-compliance was due to:

(a) illness; or
(b) mechanical breakdown; or
(c) other immediate emergency (section 77(5)).

Complaint to magistrates' court

In addition to seeking a fine for non-compliance with the directive, the local authority may apply to the magistrates' court for an order requiring the removal of the vehicles and other property (section 78(1)).

Any order made may authorise the local authority to enforce it by entering the land and removing the vehicles etc., but in relation to occupied land, at least 24 hours' notice must be given to the owner and lawful occupiers (section 78(3)).

There is no provision for the recovery of the local authority costs in enforcing these orders.

Provisions for the service of summons in relation to a local authority complaint are similar to the service of directions above.

To obtain an order for removal from the magistrates' court, it is necessary to obtain reports on the following:

- availability of sites in local gypsy sites;
- availability of school places in the locality for children in the group subject to the order;
- social services about the needs of the group subject to the order (Guidance).

Definition

Land means land in the open air.

FC112 Removal of unauthorised campers – section 77 Criminal Justice and Public Order Act 1994

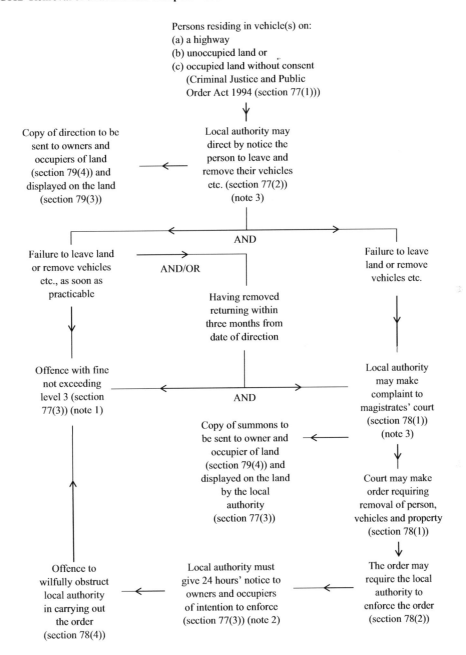

Persons residing in vehicle(s) on:
(a) a highway
(b) unoccupied land or
(c) occupied land without consent
(Criminal Justice and Public
Order Act 1994 (section 77(1)))

Local authority may direct by notice the person to leave and remove their vehicles etc. (section 77(2)) (note 3)

Copy of direction to be sent to owners and occupiers of land (section 79(4)) and displayed on the land (section 79(3))

AND

Failure to leave land or remove vehicles etc., as soon as practicable

AND/OR

Failure to leave land or remove vehicles etc.

Having removed returning within three months from date of direction

Offence with fine not exceeding level 3 (section 77(3)) (note 1)

AND

Local authority may make complaint to magistrates' court (section 78(1)) (note 3)

Copy of summons to be sent to owner and occupier of land (section 79(4)) and displayed on the land by the local authority (section 77(3))

Court may make order requiring removal of person, vehicles and property (section 78(1))

Offence to wilfully obstruct local authority in carrying out the order (section 78(4))

Local authority must give 24 hours' notice to owners and occupiers of intention to enforce (section 77(3)) (note 2)

The order may require the local authority to enforce the order (section 78(2))

Notes
1. It is a defence to show that failure to leave or remove vehicles etc. was due to illness, mechanical breakdown or other immediate emergencies (section 77(5)).
2. Notice is not required where, after reasonable enquiry, the local authority is unable to obtain the names and addresses of the person(s) subject to the court direction (section 78(3)).
3. For service of notice of direction and summons of complaint, see text.
4. For police powers to remove unauthorised camping where there is adequate provision elsewhere, see part 7 of the Anti-Social Behaviour Act 2003.

Chapter 7

PEST CONTROL

CONTROL OF RATS AND MICE

Reference

Prevention of Damage by Pests Act 1949, sections 2 to 7

Extent

This procedure applies in England, Wales and, with amendments, to Scotland but not in Northern Ireland (sections 1 and 29).

Duty of local authority

Each local authority is under a statutory duty to take such steps as may be necessary to secure, so far as practicable, that its district is free from rats and mice including:

(a) carrying out inspections;
(b) destroying rats and mice on premises which the local authority occupies and keeping the premises free from infestation;
(c) enforcing duties on owners and occupiers under Prevention of Damage by Pests Act 1949 (section 2).

Occupier's duty to notify local authority

When an occupier of any land becomes aware of an infestation by rats or mice in substantial numbers, he must notify the local authority in writing but this does not apply to agricultural land, or where food in certain categories of premises is infested when notification is then given under Part 2 of the act to the relevant minister or, in Scotland, the Secretary of State (section 3).

Notices

Notices, which must be in writing, may be served on the owner or the occupier or, where the owner is not the occupier, on both. The notice may require either or both of the following:

(a) treatment at times which may be prescribed;
(b) the carrying out of structural or other works.

The time allowed for compliance must be reasonable and should be not less than the period allowed for appeal, i.e. 21 days (section 4 and sections 290, 300 and 302, Public Health Act 1936). Works in default of the notice are allowed under section 5.

Block treatment

If rats or mice are found in substantial numbers in a group of premises in different occupation, the local authority may take such steps as it considers necessary (other than carrying out structural works), after giving at least seven days' notice to each occupier specifying the steps which it intends to take, and may subsequently recover its costs (section 6).

Power of entry

Authorised officers of local authorities, producing if required evidence of authority, may enter any land after, in the case of occupied land, giving at least 24 hours' notice, for the following purposes:

(a) inspection;
(b) compliance with notices or other requirements placed on owners or occupiers;
(c) enforcing notices served under sections 5 and 6 including the carrying out of treatment of works (section 22).

Any person who wilfully obstructs is liable on summary conviction to a fine not exceeding in the case of a first offence £5, level 1 on the standard scale, and in the case of a second or any subsequent offence £20 level 1 on the standard scale.

If any land is damaged in the exercise of a power of entry, compensation may be recovered by any person interested in the land from the local authority on whose behalf the entry was effected (section 22).

Definitions

Land includes land covered with water and any building or part of a building (section 28).

FC113 Control of rats and mice – sections 2 to 7 Prevention of Damage by Pests Act 1949

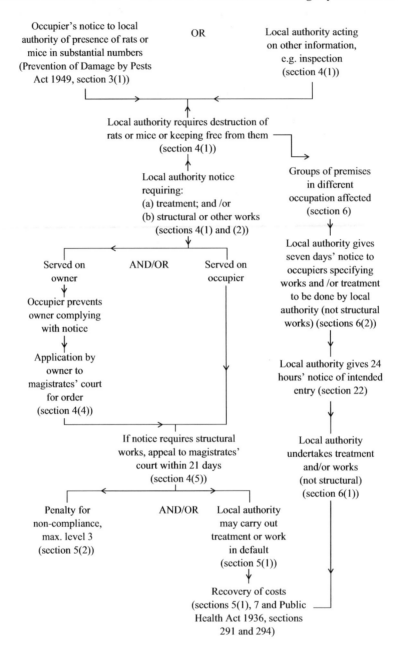

Notes

1. The following general provisions of the Public Health Act 1936 are applicable to this procedure:

   (a) authentication and service of notices (sections 284–286);

   (b) appeals against notices (only where structural works are required) (section 290(3)–(5) inclusive and 300–302 inclusive);

   (c) recovery of expenses (sections 291 and 294). These provisions are set out on p. 549.

2. This procedure applies with certain modifications to non-seagoing ships and to aircraft (section 23).

Chapter 8

PORT HEALTH

INFECTIOUS DISEASES CONTROL – AIRCRAFT

References

Public Health (Aircraft) Regulations 1979 (as amended)
The Public Health (Aircraft) (Amendment) (Wales) Regulations 2007
Public Health (Aircraft) (Scotland) Regulations 1971 (as amended)
The Public Health (Aircraft) Regulations (Northern Ireland) 2008

Extent

These provisions apply in England and Wales only. Similar provisions apply in Scotland and Northern Ireland.

Scope

The purpose of the legislation is the prevention of spread of disease of international concern and to control chemical, biological and radiological risk. The national legislation implements the World Health Organization International Health Regulations 2005.

Inspection

An authorised officer may inspect any aircraft at a customs airport.

If an authorised officer or medical officer has reasonable grounds to suspect there is case of infectious disease on board or if the commander of an aircraft notifies of a suspected infectious disease, the authorised officer must inspect the aircraft on arrival. The authorised officer can take samples of food or water for analysis or examination for the purpose of determining treatment of persons affected with any epidemic, endemic or infectious disease and for preventing the spread of such diseases or for preventing other danger to public health (regulation 7).

Examination of persons on aircraft

An authorised officer may detain people for examination and require their clothing and other articles belonging to them to be disinfected and, where necessary, disinsected, and any person found to be verminous to be disinsected. He may also prohibit any person so examined from leaving the aircraft or airport, or permit him to leave it on certain conditions. He can also require the commander to take steps reasonably necessary for preventing the spread of infection, for disinfection and the destruction of vermin, and for the removal of conditions on the aircraft likely to convey infection, including conditions the existence of which might facilitate the harbouring of insects or vermin (regulation 8).

Notice to customs officers by authorised officer

The authorised officer must inform the customs officer of any measure applied by him or at his direction, to an aircraft, any person on it, or its stores, equipment or cargo (regulation 10).

Supply of information by commanders

Commanders of aircraft must answer questions on health conditions on the aircraft put to him by an authorised officer. Commanders must also notify the authorised officer of any deaths and any cases of, or conditions likely to cause the spread of, infectious disease, including tuberculosis (regulation 11).

Notification of infectious disease on board

Where a radio message or other communication is sent to the authorised officer about the potential or existence of infectious disease on board an aircraft, he shall immediately notify the customs officer of its contents. This also applies to the person in control of the airport and the customs officer. The health part of the Aircraft General Declaration must be delivered to the authorised officer by the commander of the aircraft, or by a member of the crew acting on his behalf (regulation 12).

(In practice, if the local authority is contacted by an airline, they will liaise with the port medical officer (PMO). The PMO does a risk assessment on the information provided and will direct the environmental health officer as to what type of response is required, very much in the same way as infectious diseases are dealt with inland.)

Deratting and disinfection of aircraft

An authorised officer can require the aircraft to be deratted or disinfected if the presence of rodents is suspected on board or an infected animal, or suspected of being infected, is found on board on arrival. The methods used must be approved by the World Health Organization, unless the authorised officer decides that other measures are as safe and reliable. Additional health measures can be applied for preventing danger to public health or the spread of infection, including isolation of the aircraft. The commander of the aircraft must inform the authorised officer of the arrangements made for the disposal of the rodents (regulation 13).

Detention of aircraft

The medical officer can detain an aircraft for medical inspection if he suspects there is infectious disease on board, and he must inform the person in charge of the customs airport and the customs officer (regulation 14). Customs officers and medical officers have similar duties of notification and powers of detention (regulations 15 to 18). Only three hours is allowed for this inspection before the aircraft must be released (regulation 17).

Release of aircraft

When the authorised officer releases an aircraft from detention, he must give notice in writing to the customs officer, to the commander of the aircraft and to the person in charge of the customs airport that the aircraft is free to proceed at or after a date and time stated in the notice (regulation 19).

Removal of infected persons from aircraft where required by commander

A commander of an aircraft may require the medical officer to remove any infected person or person suffering from tuberculosis from the aircraft (regulation 21).

Removal to airport able to apply measures

If an authorised officer considers that the airport cannot apply the necessary measures, he may direct the aircraft or the person to proceed to a customs airport that is able to apply the measures. He must give the commander the reasons for the direction in writing (regulation 22). In applying any measures, the authorised officer must have regard to the need for freeing aircraft from control as quickly as possible (regulation 24).

Outgoing aircraft

The medical officer has powers of examination, prohibition from embarking, surveillance and a duty to notify medical personnel in the destination country (regulation 27).

Examination of person proposing to embark

Regulation 28 provides for the declaration of infected areas where certain diseases are present or suspected, giving the medical officer and authorised officer powers to control infection spread.

Responsibilities

It is the duty of every person subject to direction, requirement or conditions given, made or imposed by an authorised officer or customs officer under these regulations, to comply and they must furnish required information that officer may reasonably require (including information about his name and intended destination and address to which he is going on leaving an aerodrome) (regulation 29). Currently there is no penalty for non-compliance with this duty.

Persons placed under surveillance must also comply with requests for medical examination and questions about personal health.

Charges

A responsible authority may charge the commander of an aircraft for a service to apply measures for preventing danger to public health from an aircraft arriving or the spread of infection from an aircraft leaving an airport where international flights arrive or depart. Charges must not exceed the actual cost of the service rendered and be published at least ten days in advance of being levied.

The charges may be required to be paid or deposited with them before the service is performed. If the commander or person to whom the measures have been taken request, the responsible authority must provide the commander of the aircraft with particulars of the measures taken and reasons why in writing free of charge (regulation 32). Charges are recoverable either summarily as a civil debt or as a simple contract debt in any court of competent jurisdiction (regulation 33).

Commanders can refuse to comply with an authorised officer's instructions provided the commander informs the authorised officer. The authorised officer may then require the commander to remove the aircraft immediately from the aerodrome. If before leaving the aerodrome the commander wishes to discharge cargo or disembark passengers or to take on board fuel, water or stores, the authorised officer shall permit him to do so but may impose conditions the authorised officer considers necessary. When the authorised officer has required the removal of an aircraft from the aerodrome, it cannot during its voyage alight at any other place in England or Wales (regulation 36).

Note: A Port Health Plan is usually produced, which identifies the relevant organisations involved, along with their roles and responsibilities.

Definitions

Authorised officer – the medical officer, the proper officer or any other officer authorised by the responsible authority.

Commander – the person for the time being in command of an aircraft.

Disinsecting – the operation in which measures are taken to kill the insect vectors of human disease.

IHR – the International Health Regulations (2005) of the WHO adopted by the 58th World Health Assembly on 23 May 2005.

Infected aircraft:

(a) an aircraft which has on board on arrival a case of plague, cholera, yellow fever, smallpox, rabies or viral haemorrhagic fever; or

(b) an aircraft on which a plague-infected rodent is found on arrival; or

(c) an aircraft which has had a case of smallpox on board during its voyage and which has not before arrival been subjected in respect of such case to appropriate measures equivalent to those provided in these regulations.

Infected person – a person who is suffering from plague, cholera, yellow fever or smallpox or who is considered by the medical officer to be infected with one of the diseases or with some other infectious or contagious disease other than venereal disease or tuberculosis.

Infectious disease – any infectious or contagious disease other than venereal disease or tuberculosis.

Responsible authority – in relation to an aerodrome or other place, the authority charged with the duty of enforcing and executing these regulations (regulation 2).

INFECTIOUS DISEASES CONTROL – TRAINS

Reference

Public Health (International Trains) Regulations 1994

FC114 Infectious disease control – international travel – Public Health (Aircraft) Regulations 1979

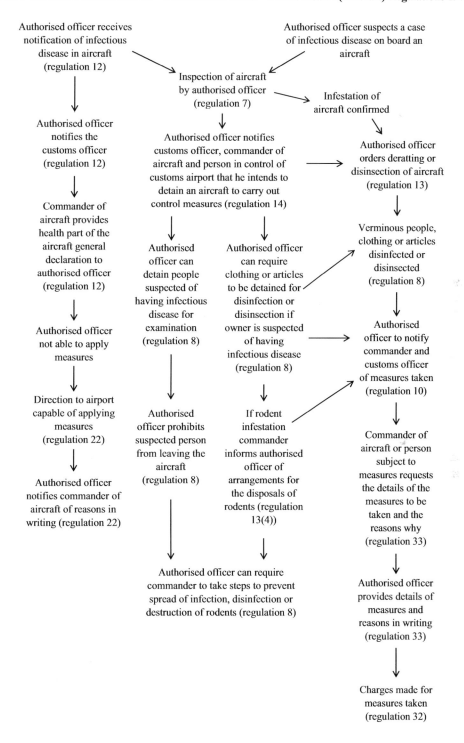

Notes
1. Samples can be taken by authorised officer (regulation 7).
2. Authorised officer must have regard to the need to free the aircraft as quickly as possible (regulation 24).

Extent

These provisions apply to England and Wales.

Stowaway animals on incoming trains

Where a train manager reports to the local authority the presence or suspected presence of a stowaway animal on board his train and the animal is or was capable of carrying rabies, plague or viral haemorrhagic fever, or is or may be contaminated, the enforcement authority may require that the train, together with such of its contents as they may reasonably specify, be deratted, disinfested or decontaminated, or be subject to other methods of control (regulation 6(1) to (3) and (5)).

The enforcement authority may require any or all of the measures to be taken at a designated customs-approved area, or a designated shuttle control area if the enforcement authority for that area agrees.

When control measures have been taken, the train operator must inform the enforcement authority for each area at which the control measures were undertaken of the arrangements made for the disposal of the stowaway animal(s) (regulations 6(6) and 6(7)).

Sick travellers on incoming trains

Train managers of an international train must notify the enforcement authority if they are aware of a sick traveller who is still on board the train, including details of the sick traveller's whereabouts and, if the sick traveller is no longer on board the train, details of the circumstances in which the traveller was identified as sick and where the traveller alighted from or was removed from the train.

The enforcement authority may require the disinfestation or decontamination of the train (regulation 8).

The enforcement authority has the power to ask travellers questions where there are reasonable grounds for suspecting that there is a significant danger to public health (regulation 9).

INFECTIOUS DISEASES CONTROL – SHIPS

References

Public Health (Ships) Regulations 1979 (as amended)
The Public Health (Ships) (Scotland) Regulations 1971 (as amended)
The Public Health (Ships) Regulations (Northern Ireland) 2008

Extent

These provisions apply in England and Wales only. Similar provisions apply in Scotland and Northern Ireland.

Scope

The purpose of the legislation is the prevention of spread of disease of international concern and to control chemical, biological and radiological risk. The national legislation implements the World Health Organization International Health Regulations 2005.

Incoming ships – inspection and notification of infectious disease

An authorised officer may inspect any ship on arrival or already in the district. The authorised officer must also inspect the ship if he receives a message of notification from the master of the ship that there has been:

(a)   (i)   the death of a person not being the result of an accident; or

      (ii)   illness where the person who is ill has or had a temperature of 38°C or greater accompanied by a rash, glandular swelling or jaundice, or where such temperature persisted for more than 48 hours; or

      (iii)   illness where the person has or had diarrhoea severe enough to interfere with work or normal activities;

(b)   the presence on board of a person who is suffering from or who has symptoms of an infectious disease or tuberculosis;

(c)   any other circumstances on board which are likely to cause the spread of infectious disease or other danger to public health; and

(d)   the presence of animals or captive birds, and the occurrence of mortality or sickness amongst such animals or birds (regulations 7 and 13).

The authorised officer must also inspect the ship if he has reasonable grounds for believing that there is on board a case or suspected case of infectious disease (regulation 7(1)(b)).

The authorised officer can also take samples of food or water for analysis or examination (regulation 7(3)).

This analysis or examination must be:

(a)   with a view to the treatment of persons affected with any epidemic, endemic or infectious disease and for preventing the spread of such diseases; or

(b)   for preventing other danger to public health (regulation 7(4)).

An authorised officer may require a ship on arrival or already in the district to be brought to, and if necessary moored or anchored at, some safe and convenient place for the purpose of medical inspection (regulation 8).

The medical officer of health has the power to examine people suspected of suffering from an infectious disease etc. (regulation 9(1)).

Examination of persons on ships

An authorised officer may:

(a)   detain persons suspected of suffering from an infectious disease for examination by the medical officer of health, either on the ship or at some place on shore appointed for the purpose;

(b)   require the clothing and other articles belonging to any such person to be disinfected and, where necessary, disinsected, and any person found to be verminous to be disinsected;

(c)   prohibit any such person from leaving the ship, or permit him to leave it subject to conditions the authorised officer considers reasonably necessary for preventing the spread of infection; and

(d)   require the master to take or assist in taking such steps as in the opinion of the authorised officer are reasonably necessary for preventing the spread of infection, for disinfection and the destruction of vermin, and for the removal of conditions on the ship likely to convey infection, including conditions the existence of which might facilitate the harbouring of insects or vermin (regulation 9).

The medical officer of health has the power to remove a suspected person to hospital, place them under surveillance or prevent them from leaving the ship (regulation 10).

The master of a ship on arrival or already in a district must

(a) answer all questions about the health conditions on board put to him by a customs officer or an authorised officer and furnish the officer with all information and assistance as he may reasonably require;

(b) notify the authorised officer immediately of any circumstances on board which are likely to cause the spread of infectious disease, including in his notification particulars as to the sanitary condition of the ship and the presence of animals or captive birds of any species, or mortality or sickness among such animals or birds, on the ship;

(c) comply with these regulations, and with any directions or requirements of an authorised officer or customs officer (regulation 11).

An authorised officer may, when he is satisfied by information received by radio from a ship from a foreign port before arrival in his district, or by any other information, that the arrival of the ship will not result in or contribute towards the spread of infectious disease, transmit free pratique to the master by radio or otherwise (regulation 12).

Maritime declaration of health

On the arrival of a ship, the master has a report to make about infectious disease on board or is directed by the medical officer to complete a maritime declaration of health. He must complete the declaration and deliver it to the authorised officer.

If, within four weeks after the master of a ship has delivered a maritime declaration of health, the ship arrives in a district or calls at another district, the master must report to the authorised officer any case or suspected case of infectious disease or tuberculosis which has occurred on board since the declaration was delivered and which has not already been reported (regulations 15 and 16).

Restriction on boarding or leaving of ships

An authorised officer can restrict the movement of persons on board or prevent them from leaving a ship until free pratique has been granted, and the master shall take all reasonable steps to secure compliance with this restriction etc.

Before granting permission to a person to leave the ship, the authorised officer may require him to state his name and his intended destination and address, and to give any other information which the authorised officer may think necessary for transmission to the medical officer for the area in which the intended destination of the person is situated.

If such a person cannot state his intended destination and address or arrives, within a period not exceeding 14 days after landing, to be specified to him by the authorised officer, at an address other than that which he has so stated, he shall immediately after his arrival at that address send particulars thereof to the authorised officer of the port where he left the ship (regulation 17).

Application for a ship sanitation control exemption certificate or ship sanitation control certificate

Where an owner or master of a ship applies in writing for a sanitation certificate, the authorised officer must inspect the ship to prevent danger to public health or spread of infection. The authorised officer may also take control measures with a view to issuing the ship sanitation certificate. If the ship is not in the area of an authorised port, an authorised officer must consult with a customs officer and direct the ship to an area of an authorised port convenient to the ship and the customs officer. If the authorised officer is satisfied that the ship is exempt from control measures, he must issue a ship sanitation control exemption certificate. If control measures have been completed to his satisfaction, he must issue a ship sanitation control certificate and note on the certificate the evidence found and the control measures taken.

A ship sanitation control exemption certificate can usually only be issued if the inspection of the ship was carried out when the ship and holds were empty, or a thorough inspection of the holds is possible (regulation 18A).

Ship sanitation certificate

If the master of a ship cannot produce a valid ship sanitation certificate, the authorised officer can inspect the ship for evidence of danger to public health or infection with a view to issuing a ship sanitation certificate or, if not in the area of the port, consult with a customs officer and direct the ship to proceed to an area of an authorised port.

The authorised officer must issue a ship sanitation control exemption certificate if he is satisfied that the ship is exempt from control measures.

If the authorised officer is not satisfied that the ship is exempt from control measures, he must order control measures necessary for the control of danger to public health or the spread of infection. If the ship is not within the area of an authorised port, the authorised officer must consult with a customs officer and direct the ship to proceed to a specified area of an authorised port.

When control measures have been completed to the satisfaction of an authorised officer, the authorised officer must:

(a) issue a ship sanitation control certificate; and
(b) note on the certificate the evidence found and the control measures taken (regulation 18B).

Control measures must be those advised by the World Health Organization. The authorised officer has the power to take other equally effective measures but must report them to the national IHR (International Health Regulation) Focal Point (regulation 18C).

Ship sanitation certificates: form, period of validity, extension and retention

A ship sanitation control exemption certificate or ship sanitation control certificate must conform to the model schedule 3 of the regulations and is valid for six months beginning with the date of issue.

An authorised officer may extend the period of validity of a ship sanitation certificate by one month if:

(a) any inspection or control measures required cannot be carried out at the port;
(b) there is no evidence of danger to public health or infection; and
(c) the port is authorised to extend the validity of a ship sanitation certificate.

The local authority must retain a copy of any ship sanitation certificate for a period of one year beginning with the date of issue (regulation 18D).

Detention of ships and ships to be taken to mooring stations

Where an infected ship or a ship suspected of being infected (within last four weeks) has not had control measures, the master must take the ship to a mooring station (regulation 21).

The authorised officer can require the ship to be subject to a medical inspection, in writing to the master (regulation 22).

Where on the arrival of a ship from a foreign port it appears to a customs officer, from information in the maritime declaration of health or otherwise, that the ship has during its voyage been in an infected area; or is one to which regulation 21 applies, he shall direct the master to take it to a mooring station for detention there unless an authorised officer otherwise allows or directs (regulation 23).

If there is a case or suspected case of, or animal infected with, plague, cholera, yellow fever, smallpox, rabies or viral haemorrhagic fever on board, the authorised officer may direct the master to take the ship to a mooring station and it must stay there until inspected by a medical officer (regulations 24 and 25).

The authorised officer may detain any ship for medical inspection at its place of mooring or at its place of discharge or loading (regulation 26).

The medical officer must inspect the ship and persons on board as soon as possible after it has been taken or directed to or detained at a mooring station. The authorised officer may apply control measures when the ship is at the mooring station (regulation 28).

The authorised officer may require the master of a ship to take all practicable measures to prevent the escape of rodents from the ship (regulation 29).

Removal of infected persons from ships where required by master

There are special control measures for specific diseases in schedule 4 of the regulations.

Outgoing ships – examination etc. of persons proposing to embark

There are similar powers available to the medical officer for ships departing from a port (regulation 33).

Infected places in England and Wales

Regulation 34 provides for the declaration of infected areas by the Secretary of State where certain diseases are present or suspected, giving the medical officer and authorised officer powers to control infection spread.

Charges

Local authorities can charge for services where there is a danger to public health from a ship arriving in its district, or the spread of infection from a ship leaving its district. Services include inspection and applying control measures in connection with issuing a ship sanitation certificate but must not exceed the actual cost of the service rendered.

At the request of the master or the person in relation to whom the measures have been taken, a local authority must provide him with particulars, free of charge, of measures taken for which a charge is made and the reasons why the measures were taken (regulation 38).

Every charge is recoverable either summarily as a civil debt or as a simple contract debt in any court of competent jurisdiction (regulation 39).

Note: A Port Health Plan is usually produced, which identifies the relevant organisations involved, along with their roles and responsibilities.

Definitions

Authorised officer – the medical officer, the proper officer or any other officer authorised by the local authority under regulation 4 to enforce and execute any of these regulations.

Authorised port – a port authorised to offer:

(a) the issuance of a ship sanitation control certificate and the provision of the services referred to:

    (i) in Annex 1 to the IHR; and
    (ii) the form reproduced at schedule 3 to these regulations;

(b) the issuance of a ship sanitation control exemption certificate following inspection of the ship, including a thorough inspection of the hold; or

(c) the extension of a ship sanitation certificate for a period of one month.

Disinsecting – the operation in which measures are taken to kill the insect vectors of human disease.

Free pratique – permission for a ship to disembark and commence operation.

IHR – the International Health Regulations (2005) of the WHO adopted by the 58th World Health Assembly on 23 May 2005.

Infected person – a person who is suffering from plague, cholera, yellow fever or smallpox or who is considered by the medical officer to be infected with such a disease or with some other infectious or contagious disease other than venereal disease or tuberculosis.

Infected ship:

(a) a ship which has on board on arrival a case of plague, cholera, yellow fever, smallpox, rabies or viral haemorrhagic fever; or

(b) a ship on which a plague-infected rodent is found on arrival; or

(c) a ship which has had on board during its voyage:

    (i) a case of human plague which developed more than six days after the embarkation of the person affected; or
    (ii) a case of cholera within five days before arrival; or
    (iii) a case of yellow fever or smallpox;

and which has not before arrival been subjected in respect of such case to appropriate measures equivalent to those provided for in these regulations.

Maritime declaration of health – means a declaration in the form set out in schedule 2 of the regulations.

Master – the person for the time being in charge of or in command of a ship.

Mooring station – a place, situated within the waters of a district, which is specified by the local authority, with the consent of the customs officer for the area in which the district is situated and the harbour master, or in such other district as the Secretary of State may allow, for the mooring of ships for medical inspection so that they do not come into contact with other ships or the shore.

Ship sanitation certificate – a ship sanitation control certificate or a ship sanitation control exemption certificate.

Ship sanitation control certificate – a certificate that conforms to the model in Annex 3 to the IHR reproduced at schedule 3 to these regulations that is issued in accordance with article 39 of the IHR (ship sanitation certificates).

Ship sanitation control exemption certificate – a certificate that conforms to the model in Annex 3 to the IHR reproduced at schedule 3 to these regulations that is issued in accordance with article 39 of the IHR (regulation 2).

FC115 Infectious disease control – Public Health (Ships) Regulations 1979 (as amended)

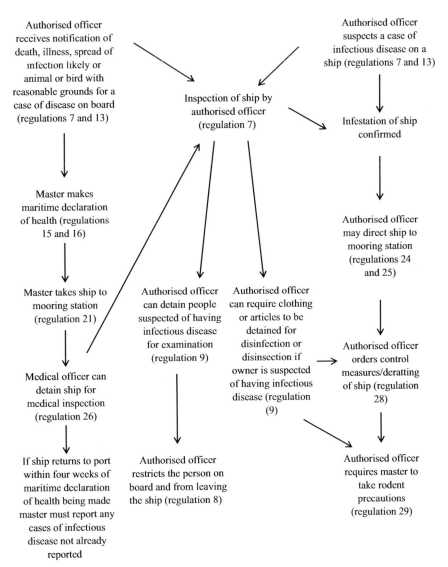

Notes

1. Local authorities can charge for measures taken (regulation 38).
2. Authorised officer must have regard to the need to free the ship as quickly as possible (regulation 27).

FC116 Ship sanitation control certificates – Public Health (Ships) Regulations 1979 (as amended) Regulations 18A to 18D

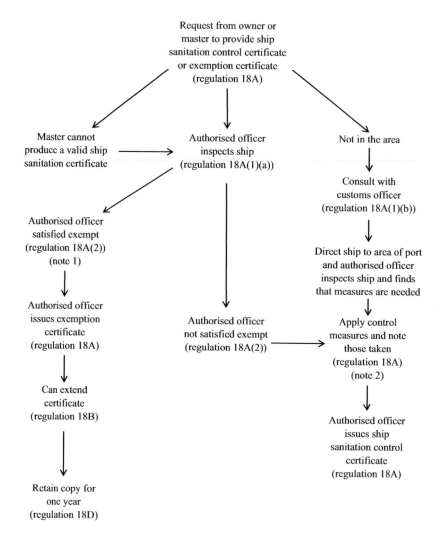

Notes

1. Holds should be empty for ship sanitation exemption certificate to be issued (regulation 18A(3)).
2. Charges can be made for inspection and control measures taken (regulation 39).

Chapter 9

PUBLIC HEALTH FUNERALS

PUBLIC HEALTH FUNERALS

References

Public Health (Control of Disease) Act 1984, section 46
Burial and Cremation (Scotland) Act 2016, section 87
Welfare Service Act (Northern Ireland) 1971, section 25

Extent

These provisions apply to England and Wales. Similar provisions apply in Scotland and Northern Ireland.

Scope

Local authorities have a duty to dispose of a dead body where the deceased's estate or their family cannot pay for the deceased's burial or cremation. They may pay for and arrange funeral and trace next of kin to recover costs.

Most local authorities will receive a referral via one of these:

- coroner's office
- police
- local hospital
- care home
- GP

The authority may decide that it wishes to make use of free services offered by genealogists to trace any next of kin before moving forward.

By doing this, the authority minimises the risk of family not being traced and opens up the possibility that the arrangements can be passed on, thus sparing cost to the authority and time spent in registering a death and arranging a funeral.

Local authorities also carry out probate searches to find next of kin and if the person has died intestate and no next of kin can be found they can pass the case onto the Government's Bona Vacantia department for it to deal with the case.

Alternatively, if an authority believes that there are no known next of kin and that there is no estate, they can apply to Finders International for a payment from the unique Finders International Funeral Fund. The FIFF will pay all or part of the costs of the funeral should Finders confirm that there is no next of kin and that the estate has no funds.

Authorities are often approached by families for financial help. The only state aid for funerals is provided by the Department for Works and Pensions and individuals have to be in receipt of (or have recently claimed) the following benefits:

- Income Support
- income-based Jobseeker's Allowance
- income-related Employment and Support Allowance
- Pension Credit
- Housing Benefit
- the disability or severe disability element of Working Tax Credit
- Child Tax Credit
- Universal Credit

The DWP advises that the payment may not cover the full cost of a funeral and that any monies held by the deceased should be used first to make any payment.

Burial and cremation

Local authorities have a duty to buried or cremate the body of any person who has died or been found dead in their area, where it appears to the authority that no suitable arrangements for the disposal of the body have been or are being made (section 46(1)).

The authority may bury or cremate the body of any deceased person who immediately before his death was being provided with accommodation under Parts 1 or 4 of the Care Act 2014 (section 46(2) and (2A)).

The authority cannot cremate the body where they have reason to believe that cremation would be contrary to the wishes of the deceased (section 46(3)).

The authority may recover from the estate of the deceased person expenses in burial or cremation (section 46(5)). The sum due to an authority is recoverable summarily as a civil debt by proceedings brought within three years after the sum becomes due (section 46(6)).

Powers of entry

Section 61 of the Public Health (Control of Disease) Act 1984 gives powers of entry to all provisions of the act that require action.

FC117 Public health funerals – Public Health (Control of Disease) Act 1984, section 46

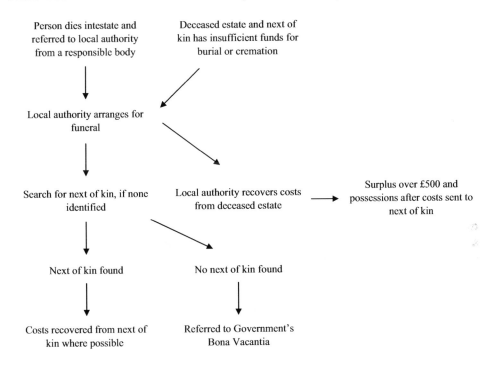

Person dies intestate and referred to local authority from a responsible body

Deceased estate and next of kin has insufficient funds for burial or cremation

Local authority arranges for funeral

Search for next of kin, if none identified

Local authority recovers costs from deceased estate

Surplus over £500 and possessions after costs sent to next of kin

Next of kin found

No next of kin found

Costs recovered from next of kin where possible

Referred to Government's Bona Vacantia

Chapter 10

SANITATION AND DRAINAGE

GENERAL PROCEDURAL PROVISIONS: PUBLIC HEALTH ACTS 1936 AND 1961

The following general provisions are those which are normally applicable to actions under the Public Health Acts 1936 and 1961 and are also generally applicable to procedures being taken under the Clean Air Act 1993 and the Prevention of Damage by Pests Act 1949. Unless otherwise indicated, they will apply to the procedures in this chapter.

Notices requiring the execution of works

These provisions (known as the Part 12 provisions) apply to any procedures where the section concerned specifically indicates this to be the case and this is shown in the procedures where appropriate.

(a) **Content.** The notice must indicate the nature of the works required and state the time within which they are to be executed. The time allowed should not be less than the period for appeal, i.e. 21 days (section 290).

(b) **Appeals.** Appeal against these notices may be made to a magistrates' court within 21 days of service on any of the following grounds:

    (i)   the notice is not justified;

   (ii)   there has been some material informality, defect or error in, or in connection with, the notice;

  (iii)   the local authority has refused unreasonably to approve alternative works;

  (iv)   the works required are unreasonable in character or extent or are unnecessary;

   (v)   the time allowed for compliance is not reasonable;

  (vi)   it would have been equitable for the notice to have been served on the occupier instead of the owner or vice versa;

 (vii)   that some other person ought to contribute towards the cost of compliance.

       Unless a longer period is specified, any appeals to magistrates' courts are to be made by way of complaint for an order within 21 days from the date on which the council's requirement, notice, etc. was served. Notices and other documents must state the rights of appeal and the time within which appeal may be made (section 300).

       Persons aggrieved by decisions of the magistrates' court may appeal to the Crown Court, except where there is a provision for arbitration (Public Health Act 1936, section 301).

(c) **Enforcement.** Following failure to comply with a notice to execute works within the time specified, subject to appeal, the local authority may, at its discretion, carry out the work itself and recover expenses reasonably incurred from the person on whom the notice was served. In addition, the person concerned is liable to a fine not exceeding level 4 on the standard scale and a maximum daily penalty of £2 for a continuing offence (sections 290 and 300).

Recovery of costs

Costs incurred by local authorities in carrying out works in default of an owner are recoverable both by proceeding in the County Court and by the debt becoming a charge on the property. They may be recovered from an owner by instalments and with interest (section 291).

Power of a local authority to execute work on behalf of owners or occupiers at their request

A local authority, by agreement with an owner or occupier of any premises, may undertake on his behalf works which it has required him to carry out under the Act or undertake sewer or drain construction, alteration or repair which the owner or occupier is entitled to execute. The costs are chargeable to the owner or occupier (section 275).

Power of local authority to sell certain materials

A local authority may sell any materials which have been removed by them from any premises, including any street, when executing works under the Act, which are not claimed by the owner and taken away by him within three days of the work.

Where a local authority sells any materials, they must pay the proceeds to the person to whom the materials belonged after deducting expenses (section 276).

GENERAL PROCEDURAL PROVISIONS: BUILDING ACT 1984

You will notice that these provisions are almost identical to those of the provisions in the Public Health Act 1936.

Notices requiring the execution of works

Where specifically indicated by the particular provisions relating to a procedure, the following apply:

(a) **Content.** The notice must indicate the nature of the works required and state the time within which they are to be executed. The time allowed should not be less than the period for appeal, i.e. 21 days (section 99(1)).
(b) **Appeals.** Appeal against these notices may be made to a magistrates' court within 21 days on any of the following grounds:

    (i) the notice is not justified;
    (ii) there has been some material informality, defect or error in, or in connection with, the notice;

    (iii) the local authority has refused unreasonably to approve alternative works;

    (iv) the works required are unreasonable in character or extent or are unnecessary;

    (v) the time allowed for compliance is not reasonable;

    (vi) it would have been equitable for the notice to have been served on the occupier instead of the owner or vice versa;

    (vii) that some other person ought to contribute towards the cost of compliance (sections 102 and 103).

        Any appeals to magistrates' courts are to be made by way of complaint for an order within 21 days from the date on which the council's requirement, notice, etc., was served (unless a longer period is specified). Notices and other documents must give the rights of appeal and state the time within which appeal may be made (section 103).

        Persons aggrieved by decisions of the magistrates' court may appeal to the Crown Court, except where there is a provision for arbitration (section 86).

(c) **Enforcement.** Following failure to comply with a notice to execute works within the time specified, subject to appeal, the local authority may, at its discretion, carry out the work itself and recover expenses reasonably incurred from the person on whom the notice was served. In addition, the person concerned is liable to a fine not exceeding level 4 and a maximum daily penalty of £2 for a continuing offence (section 99(2)).

Recovery of costs

Costs incurred by local authorities in carrying out works in default of an owner are recoverable either by proceedings in the County Court or by becoming a charge on the property. They may be recovered from an owner by instalments and with interest (sections 107(1) and 108).

Power of local authority to execute work on behalf of owners or occupiers

A local authority, by agreement with an owner or occupier of any premises, may undertake on his behalf works which it has required him to carry out under the Act or undertake sewer construction, alteration or repair which the owner or occupier is entitled to execute. The costs are chargeable to the owner or occupier (section 97).

Power to sell materials

A local authority may sell any materials that have been removed by them from any premises, including a street, when executing works and are not claimed by the owner and taken away by him after three days of the works.

    Where a local authority sells materials, they must pay the proceeds to the person to whom the materials belonged, after deducting any expenses (section 100).

DEFECTIVE SANITARY CONVENIENCES

Reference

Public Health Act 1936, section 45 and part 12

FC118 Defective sanitary conveniences – section 45 Public Health Act 1936

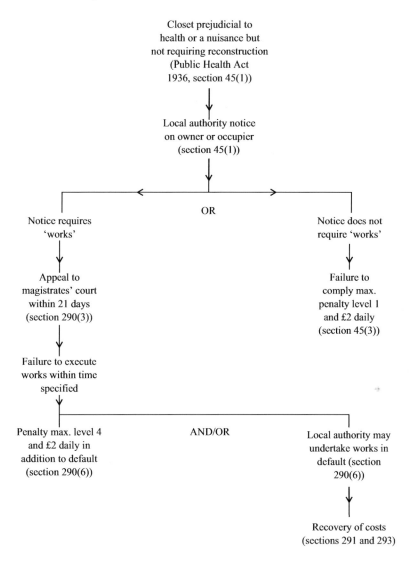

Closet prejudicial to
health or a nuisance but
not requiring reconstruction
(Public Health Act
1936, section 45(1))

Local authority notice
on owner or occupier
(section 45(1))

OR

Notice requires
'works'

Notice does not
require 'works'

Appeal to
magistrates' court
within 21 days
(section 290(3))

Failure to
comply max.
penalty level 1
and £2 daily
(section 45(3))

Failure to execute
works within time
specified

Penalty max. level 4
and £2 daily in
addition to default
(section 290(6))

AND/OR

Local authority may
undertake works in
default (section
290(6))

Recovery of costs
(sections 291 and 293)

Notes
1. For closets prejudicial to health or a nuisance but requiring reconstruction, see Building Act 1984, section 64
    (FC121).
2. At any time after service of the local authority notice, the local authority may undertake the required works at
    the request of the owner or occupier and charge the costs (section 275).

Extent

This procedure applies to England and Wales only.

Scope

This procedure deals with closets provided in a building which are in such a state as to be prejudicial to health or a nuisance and can, without reconstruction, be put into a satisfactory condition. This includes both repair and cleansing (Public Health Act 1936, section 45(1)).

It does not apply to closets in a factory, workshop or workplace (section 45(4)). In such cases, the provisions of the Health and Safety at Work etc. Act 1974 can be used (FC48).

Notices etc.

A local authority can serve a notice on the owner or occupier requiring defective sanitary conveniences to be repaired. The provisions of Part 12 of the Public Health Act 1936 apply to appeals against and the enforcement of notices requiring repairs but not to those requiring cleansing only (see p. 548).

Local authorities can carry out works in default of the notice through section 290(6).

Appeals

In relation to a notice not requiring works, i.e. requiring cleansing etc., the defendant in subsequent proceedings for penalty may question the reasonableness of the requirements and the decision to address the notice to him and not the owner or occupier as the case may be (section 45(3)).

Penalties

If the person who is served with the notice fails to comply with the notice, he is liable to a fine not exceeding level 1 on the standard scale and to a further fine not exceeding £2 for each day on which the offence continues after conviction (section 45(4)).

Definitions

Closet includes privy (section 90(1)). It is suggested that the word includes the structure in which the appliance is housed as well as the appliance itself.
Prejudicial to health means injurious, or likely to be injurious to health.

SANITARY CONVENIENCES: PROVISION/REPLACEMENT

Reference

Building Act 1984, sections 64 and 65

Extent

This procedure applies to England and Wales only.

FC119 Sanitary conveniences: provision/replacement – sections 64 and 65 Building Act 1984

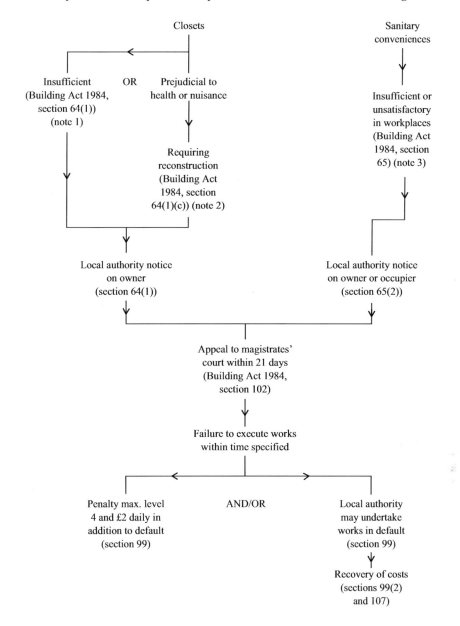

Notes

1. This relates to buildings, or parts of buildings occupied as a separate dwelling, which are without sufficient closet accommodation but does not include workplaces.
2. The procedure for dealing with existing closets of buildings which are prejudicial to health or a nuisance is split between those which can be put into satisfactory condition without construction (section 45, Public Health Act 1936, FC118) and those which cannot (section 64, Building Act 1984).
3. This section covers all workplaces (does not include factories). Alternatively the situation could be dealt with by using the enforcement procedures of the Health and Safety at Work etc. Act 1974, see FC48.
4. For sanitary conveniences in a place of public entertainment etc., see FC120.
5. For the power of a local authority to require replacement of earth closets by water closets at the joint expense of the owner and local authority, see FC121.

Scope

The procedure applies to:

(a) insufficient or replacement accommodation in buildings (other than workplaces) where the existing closet accommodation is insufficient or is prejudicial to health or a nuisance and requires reconstruction. The notices may require the provision of additional or replacement closets, but water closets may not be required unless a sufficient water supply and sewer are available (section 64):

   (i) sewer available – must be within 100 ft of the building at a level which makes connection reasonably practicable, is a sewer which the owner of the building is entitled to use and the intervening land is land through which he is entitled to construct a drain;

   (ii) water supply available – water is laid on or can be laid on from a point within 100 ft of the building and the intervening land is land through which the owner of the building is entitled to lay a pipe (section 125);

(b) sanitary conveniences in workplaces. In deciding whether or not sanitary conveniences for buildings used as workshops are sufficient and satisfactory, regard must be paid to the number of persons employed in or in attendance at the building and the need to provide separate accommodation for the sexes, although the local authority may waive this latter provision if it so wishes (section 65(1)).

Notices, appeals, etc.

The provisions of sections 99 and 102 apply to notices used in this procedure (p. 549).

Definition (see also pp. 552, 558, 562, 564)

Closet includes privy.

Drain means a drain used for the drainage of one building or of buildings or yards appurtenant to buildings within the same curtilage, and includes any manholes, ventilating shafts, pumps or other accessories belonging to the drain.

Sewer does not include a drain, but otherwise it includes all sewers and drains used for the drainage of buildings and yards appurtenant to buildings, and any manholes, ventilating shafts, pumps or other accessories belonging to the sewer (section 126).

SANITARY CONVENIENCES AT PLACES OF ENTERTAINMENT ETC.

References

Local Government (Miscellaneous Provisions) Act 1976, sections 20 and 21
Chronically Sick and Disabled Persons Act 1970, section 6
Public Health and Local Government (Miscellaneous Provisions) Act (Northern Ireland) 1955

Extent

This procedure applies to England and Wales (section 83). Similar provisions apply in Scotland or Northern Ireland.

FC120 Sanitary conveniences at places of entertainment etc. – section 20 Local Government (Miscellaneous Provisions) Act 1976

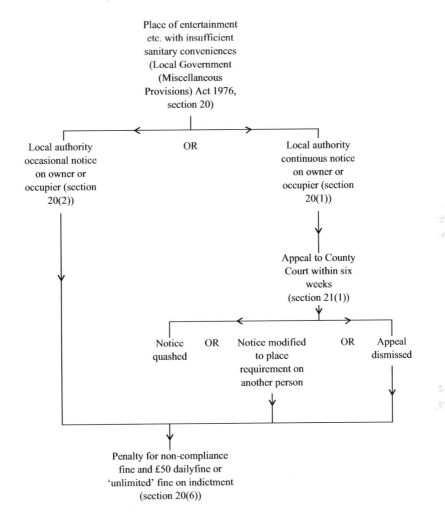

Notes

1. Notices under this section may be served by a district council, London borough council, the Common Council and the Council of the Isles of Scilly (Local Government (Miscellaneous Provisions) Act 1976, sections 20(1) and 44(1)).
2. There is no provision for appeal against 'occasional' notices.
3. The notice must be in writing (Local Government (Miscellaneous Provisions) Act 1976, section 44(1)).

Scope

The premises covered by these provisions are termed 'relevant places' and are premises used, or proposed to be used, occasionally or permanently, for the following purposes:

(a) the holding of any entertainment, exhibition or sporting event to which the public is admitted as spectators or otherwise;
(b) the sale of food and drink to the public for consumption on the premises;
(c) a betting office (section 20(9)).

In respect of (a), a licence for the holding of entertainment may also be required under the Licensing Act 2003.

It is worth noting that the provision of welfare facilities required by the Health and Safety at Work Act 1974 is likely to override this provision. Care must be taken in deciding whether this provision is relevant to a situation which may involve the Health and Safety at Work Act 1974 provisions.

Notices

Notices must be in writing on either the owner or the occupier of the premises and can be served in anticipation of the use of a premises as a 'relevant place'. Notices may require:

(a) the provision of a specified number and type of sanitary appliances at specified positions within a defined period, which should be not less than six weeks;
(b) the maintenance and cleaning of the appliances;
(c) the provision and maintenance of washing facilities, including hot and cold water and other facilities for use in connection with the sanitary appliances, e.g. hand drying facilities etc.;
(d) that the appliance and associated facilities should be made available to the public resorting to the premises, if required by the notice, free of charge.

The notice may require that existing sanitary appliances must be made available but cannot require either movable sanitary appliances at a betting office or, in the case of new buildings, provisions in excess of requirements of the building regulations (section 20(1), (2) and (3)).

Occasional notices

As an alternative to permanent provision, the notice may require provision of the specified appliances and associated facilities only on occasions which are specified in the notice (section 20(2)).

Disabled persons

A person upon whom a section 20 notice is served must make provision for the needs of the disabled so far as is practicable and reasonable (Chronically Sick and Disabled Persons Act 1970, section 6). The notice served under section 20 must draw attention to the requirements of the Chronically Sick and Disabled Persons Act 1970 and to BS 5810/1979, which contains a code at practice relating to Access for the Disabled to Buildings (sections 20(11) and 20(12)).

Appeals

So far as notices requiring permanent provision of facilities are concerned, appeal may be made within six weeks of service to the County Court on the following grounds:

(a) a requirement is unreasonable;

(b) it would have been fairer to have served the notice on the owner or occupier as the case may be (section 21).

There is no appeal against an occasional notice but in any subsequent proceedings it will be a defence to prove situations as in (a) and (b) above.

Defences

In proceedings for non-compliance with notices served under section 20, it will be a defence to prove:

(a) that at the time of the failure the person on whom the notice was served was neither the owner nor occupier and that he did not cease to be the owner or occupier with a view to avoiding compliance with the notice; or

(b) where the contravention relates to a particular day, that the relevant place was closed to members of the public or not used as a relevant place; or

(c) in respect of occasional notices only that:

   (i)   the requirement was unreasonable; or

   (ii) that it would have been fairer to have served the notice on another person (section 20(7) and (8)).

Power of entry

Authorised officers producing evidence of such may enter a 'relevant place' at any reasonable time to determine if a notice should be served or is being complied with. Penalty for obstruction is max. level 3 on the standard scale (section 20(5)).

Definitions

Betting office means premises, other than a track, for which a betting office licence within the meaning of the Gambling Act 2005, is in force (section 20(9)).

Owner in relation to any land, place or premises, means a person who, either on his own account or as agent or trustee for another person, is receiving the rackrent of the land, place or premises or would be entitled to receive it if the land, place or premises were let at a rackrent and 'owned' shall be construed accordingly (section 44(1)).

Sanitary appliances means water closets, other closets, urinals and wash basins (section 20(9)).

CONVERSION OF EARTH CLOSETS ETC. TO WATER CLOSETS

Reference

Building Act 1984, section 66

Extent

This procedure applies to England and Wales only.

Scope

The procedure allows a local authority to secure the replacement of any **closet, other than a water closet**, by a water closet, even though the existing closet may be sufficient and not prejudicial to health or a nuisance, provided that the costs are equally shared between the local authority and the owner of the building and that a **water supply and sewer are available**.

This procedure is similar to the replacement of sanitary conveniences under sections 64 and 65, Building Act 1984.

It is unlikely that there are any earth closets still in existence in the UK, therefore it is suggested that this power is redundant. However, because the wording allows 'closets, other than a water closet' to be replaced and the emergence of compost and waterless toilets, it may still have a role!

Water supply and sewer available

Sewer available – this must be within 100 ft of the building at a level which makes connection reasonably practicable and be a sewer which the owner of the building is entitled to use, and the intervening land should be land through which he is entitled to construct a drain.

Water supply available – water is laid on or can be laid on from a point within 100 ft of the building and the intervening land should be land through which the owner of the building is entitled to lay a pipe (section 125).

The limit of 100 ft does not apply if the local authority undertakes to bear so much of the expenses reasonably incurred in constructing, and in maintaining and repairing, a drain to communicate with a sewer or, as the case may be, in laying and in maintaining and repairing a pipe for the purpose of obtaining a supply of water, even though the distance of the sewer or water supply exceeds 100 ft (section 125(3)).

The availability of a sewer or water supply in these terms is not necessary for the local authority to contribute towards the costs of an owner converting by agreement.

In practice it the local authority may come to an arrangement whereby a septic tank may be provided to deal with the products of the new facilities.

Notices etc.

The provisions of sections 99, 102 and 66(5) (p. 549) apply to appeals against, and the enforcement of notices served under, this procedure. Note that there is no appeal that the works are unnecessary (section 66(5)). Notices to be served on owner. Section 66(3) allows half of the expenses to be shared between the owner and the local authority.

Definitions (also p. 552)

Closet includes privy (section 126).
Water closet means a closet which has a separate fixed receptacle connected to a drainage system and separate provision for flushing from a supply of clean water, either by the operation of mechanism or by automatic action (section 126).

FC121 Conversion of earth closets etc. to water closets – section 66 Building Act 1984

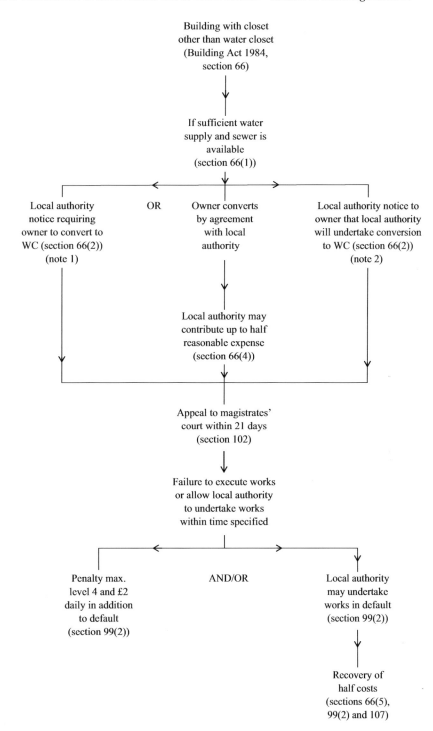

Notes

1. In this case, the owner may recover half of the reasonable cost from the local authority.
2. In this case, the local authority may recover half of the reasonable cost from the owner.

BLOCKED PRIVATE SEWERS

Reference

Local Government (Miscellaneous Provisions) Act 1976, section 35

Extent

This procedure applies to England and Wales only.

Scope

This procedure allows a notice requiring the unblocking of a private sewer to be served on owners or occupiers of all premises draining into it (section 35(1)).

It must be noted that most private sewers, lateral drains and pumping stations that form part of the sewer or lateral drain that connect to the public sewer network have been transferred to the ownership of the regulated sewerage companies in England and Wales. It is suggested that this means that local authorities can transfer responsibility for maintaining private sewers to the water companies making this provision largely redundant.

Notices

Notice may be served on each of the persons who is an owner or occupier of premises served by the sewer or on any of these as the local authority sees fit.

The time specified within which the works are required must not be less than 48 hours but can be longer at the discretion of the local authority (section 35(1)). Notices must be in writing (Public Health Act 1936, section 283).

Notices apportioning costs

Following works in default (section 35(2)), the local authority must specify by notice to those persons in receipt of the notice to clear the blockage, the amount to be recovered from each and which other persons are to be charged and the amount. In apportioning cost, the local authority may have regard to any matters relating to the cause of the blockage and to any agreements relating to the cleansing of that private sewer (section 35(3)). Particular regard should be paid to any determinations under section 22 of the Building Act 1984, dealing with the drainage of buildings in combination.

Appeals against notices to recover costs

In considering an appeal that whole or part should be paid by some other person, the County Court may either dismiss the appeal or order the whole or part to be paid by other owners or occupiers of premises served by the sewer, provided that such owners or occupiers have been given eight days' notice of the appeal (section 35(5)).

Definitions (see also pp. 554, 562, 564)

Notice means a notice in writing (Local Government (Miscellaneous Provisions) Act 1976, section 44).

FC122 Blocked private sewers – section 35 Local Government (Miscellaneous Provisions) Act 1976

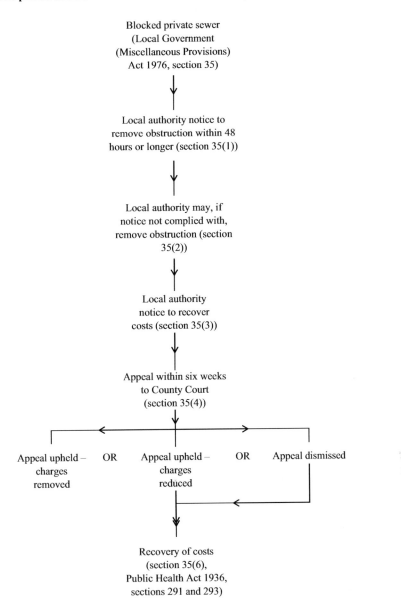

Blocked private sewer
(Local Government
(Miscellaneous Provisions)
Act 1976, section 35)

Local authority notice to
remove obstruction within 48
hours or longer (section 35(1))

Local authority may, if
notice not complied with,
remove obstruction (section
35(2))

Local authority
notice to recover
costs (section 35(3))

Appeal within six weeks
to County Court
(section 35(4))

Appeal upheld –   OR   Appeal upheld –   OR   Appeal dismissed
charges   charges
removed   reduced

Recovery of costs
(section 35(6),
Public Health Act 1936,
sections 291 and 293)

Notes
1. There is no appeal against the notice to remove the blockage under section 35(1) and neither is there a penalty for non-compliance.
2. For defective private sewers, see FC123 and FC124.

Owner in relation to any land, place or premises means a person who, either on his own account or as agent or trustee for another person, is receiving the rackrent of the land, place or premises or would be entitled to receive it if the land, place or premises were let at a rackrent (Local Government (Miscellaneous Provisions) Act 1976, section 44).

Private sewer means a sewer which is not a public sewer.

Sewer does not include a drain (see definition above) . . . but . . . includes all sewers and drains used for the drainage of buildings and yards appurtenant to buildings (Public Health Act 1936, section 343).

In practical terms a private sewer is a building sewer that receives the discharge from more than one building drain and conveys it to a public sewer, private sewage disposal system, or other point of disposal.

STOPPED-UP DRAINS, PRIVATE SEWERS ETC.

Reference

Public Health Act 1961, section 17(3) (as substituted by section 27 of the Local Government (Miscellaneous Provisions) Act 1982)

Extent

This procedure applies to England and Wales only.

Scope

This procedure may be applied to any drain, private sewer, water closet, waste pipe or soil pipe on any premises which is stopped-up. There is no exclusion here for premises owned by statutory undertakers as there is in relation to other parts of section 17 (see FC125).

Because section 17 appears to limit the service of the notice to the owner or occupier of the premises at which the drain etc. is stopped-up, blocked private sewers could also be dealt with under section 35 (Local Government (Miscellaneous Provisions) Act 1976) since owners of other premises draining into the sewer may be involved (FC122). Please note the comments about the practical viability of section 35 use (p. 560).

Notices

Notices requiring remedy of the defect within 48 hours (this period is specified by the section and should not be shorter or longer) must be in writing, but are not subject to section 290 of the Public Health Act 1936 provisions and are to be served on either the owner or the occupier of the premises on which the drain, private sewer, etc. is stopped-up (section 17(3)).

Recovery of costs

If the cost to the local authority of undertaking the works in default does not exceed £10, recovery may be waived, otherwise costs are recoverable from persons on whom the notices

FC123 Stopped-up drains, private sewers, etc. – section 17(3) Public Health Act 1961

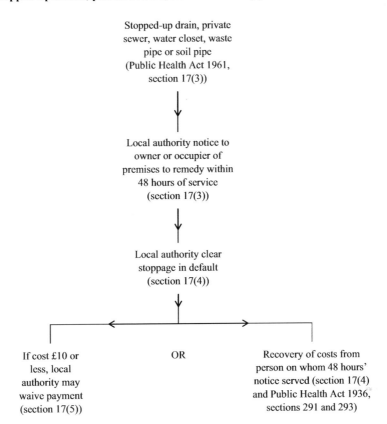

Stopped-up drain, private
sewer, water closet, waste
pipe or soil pipe
(Public Health Act 1961,
section 17(3))

↓

Local authority notice to
owner or occupier of
premises to remedy within
48 hours of service
(section 17(3))

↓

Local authority clear
stoppage in default
(section 17(4))

If cost £10 or
less, local
authority may
waive payment
(section 17(5))

OR

Recovery of costs from
person on whom 48 hours'
notice served (section 17(4)
and Public Health Act 1936,
sections 291 and 293)

Notes

1. At the request of the owner or occupier, a local authority may cleanse or repair drains, water closets, sinks or gullies and recover costs. This could be done before the expiration of the 48 hours' notice by agreement (Public Health Act 1961, section 22).
2. There is no appeal against a notice served under section 17, neither is there a penalty for non-compliance.
3. For defective drainage, see FC124 and FC125.
4. For alternative procedure for blocked private sewers, see Local Government (Miscellaneous Provisions) Act 1976, section 35, at FC122.

were served (section 17(5)). In any proceedings to recover expenses, the magistrates' court may inquire:

(a) whether any requirement in the notice was reasonable; and
(b) whether the expenses ought to be borne wholly or in part by someone other than the defendant, but due notice of the proceedings and an opportunity to be heard must be given to these other persons (section 17(6)).

Definitions

Drain means a drain used for the drainage of one building or any buildings or yards appurtenant to buildings within the same curtilage.

Owner means the person for the time being receiving the rackrent of the premises . . . whether on his own account or as agent or trustee for any other person, or who would so receive the same if those premises were let at a rackrent.

Premises includes messuages, buildings, lands, easements and hereditaments of any tenure.

Private sewer means a sewer which is not a public sewer.

Sewer does not include a drain (see definition above) . . . but . . . includes all sewers and drains used for the drainage of buildings and yards appurtenant to buildings (Public Health Act 1936, section 343).

DEFECTIVE DRAINAGE TO EXISTING BUILDINGS

Reference

Building Act 1984, sections 59 and 60

Extent

This procedure applies to England and Wales only.

Scope

The procedure is applied to buildings which have:

(a) unsatisfactory provision for drainage (section 59(1)(a));
(b) cesspools, private sewers, drains, soil pipes, rainwater pipes, spouts, sinks or other necessary appliances which are insufficient or, in the case of a private sewer or drain communicating with a public sewer, is so defective as to admit subsoil water (section 59(1)(b));
(c) cesspools etc. as detailed in (b) above, in such a condition as to be prejudicial to health or a nuisance, and this also covers cesspools, private sewers and drains no longer in use (section 59(1)(c) and (d));
(d) rainwater pipes being used for foul waste, soil pipes from water closets not properly ventilated and surface water pipes acting as vents to foul drains or sewers (section 60).

In relation to (a), unsatisfactory drainage is not defined.

FC124 Defective drainage to existing buildings – sections 59 and 60 Building Act 1984

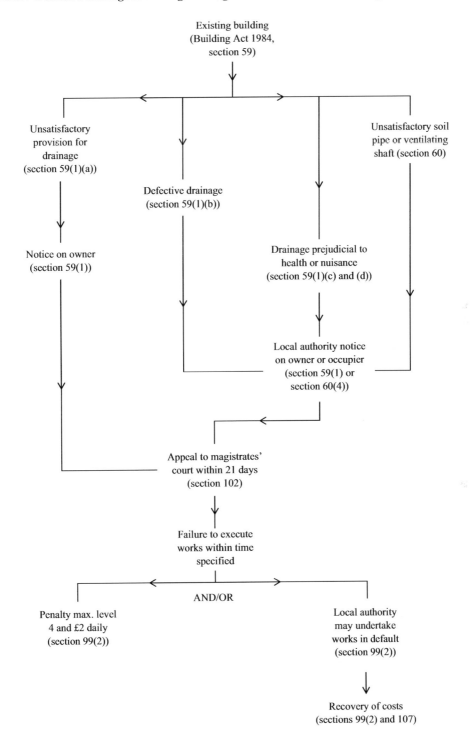

Notes

1. At the request of an owner or occupier, a local authority may cleanse or repair drains, water closets, sinks or gullies and recover costs. This could be done by agreement before the expiration of any notice served under section 59 and whether or not a notice has been served (Public Health Act 1961, section 22).
2. See also Public Health Act 1961, section 17 as amended (FC125).

Notices etc.

The provisions on p. 549 apply to appeals against and the enforcement of notices served under this procedure. Notices are to be served on the owner of the building concerned. Appeal provisions are included in section 102 of the Building Act 1984. Works and default powers, penalties and recovery of costs are included in sections 99 and 107 of the Building Act 1984.

Testing of drains etc.

Local authorities have the power to examine sanitary conveniences, drains, private sewers or cesspools and to apply tests, other than tests by water under pressure, where they suspect deficiencies under section 59. This includes exposing the pipes by opening the ground (section 48, Public Health Act 1936).

Notice of intention to repair drains etc.

Before carrying out repairs to drains or private sewers, the person undertaking the work must give at least 24 hours' notice to the local authority, except if the work is required in an emergency, in which case he must not cover over the drain or sewer without such notice (Building Act, section 61).

Definitions (see also pp. 552, 558)

Surface water includes water from roofs.

Water closet means a closet that has a separate fixed receptacle connected to a drainage system and separate provision for flushing from a supply of clean water, either by the operation of a mechanism or by automatic action (section 126).

Power of local authority to examine and test drains, &c., believed to be defective – section 48 Public Health Act 1936

If a local authority has reasonable grounds for believing that a sanitary convenience, drain, private sewer or cesspool is in such a condition as to be prejudicial to health or a nuisance, or that a drain or private sewer communicating indirectly with a public sewer is so defective as to admit subsoil water, they may examine its condition, and may apply any test, other than a test by water under pressure, and, if they deem it necessary, open the ground (section 48(1)).

If on examination the convenience, drain, sewer or cesspool is found to be in proper condition, the authority must reinstate any ground which has been opened by them and make good any damage done by them (section 48(2)).

REPAIR OF DRAINS, PRIVATE SEWERS ETC.
BY LOCAL AUTHORITIES

Reference

Public Health Act 1961, section 17(1) (as substituted by section 27 of the Local Government (Miscellaneous Provisions) Act 1982)

FC125 Repair of drains, private sewers, etc. by local authorities – section 17 Public Health Act 1936

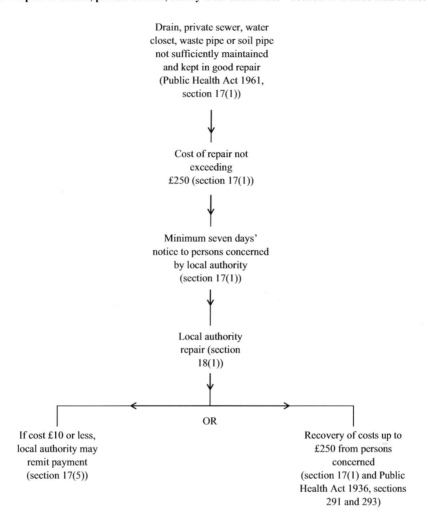

Drain, private sewer, water
closet, waste pipe or soil pipe
not sufficiently maintained
and kept in good repair
(Public Health Act 1961,
section 17(1))

Cost of repair not
exceeding
£250 (section 17(1))

Minimum seven days'
notice to persons concerned
by local authority
(section 17(1))

Local authority
repair (section
18(1))

OR

If cost £10 or less,
local authority may
remit payment
(section 17(5))

Recovery of costs up to
£250 from persons
concerned
(section 17(1) and Public
Health Act 1936, sections
291 and 293)

Notes

1. For alternative procedure, see Building Act 1984, section 59, FC124.
2. For blocked drains and private sewers, see FC123.
3. There is no appeal against a notice served under section 17, neither is there a penalty for non-compliance.
4. Local authorities may clear obstructions in drains or private sewers by agreement before the expiration of the notice (Public Health Act 1961, section 22).

Extent

This procedure applies to England and Wales only.

Scope

The procedure applies to any drain, private sewer, water closet, waste pipe or soil pipe which is:

(a) not sufficiently maintained and kept in good repair; and
(b) can be sufficiently repaired at a cost not exceeding £250 (section 17(1)).

Works by local authorities on land belonging to statutory undertakers and held or used by them for those purposes is not authorised unless it affects houses or buildings used as offices or showrooms, other than those forming part of a railway station (section 17(10) and (11)).

It is suggested that the limit of £250 to repair an element of a sewerage system makes this provision largely redundant and only in the simplest cases would it be used.

The procedure under section 17 to deal with blocked or stopped-up (as opposed to defective) drains, private sewers etc. is covered separately in FC123 (p. 563).

Persons concerned

Notices are to be served on the person or persons concerned, which means:

(a) in relation to a water closet, waste pipe or soil pipe, the owner or occupier of the premises; and
(b) in relation to a drain or private sewer, any person owning any premises drained by means of it and also, in the case of a sewer, the owner of the sewer (section 17(2)).

Notices

Notices must be in writing (Public Health Act 1936, section 283) but are not subject to the provisions of section 290 (Appeals against, and the enforcement of, notices requiring execution of works).

Recovery of costs

Costs not exceeding £250 incurred by the local authority are recoverable from the persons concerned (as defined above) in such proportions, if there is more than one person concerned, as the local authority may determine. Costs of £10 or less may be waived (section 17(4) and (5)).

In any proceedings to recover the local authority's expenses of up to £250, the court must consider whether or not the local authority was justified in concluding that the drain, private sewer, etc. was not sufficiently maintained and kept in good repair and decide whether any apportionment between different persons is fair (section 17 (6)).

PAVING OF YARDS AND PASSAGES

References

Building Act 1984, section 84
Housing (Scotland) Act 2006

FC126 Paving of yards and passages – section 84 Building Act 1984

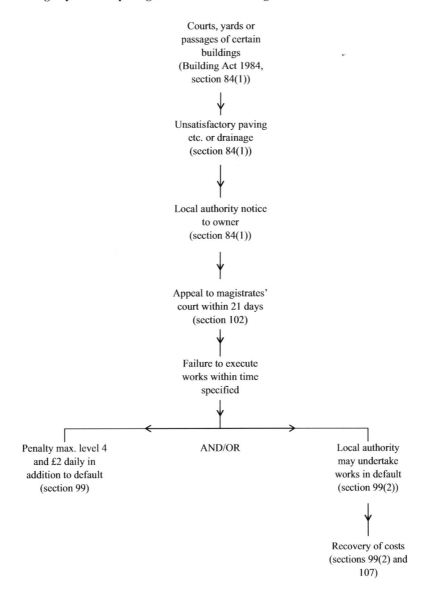

Courts, yards or
passages of certain
buildings
(Building Act 1984,
section 84(1))

Unsatisfactory paving
etc. or drainage
(section 84(1))

Local authority notice
to owner
(section 84(1))

Appeal to magistrates'
court within 21 days
(section 102)

Failure to execute
works within time
specified

AND/OR

Penalty max. level 4
and £2 daily in
addition to default
(section 99)

Local authority
may undertake
works in default
(section 99(2))

Recovery of costs
(sections 99(2) and
107)

Extent

This procedure applies to England and Wales only. Similar provisions apply in Scotland.

Scope

The section applies to any court or yard appurtenant to, or any passage giving access to, buildings as defined below, including those used in common by two or more houses provided it is not repairable by the inhabitants at large as a highway (section 84). It does not apply to garden paths.

The procedure applies to buildings which are houses, industrial or commercial buildings.

Paving and drainage

Courts, yards and passages must be so formed, flagged, asphalted or paved, or provided with such works on, above or below its surface, as to allow satisfactory drainage of its surface or subsoil to a proper outfall (section 84(1)). Regard should be made to the need to remove water from the court, yard or passage and also to the need to dispose of it satisfactorily in the course of or after its removal.

Notices

The local authority may by notice require any person who is the owner of any of the buildings to execute all such works as may be necessary to remedy the defect.

The provision relating to enforcement of notices and appeals on p. 549 applies to this procedure.

Notices, which are to be served on the owner, must indicate the nature of the works required and state the time within which they must be executed. The time allowed must be reasonable and not less than the period for appeal, i.e. 21 days (sections 99 and 102).

Grounds of appeal

These are as in section 102 set out on p. 549.

OVERFLOWING AND LEAKING CESSPOOLS

References

Public Health Act 1936, section 50 and part 12
Sewerage (Scotland) Act 1968
The Water (Northern Ireland) Order 1999

Extent

This procedure applies to England and Wales only. Similar provisions apply in Scotland and Northern Ireland.

FC127 Overflowing and leaking cesspools – section 50 Public Health Act 1936

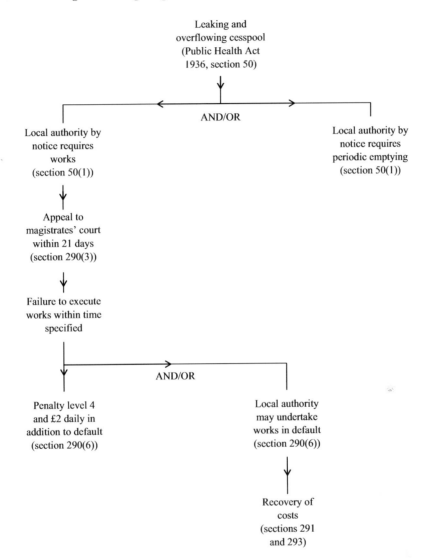

Notes
1. There is no provision for appeal against notices not requiring works under this section (e.g. periodic emptying), but it is open to the defendant to question the reasonableness of the notice in subsequent proceedings.
2. The local authority may at any time undertake any works which it has requested if asked to do so by the owner or occupier (Public Health Act 1936, section 275).

Scope

The section does not apply in relation to the effluent from a properly constructed tank for the reception and treatment of sewage provided that the effluent is not prejudicial to health or a nuisance. It applies to any leaking or overflowing cesspool even though the situation may not be prejudicial to health or a nuisance.

The local authority may by notice require the person by whose act, default or sufferance the soakage or overflow occurred or continued, to execute such works, or to take such steps by periodically emptying the cesspool or otherwise, for preventing the soakage or overflow (section 50(1)).

Notices etc.

Part 12 of the Public Health Act 1936 applies to appeals against the enforcement of notices served under this procedure (p. 548).

Appeals against notices

The grounds of appeal against notices requiring works are set out on p. 548. Notices which require periodic emptying are not subject to the provisions of section 290 and no appeal is provided for. In defending any proceedings for penalty because of non-compliance with this form of notice, it is open to the defendant to question the reasonableness of the requirements (section 50(3)).

Definition

Cesspool includes a settlement tank or other tank for the reception or disposal of foul matter from buildings (section 90(1)).

Chapter 11

STATUTORY NUISANCES

STATUTORY NUISANCES

References

Environmental Protection Act 1990, sections 79–82 and schedule 3 (as amended by the Noise and Statutory Nuisance Act 1993, the Environment Act 1995, the Pollution Prevention and Control Act 1999 and the Clean Neighbourhoods and Environment Act 2005)

The Statutory Nuisance (Appeals) Regulations 1995 (as amended 2006)

The Statutory Nuisance (Insects) Regulations 2006

The Statutory Nuisance (Artificial Lighting) (Designation of Relevant Sports) (England) Order 2006

Statutory Nuisance from Insects and Artificial Light – Guidance on sections 101 to 103 of the Clean Neighbourhoods and Environment Act 2005

Public Health etc. (Scotland) Act 2008

Clean Neighbourhoods and Environment Act (Northern Ireland) 2011

Extent

The 1990 Act applies in England and Wales and, with certain amendments, in Scotland. Similar provisions apply in Northern Ireland.

Statutory nuisances

These are nuisances to which the abatement procedures of Part 3 of the Environmental Protection Act 1990 have been applied and are as follows:

1. Any premises in such a state as to be prejudicial to health or a nuisance (section 79(1)(a)). (For an expedited procedure to deal with the defective premises, see Building Act 1984, section 76, FC62.)
2. Smoke emitted from premises so as to be prejudicial to health or a nuisance; but this does not apply to:

   (a) premises occupied by the Crown for military or Ministry of Defence purposes;
   (b) smoke emitted from the chimney of a house within a smoke control area;
   (c) dark smoke from the chimney of a building or of a furnace attached to a building or installed on any land;
   (d) smoke from a railway locomotive steam engine;
   (e) dark smoke from any industrial or trade premises (section 79(1)(b), (2) and (3)).

3. Fumes or gases emitted from private dwellings so as to be prejudicial to health or a nuisance (section 79(1)(c) and (4)).

4. Any dust, steam (other than from a railway locomotive engine), smell or other effluvia arising on industrial, trade or business premises and being prejudicial to health or a nuisance (section 79(1)(d) and (5)).

5. Any accumulation or deposit which is prejudicial to health or a nuisance (section 79(1)(e)).

6. Any animal kept in such a place or manner as to be prejudicial to health or a nuisance (section 79(1)(f)).

7. Noise (except that from aircraft other than model aircraft) emitted from premises so as to be prejudicial to health or a nuisance; but this does not apply to Crown premises used for military or Ministry of Defence purposes (section 79(1)(g), (2) and (6) as amended).

8. Noise that is prejudicial to health or a nuisance and is emitted from or caused by a vehicle, machinery or equipment in a street (other than noise made by traffic, by any military force or by political demonstration or a demonstration supporting or opposing a cause or campaign) (section 79(1)(ga) and (6A)).

9. Insects (other than those listed in schedule 45 of the Wildlife and Countryside Act 1981) emanating from relevant industrial, trade or business premises (certain types of farmland and land within sites of special scientific importance and certain waters are excluded – see the Statutory Nuisances (Insects) Regulations 2006) and being prejudicial to health or a nuisance (section 79 (1)(fa)).

10. Artificial light emitted from industrial or trade premises or from lights used only to illuminate outdoor relevant sports facilities so as to be prejudicial to health or a nuisance. Certain types of premises are excluded, e.g. those used for defence, some transport-related premises, airports and prisons (section 79(1)(fb) and 80(aza)(not yet in force)).

11. Any other matter declared by any enactment to be a statutory nuisance (section 79(1)(h)) and these include:

    (a) any well, tank, cistern or water butt used for the supply of water for domestic purposes which is so placed, constructed or kept as to render the water therein liable to contamination prejudicial to health (Public Health Act 1936, section 141);

    (b) any pond, pool, ditch, gutter or watercourse which is so foul or in such a state as to be prejudicial to health or a nuisance (Public Health Act 1936, section 259(1)(a));

    (c) any part of a watercourse, not being a part ordinarily navigated by vessels employed in the carriage of goods by water, which is so choked or silted up as to obstruct or impede the proper flow of water and thereby to cause a nuisance or give rise to conditions prejudicial to health (Public Health Act 1936, section 259(1)(b));

    (d) a tent, van, shed or similar structure used for human habitation:

        (i) which is in such a state, or so overcrowded, as to be prejudicial to the health of the inmates; or

        (ii) the use of which, by reason of the absence of proper sanitary accommodation, or otherwise, gives rise, whether on the site or on other land, to a nuisance or to conditions prejudicial to health (Public Health Act 1936, section 268(2));

    (e) a shaft or outlet of certain abandoned or disused mines where:

        (i) it is not provided with a properly maintained device designed and constructed to prevent persons from accidentally falling down the shaft or accidentally entering the outlet; or

    (ii) by reason of its accessibility from a highway or a place of public resort, it constitutes a danger to the public (Mines and Quarries Act 1954, section 151);

(f) a quarry which is not provided with an efficient and properly maintained barrier so designed and constructed as to prevent persons from accidentally falling into it and which, by reason of its accessibility from a highway or place or public resort, constitutes a danger to the public (Mines and Quarries Act 1954, section 151).

In implementing the statutory nuisance procedures, local authorities may not have regard to the radioactive state of any substance, article or premises. These are matters to be dealt with under the Environmental Permitting (England and Wales) Regulations 2016.

    Unless the Secretary of State has granted consent, a local authority may not bring summary proceedings for causing a statutory nuisance where proceedings may be brought under the Integrated Pollution Prevention Control regime under the Pollution Prevention and Control Act 1999 – see FC27 (section 79(10)). The courts have determined that such consent is only needed for prosecuting for the offence of non-compliance with an abatement notice, not for the service of such a Notice nor other enforcement proceedings such as a High Court Injunction.

    However, activities that are not covered by Integrated Pollution Prevention Control, even though they are on the same site as an Integrated Pollution Prevention Control installation, may be regulated by the statutory nuisance procedures. The restriction also does not affect the right of members of the public to use private proceedings under section 82 of the Environmental Protection Act 1990 (see p. 580).

Abatement notices

Where the local authority (definition on p. 40) is satisfied that a statutory nuisance:

(a) exists; or
(b) is likely to occur; or
(c) is likely to recur,

it is required to serve an abatement notice:

(a) requiring the abatement of the nuisance or prohibiting or restricting its occurrence;
(b) requiring the execution of such works or steps as necessary for those purposes;
(c) specifying the time or times within which the notice is to be complied with (section 80(1)).

The notice must also indicate the rights for and times of appeal (schedule 3, paragraph 6). The period allowed for compliance must be reasonable ((d) under 'Appeals' below) but may nevertheless be short, e.g. to deal with noise from a party. Where the period allowed is less than the time allowed for appeal, the notice is suspended in certain circumstances ('Suspension of notices' below). For the service of notices, see p. 43.

    The notice may be served by a local authority to deal with a nuisance which exists or has occurred from outside its area (section 81(2)).

Deferral of duty to serve abatement notice

In dealing with a statutory nuisance caused by noise, the local authority may defer the service of the abatement notice for seven days while it takes other appropriate action to persuade the person who would receive the notice to abate the nuisance. After the seven days, or before if the local authority judges it right, it may proceed by way of the usual abatement notice (section 80(2A)).

Persons responsible

The abatement notice is to be served:

(a) on the person/s responsible for the nuisance except as in (b) and (c) below;
(b) on the owner where the nuisance arises from any defect of a structural character;
(c) where the person responsible cannot be found or the nuisance has not yet occurred, on the owner or occupier of the premises (section 80(2)).

Where more than one person is responsible, the notice must be served on each (section 81(1)).
For noise nuisances from vehicles, machinery or plant in streets, the notice is to be served:

(a) on the person responsible for that vehicle etc.; or
(b) where that person cannot be found, by fixing it to the vehicle etc. (section 80A(2)).

Appeals

Any person served with an abatement notice may appeal to a magistrates' court within 21 days of the date of service (section 80(3), schedule 3, paragraphs 1–4) on one or more of the following grounds:

(a) that the abatement notice is not justified by section 80 of the 1990 Act;
(b) that there has been some material informality, defect or error in, or in connection with, the abatement notice or in, or in connection with, any copy of the abatement notice served under section 80A(3) (which relates to notices in respect of vehicles, machinery or equipment);
(c) that the authority has refused unreasonably to accept compliance with alternative requirements, or that the requirements of the abatement notice are otherwise unreasonable in character or extent, or are unnecessary;
(d) that the time, or, where more than one time is specified, any of the times, within which the requirements of the abatement notice are to be complied with is not reasonably sufficient for the purpose;
(e) where the nuisance to which the notice relates:

 (i) is a nuisance falling within (1) premises, (4) dust etc., (5) accumulations etc., (6) animals, (7) noise, (8) noise in streets, (9) insects or (10) artificial lighting in 'Statutory nuisances' above, and arises on industrial, trade or business premises or, in the case of a vehicle, equipment or machinery, is being used for industrial, trade or business purposes; or
 (ii) is a nuisance falling within (2) smoke etc. of the same section above and the smoke is emitted from a chimney;
      that the best practicable means were used to prevent, or to counteract the effects of, the nuisance;

(f) that, in the case of a nuisance under (7) noise or (8) noise in streets, the requirements imposed by the abatement notice are more onerous than the requirements for the time being in force, in relation to the noise to which the notice relates, of:

    (i) any notice served under section 60 or 66 of the Control of Pollution Act 1974; or

    (ii) any consent given under section 61 or 65 of the Control of Pollution Act 1974; or

    (iii) any determination made under section 67 of the Control of Pollution Act 1974; or

    (iv) in relation to noise in streets, of any condition of consent given under paragraph 1 of schedule 2 of the Environmental Protection Act 1990;

(g) that the abatement notice should have been served on some person instead of the appellant, being:

    (i) the person responsible for the nuisance; or

    (ii) the person responsible for the vehicle, machinery or equipment; or

    (iii) in the case of a nuisance arising from any defect of a structural character, the owner of the premises; or

    (iv) in the case where the person responsible for the nuisance cannot be found or the nuisance has not yet occurred, the owner or occupier of the premises;

(h) that the abatement notice might lawfully have been served on some person instead of the appellant being:

    (i) in the case where the appellant is the owner of the premises, the occupier of the premises; or

    (ii) in the case where the appellant is the occupier of the premises, the owner of the premises;

      and that it would have been equitable for it to have been so served (copy of the appeal notice to be served on the person/s to be implicated);

(i) that the abatement notice might lawfully have been served on some person in addition to the appellant, being:

    (i) a person responsible for the nuisance; or

    (ii) a person who is also an owner of the premises; or

    (iii) a person who is also an occupier of the premises; or

    (iv) a person who is also the person responsible for the vehicle, machinery or equipment (copy of appeal notice to be served on the person/s to be implicated);

and that it would have been equitable for it to have been so served (copy of the appeal notice to be served on the person/s to be implicated) (regulation 2 of the Statutory Nuisance (Appeals) Regulations 1995).

Any party may appeal to the Crown Court against any decision of the magistrates' court (schedule 3, paragraph 1(3)). The court may either:

(a) quash the notice; or

(b) vary it; or

(c) dismiss the appeal and may make orders about the responsibility of persons for costs of workers.

Suspension of notices

Where the nuisance:

(a) is injurious to health; or
(b) is likely to be of limited duration; or
(c) any expenditure incurred in compliance with the notice would not be disproportionate to the public benefit expected; and
(d) the notice has specified the existence of these circumstances,

the abatement notice is *not* suspended pending the hearing of any appeal. In any other case where:

(a) expenditure is involved in attaining compliance; or
(b) in the case of noise nuisance, the noise is caused in the course of the performance of a duty imposed by the local authority,

the notice is suspended pending the determination of the appeal (regulation 3).

Offences

If a person on whom an abatement notice is served, without reasonable excuse, contravenes or fails to comply with any requirement or prohibition imposed by the notice, he is guilty of an offence and is liable on summary conviction to an unlimited fine together with a further fine of an amount equal to one-tenth of the greater of £5,000 or level 4 on the standard scale for each day on which the offence continues after the conviction. Upon conviction for non-compliance with the requirements of an abatement notice on industrial, trade or business premises, a person is liable on summary conviction to a fine (section 80(5) and (6)).

Defences

It will be a defence to prove that the best practicable means have been used to prevent or counteract the effects of the nuisance except:

(a) where the nuisance arises on *other than* industrial, trade or business premises *and* involves nuisance premises, dust etc., accumulations etc., animals and insects, noise;
(b) in the case of a smoke nuisance, except where it arises from smoke emitted from a chimney;
(c) in the case of artificial light except where –

   (i) the artificial light is emitted from industrial, trade or business premises, or
   (ii) the artificial light illuminating an outdoor relevant sports facility;

(d) in the case of a noise nuisance from vehicles etc except where the noise is emitted from or caused by a vehicle, machinery or equipment being used for industrial, trade or business purposes;
(e) in relation to category (3) fumes etc. and (8) statutory nuisances declared by Acts other than the Environmental Protection Act 1990 (sections 80(7) and (8)).

In relation to noise nuisances only, including noise in streets, it will be a defence to prove that:

(a) the situation was covered by a notice under section 60 or a consent under sections 61 or 65 of the Control of Pollution Act 1974 relating to construction sites; or

(b) where a section 66 Control of Pollution Act 1974 noise reduction notice was in force, the level of noise was below that specified in the notice; or

(c) although a section 66 notice was not in force, there was a section 67 notice (Control of Pollution Act 1974) relating to new buildings liable to an abatement order and the noise was less than the specified level (section 80(9)).

Defaults

When an abatement notice has not been complied with, the local authority may, in addition to prosecuting for non-compliance, do whatever is necessary in the terms of the notice to abate or prevent the nuisance (section 81(3)). This may include the confiscation of sound amplification equipment (FC25). The special procedures for dealing with noise from vehicles, machinery and plant in streets are also dealt with in the Noise section (p. 135).

The costs of abating or preventing the nuisance are recoverable from the person/s by whose act or default the nuisance was caused (section 81(4)). Where the owner of any premises is the person responsible for the nuisance, the local authority may, with prior notification to that person and subject to appeal, recover costs in executing notices by a charge on the property (section 81A). A local authority may also recover its costs by instalments (section 81B).

High Court proceedings

If a local authority considers that the taking of summary proceedings (service of abatement notice etc.) would afford an inadequate remedy in relation to the abatement, prohibition or restriction of any statutory nuisance, it may take proceedings in the High Court or, in Scotland, in any court of competent jurisdiction (section 81(5)). The courts have determined that an abatement notice must have been served first, and the local authority must formally resolve that other remedies would be inadequate. It is not necessary to prosecute for non-compliance first and it does not matter that there is an outstanding appeal against the Notice.

In cases involving noise nuisance, it will be a defence to show that the noise was authorised by either a notice under section 60 or a consent under section 61 of the Control of Pollution Act 1974 dealing with construction sites (see FC23). Otherwise the defences identified above relating to summary proceedings are not applicable to the High Court procedure (section 81(6)).

Powers of entry

Specific powers relating to power of entry to deal with statutory nuisances are provided in schedule 3 to the Environmental Protection Act 1990.

Authorised officers of the local authority, upon producing if required their authority, may enter premises at any reasonable time to see whether or not a statutory nuisance exists or to execute works. Unless in an emergency, 24 hours' notice of entry is required to the occupier of residential premises. Where admission is refused or apprehended, where premises are unoccupied, in an emergency or where application for entry would defeat the object, application may be made to the magistrates' court for a warrant. The maximum penalty for obstruction is level 3 (schedule 3, paragraphs 2, 2A and 3).

Summary proceedings by persons aggrieved (FC129)

It is possible for an aggrieved person or persons to proceed directly to abate or prevent the recurrence of a statutory nuisance by laying information to the magistrates' court. For example, this would include the tenant of a house, local authority or privately owned, seeking a remedy to disrepair. At least 21 days' notice shall be given to the person responsible for the nuisance (three days for noise nuisances, including noise from vehicles, machinery or equipment in streets).

If the court is satisfied, it may make an order for either or both the abatement or prohibition of the nuisance by the carrying out of specified works within the time specified. The court may also impose an unlimited fine on the defendant. Because the courts have determined that the ability to fine at this stage makes this a criminal procedure, it is correctly started by laying information not by complaint as stated and the aggrieved person has to prove the relevant facts beyond reasonable doubt. The defences available parallel those where the local authority serves the abatement notice, as identified above.

In the event of non-compliance with the order, the court may impose an unlimited fine and one-tenth of £5000 or the maximum of level 4 on the standard scale daily for its continuation. The court may also, after giving the local authority an opportunity of being heard, order the local authority to undertake whatever works or steps are necessary to abate or prohibit the nuisance.

At the discretion of the court and upon the making of an order, compensation may be payable by the defendant to the aggrieved persons making the complaint (section 82).

Definitions

Best practicable means is to be interpreted by reference to the following provisions:

(a) practicable means reasonably practicable, having regard among other things to local conditions and circumstances, to the current state of technical knowledge and to the financial implications;
(b) the means to be employed include the design, installation, maintenance and manner and periods of operation of plant and machinery, and the design, construction and maintenance of buildings and structures;
(c) the test is to apply only so far as it is compatible with any duty imposed by local authority;
(d) the test is to apply only so far as it is compatible with safety and safe working conditions, and with the exigencies of any emergency or unforeseeable circumstances;

and, in circumstances where a Code of Practice under section 71 of the Control of Pollution Act 1974 (noise minimisation) is applicable, regard shall also be had to guidance given in it (section 79(9)).

Chimney includes structures and openings of any kind from or through which smoke may be emitted.

Dust does not include dust emitted from a chimney as an ingredient of smoke.

Equipment includes a musical instrument.

Fumes means any airborne solid matter smaller than dust.

Gas includes vapour and moisture precipitated from vapour.

Industrial, trade or business premises means premises used for any industrial, trade or business purposes or premises not so used on which matter is burnt in connection with any industrial, trade or business process, and premises are used for industrial purposes where they are used for the purpose of any treatment or process as well as where they are used for the purposes of manufacturing.

Noise includes vibration.

Person responsible:

(a) in relation to a statutory nuisance, means the person to whose act, default or sufferance the nuisance is attributable;

(b) in relation to a vehicle, includes a person in whose name the vehicle is for the time being registered under the Vehicles (Excise) Act 1994 and any other person who is for the time being the driver of the vehicle;

(c) in relation to machinery or equipment, includes any person who is for the time being the operator of the machinery or equipment.

Prejudicial to health means injurious, or likely to cause injury, to health.

Premises includes land and any vessel (other than one powered by steam reciprocating machinery).

Private dwelling means any building, or part of a building, used or intended to be used as a dwelling.

Smoke includes soot, ash, grit and gritty particles emitted in smoke.

Street means a highway and any other road, footway, square or court that is for the time being open to the public (section 79(7) and (8)).

FC128 Statutory nuisances: local authority action – section 80 Environmental Protection Act 1990

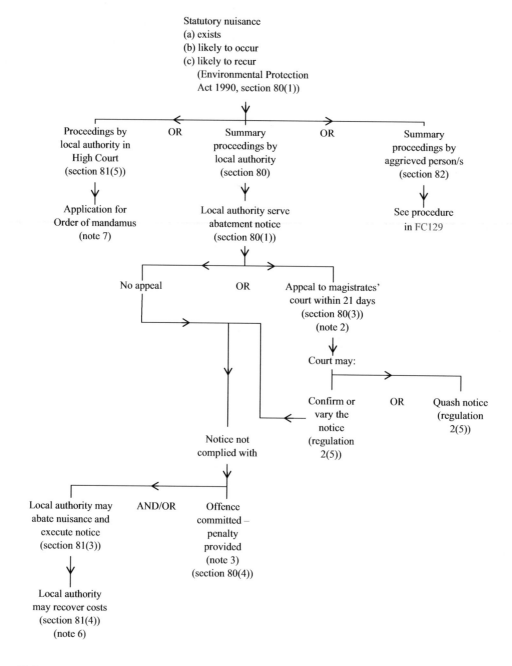

Statutory nuisance
(a) exists
(b) likely to occur
(c) likely to recur
(Environmental Protection
Act 1990, section 80(1))

Proceedings by local authority in High Court (section 81(5))   OR   Summary proceedings by local authority (section 80)   OR   Summary proceedings by aggrieved person/s (section 82)

Application for Order of mandamus (note 7)

Local authority serve abatement notice (section 80(1))

See procedure in FC129

No appeal   OR   Appeal to magistrates' court within 21 days (section 80(3)) (note 2)

Court may:

Confirm or vary the notice (regulation 2(5))   OR   Quash notice (regulation 2(5))

Notice not complied with

Local authority may abate nuisance and execute notice (section 81(3))   AND/OR   Offence committed – penalty provided (note 3) (section 80(4))

Local authority may recover costs (section 81(4)) (note 6)

Notes
1. Regulation numbers refer to the Statutory Nuisance (Appeals) Regulations 1995.
2. For the suspension of some notices pending hearing of the appeal, see p. 578.
3. Details of penalties are given on p. 578.
4. For expedited procedure to deal with the defective premises, see FC62.
5. For service of notices under the Environmental Protection Act 1990, see p. 43.
6. Or local authority may put a charge on the property (section 81A).
7. On Judicial Review of any relevant decision by a lower court or a public authority, the High Court may order them to do or refrain from doing some act which that body is obliged to do under the law.

FC129 Statutory nuisances: action by aggrieved persons – section 82 Environmental Protection Act 1990

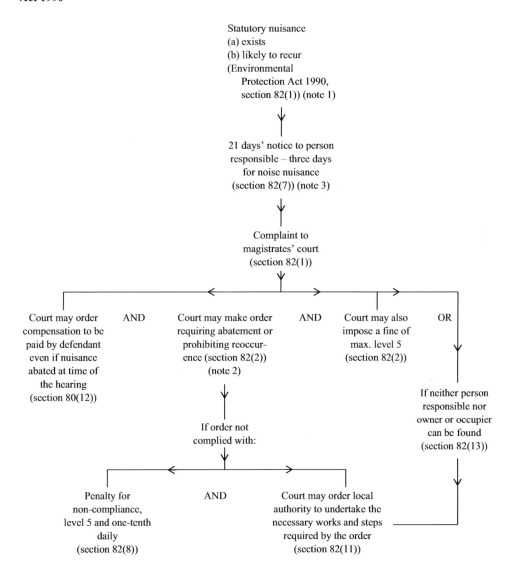

Notes

1. Use of this procedure is not possible to take action to prohibit an anticipated nuisance which does not exist or has not already occurred.
2. The court may also order the closure of a premises rendered unfit for human habitation by the nuisance until they are rendered fit (section 82(3)).
3. This includes noise nuisances for vehicles, machinery or plant in streets.

Chapter 12

WASTE

WASTE ON LAND

Definitions – General – Environmental Protection Act 1990

Environment consists of all, or any, of the following media, namely land, water and the air.

Pollution of the environment means pollution of the environment due to the release or escape (into any environmental medium) from:

(a) the land on which controlled waste is treated;
(b) the land on which controlled waste is kept;
(c) the land in or on which controlled waste is deposited;
(d) fixed plant by means of which controlled waste is treated, kept or disposed of,

of substances or articles constituting or resulting from the waste, and capable (by reason of the quantity or concentrations involved) of causing harm to man or any other living organisms supported by the environment.

This applies in relation to mobile plant by means of which controlled waste is treated or disposed of as it applies to plant on land by means of which controlled waste is treated or disposed of.

For the purposes of the above, **harm** means harm to the health of living organisms or other interference with the ecological systems of which they form part, and in the case of humans includes offence to any of his senses or harm to his property.

Disposal of waste includes its disposal by way of deposit in or on land and, subject to the paragraph below, waste is **'treated'** when it is subjected to any process, including making it re-usable or reclaiming substances from it, and **recycle** is construed accordingly.

Land includes land covered by waters where the land is above the low water mark of ordinary spring tides and references to land on which controlled waste is treated, kept or deposited are references to the surface of the land (including any structure set into the surface).

Substance means any natural or artificial substance, whether in solid or liquid form or in the form of a gas or vapour (section 29).

The definition of waste

Waste means anything that is waste within the meaning of Article 3(1) of Directive 2008/98/EC of the European Parliament and of the council on waste. That is, 'waste' means any substance or object which the holder discards or intends or is required to discard.

Controlled waste means household, industrial and commercial waste or any such waste.

Household waste means waste from:

(a) domestic property, that is to say, a building or self-contained part of a building which is used wholly for the purposes of living accommodation;
(b) a caravan (as defined in section 29(1) of the Caravan Sites Control and Development Act 1960) which usually and for the time being is situated on a caravan site (within the meaning of that Act);
(c) a residential home;
(d) premises forming part of a university or school or other educational establishment;
(e) premises forming part of a hospital or nursing home.

Industrial waste means waste from any of the following premises:

(a) any factory (within the meaning of the Factories Act 1961);
(b) any premises used for the purposes of, or in connection with, the provision to the public of transport services by land, water or air;
(c) any premises used for the purposes of, or in connection with, the supply to the public of gas, water or electricity or the provision of sewerage services; or
(d) any premises used for the purposes of, or in connection with, the provision to the public of postal or telecommunications services.
(e) any mine or quarry or any premises used for agriculture.

Commercial waste means waste from premises used wholly or mainly for the purposes of a trade or business or the purposes of sport, recreation or entertainment excluding:

(a) household waste;
(b) industrial waste;
(c) waste of any other description prescribed by regulations made by the Secretary of State for the purposes of this paragraph (section 75).

The Controlled Waste (England and Wales) Regulations 2012, the Hazardous Waste (England and Wales) Regulations 2005 and Guidance on the classification and assessment of waste, Technical Guidance WM3, clarify the definitions of household, commercial and industrial waste in relation to the application of the duty of care, the obligations of collection authorities and the charging regime.

Guidance on the legal definition of waste and its application, August 2012 gives further guidance.

Waste collection and regulation authorities

The waste collection authorities are:

(a) for any district in England not within Greater London, the council of the district;
(b) in Greater London, the following:

    (i) for any London borough, the council of the borough;
    (ii) for the City of London, the Common Council;
    (iii) for the Temples, the Sub-Treasurer of the Inner Temple and the Under-Treasurer of the Middle Temple, respectively;

(c) for any county or county borough in Wales, the council of the county or county borough;
(d) in Scotland, a council constituted under the Local Government (Scotland) Act 1994 (section 30(3)).

Waste regulation authority:

a) in relation to England, is a reference to the Environment Agency;
b) in relation to Wales, the Natural Resources Body for Wales and
c) in relation to Scotland, is a reference to the Scottish Environment Protection Agency (section 30(1)).

Duty of care etc. on waste

There is a duty on any person who imports, produces, carries, keeps, treats or disposes of controlled waste (and extractive waste) or, as a dealer or broker, has control of such waste, to take all such measures to prevent any waste related contravention in the Environmental Protection Act and Environmental Permitting Regulations or of a condition of an environmental permit; to prevent the escape of the waste from his control or that of any other person; and on the transfer of the waste, to secure that the transfer is only to an authorised person or to a person for authorised transport purposes; and a written description of the waste enabling others to avoid a contravention (section 34(1) and (1A)).

This does not apply to an occupier of domestic property concerning the household waste produced on the property (section 34(2)).

It is the duty of the occupier of any domestic property to take all such measures to secure that any transfer by him of household waste produced on the property is only to an authorised person or someone authorised to transport the waste (section 34(2A)).

Any person who fails to comply with the duty is liable on summary conviction, to a fine not exceeding the statutory maximum; and on conviction on indictment, to a fine (section 34(6)).

Fixed penalty notices

Where a person has failed to comply with a duty to produce documents the authority may serve a notice on that person offering him the opportunity of discharging any liability to conviction for an offence by payment of a fixed penalty (section 34A(1) and (2)).

Where a person is given a notice no proceedings may be instituted for that offence within 14 days of the notice being served; and he cannot be convicted of that offence if he pays the fixed penalty before the end of the period (section 34A(3)).

A notice must give particulars of the alleged circumstances of the offence to give reasonable information of the offence (section 34A(4)).

A notice must also state the period during which proceedings will not be taken for the offence; the amount of the fixed penalty; and the person to whom and the address at which the fixed penalty may be paid (section 34A(5)).

The fixed penalty payable to an enforcement authority is £300 (section 34A(9)), but a different lower amount can be set (section 34A(10)).

The authority to which a fixed penalty is payable may treat it as having been paid if a lesser amount is paid before the end of a period specified by the authority (section 34A(11)).

In any proceedings a certificate which purports to be signed on behalf of the chief finance officer of the enforcement authority, and states that payment of a fixed penalty was or was not received by a date specified in the certificate, is evidence of the facts stated (section 34A(13)).

LITTER

Definitions – Environmental Protection Act 1990

Principal litter authority (PLA) (also see p. 596)

In England and Wales, the following are PLAs:

- (a) a county council;
- (b) a county borough council;
- (c) a district council;
- (d) a London borough council;
- (e) the Common Council of the City of London; and
- (f) the Council of the Isles of Scilly.

This definition includes the unitary authorities in England and Wales. The Secretary of State may, by order, designate other descriptions of local authorities as litter authorities for the purposes of this part; and any such authority shall also be a PLA (section 86(2)).

In Scotland, the PLAs are councils constituted under the Local Government etc. (Scotland) Act 1994, i.e. unitary councils, and joint boards.

Litter

Litter is defined as including:

- (a) the discarded ends of cigarettes, cigars and like products; and
- (b) discarded chewing-gum and the discarded remains of other products designed for chewing (section 98(5A)).

Refuse on land

The Litter (Animal Droppings) Order 1991 includes as refuse, for the whole of Part 4 of the Environmental Protection Act 1990, dog faeces on land of the following description which is not heath or woodland used for the grazing of animals:

- (a) any public walk or pleasure ground;
- (b) any land, whether enclosed or not, on which there are no buildings or of which no more than one-twentieth part is covered with buildings, and the whole or the remainder of which is laid out as a garden or is used for the purposes of recreation;
- (c) any part of the seashore (that is to say, every cliff, bank, barrier, dune, beach, flat or other land adjacent to and above the place to which the tide flows at mean high water springs) which is:
  - (i) frequently used by large numbers of people; and
  - (ii) managed by the person having direct control of it as a tourist resort or recreational facility;
- (d) any esplanade or promenade which is above the place to which the tide flows at mean high water springs;

(e) any land not forming part of the highway, or, in Scotland, a public road, which is open to the air, which the public are permitted to use on foot only, and which provides access to retail premises;

(f) a trunk road picnic area provided by the minister under section 112 of the Highways Act 1980 or, in Scotland, by the Secretary of State under section 55 of the Roads (Scotland) Act 1984;

(g) a picnic site provided by a local planning authority under section 10(2) of the Countryside Act 1968 or, in Scotland, a picnic place provided by an islands or district council or a general or district planning authority under section 2(2)(a)(i) of the Local Government (Development and Finance) (Scotland) Act 1964;

(h) land (whether above or below ground and whether or not consisting of or including buildings) forming or used in connection with off-street parking places provided in accordance with section 32 of the Road Traffic Regulation Act 1984 (section 86).

Relevant land

The duties placed on various public bodies, statutory undertakers and others to keep their land clear, as far as practicable, from litter and refuse relate to the 'relevant land' of those organisations and individuals. Relevant land is defined in section 86 as:

(4) Subject to subsection (8) below, land is 'relevant land' of a principal litter authority if, not being relevant land falling within subsection (7) below, it is open to the air and is land (but not a highway or in Scotland a public road) which is under the direct control of such an authority to which the public are entitled or permitted to have access with or without payment *(see note at (8) below)*.

(5) Land is Crown Land if it is land:

(a) occupied by the Crown Estate Commissioners as part of the Crown Estate;

(b) occupied by or for the purposes of a government department or for naval, military or air force purposes; or

(c) occupied or managed by anybody acting on behalf of the Crown; and is 'relevant Crown land' if it is Crown land which is open to the air and is land (but not a highway or in Scotland a public road) to which the public are entitled or permitted to have access with or without payment; and 'the appropriate Crown authority' for any Crown land is the Crown Estate Commissioners, the minister in charge of the government department or the body which occupies or manages the land on the Crown's behalf, as the case may be *(see note at (8) below)*.

(6) Subject to subsection (8) below, land is 'relevant land' of a designated statutory undertaker if it is the land which is under the direct control of any statutory undertaker of any description which may be designated by the Secretary of State, by order, for the purposes of this Part, being land to which the public are entitled or permitted to have access with or without payment or, in such cases as may be prescribed in the designation order, land in relation to which the public have no such right or permission. *(See the Litter (Statutory Undertakers) (Designation and Relevant Land) Order 1991 (as amended 1992).)*

(7) Subject to subsection (8) below, land is 'relevant land' of a designated educational institution if it is open to the air and is land which is under the direct control of the governing body of or, in Scotland, of such body or of the educational authority responsible for the management of, any educational institution or educational institution of

any description which may be designated by the Secretary of State, by order, for the purposes of this Part. *(The Litter (Designated Educational Institutions) Order 1991.)*

(8) The Secretary of State may, by order, designate descriptions of land which are not to be treated as relevant Crown land or as relevant land of Principal Litter Authorities, of designated statutory undertakers or of designated educational institutions or of any description of any of them. *NB. The Litter (Relevant Land of Principal Litter Authorities and Relevant Crown Land) Order 1991 excludes from the interpretation of relevant land as respects both PLAs and the Crown, land below the high water spring tide level.*

(9) Every highway maintainable at the public expense other than a trunk road which is a special road is a 'relevant highway' and the local authority which is, for the purposes of this Part, 'responsible' for so much of it as lies within its area is, subject to any order under subsection (11) below *(The Highway Litter Clearance and Cleaning (Transfer of Duties) Order 1991):*

(a) in Greater London, the Council of the London borough or the Common Council of the City of London;

(b) in England, outside Greater London, the Council of the district;

(c) in Wales, the Council of the County or County borough; and

(d) the Council of the Isles of Scilly.

(10) In Scotland, every public road other than a trunk road which is a special road is a 'relevant road' and the local authority which is, for the purposes of this Part, 'responsible' for so much of it as lies within its area is, subject to any order under subsection (11) below, the council constituted under section 2 of the Local Government etc. (Scotland) Act 1994.

(11) The Secretary of State may, by order, as respects relevant highways or relevant roads, relevant highways or relevant roads of any class or any part of a relevant highway or relevant road specified in the order, transfer the responsibility for the discharge of the duties imposed by section 89 below from the local authority to the highway or roads authority; but he shall not make an order under this subsection unless:

(a) (except where he is the highway or roads authority) he is requested to do so by the highway or roads authority;

(b) he consults the local authority; and

(c) it appears to him to be necessary or expedient to do so in order to prevent or minimise interference with the passage or with the safety of traffic along the highway or, in Scotland, road in question; and where, by an order under this subsection, responsibility for the discharge of those duties is transferred, the authority to which the transfer is made is, for the purposes of this Part, 'responsible' for highway, road or part specified in the order. *(See various Highway Litter Clearance and Cleaning (Transfer of Duties) Orders.)*

(12) A place on land shall be treated as 'open to the air' notwithstanding that it is covered if it is open to the air on at least one side.

(13) The Secretary of State may, by order, apply the provisions of this Part which apply to refuse to any description of animal droppings in all or any prescribed circumstances subject to such modifications as appear to him to be necessary. *(See the Litter (Animal Droppings) Order 1991.)*

(14) Any power under this section may be exercised differently as respect different areas, different descriptions of land or for different circumstances.

(section 86)

Statutory undertaker

Statutory undertaker means:

(a) any person authorised by any enactment to carry on any railway, light railway, tramway or road transport undertaking;

(b) any person authorised by any enactment to carry on any canal, inland navigation, dock, harbour or pier undertaking; or

(c) any relevant airport operator (within the meaning of Part V of the Airports Act 1986) (section 98(6)).

LITTER OFFENCES

References

Environmental Protection Act 1990, sections 87 and 88, as amended by the Clean Neighbourhoods and Environment Act 2005 (CNEA 2005)
The Environmental Offences (Fixed Penalties) (Miscellaneous Provisions) Regulations 2007
The Environmental Offences (Use of Fixed Penalty Receipts) Regulations 2006
Code of Practice on Litter and Refuse, DEFRA, April 2006
The Litter (Northern Ireland) Order 1994
Clean Neighbourhoods and Environment Act (Northern Ireland) 2011

Extent

These provisions apply to England, Wales and Scotland. Similar provisions apply in Northern Ireland.

Scope

These procedures allow a litter authority to prosecute or apply a fixed penalty to people who drop litter on land and from a vehicle.

Litter offence

Any person who throws down, drops or otherwise deposits litter (definition on p. 597) and leaves anything which causes the defacement of a place by litter is guilty of an offence (section 87(1)).

This applies to land anywhere in the open air regardless of ownership, except in a covered place open to the air on at least one side and to which the public does not have access. It also applies to dropping litter into water such as rivers or lakes and on beaches above the low water mark (section 87).

Fixed penalty schemes

As an alternative to proceeding summarily for the littering offence, an authorised officer of a litter authority may serve a fixed penalty notice on the offender (section 88(1)), which requires the payment of the fine to the local authority within 14 days (section 88(2)). The level of the fixed penalty may be set by the local authority at between £50 and £80, with discounts for early payment and, if paid within 14 days, no further action can be taken. Where an authorised officer proposes to issue a fixed penalty notice, he may require the person to give his name and address and it is an offence not to do so or to give false information (section 88).

LITTERING FROM VEHICLES

A littering offence is committed in respect of a vehicle occurs as a result of litter being thrown, dropped or otherwise deposited from the vehicle (whether or not by the vehicle's keeper) (section 88A(2)).

The local authority may authorise a person to give a penalty notice for a littering offence committed in respect of a vehicle (section 88A(4)).

No proceedings may be instituted for a littering offence in respect of a vehicle if:

(a) a penalty notice has been given to the keeper of the vehicle in respect of which the offence was committed, and
(b) the fixed penalty has been paid or recovered in full (section 87(4D))

Littering from vehicles outside London (Keepers: Civil Penalties) Regulations 2018

Penalty notices

A litter authority may give a penalty notice to a person who is the keeper of a vehicle if the litter authority has reason to believe that a littering offence has been committed in respect of the vehicle on the authority's land (regulation 4(1) and (3)).

A penalty notice is a written notice requiring the person to pay a fixed penalty (regulation 4(2)).

A penalty notice must not be given:

(a) more than 35 days after the day on which the littering offence in question occurred,
(b) if a notice under section 88(1) of the EPA 1990 (which relates to fixed penalty notices for leaving litter) has been given to a person in respect of the same offence (whether or not the person is the vehicle's keeper), or
(c) if a prosecution has been brought against a person under section 87 of the EPA 1990 (offence of littering) in respect of the same offence (whether or not the person is the vehicle's keeper and whether or not the prosecution has concluded or was successful) (regulation 4(5)).

A litter authority may cancel a penalty notice at any time by informing the recipient in writing (regulation 4(6)).

Content of penalty notices

A penalty notice must state:

(a) the circumstances alleged to constitute the littering offence in question, including the registration mark (if known) of the vehicle concerned;
(b) the fixed penalty payment period;

(c) the amount of the fixed penalty if paid within that period;

(d) that the amount of the fixed penalty increases by 100% if not paid within that period;

(e) that the litter authority may recover any fixed penalty not paid within the fixed penalty payment period in court;

(f) any lesser amount;

(g) the date by which the lesser amount must be paid in order for it to be treated as discharging the liability to pay the fixed penalty

(h) the name and address of the person to whom the fixed penalty must be paid and the permissible methods of payment;

(i) that the person to whom the notice is addressed has a right to make representations to the litter authority;

(j) the grounds on which, and the manner in which, representations may be made and the date by which they must be made and

(k) in general terms, the form and manner in which an appeal to an adjudicator may be made (regulation 5(1)).

Penalty amount and payment

The amount of a fixed penalty is the amount specified by the litter authority under section 88(6A)(a) of the EPA 1990 (regulation 6(1)).

But if no amount is specified by the litter authority, the amount of the fixed penalty is £100 (regulation 6(2)).

If a fixed penalty is not paid in full within the fixed penalty payment period, the amount of the fixed penalty increases by 100% with effect from the day after the last day of the fixed penalty payment period (regulation 6(3)).

The fixed penalty payment period is usually 28 days unless there is an appeal (regulation 6(4) and (5)).

An authority may make provision for treating the amount in paragraph (1) or (2) as having been paid in full if a lesser amount is paid within 14 days (regulation 6(6) and (7)). The lesser amount must not be less than £50 (regulation 6(8)).

Recovery of unpaid amounts

Where a litter authority has given a person a penalty notice and the person has not paid the fixed penalty in full within the payment period the litter authority may recover any unpaid amount (being the increased amount referred to in regulation 6(3)) and any related costs awarded by an adjudicator as a civil debt, or as if payable under a County Court order (regulation 7(2)).

Cancelling notices

Where a litter authority cancels or is deemed to have cancelled a penalty notice, the authority must as soon as practicable refund any amount paid in respect of the notice (regulation 9(1)).

This does not apply where an adjudicator has:

(a) given directions to a litter authority requiring the cancellation of a penalty notice under regulation 16(6), and

(b) the directions include directions about the refund of any amount paid in respect of the penalty notice (regulation 9(2)).

A cancellation or deemed cancellation of a penalty notice does not prevent the litter authority from giving a further penalty notice in respect of the same littering offence (regulation 9(3)).

Use of receipts by litter authorities

Sums received by a litter authority may be used by the authority for the purposes of any of its functions which are:

(a) listed in section 96(4)(a) to (c) of the Clean Neighbourhoods and Environment Act 2005, or

(b) specified in regulations under section 96(4)(d) of that Act.

Public service vehicles and licensed taxis etc.

The keeper of a public service vehicle, a hackney carriage or a private hire vehicle is not liable to pay a fixed penalty for a littering offence if the person who committed the offence was a passenger in the vehicle (regulation 12).

Discharge of liability where action taken against person who littered

The liability of a person who is the keeper of a vehicle to pay a fixed penalty for a littering offence in respect of the vehicle is discharged if –

(a) a notice under section 88(1) of the EPA 1990 is subsequently given to a person in respect of the same offence (whether or not the person is the vehicle's keeper), or

(b) a prosecution is subsequently brought against a person under section 87 of the EPA 1990 in respect of the same offence (whether or not the person is the vehicle's keeper and whether or not the prosecution is successful) (regulation 13).

Representations against penalty notice

A person to whom a penalty notice is given may make written representations to the litter authority within 28 days if it appears to the person that one or more of grounds A to L apply (regulation 14(1) and (2)).

Ground A is that the littering offence in question did not occur.

Ground B is that the person was not the keeper of the vehicle at the time of the littering offence because the person became the keeper of the vehicle after the littering offence occurred.

Ground C is that the person was not the keeper of the vehicle at the time of the littering offence because the person had disposed of the vehicle to another person before the littering offence occurred.

Ground D is that the person was not the keeper of the vehicle at the time of the littering offence because the vehicle was a stolen vehicle when the littering offence occurred.

Ground E is that the person –

(a) was engaged in the hiring of vehicles in the course of a business at the time of the littering offence, and

(b) was not the keeper of the vehicle at that time by virtue of a vehicle hire agreement

Ground F is that the person was not the keeper of the vehicle at the time of the littering offence for a reason not mentioned in grounds B to E.

Ground G is that the litter authority was not, by virtue of regulation 4(5), authorised to give the person a penalty notice.

Ground H is that the person is not liable to pay the fixed penalty by virtue of regulation 12.

Ground I is that liability to pay the fixed penalty has been discharged in the circumstances set out in regulation 13.

Ground J is that the fixed penalty exceeds the amount payable under these Regulations.

Ground K is that the litter authority has failed to observe any requirement imposed on it by these Regulations in relation to the imposition or recovery of the fixed penalty.

Ground L is that there are compelling reasons why, in the particular circumstances of the case, the penalty notice should be cancelled (whether or not any of grounds A to K apply) (regulation 14 (3) to (14)).

If ground B applies, the representations must include the name and address of the other person from whom the vehicle was acquired (if known) (regulation 14(15)).

If ground C applies, the representations must include:

(a) the name and address of the other person to whom the vehicle had been disposed of (if known), or

(b) a statement that the name and address of that person is not known (regulation 14(16)).

If ground D applies, the representations must include the crime reference number, insurance claim reference or other evidence of the vehicle's theft (regulation 14(17)).

If ground E applies, the representations must include:

(a) a statement signed by or on behalf of the person to the effect that at the time of the littering offence the vehicle was hired to a named person under a vehicle hire agreement with the person, and

(b) a copy of the vehicle hire agreement (regulation 14(18)).

Functions of litter authority following representations

A litter authority which receives representations must:

(a) consider them and any supporting evidence which the person making the representations provides, and

(b) decide whether or not it accepts that one or more of the grounds in regulation 14 applies (regulation 15(1)).

If the litter authority accepts that one or more of the grounds applies, it must cancel the penalty notice and inform the person who made the representations of the cancellation in writing (regulation 15(2)).

If the litter authority does not accept that one or more of the grounds applies, it must give a notice of rejection to the person who made the representations (regulation 15(3)).

A notice of rejection informs the person who made the representations that the litter authority does not accept that one or more grounds applies (regulation 15(4)).

The notice of rejection must state:

(a) the litter authority's decision and the reasons for it,
(b) that the person has a right to appeal to an adjudicator within the period of 28 days beginning with the day on which the notice of rejection is given,
(c) in general terms, the form and manner in which an appeal to an adjudicator may be made, and
(d) that an adjudicator has power to award costs against a person appealing against the decision set out in the notice of rejection (regulation 15(5)).

The litter authority must carry out these functions within 56 days of representations being received (regulation 15(6)).

If a litter authority fails to comply with this regulation, it is deemed to have:

(a) decided that it accepts that one or more of the grounds applies, and
(b) cancelled the penalty notice (regulation 15(7)).

Appeals against notice of rejection

A person who is given a notice of rejection may appeal to an adjudicator against it (regulation 16(1) and (2)).

The appeal must be made within 28 days of the notice of rejection being given (regulation 16(3) and (4)).

If the adjudicator concludes that one or more of the grounds applies, the appeal must be allowed (regulation 16(5)).

Where an appeal is allowed, the adjudicator may give written directions to the litter authority which the adjudicator considers appropriate for the purpose of giving effect to the decision (regulation 16(6)).

Despite not allowing an appeal, an adjudicator may give a written recommendation to the litter authority that it cancel the penalty notice if the adjudicator is satisfied that there are compelling reasons why the penalty notice should be cancelled (regulation 16(7) and (8)).

An adjudicator must dismiss an appeal if the adjudicator concludes that:

(a) none of the grounds apply, and
(b) there are no compelling reasons why the penalty notice should be cancelled (regulation 16(9)).

Functions of litter authority following adjudication

A litter authority must comply with any direction given to it as soon as practicable (regulation 17(1)).

A litter authority which is the subject of a recommendation must reconsider whether to cancel the penalty notice, taking account of any observations made by the adjudicator (regulation 17(2)).

The authority must inform the appellant and the adjudicator in writing within 35 days:

(a) whether or not it accepts the adjudicator's recommendation,
(b) if it does accept the adjudicator's recommendation, that the penalty notice is cancelled, and

(c) if it does not accept the adjudicator's recommendation, of the reasons for its decision (regulation 17(3)).

No further appeal is allowed (regulation 17(4)).

If a litter authority fails to comply with paragraph (3), it is deemed to have accepted the adjudicator's recommendation and to have cancelled the penalty notice (regulation 17(5)).

Appeal procedure

The Schedule to the Road User Charging Schemes (Penalty Charges, Adjudication and Enforcement) (England) Regulations 2013 applies in respect of appeals made under these regulations (regulation 19(1)).

Evidence produced by a recording device

Evidence may be given by the production of a record produced by a recording device, and a certificate stating the circumstances in which the record was produced, signed by a person authorised to do so by the litter authority which installed the device (regulation 20(1)).

Definitions

Authorised officer means an officer of a litter authority or someone not an employee of the local authority who is authorised in writing by the authority for the purpose of issuing notices under this section (section 88(10)).

Keeper, in relation to a vehicle, means the person by whom the vehicle is kept at the time when the littering offence in question occurs, which in the case of a registered vehicle is to be presumed, unless the contrary is proved, to be the registered keeper (section 88A(9)).

Litter authority for the purpose of this procedure is:

(a) any principal litter authority, other than an English county council or a joint board;

(b) any English county council or joint board designated by the Secretary of State, by order (no such orders have been made), in relation to such area as is specified in the order (not being an area in a National Park);

(c) the Broads Authority (section 88(9)).

Recording device is a camera or other device capable of producing a record of –

(a) the presence of a particular vehicle on the litter authority's land, and

(b) the date and time at which the vehicle is present,

and includes any equipment used in conjunction with the camera or other device for the purpose of producing such a record (regulation 20(2)).

Registered keeper, in relation to a registered vehicle, means the person in whose name the vehicle is registered.

Registered vehicle means a vehicle which is for the time being registered under the Vehicle Excise and Registration Act 1994.

Vehicle means a mechanically propelled vehicle or a vehicle designed or adapted for towing by a mechanically propelled vehicle (section 88A(9)).

FC130 Litter offences – sections 87 and 88 Environmental Protection Act 1990

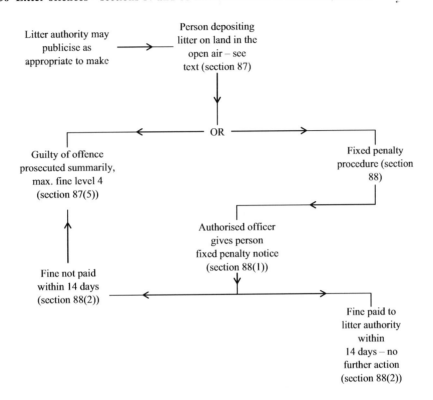

Note

1. Publicity is not a prerequisite to action under sections 87 or 88.

FC131 Littering from vehicles – The Littering From Vehicles Outside London (Keepers: Civil Penalties) Regulations 2018

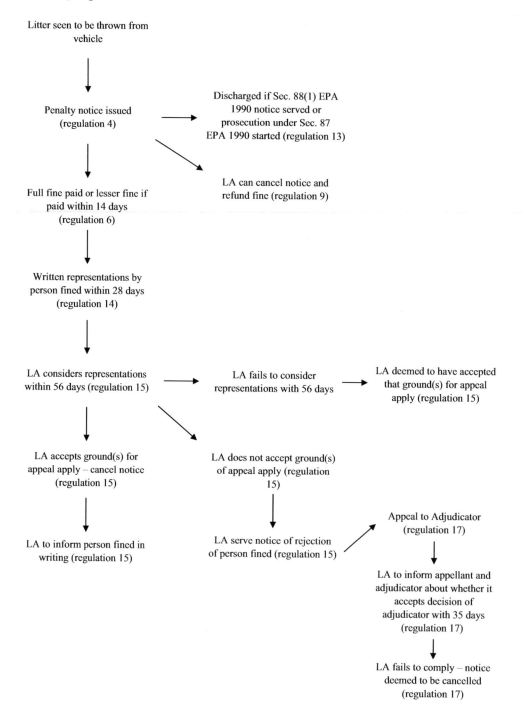

Litter seen to be thrown from vehicle

Penalty notice issued (regulation 4)

Discharged if Sec. 88(1) EPA 1990 notice served or prosecution under Sec. 87 EPA 1990 started (regulation 13)

LA can cancel notice and refund fine (regulation 9)

Full fine paid or lesser fine if paid within 14 days (regulation 6)

Written representations by person fined within 28 days (regulation 14)

LA considers representations within 56 days (regulation 15)

LA fails to consider representations with 56 days

LA deemed to have accepted that ground(s) for appeal apply (regulation 15)

LA accepts ground(s) for appeal apply – cancel notice (regulation 15)

LA does not accept ground(s) of appeal apply (regulation 15)

LA to inform person fined in writing (regulation 15)

LA serve notice of rejection of person fined (regulation 15)

Appeal to Adjudicator (regulation 17)

LA to inform appellant and adjudicator about whether it accepts decision of adjudicator with 35 days (regulation 17)

LA fails to comply – notice deemed to be cancelled (regulation 17)

LITTER ABATEMENT ORDERS

References

Environmental Protection Act 1990, sections 89 and 91
Code of Practice on Litter and Refuse, DEFRA, April 2006
Clean Neighbourhoods and Environment Act (Northern Ireland) 2011

Extent

This provision applies in England, Wales and Scotland. Similar provisions apply in Northern Ireland.

Duty to keep land clear of litter

This procedure allows the public to bring to the attention of a court the fact that a responsible body is not complying with its responsibilities under section 89.

The following are placed under a duty to keep their relevant land (definition on p. 588) clear of litter and refuse, as far as is practicable:

   (a) local authorities as respects highways;
   (b) the Secretary of State as respects trunk roads or other highways for which he is responsible;
   (c) principal litter authorities (PLAs) as respects their relevant land;
   (d) the Crown as respects its relevant land;
   (e) designated statutory undertakers as respects their relevant land;
   (f) governing bodies as designated educational institutions as respects their relevant land (section 89(1)).

In respect of (a) and (b), each local authority and the Secretary of State is also required to keep highways and roads clean, as far as is practicable (section 89(2)).

Litter standards

Regard must be had to the character and use of the land in question as well as to the measures which are practicable in the circumstances (section 89(3)).

The Secretary of State has prescribed standards to which the responsible bodies must have regard in discharging their duties in a Code of Practice made under section 89(7). These establish what are reasonable and generally acceptable levels of cleanliness to be met and are admissible in court in any proceedings for a litter abatement order (section 91(11)). See Code of Practice on Litter and Refuse DETR under section 89 of Environ mental Protection Act, April 2006.

Litter abatement order

Any person aggrieved (other than a PLA) by the defacement by litter or refuse of any of the land set out under 'Duty to keep land clear of litter' (above) or, where appropriate, aggrieved by a highway not being clear may complain to a magistrates' court (section 91(1) and (2)).

FC132 Litter abatement orders (LAOs) – sections 89 and 91 Environmental Protection Act 1990

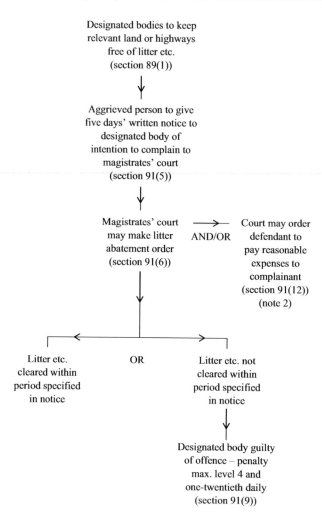

Designated bodies to keep
relevant land or highways
free of litter etc.
(section 89(1))

Aggrieved person to give
five days' written notice to
designated body of
intention to complain to
magistrates' court
(section 91(5))

Magistrates' court   →   Court may order
may make litter   AND/OR   defendant to
abatement order   pay reasonable
(section 91(6))   expenses to
                                  complainant
                                  (section 91(12))
                                  (note 2)

Litter etc.   OR   Litter etc. not
cleared within   cleared within
period specified   period specified
in notice   in notice

Designated body guilty
of offence – penalty
max. level 4 and
one-twentieth daily
(section 91(9))

Notes

1. This procedure allows the public to bring to the attention of a court the fact that a responsible body is not complying with its responsibilities under section 89.

2. Even if a litter abatement order is not made, the court may require the complainant to be compensated if, although the litter etc. had been cleared away, it had been present at the time the complaint was made and there were reasonable grounds for bringing the complaint (section 91(12)).

The person must give the responsible person or body five days' written notice of his intention to make the complaint (section 91(5)).

If the court is satisfied that the land or highway is defaced by litter and refuse or the highway is not clear and that the person has not complied with his statutory duty, taking due account of the Code of Practice, it may make a litter abatement order (section 91(6)).

Effect of a litter abatement order

The litter abatement order requires the litter or refuse to be cleared away or cleaning to be undertaken within a specified period, failing which the person or body to whom the order is addressed is guilty of an offence (section 91(9)).

Payment of expenses

Whether or not the court makes a litter abatement order, it may order the defendant authority to pay expenses to the complainant where it feels that there were reasonable grounds for bringing the complaint and that the defacement or uncleanliness existed at the time the complaint was made (section 91(12)).

ACCUMULATIONS OF RUBBISH

References

Public Health Act 1961, section 34
Environmental Protection Act 1990, section 33

Extent

These provisions apply in England and Wales. Similar provisions apply in Scotland and Northern Ireland.

Scope

The procedure deals with any rubbish which is in the open air and which is seriously detrimental to the amenities of the neighbourhood (section 34(1)).

Notices

Notices must be in writing (Public Health Act 1936, section 283), specify the steps which the local authority proposes to take and indicate the provisions for counter-notice and appeals.

The section makes provision for the service of a counter-notice from the person initially served with the notice. This should state that the person will do the work required to comply with the notice. The local authority can take no action until 28 days from the service of the notice to allow the potential service of a counter-notice or an appeal to the magistrates' court (section 34(2)).

FC133 Accumulations of rubbish – section 34 Public Health Act 1961

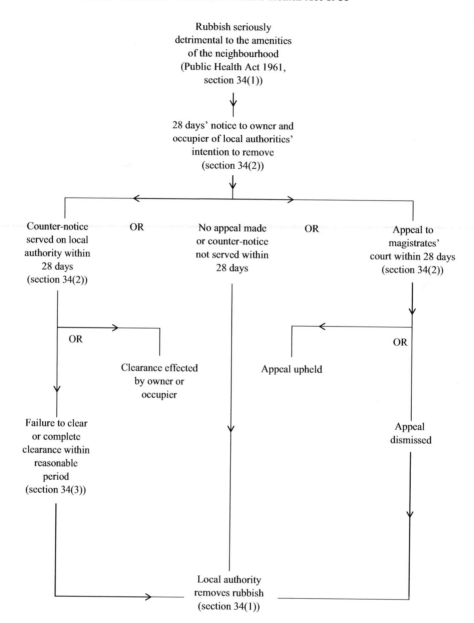

Notes

1. Where rubbish constitutes a statutory nuisance, see FC128.
2. For proper officer's power relating to noxious matter, see Public Health Act 1936, section 79, FC136.
3. For local authority power relating to removal of manure, see Public Health Act 1936, section 80.
4. For removal of rubbish resulting from demolition, see FC71.
5. There is no power for the local authority to recover its costs.
6. The Town and Country Planning Act 1990, section 215, allows a local authority to deal with a site which adversely affects the neighbourhood.

Appeals

Appeal is to the magistrates' court within 28 days of service on the grounds that the local authority was not justified in taking action under this section, or that the steps which it proposes to take are unreasonable (section 34(2)(b)). There is no power for the local authority to recover costs.

Definition

Rubbish means rubble, waste paper, crockery and metal, and any other kind of refuse (including organic matter), but does not include any material accumulated for, or in the course of, any business.

REMOVAL AND DISPOSAL OF ABANDONED REFUSE

References

Refuse Disposal (Amenity) Act 1978, section 6
Removal of Refuse Regulations 1967
Clean Neighbourhoods and Environment Act (Northern Ireland) 2011
The Waste Regulations (Northern Ireland) 2011

Extent

This procedure applies in England, Wales and Scotland (section 13(5)). Similar provisions apply in Northern Ireland.

Scope

These provisions deal with anything (other than motor vehicles) abandoned without lawful authority on any land in the open air or on any other land forming part of a highway (section 6(1)).

Unlike the provisions relating to abandoned cars etc., there is no duty placed here on the local authority to act; the powers are discretionary.

Unoccupied land

Where the land in question appears to be unoccupied, the local authority may immediately remove the articles without notice (section 6(1)).

Occupied land

In this situation, the local authority is first required to serve notice on the occupier stating its intention to remove the abandoned articles. If the occupier objects in writing to the proposal within 15 days of service of the notice, the local authority cannot proceed further unless other legislative procedures are available for the particular circumstances (notes to FC134). If no objection within the 15-day period (see Removal of Refuse Regulations 1967, regulation 4) is received, the local authority may remove the articles (section 6(2)).

FC134 Removal and disposal of abandoned refuse – section 6 Refuse Disposal (Amenity) Act 1978

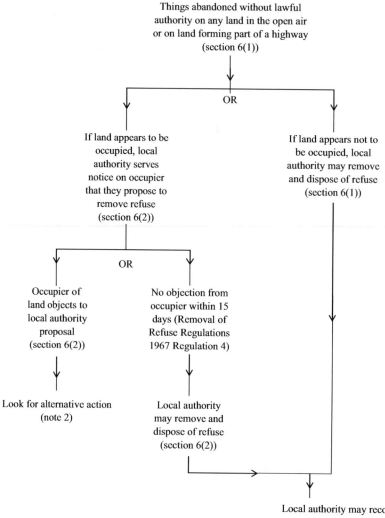

Things abandoned without lawful
authority on any land in the open air
or on land forming part of a highway
(section 6(1))

OR

If land appears to be
occupied, local
authority serves
notice on occupier
that they propose to
remove refuse
(section 6(2))

If land appears not to
be occupied, local
authority may remove
and dispose of refuse
(section 6(1))

OR

Occupier of
land objects to
local authority
proposal
(section 6(2))

No objection from
occupier within 15
days (Removal of
Refuse Regulations
1967 Regulation 4)

Look for alternative action
(note 2)

Local authority
may remove and
dispose of refuse
(section 6(2))

Local authority may recover cost of
removal and disposal from:
(a) any person by whom it was put
in the place from which it was
removed; or
(b) any person convicted of an
offence under section 2, Refuse
Disposal (Amenity) Act 1978 of
unlawful dumping of refuse
(sections 6(3) and 5(2) and (3))

Notes
1. This procedure does not apply to abandoned cars – see FC138.
2. The provisions do not provide for any action beyond this point but see following notes for alternative procedures.
3. For accumulations constituting a statutory nuisance, see FC128.
4. For removal of rubbish resulting from demolition, see FC71.
5. For removal of rubbish seriously detrimental to the amenities of a neighbourhood, see FC133.
6. This procedure applies in Scotland but not in Northern Ireland.

Notice

The form of the notice is specified in the Removal of Refuse Regulations 1967, which, although made under the Civic Amenities Act 1967 by a provision now repealed, has been continued in force by section 12(3) of the Refuse Disposal (Amenity) Act 1978. With appropriate rewording, this form should be used.

Recovery of costs

The local authority is entitled to recover its costs of removal from either:

(a) the person who left the articles on the land; or

(b) any person convicted under section 2(1) of the 1978 Act for illegally abandoning the articles (section 6(4)).

Sums are recoverable as a simple contract debt (section 5(2)) and courts making a conviction under section 2(1) may, on application by the local authority, order the person convicted to reimburse to the local authority their costs of removal (section 6(6)).

Penalties

This procedure itself carries no penalties but prosecution for abandoning the articles may be taken under section 2(1)(b) of the 1978 Act, with penalties not exceeding level 4 on the standard scale or imprisonment for up to three months, or both.

REMOVAL OF CONTROLLED WASTE ON LAND

References

Environmental Protection Act 1990, section 59, as amended by the Anti-Social Behaviour Act 2003 and the Clean Neighbourhoods and Environment Act 2005 (CNEA 2005)
Clean Neighbourhoods and Environment Act (Northern Ireland) 2011
The Waste Regulations (Northern Ireland) 2011

Extent

This provision applies to England, Wales and Scotland. Similar provisions apply in Northern Ireland.

Scope

These provisions apply whenever controlled waste (see 'The definition of waste' on p. 584) has been deposited on land without the authority of a waste management licence under section 35, and may be operated by either a waste collection authority (WCA) or the Environment Agency as the waste regulation authority (WRA) (section 59(1)). Section 59A enables the Secretary of State to issue directions as to which categories of waste should be given priority under this procedure.

FC135 Removal of controlled waste on land – section 59 Environmental Protection Act 1990

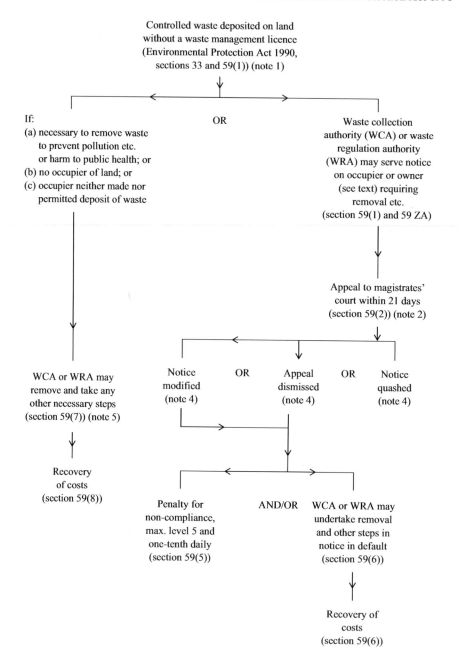

Notes
1. For definition of controlled waste, see p. 584.
2. The notice is suspended pending determination of the appeal (section 59(4)).
3. For procedure for removing abandoned vehicles, see FC138A and FC138B.
4. Either party may appeal to the Crown Court against the decision of the magistrates' court (section 73(1)).
5. Once removed, the waste belongs to the WCA or WRA and may be dealt with accordingly (section 59(9)).
6. For service of notices, see p. 40.

Removal by notice

The WCA or WRA may serve written notice on the occupier of the land requiring:

(a) removal of the waste within a specified period of not less than 21 days; and/or
(b) the taking of specified steps to eliminate or reduce the consequences of the deposit (section 59(1)).

Where:

(a) there is no occupier; or
(b) the occupier cannot be found; or
(c) the occupier has successfully appealed against a notice served on him,

the notice may be served on the owner (section 59ZA).

Where there is non-compliance, the WCA or WRA may remove the waste and/ or prosecute. Costs of acting in default may be recovered (section 59(5), (6) and (7)).

Removal etc. by a WCA or WRA

This may be effected, without recourse to the notice procedure, where:

(a) there is no occupier of the land affected; or
(b) the occupier neither made nor knowingly permitted the deposit; or
(c) removal, and/or steps to eliminate or reduce the consequences of the deposit is necessary forthwith to remove or prevent pollution of land, water or air or harm to human health – see definitions, p. 584 (section 59(7)).
(d) Recovery of costs both of removing the waste and of taking steps to eliminate or reduce the consequences of the deposit is possible from any person who caused, or knowingly permitted, the deposit (section 59(8)).

Appeals

Appeal is to the magistrates' court within 21 days of service and no particular grounds of appeal are specified. The court is required to quash the notice if either:

(a) the appellant neither deposited nor knowingly caused or permitted the deposit; or
(b) there is a material defect in the notice (section 59(3)).

The notice is suspended pending determination of the appeal (section 59(4)).

REMOVAL OF NOXIOUS MATTER

References

Public Health Act 1936, section 79
Note: Although provision was made in the Control of Pollution Act 1974, schedule 4 for section 79 of the Public Health Act 1936 to be repealed, this has not been implemented and this procedure remains available.

FC136 Removal of noxious matter – section 79 Public Health Act 1936

Noxious matter on any premises
(Public Health Act 1936, section
79(1))

Notice from proper officer to
owner or occupier requiring
removal within 24 hours
(section 79(1))

If notice not
complied with

Proper officer may remove
accumulation
(section 79(1))

Local authority may recover
expenses from owner or
occupier (section 79(2))

Notes

1. For accumulations constituting a statutory nuisance, see FC128.
2. For removal of rubbish resulting from demolition, see FC71.
3. For removal of abandoned articles, see FC134.
4. For removal of abandoned cars, see FC138A and 138B.

Environmental Protection Act 1990
Clean Neighbourhoods and Environment Act (Northern Ireland) 2011
The Waste Regulations (Northern Ireland) 2011

Extent

This provision applies to England and Wales. Similar provisions apply in Scotland and Northern Ireland.

Scope

The procedure may be used where the proper officer of the local authority considers that any accumulation of noxious matter ought to be removed. The word 'noxious' is not defined, but according to the Concise Oxford Dictionary means 'harmful, unwholesome' (section 79(1)).

Notice

The notice is served by the proper officer on either the owner or the occupier of the premises concerned, requiring removal of the accumulation within 24 hours (section 79(1)). Notices must be in writing (Public Health Act 1936, section 283). There is no provision for appeal or for penalties.

Default

If the accumulation is not removed within 24 hours from the service of the notice, the proper officer may have it removed (section 79(1)).

Recovery of costs

The local authority can recover from either the owner or the occupier the expenses of any reasonable actions taken by its proper officer in acting in default (section 79(2)).

Definition

Proper officer means the specified officer in the Act (Local Government Act 1972, schedule 29, part 1).

ABANDONED TROLLEYS

References

Environmental Protection Act 1990, section 99 and schedule 4, as amended by the Clean Neighbourhoods and Environment Act 2005
DEFRA Guidance on the Clean Neighbourhoods and Environment Act 2005 – Abandoned Trolleys 2006
Clean Neighbourhoods and Environment Act (Northern Ireland) 2011

FC137 Abandoned trolleys – section 99 Environmental Protection Act 1990

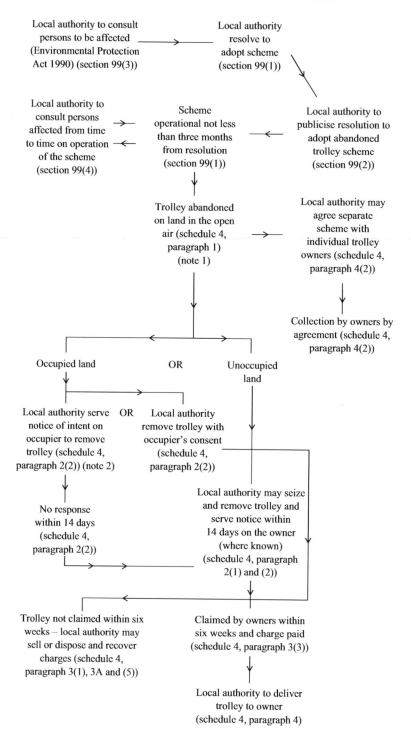

Local authority to consult persons to be affected (Environmental Protection Act 1990) (section 99(3))

Local authority resolve to adopt scheme (section 99(1))

Local authority to consult persons affected from time to time on operation of the scheme (section 99(4))

Scheme operational not less than three months from resolution (section 99(1))

Local authority to publicise resolution to adopt abandoned trolley scheme (section 99(2))

Trolley abandoned on land in the open air (schedule 4, paragraph 1) (note 1)

Local authority may agree separate scheme with individual trolley owners (schedule 4, paragraph 4(2))

Collection by owners by agreement (schedule 4, paragraph 4(2))

Occupied land   OR   Unoccupied land

Local authority serve notice of intent on occupier to remove trolley (schedule 4, paragraph 2(2)) (note 2)

OR

Local authority remove trolley with occupier's consent (schedule 4, paragraph 2(2))

No response within 14 days (schedule 4, paragraph 2(2))

Local authority may seize and remove trolley and serve notice within 14 days on the owner (where known) (schedule 4, paragraph 2(1) and (2))

Trolley not claimed within six weeks – local authority may sell or dispose and recover charges (schedule 4, paragraph 3(1), 3A and (5))

Claimed by owners within six weeks and charge paid (schedule 4, paragraph 3(3))

Local authority to deliver trolley to owner (schedule 4, paragraph 4)

Notes
1. For exceptions, see text.
2. For service of notices under Environmental Protection Act 1990, see p. 40.

Extent

This section applies to England, Scotland and Wales. Similar provisions apply in Northern Ireland.

Scope

This procedure is discretionary and the powers may only be used after the scheme has been adopted by resolution of the local authority (section 99(1)). The powers adopted allow the local authority to seize and deal with abandoned shopping trolleys.

Consultation

Before passing a resolution adopting the scheme, the local authority must consult with persons likely to be affected or their representatives, e.g. supermarket operators (section 99(3)).

There is also a duty on the local authority once a scheme has been adopted to consult the same people from time to time on the operation of the scheme (section 99(4)).

Publicity

Following the resolution of adoption, the scheme must be publicised in at least one local newspaper (section 99(2)).

Voluntary schemes

The local authority is able to reach separate agreements with each owner of trolleys whereby the operators agree to arrange collection themselves outside of the statutory scheme. In such cases, if abandoned trolleys are collected by the local authority, for any trolleys removed by them within the times specified in the statutory scheme, no charge is payable (schedule 4, paragraph 4(2)).

Abandoned trolleys

Trolleys which may be dealt with by the local authority under a scheme are those found to be abandoned in the open air except:

(a) on land in which the trolley owner has a legal estate;
(b) in a trolley park in an off-street parking place;
(c) in any other trolley park designated by the local authority; or
(d) on land used for the purposes of the undertaking by the owner of the luggage trolley (schedule 4, paragraph 1).

Removal

On unoccupied land, the local authority may remove the trolley immediately. However, on occupied land a notice of intention must be served on the occupier, who has 14 days to respond. In the absence of a response within that period, the local authority may remove the trolley (schedule 4, paragraph 2(2)).

Retention, return and disposal

Once seized and removed, the local authority must serve notice within 14 days on the owner notifying him of the removal of the trolley, the place where it is being kept and notification that, if not claimed within six weeks, it will dispose of it (schedule 4, paragraph 3(2)).

If claimed within the six weeks, the trolley is delivered back to its owner by the local authority and a charge made (schedule 4, paragraph 3(3) and (4)).

If not claimed within the six-week period and the local authority has made reasonable enquiries to ascertain ownership, the local authority may sell or otherwise dispose of the trolley (schedule 4, paragraph 3(1)(b) and (5)).

Charges

Charges to be levied by the local authority should be such as to recover the costs of removing, storing and disposing of trolleys under the scheme (schedule 4, paragraph 3A and 4(2)).

Definitions

The scheme applies to both shopping and luggage trolleys defined as:

Luggage trolley means a trolley provided by a person carrying on an undertaking of any railway, light railway, tramway, road transport undertaker or airport operator to travellers for use by them for carrying their luggage to, from or within the premises used for the purpose of the undertaking, not being a trolley which is power-assisted.

Shopping trolley means a trolley provided by the owner of a shop to customers for use by them for carrying goods purchased at the shop, not being a trolley which is power-assisted (schedule 4, paragraph 5).

ABANDONED VEHICLES

References

Refuse Disposal (Amenity) Act 1978, sections 3 to 5, as amended by the Clean Neighbourhoods and Environment Act 2005

Removal and Disposal of Vehicles Regulations 1986 (as amended)

Road Traffic Regulation Act 1984

Removal, Storage and Disposal of Vehicles (Prescribed Sums and Charges) Regulations 2008

Guidance to LAs on the Clean Neighbourhoods and Environment Act 2005 – Abandoned Vehicles, DEFRA, 2006

Clean Neighbourhoods and Environment Act (Northern Ireland) 2011

Extent

This procedure applies in England and Wales (slightly different provisions for Scotland) (section 13(5)). Similar provisions apply in Northern Ireland.

Scope

Local authorities are placed under a statutory duty to remove motor vehicles unlawfully abandoned on any land in the open air or on any other land forming part of a highway except

FC138A Part 1 Abandoned vehicles: removal – sections 3 and 4 Refuse Disposal (Amenity) Act 1978

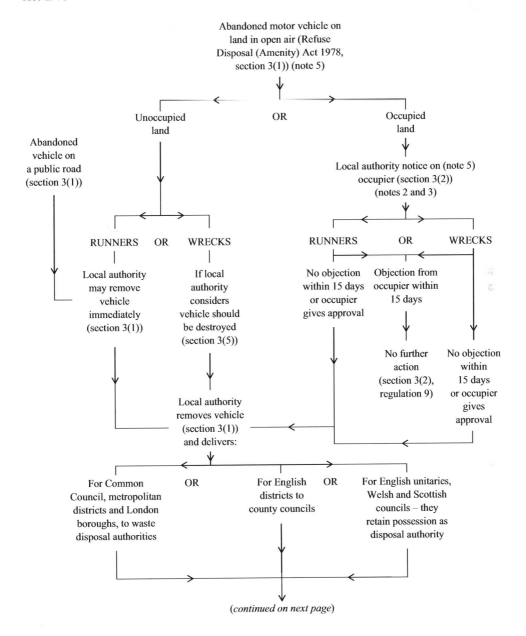

(continued on next page)

Notes
1. All regulation numbers refer to the Removal of Vehicles Regulations 1986.
2. If the land is occupied by the local authority itself, no 'occupier' notice is required.
3. For powers to remove vehicles illegally, obstructively or dangerously parked, see Road Traffic Regulation Act 1984 and Removal of Vehicles Regulations 1986 (as amended 1993).
4. For power of entry, see section 8.
5. There is a penalty for this offence and an alternative fixed penalty procedure – see text.

FC138B Part 2 Abandoned vehicles: disposal

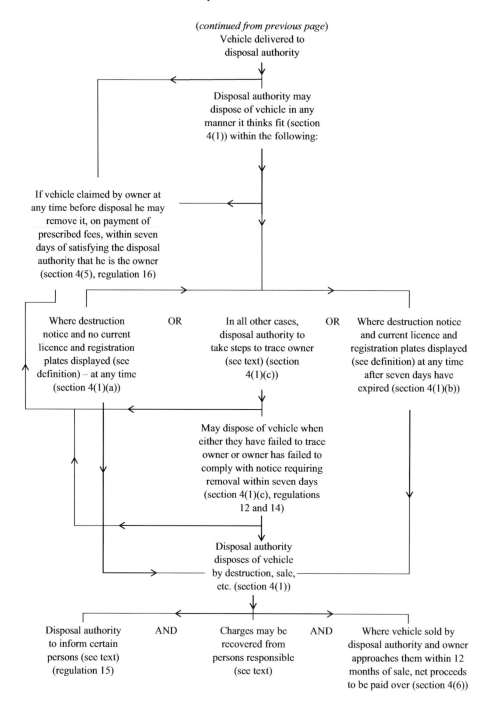

(*continued from previous page*)
Vehicle delivered to
disposal authority

Disposal authority may
dispose of vehicle in any
manner it thinks fit (section
4(1)) within the following:

If vehicle claimed by owner at
any time before disposal he may
remove it, on payment of
prescribed fees, within seven
days of satisfying the disposal
authority that he is the owner
(section 4(5), regulation 16)

Where destruction
notice and no current
licence and registration
plates displayed (see
definition) – at any time
(section 4(1)(a))

OR

In all other cases,
disposal authority to
take steps to trace owner
(see text) (section
4(1)(c))

OR

Where destruction notice
and current licence and
registration plates displayed
(see definition) at any time
after seven days have
expired (section 4(1)(b))

May dispose of vehicle when
either they have failed to trace
owner or owner has failed to
comply with notice requiring
removal within seven days
(section 4(1)(c), regulations
12 and 14)

Disposal authority
disposes of vehicle
by destruction, sale,
etc. (section 4(1))

Disposal authority
to inform certain
persons (see text)
(regulation 15)

AND

Charges may be
recovered from
persons responsible
(see text)

AND

Where vehicle sold by
disposal authority and owner
approaches them within 12
months of sale, net proceeds
to be paid over (section 4(6))

that, where the cost of removal to the nearest convenient carriageway would be unreasonably high, the procedure is discretionary (section 3(1) and (3)).

There is no legal definition of 'abandoned' but it is suggested that this would include:

(a) untaxed, with any or some of the following:
(b) no current vehicle keeper;
(c) stationary for a significant time;
(d) significantly damaged, run down or unroadworthy;
(e) burned out;
(f) lacking one or more number plates;
(g) containing waste.

The removal of vehicles broken down or illegally, obstructively or dangerously parked are matters for the police and traffic wardens under the Removal and Disposal of Vehicles Regulations 1986 (as amended). Traffic officers have powers to remove cars in certain circumstances under the Removal and Disposal of Vehicles (Traffic Officers) (England) Regulations 2008.

Unoccupied land

If the vehicle is on unoccupied land, it may be removed immediately without notice (section 3(1)).

Occupied land

If the vehicle is situated on occupied land, a further notice procedure is required. In this case, the local authority must give at least 15 days' written notice to the occupier (section 3(2)).

Vehicles abandoned on roads

No notice of intended removal is required for a vehicle abandoned on a road within the meaning of the Road Traffic Regulation Act 1984. In essence, this is any length of highway or any road to which the public has access and will therefore include all highways to which the public has a right to pass. Access roads on housing estates and car parking areas are therefore likely to be covered by this definition (section 3(2A)).

Delivery of vehicle to disposal authority

Having removed the vehicle, the local authority must deliver it by arrangement:

(a) in London and English metropolitan areas, to the waste disposal authorities; or
(b) in the rest of England, to the county council.

In England, Wales and Scotland, the unitary authority retains possession and continues to operate the procedure from that point as refuse disposal authority (section 3(7) and (8)).

Arrangements

Where two authorities are involved, delivery to the waste disposal authority is required to be in accordance with arrangements agreed between the local authority and the waste disposal authority, and these arrangements may include sharing of expenses incurred or income received. In default of agreement, there is provision for arbitration or, in the case of London, reference to the Secretary of State (sections 3(7)).

Penalties

No penalties are specified for not recovering vehicles after notices have been served but it is an offence to abandon a vehicle in the open air or on land forming part of a highway, and penalties are not exceeding level 4 on the standard scale and/or imprisonment for up to three months (section 2(1)).

Fixed penalty notices

Fixed penalty procedures in sections 2A, 2B and 2C of the Refuse Disposal (Amenity) Act 1978 are available to local authorities for the offence of abandoning a vehicle. The penalty is set at £200 but may be amended by order, and the local authority may treat the amount as having been paid if a lesser amount is paid before the end of a shorter period as it may specify.

Safe custody

For vehicles other than those intended for destruction with the appropriate notice attached, the local authority or waste disposal authority having custody of the vehicle is responsible for taking reasonable steps to ensure its safe custody (section 3(8)). In the event of this being in question, approaches may be made by the person concerned to the Secretary of State, who must hold a public inquiry (sections 1(5) and 3(7)).

Disposal of vehicles

The waste disposal authority may dispose of vehicles delivered to them in any manner it thinks fit in the timescales shown in FC138, part 2 (section 4(1)). The sale of a vehicle by the local authority, e.g. at an auction, may be an acceptable form of disposal for a suitable vehicle.

The steps to be taken to trace the owner (i.e. other than where a destruction notice is given) are specified in regulation 12 of the Removal and Disposal of Vehicles Regulations 1986.

Notices to owners requiring removal of the vehicle are to be served by either:

(a) delivering it to the owner;
(b) leaving it at the usual or last known place of abode;
(c) sending it by a pre-paid registered letter or recorded delivery addressed to him at his usual or last known abode;
(d) where the owner is an incorporated company or body, delivering it to the secretary or clerk or sending it addressed to that person by either pre-paid registered letter or recorded delivery at their registered or principal office (regulation 13).

The persons to whom information must be given following the disposal of the vehicle are detailed in regulation 15 of the Removal and Disposal of Vehicles Regulations 1986.

Charges for storage and removal

Statutory charges are prescribed from time to time by the Removal, Storage and Disposal of Vehicles (Prescribed Charges) Regulations 2008. Charges may be recovered from the person responsible:

(a) the owner at the time the vehicle was abandoned unless he shows that he was not aware of or concerned in the abandonment; or

(b) any person who put it in the place from which it was removed; or

(c) persons convicted of abandoning the vehicle under section 2(1) (section 5(4) of the Refuse Disposal (Amenity) Act 1978).

Definitions

Motor vehicle means a mechanically propelled vehicle intended or adapted for use on roads, whether or not it is in a fit state for such use, and includes any trailer intended or adapted for such use as an attachment to such a vehicle, any chassis or body, with or without wheels, appearing to have formed part of such a vehicle or trailer and anything attached to such a vehicle or trailer (section 11(1)). This definition includes caravans.

RECEPTACLES FOR COMMERCIAL AND INDUSTRIAL WASTE

References

Environmental Protection Act 1990, section 47, as amended by the Clean Neighbourhoods and Environment Act 2005 (CNEA 2005)
The Environmental Offences (Fixed Penalties) (Miscellaneous Provisions) Regulations 2007
DEFRA Guidance on the CNEA 2005 – Waste 2006 and Fixed Penalty Notices 2006
The Waste and Contaminated Land (Northern Ireland) Order 1997

Extent

This provision applies to England, Wales and Scotland. Similar provisions apply in Northern Ireland.

Scope

This procedure may be applied where:

(a) commercial or industrial waste is stored at a premises; and

(b) if not stored in receptacles of a particular kind, the waste is likely to cause nuisance or be detrimental to the amenities of the neighbourhood (section 47(2)).

Power of waste collection authority (WCA) to provide receptacles

A WCA has power to supply receptacles for commercial and industrial waste (see 'The definition of waste' on p. 584) which the WCA has been requested to collect. A charge for a receptacle for industrial waste is mandatory, but the charge for a receptacle for commercial waste is discretionary (section 47(1)). The charge may be on a purchase or rental basis.

Public health

WCAs are under a duty to collect commercial waste when asked to do so by an occupier but have discretion on the collection of industrial waste. Where collections are made, the WCA must make a charge (section 45).

Contents of notice

The notice may include requirements covering:

(a) size, construction and maintenance of the receptacle (the kind and number must be reasonable);
(b) the siting of the receptacle for emptying and access for that purpose;
(c) placing of the receptacle on the highway, with the consent of the highway authority and in accordance with arrangements for damage liability;
(d) the substances which may or may not be put into the receptacles and the precautions to be taken;
(e) the steps to be taken by occupiers to facilitate collection (section 47(3) and (4)).

The time allowed for compliance (where appropriate) must be reasonable, and should not be less than the period within which an appeal may be made, i.e. 21 days.

Appeals

Appeal may be made to the magistrates' court within 21 days of receipt of the notice on the following grounds:

(a) any requirement is unreasonable;
(b) the waste is not likely to cause nuisance or be detrimental to the amenities of the neighbourhood (section 47(7) and (8)).

Upon appeal, the notice is suspended until the appeal is determined and, in any subsequent proceedings, no question concerning the reasonability of any requirement can be entertained (section 47(9)(c)).

Penalties

A person not complying with the requirements of the notice without reasonable excuse is liable, on conviction, to a fine not exceeding level 3 (section 47(6)).

Fixed penalty notices are available to an authorised officer where a person has not complied with the notice. This offers the offender the opportunity to discharge his liability to conviction by payment of the fixed amount (section 47ZA).

The amount is at the discretion of the local authority and is either £100 with discounts for early payment (section 47ZB), or between £75 and £100 (Environmental Offences (Fixed Penalties) (Miscellaneous Provisions) Regulations 2007 as amended).

Definition

Receptacle includes a holder for receptacles (section 47(10)).

FC139 Receptacles for commercial and industrial waste – section 47 Environmental Protection Act 1990

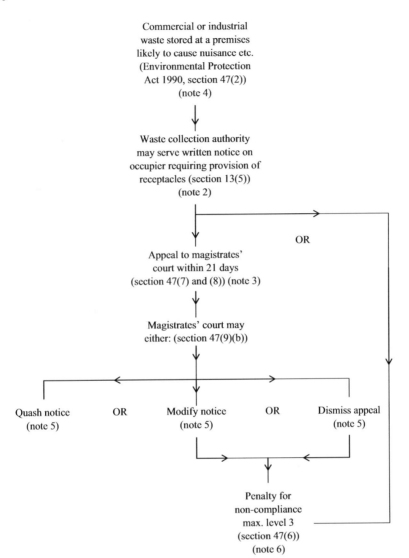

Commercial or industrial
waste stored at a premises
likely to cause nuisance etc.
(Environmental Protection
Act 1990, section 47(2))
(note 4)

Waste collection authority
may serve written notice on
occupier requiring provision of
receptacles (section 13(5))
(note 2)

OR

Appeal to magistrates'
court within 21 days
(section 47(7) and (8)) (note 3)

Magistrates' court may
either: (section 47(9)(b))

Quash notice   OR   Modify notice   OR   Dismiss appeal
(note 5)   (note 5)   (note 5)

Penalty for
non-compliance
max. level 3
(section 47(6))
(note 6)

Notes
1. For provision of receptacles for household waste, see FC140.
2. For service of notices, see p. 40.
3. The notice is suspended pending determination of the appeal (section 47(9)(a)).
4. For definition of industrial and commercial waste, see p. 585.
5. Either party may appeal to the Crown Court against the decision of the magistrates' court (section 73(1)).
6. There is a fixed penalty option – see text.

RECEPTACLES FOR HOUSEHOLD WASTE

References

Environmental Protection Act 1990, section 46, as amended by the Clean Neighbourhoods
    and Environment Act 2005 (CNEA 2005)
The Environmental Offences (Fixed Penalties) (Miscellaneous Provisions) Regulations 2007
DEFRA Guidance on the CNEA 2005 – Waste 2006 and Fixed Penalty Notices 2006
The Waste and Contaminated Land (Northern Ireland) Order 1997

Extent

This provision applies to England, Wales and Scotland. Similar provisions apply in Northern
Ireland.

Scope

This procedure may be used by a waste collection authority (WCA) to require the occupier of
any premises from which it has the duty to collect household waste ('The definition of waste'
on p. 584) to place the waste for collection in receptacles of the kind and number specified
in the notice (section 46(1)).

WCAs have a duty to collect household waste. No charge may be made except in
prescribed cases detailed in the Controlled Waste Regulations 1992 (section 45).

Content of notice

The kind and number of receptacles to be used must be reasonable, but separate receptacles
can be required for waste which is to be recycled (section 46(2)). The notice may include
requirements dealing with:

(a) the size, construction and maintenance of receptacles;
(b) the placing of them to facilitate emptying and access for that purpose;
(c) the placing of receptacles on the highway (with the consent of the highway authority
    and with an arrangement for damage liability);
(d) limitation of the waste which may or may not be put into the receptacles and the
    precautions to be taken with particular substances or articles; and
(c) the steps to be taken by occupiers to facilitate the collection of the waste (section
    46(4) and (5)).

Provision of receptacles

The WCA has a choice of the way in which the necessary receptacles are to be provided:

(a) by the WCA free of charge;
(b) with the occupier's agreement, by the WCA on payment by the occupier (single pur-
    chase or periodic payments);
(c) by the occupier, including where the occupier refuses an agreement under (b) (section
    46(3)).

Appeal

The occupier may appeal to the magistrates' court within 21 days of the notice being served or, where required to enter into an agreement under (b) above, within 21 days of the expiry of the period allowed by the WCA to enter into the agreement. The grounds of appeal are:

(a) that any requirement is unreasonable; and
(b) that the receptacles already used are adequate (section 46(7) and (8)).

Upon appeal, the notice is suspended until the appeal is determined and in any subsequent proceedings for non-compliance with the notice, no question regarding the reasonability of any requirement can be entertained (section 46(9)).

Penalties

A person not complying with the requirements of the notice without reasonable excuse is liable, on conviction, to a fine not exceeding level 3 on the standard scale (section 46(6)). Fixed penalty notices are available to an authorised officer where a person has not complied with the notice. This offers the offender the opportunity to discharge his liability to conviction by payment of the fixed amount (section 47ZA). The amount is at the discretion of the local authority between £75 and £110, with discounts for early payment (Environmental Offences (Fixed Penalties) (Miscellaneous Provisions) Regulations 2007).

Written warnings and penalties

Where an authorised officer of a waste collection authority is satisfied that:

(a) a person has failed without reasonable excuse to comply with a requirement to place waste in a receptacle for household waste, and
(b) the person's failure to comply –

  (i) has caused, or is or was likely to cause, a nuisance, or
  (ii) has been, or is or was likely to be, detrimental to any amenities of the locality,

the authorised officer may give a written warning to the person (section 46A(1) and (2)).
A written warning must:

(a) identify the requirement with which the person has failed to comply,
(b) explain the nature of the failure to comply,
(c) explain how the failure to comply has had, or is or was likely to cause a nuisance, or be detrimental to any amenities of the locality.
(d) if the failure to comply is continuing, specify the period within which the requirement must be complied with and explain the consequences of the requirement not being complied with within that period, and
(e) whether or not the failure to comply is continuing, explain the consequences of the person subsequently failing to comply (section 46A(3)).

Where a written warning is continuing, an authorised officer may require the person to pay a fixed penalty to the authority (section 46A(4)).

Where a person has been required to pay a fixed penalty and that requirement has not been withdrawn on appeal, an authorised officer of the authority may require the person to pay a further fixed penalty to the authority if satisfied that the failure to comply is still continuing one year after the written warning being given (section 46A(5)).

Further fixed penalties are available if the person fails again without reasonable excuse to comply with the requirement (section 46A(7) and (8)).

The amount of the monetary penalty is specified by the waste collection authority, or if no amount is so specified, £60 (section 46B(1)).

A lesser amount can be paid before the end of a period specified by the authority if, it resolves to (section 46B(2)).

A fixed penalty is recoverable summarily as a civil debt and as if it were payable under an order of the High Court or the County Court, if the court in question orders (section 46B(6)).

Procedure regarding notices of intent and final notices

Before requiring a person to pay a fixed penalty, an authorised officer must serve on the person notice of intention to do so (section 46C(1)).

A notice of intent must contain information about:

(a) the grounds for proposing to require payment of a fixed penalty,
(b) the amount of the penalty that the person would be required to pay, and
(c) the right to make representations (section 46C(2)).

A person on whom a notice of intent is served may make representations to the authorised officer as to why payment of a fixed penalty should not be required (section 46C(3)).

Representations must be made within 28 days of notice of intent effected (section 46C(4)).

In order to require a person to pay a fixed penalty, an authorised officer must serve on the person a 'final notice' (section 46C(5)).

A final notice cannot be served within the 28 days of the notice of intent is effected (section 46C(6)).

Before serving a final notice on a person, an authorised officer must consider any representations made by the person (section 46C(7)).

The final notice must contain information about –

(a) the grounds for requiring payment of a fixed penalty,
(b) the amount of the penalty,
(c) how payment may be made,
(d) the period within which payment is required to be made (which must not be less 28 days beginning with the day service of the final notice is effected),
(e) any provision giving a discount for early payment,
(f) the right to appeal, and
(g) the consequences of not paying the penalty (section 46C(8)).

Appeals against penalties

A person on whom a final notice is served may appeal to the First-Tier Tribunal against the decision to require payment of a fixed penalty (section 46D(1)).

On an appeal the First-Tier Tribunal may withdraw or confirm the requirement to pay the fixed penalty (section 46D(2)).

FC140 Receptacles for household waste – section 46 Environmental Protection Act 1990

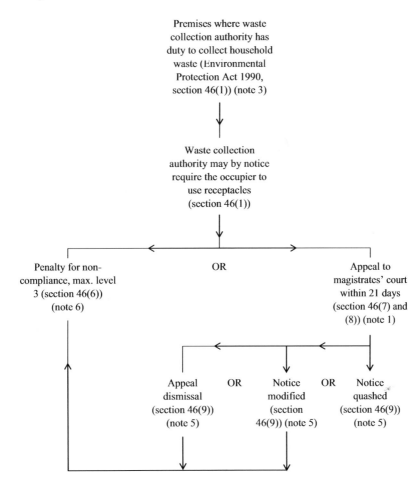

Notes

1. The notice is suspended pending determination of the appeal (section 46(9)(a)).
2. For the provision of receptacles for industrial and commercial waste, see FC139.
3. For definition of waste, see p. 584.
4. For service of notices, see p. 40.
5. Either party may appeal to the Crown Court against the decision of the magistrates' court (section 73(1)).
6. There is a fixed penalty option – see text.

FC141 Receptacles for household waste – written warnings etc.

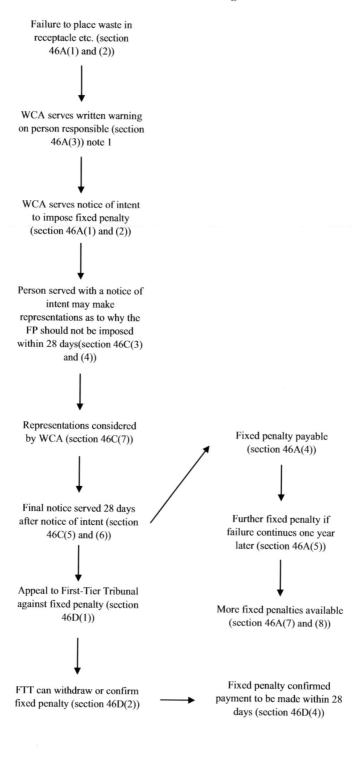

The requirement to pay the fixed penalty is suspended pending the determination or withdrawal of the appeal that is the final appeal made by the person against the decision to require payment of the penalty (section 46D(3)).

Where the requirement to pay the fixed penalty is confirmed at any stage in the proceedings on appeal, payment must be made before 28 days beginning with the day on which the requirement is confirmed unless the person makes a further appeal before the end of that period (section 46D(4)).

Definition

Receptacle includes a holder for receptacles (section 46(10)).

LICENSING OF SCRAP METAL DEALERS

References

Scrap Metal Dealers Act 2013
Scrap Metal Dealers Act 2013 (Prescribed Documents and Information for Verification of Name and Address) Regulations 2013
Scrap Metal Dealers Act 2013 (Prescribed Relevant Offences and Relevant Enforcement Action) Regulations 2013
Determining suitability to hold a scrap metal dealer's licence Home Office October 2013
Scrap Metal Dealer Act 2013: licence fee charges Home Office August 2013
Scrap Metal Dealers Act 2013: supplementary guidance Home Office October 2013
LGA Scrap Metal Dealers Act 2013: Compliance and enforcement guide
LGA Guide to the Scrap Metal Dealers Act 2013: Applications
Environment Agency Guidance Note: Scrap Metal Dealers Act 2013

Extent

These provisions apply in England and Wales. There are licensing provisions for scrap metal dealers in Scotland and Northern Ireland.

Scope

These provisions cover the licensing of scrap metal dealers to ensure scrap metal is collected in an orderly and controlled manner and tackling illegal trade in scrap metal collection.

Requirement for licence to carry on business as scrap metal dealer

Scrap metal dealers require an authorised scrap metal licence (section 1(1)). Without one they are guilty of an offence and liable on summary conviction to a fine not exceeding level 5 on the standard scale (section 1(3)).

Form and effect of licence

A scrap metal licence issued by a local authority must be either a site licence, or a collector's licence (section 2(1) and (2)).

A site licence authorises the licensee to carry on business at any site in the authority's area which is identified in the licence (section 2(3)). It must:

(a) name the licensee,
(b) name the authority,
(c) identify all the sites in the authority's area at which the licensee is authorised to carry on business,
(d) name the site manager of each site, and
(e) state the date on which the licence is due to expire (section 2(4))

A collector's licence authorises the licensee to carry on business as a mobile collector in the authority's area (section 2(5)). A collector's licence must name the licensee and the authority, and state the date on which the licence is due to expire (section 2(6)).

Regulations prescribe further requirements about the form and content of licences (section 2(8)).

A person can hold more than one licence issued by different local authorities, but cannot hold more than one licence issued by any one authority (section 2(9)).

Issue of licence

A local authority must not issue or renew a scrap metal licence unless it is satisfied that the applicant is a suitable person to carry on business as a scrap metal dealer (section 3(1)).

In determining this, the authority may have regard to any information which it considers to be relevant, including:

(a) whether the applicant or any site manager has been convicted of any relevant offence or has been the subject of any relevant enforcement action;
(b) any previous refusal of an application for the issue or renewal of a scrap metal licence or for a relevant environmental permit or registration;
(c) any previous revocation of a scrap metal licence;
(f) whether the applicant has demonstrated that he has adequate procedures to ensure that this Act is complied with (section 3(2)).

In determining whether a company is a suitable person to carry on business as a scrap metal dealer, a local authority is to have regard to whether any of the following is a suitable person –

(a) any director of the company;
(b) any secretary of the company;
(c) any shadow director of the company.

In determining whether a partnership is a suitable person to carry on business as a scrap metal dealer, a local authority is to have regard to whether each of the partners is a suitable person (section 3(5)).

The authority must also have regard to any guidance on determining suitability (section 3(6)).

The authority may consult other persons regarding the suitability of an applicant, including:

(a) any other local authority;
(b) the Environment Agency;

(c) the Natural Resources Body for Wales;

(d) an officer of a police force (section 3(7)).

If the applicant or any site manager has been convicted of a relevant offence, the authority may include times restrictions in the licence (section 3(8)).

Term of licence

A licence lasts for three years (Schedule 1 paragraph 1(1)). But if an application to renew a licence is received before the licence expires, the licence continues.

If the application is withdrawn, the licence expires on the day withdrawn. If the application is refused, the licence expires when no appeal is possible in relation to the refusal or the appeal is finally determined or withdrawn. If the licence is renewed, it lasts for three years (Schedule 1 paragraph 1(2)).

Applications

A licence is to be issued or renewed on an application, which must be accompanied by:

(a) if the applicant is an individual, the full name, date of birth and usual place of residence of the applicant,

(b) if the applicant is a company, the name and registered number of the applicant and the address of the applicant's registered office,

(c) if the applicant is a partnership, the full name, date of birth and usual place of residence of each partner,

(d) any proposed trading name,

(e) the telephone number and email address of the applicant,

(f) the address of any site in the area of any other local authority at which the applicant carries on business as a scrap metal dealer or proposes to do so,

(g) details of any relevant environmental permit or registration in relation to the applicant,

(h) details of any other scrap metal licence issued (whether or not by the local authority) to the applicant within 3 years of the date of the application,

(i) details of the bank account which is proposed to be used in order to comply with scrap metal not to be bought for cash etc, and

(j) details of any conviction of the applicant for a relevant offence, or any relevant enforcement action taken against the applicant (Schedule 1 paragraph 2(1)).

If the application relates to a site licence, it must also be accompanied by:

(a) the address of each site proposed to be identified in the licence, and

(b) the full name, date of birth and usual place of residence of each individual proposed to be named in the licence as a site manager (other than the applicant) (Schedule 1 paragraph 2(2)).

Variation of licence

A local authority may, on an application, vary a licence by changing it from one type to the other and contain particulars of the changes to be made to the licence (Schedule 1 paragraph 3(1) and (4)).

A licensee who fails to make application for variation is guilty of an offence and liable on summary conviction to a fine not exceeding level 3 on the standard scale (Schedule 1 paragraph 3(5)).

It is a defence to prove that the person took all reasonable steps to avoid committing the offence (Schedule 1 paragraph 3(6)).

Further information

The local authority may request that the applicant provide further information the authority considers relevant to consider the application (paragraph 4(1)). If an applicant fails to provide information requested, the authority may decline to proceed with the application (Schedule 1 paragraph 4(2)).

Offence of making false statement

An applicant who in an application or in response to a request for further information makes a statement knowing it be false in a material particular, or recklessly makes a statement which is false in a material particular, is guilty of an offence and liable on summary conviction to a fine not exceeding level 3 on the standard scale (Schedule 1 paragraph 5).

Fee

An application must be accompanied by a fee set by the authority following guidance (Schedule 1 paragraph 6).

Right to make representations

If a local authority proposes to refuse an application, revoke or vary a licence the authority must give the applicant or licensee a notice which sets out what the authority proposes to do and the reasons for it (Schedule 1 paragraph 7(1)).

A notice must also state that, within the period specified (more than 14 days) in the notice, applicant or licensee (A) may either:

(a) make representations about the proposal, or
(b) inform the authority that A wishes to do so (Schedule 1 paragraph 7(3) and (4)).

The authority may refuse the application, or revoke or vary the licence, if:

(a) within the period specified in the notice, A informs the authority that A does not wish to make representations, or
(b) the period specified in the notice expires and A has neither made representations nor informed the authority that A wishes to do so (Schedule 1 paragraph 7(5)).

If, within the period specified in the notice, A informs the authority that A wishes to make representations, the authority:

(a) must allow A a further reasonable period to make representations, and
(b) may refuse the application, or revoke or vary the licence, if A fails to make representations within that period (Schedule 1 paragraph 7(6)).

If A makes representations, the authority must consider them (Schedule 1 paragraph 7(7)).

If A informs the authority that A wishes to make oral representations, the authority must give A the opportunity of appearing before, and being heard by, a person appointed by the authority (Schedule 1 paragraph 7(8)).

Notice of decision

If the authority refuses the application, revokes or varies the licence, it must give A notice setting out the decision and the reasons for it (Schedule 1 paragraph 8(1)).

A notice must also state:

(a) that A may appeal against the decision,
(b) the time within which such an appeal may be brought, and
(c) in the case of a revocation or variation, the date on which the revocation or variation is to take effect (Schedule 1 paragraph 8(2)).

Appeals

An applicant may appeal to a magistrates' court against the refusal of an application (Schedule 1 paragraph 9(1)).

A licensee may appeal to a magistrates' court against:

(a) the inclusion in a licence of a condition or
(b) the revocation or variation of a licence (Schedule 1 paragraph 9(2)).

An appeal must be made within 21 days on which notice was given (Schedule 1 paragraph 9(3)).

An appeal is to be by way of complaint for an order (Schedule 1 paragraph 9(4)).

The magistrates' court may:

(a) confirm, vary or reverse the authority's decision, and
(b) give directions it considers appropriate (Schedule 1 paragraph 9(6)).

The authority must comply with any directions given by the magistrates' court (Schedule 1 paragraph 9(7)).

The authority need not comply with any such directions:

(a) until the time for making an application by way of case stated has passed, or
(b) if such an application is made, until the application is finally determined or withdrawn (Schedule 1 paragraph 9(8)).

Revocation of licence and imposition of conditions

The authority may revoke a licence

- if it is satisfied that the licensee does not carry on business at any of the sites identified in the licence (section 4(1)) or
- if it is satisfied that a site manager named in the licence does not act as site manager at any of the sites in the licence (section 4(2)).

- if it is no longer satisfied that the licensee is a suitable person to carry on business as a scrap metal dealer (section 4(3)).

If the licensee or any site manager named in a licence is convicted of a relevant offence, the authority may vary the licence by adding conditions (section 4(5)).

A revocation or variation comes into effect when no appeal is possible, or when any such appeal is finally determined or withdrawn (section 4(6)).

But if the authority considers that the licence should not continue in force without conditions, it may by notice provide:

(a) that, until a revocation comes into effect, the licence is subject to one or both of the conditions set out in section 3(8), or
(b) that a variation comes into effect immediately (section 4(7)).

Supply of information by authority

The local authority must supply any information to any other local authority; the Environment Agency; the Natural Resources Body for Wales or an officer of a police force (section 6).

Register of licences

The Environment Agency and the Natural Resources Body for Wales must maintain a register of scrap metal licences issued by authorities in England and Wales respectively. Each entry in the registers must record the name of the authority which issued the licence, the name of the licensee, any trading name of the licensee, the address of any site identified in the licence, the type of licence, and the date on which the licence is due to expire.

The registers are to be open for inspection to the public (section 7).

Notification requirements

An applicant for a scrap metal licence, or the renewal or variation of a licence, must notify the authority to which the application was made of any changes which materially affect the accuracy of the information which the applicant has provided in the application (section 8(1)).

A licensee who is not carrying on business as a scrap metal dealer in the area of the authority which issued the licence must notify the authority (section 8(2)).

This notification must be given within 28 days (section 8(3)).

If a licensee carries on business under a trading name, the licensee must notify the authority which issued the licence within 28 days of any change to that name (section 8(4) and (5)).

An authority must notify the relevant environment body of any notification given to it, any variation and revocation by the authority of a licence (section 8(6)). This notification must be given within 28 days (section 8(7)).

Where an authority notifies the relevant environment body, the body must amend the register accordingly (section 8(8)).

An applicant or licensee who fails to comply with this is guilty of an offence and is liable on summary conviction to a fine not exceeding level 3 on the standard scale (section 8(9)).

It is a defence for a person charged with this offence to prove that the person took all reasonable steps to avoid committing the offence (section 8(10)).

Closure of unlicensed sites

Schedule 2 makes provision for the closure of sites at which a scrap metal business is being carried on without a licence (section 9).

Closure notice

Where a constable or the local authority is satisfied that premises used by a scrap metal dealer is not licensed (not a residential premises) (paragraph 2(1) and (2)) they may issue a closure notice which:

(a) states that the constable or authority is satisfied above,
(b) gives the reasons for that,
(c) states that the constable or authority may apply to the court for a closure order and
(d) specifies the steps which may be taken to ensure that the alleged use of the premises ceases (paragraph 2(3)).

The constable or authority must give the closure notice to:

(a) the person who appears to be the site manager of the premises, and
(b) any person who appears to be a director, manager or other officer of the business and to any person who has an interest in the premises (paragraph 2(4) to (7)).

Cancellation of closure notice

A closure notice may be cancelled by a cancellation notice issued by a constable or the local authority (paragraph 3(1)).

A cancellation notice takes effect when it is given to any one of the persons to whom the closure notice was given (paragraph 3(2)).

The cancellation notice must also be given to any other person to whom the closure notice was given (paragraph 3(3)).

Application for closure order

Where a closure notice has been given, a constable or the local authority may make a complaint to a Justice of the Peace for a closure order (paragraph 4(1)).

A complaint must be made between seven days and six months after the closure notice was given (paragraph 4(2)).

A complaint cannot be made if the constable or authority is satisfied that the premises are not (or are no longer) being used by a scrap metal dealer, and there is no reasonable likelihood that the premises will be so used in the future (paragraph 4(3)).

Where a complaint has been made, the Justice may issue a summons to answer to the complaint (paragraph 4(4)). The summons must be directed to any person to whom the closure notice was given (paragraph 4(5)).

If a summons is issued, notice of the date, time and place at which the complaint will be heard must be given to all the persons to whom the closure notice was given (paragraph 4(6)).

Closure order

If the court is satisfied that the closure notice was given and that the premises continue to be used by a scrap metal dealer in the course of business, or there is a reasonable likelihood that the premises will be so used in the future they may make a closure order (paragraph 5(1) and (2)).

A closure order may, in particular, require:

(a) that the premises be closed immediately to the public and remain closed until a constable or the local authority makes a certificate under paragraph 6;

(b) that the use of the premises by a scrap metal dealer in the course of business be discontinued immediately;

(c) that any defendant pay into court (to a designated officer) a sum the court determines and the sum will not be released by the court to that person until the other requirements of the order are met (paragraph 5(3) and (7)).

A closure order including a requirement on immediate closure may include conditions the court considers appropriate:

(a) the admission of persons onto the premises;

(b) the access by persons to another part of any building (paragraph 5(4)).

A closure order may include provisions the court considers appropriate for dealing with the consequences if the order should cease to have effect under paragraph 6 (paragraph 5(5)).

As soon as practicable after a closure order is made, the complainant must fix a copy of it in a conspicuous position on the premises (paragraph 5(6)).

Termination of closure order by certificate of constable or authority

Where a constable or the local authority is satisfied that the need for the order has ceased the constable or authority may make a certificate to that effect (paragraph 6(1) and (2)). The closure order ceases to have effect when the certificate is made (paragraph 6(3)).

Any sum paid into court under the order is to be released by the court to the defendant (paragraph 6(4)).

As soon as practicable after making a certificate, the constable or authority must:

(a) give a copy of it to any person against whom the closure order was made,

(b) give a copy of it to the designated officer for the court which made the order, and

(c) fix a copy of it in a conspicuous position on the premises in respect of which the order was made (paragraph 6(5)).

(d) give a copy of the certificate to any person who requests one (paragraph 6(6)).

Discharge of closure order by court

Any person given the closure notice or with an interest in it may make a complaint to a Justice of the Peace for a discharge order (paragraph 7(1)).

The court may not make a discharge order unless it is satisfied that there is no longer a need for the closure order (paragraph 7(2)).

The Justice may issue a summons directed to a constable the Justice considers appropriate, or the local authority, requiring that person to appear before the magistrates' court to answer to the complaint (paragraph 7(3)).

If a summons is issued notice of the date, time and place at which the complaint will be heard must be given to all the persons to whom the closure notice was given (paragraph 7(4)).

Appeals

An appeal may be made to the Crown Court against:

    (a) a closure order;
    (b) a decision not to make a closure order;
    (c) a discharge order;
    (d) a decision not to make a discharge order (paragraph 8(1)).

Any appeal must be made within 21 days of the order or decision in question was made (paragraph 8(2)).

An appeal against a closure order or a decision not to make a discharge order may be made by any person given the closure notice or with an interest in the premises (paragraph 8(3)).

An appeal against a decision not to make a closure order or against a discharge order may be made by a constable or the local authority (paragraph 8(4)).

On appeal the Crown Court may make an order it considers appropriate (paragraph 8(5)).

Enforcement of closure order

A person is guilty of an offence if the person, without reasonable excuse:

    (a) permits premises to be open in contravention of a closure order, or
    (b) otherwise fails to comply with, or is in contravention of, a closure order (paragraph 9(1)).

If a closure order has been made, a constable or an authorised person may (if necessary using reasonable force):

    (a) enter the premises at any reasonable time, and
    (b) having entered the premises, do anything reasonably necessary for the purpose of securing compliance with the order (paragraph 9(2)).

Where a constable or an authorised person seeks to exercise his powers and if the owner, occupier requires evidence of the officer's identity or authority to exercise those powers, the officer must produce that evidence (paragraph 9(3) and (4)).

A person who intentionally obstructs a constable or an authorised person in the exercise of powers is guilty of an offence (paragraph 9(5)) and is liable on summary conviction to a fine not exceeding level 5 on the standard scale (paragraph 9(6)).

Definitions

Carrying on business as a scrap metal dealer

If the person:

(a) carries on a business which consists wholly or partly in buying or selling scrap metal, whether or not the metal is sold in the form in which it was bought, or
(b) carries on business as a motor salvage operator (section 21(2)).

Scrap metal includes –

(a) any old, waste or discarded metal or metallic material, and
(b) any product, article or assembly which is made from or contains metal and is broken, worn out or regarded by its last holder as having reached the end of its useful life (section 21(6)).

FC142A Scrap metal dealers – Scrap Metal Dealers Act 2013 – Site and collectors' licences (1)

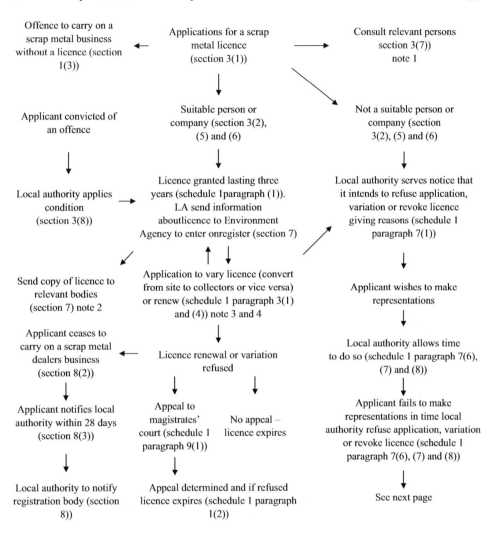

FC142B Scrap metal dealers – Scrap Metal Dealers Act 2013 – Site and collectors' licences (2)

From previous page

↓

Local authority refuses, revokes or varies licence must give notice of decision to applicant with reasons (schedule 1 paragraph 8(1))

Local authority may revoke scrap metal dealer's licence (section 4(1)(2) and (3))

↓

Applicant may appeal to magistrates' court within 21 days (schedule 1 paragraph 9(1) and (3))

No appeal – revocation comes into effect (section 4(6))

↓

Magistrates court may confirm, vary or reverse local authority decision and give directions (schedule 1 paragraph 9 (6))

↓

Local authority must comply with court directions (schedule 1 paragraph 9(7))

FC143 Closure of unlicensed scrap metal sites – section 9 and Schedule 2 Scrap Metal Dealers Act 2013

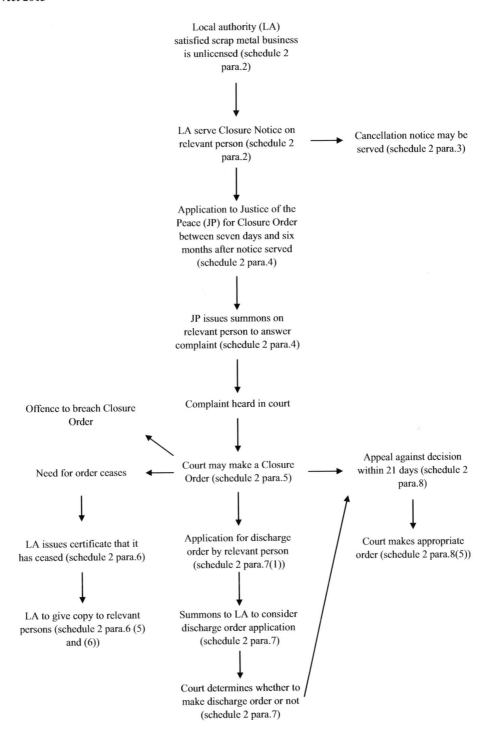

Chapter 13

WATER SUPPLY

GENERAL PROCEDURAL PROVISIONS: WATER INDUSTRY ACT 1991

The following provisions are applicable to the use by local authorities of all procedures dealt with in this section.

Information

A local authority may by notice require any person to provide such information as it may reasonably require to fulfil its powers and duties with respect to both public and private water supplies. The maximum penalty for non-compliance with such notices is level 5 (section 85).

Power of entry etc.

A person designated in writing by the local authority may:

(a) enter premises supplied with private or public water to see if any action is required of the local authority;

(b) carry out such inspections, measurements and tests, including the taking of samples of water, land and articles, as is considered appropriate.

Unless the situation is one of emergency, power of entry cannot be required unless it is requested at a reasonable time and after having given at least 24 hours' notice to the occupier in relation to non-business premises. Persons convicted of obstructing a local authority officer are subject to penalties not exceeding level 5 (section 84(3) and schedule 6).

Where the local authority shows that entry has been refused, or is anticipated, or the premises are unoccupied or the case is one of emergency, a Justice of the Peace may issue a warrant to authorise entry, if need be by force (schedule 6, paragraph 2).

Definitions

Owner in relation to any premises, means the person who:

(a) is for the time being receiving the rackrent of the premises whether on his own account or as the agent or trustee for another person; or

(b) would receive the rackrent if the premises were let at a rackrent (section 219(1)).

Private supply means:

1. subject to subsection (2) below, a supply of water provided otherwise than by a water undertaker or by a licensed water supplier (including a supply provided for the purposes of the bottling of water), and cognate expressions shall be construed accordingly;
2. for the purposes of any reference . . . to a private supply, or to supplying water by means of a private supply, water shall be treated as supplied to any premises not only where it is supplied from outside those premises, but also where it is abstracted, for the purpose of being used or consumed on those premises, from a source which is situated on the premises themselves; and for the purposes of this subsection water shall be treated as used on any premises where it is bottled on those premises for use or consumption elsewhere.

(section 93(1) and (2))

Water undertaker – a company may be appointed:

(a) by the Secretary of State; or
(b) with the consent of or in accordance with a general authorisation given by the Secretary of State, by the Director General of Water Services, to be the water under-taker . . . for any area of England and Wales (section 6(1)).

CONTROL OVER PUBLIC WATER SUPPLIES

References

Water Industry Act 1991, sections 77 and 78
Water Supply (Water Quality) Regulations 2018
The Public Water Supplies (Scotland) Regulations 2014
The Water Supply (Water Quality) Regulations (Northern Ireland) 2017

Extent

These provisions apply to England and Wales. Similar provisions apply in Scotland and Northern Ireland.

Scope

Public supplies are those provided by an appointed water undertaker (WU) and licensed water supplier, and each is placed under a statutory duty to supply water for domestic or food production purposes which is wholesome and to ensure that, in general, there is no deterioration in the quality of water from any of its sources (section 68(1)).

In this connection, domestic purposes include drinking, washing, cooking, central heating and sanitary purposes (section 218(1)). 'Food production purposes' means the manufacturing, processing, preserving or marketing purposes with respect to food or drink for which water supplied to food production premises may be used; and 'food production premises' means premises used for the purposes of a business of preparing food or drink for consumption otherwise than on premises (section 93).

Enforcement

Those duties placed on water undertakers are enforceable by the Secretary of State and the Director General of Water Services through enforcement orders (section 18).

Wholesomeness

Water is to be judged as being wholesome if it satisfies the requirements of regulation 4 of the Water Supply (Water Quality) Regulations 2018 or regulation 4 of the Private Water Supplies (England) Regulations 2016 as appropriate (section 67(1)).

Local authority role

Local authorities are placed under a statutory duty to take all such steps as they consider appropriate for keeping themselves informed about the wholesomeness and sufficiency of all public water supplies (section 77(1)). In this connection, local authorities:

(a) must make arrangements with the specified relevant supplier to be notified of events giving rise to significant risks to health; and

(b) may take and have analysed by designated officers such public water samples as they may reasonably require (regulations 37 Water Supply (Water Quality) Regulations 2018).

Powers of the local authority

There are three situations which the local authority deals with in relation to public supplies:

1. Where a local authority believes that:

    (a) the water is, or is likely to become, unwholesome;
    (b) the water is insufficient for domestic purposes;
    (c) the water is, or is likely to cause, a danger to life or health because of unwholesomeness or insufficiency; or
    (d) the quality of sources of water are deteriorating,

    the local authority is required to inform the water undertaker or water supply licensee. Where the local authority is not satisfied that all appropriate remedial action is being taken by either, it must notify the Secretary of State, who may then use his powers of enforcement (section 78).

2. The local authority is required to notify the water undertaker or wherever it (the water undertaker) is satisfied that it is not reasonably practicable at reasonable cost to provide, or to maintain, a piped supply of wholesome water that is sufficient for domestic purposes, and to request the water undertaker to provide a supply of water to particular premises by means other than pipes where:

    (a) this is practicable at reasonable cost; and
    (b) the unwholesomeness or insufficiency is such as to cause danger to life and health.

    In these circumstances, the local authority is required to pay to the water undertaker the appropriate water charges, but may recover these from the owners or occupiers of the premises benefiting from the supply. The water undertaker is under a duty to comply with a local authority request which fulfils the criteria outlined above (section 79).

3. Where the local authority is of the opinion that a supply is so polluted as to be prejudicial to health, it may apply to a magistrates' court for an order closing or restricting that supply. This procedure is detailed in FC147 (Public Health Act 1936, section 140).

Authorisations for quality standards for public water supplies

The Secretary of State deals with authorisation of the temporary supply of public water that is not wholesome. Such authorisations are time related and cannot extend beyond three years.

In making an application to relax the standards of the Water Supply (Water Quality) Regulations 2018, the applicant must send a copy to the local authority, which may make representations to the Secretary of State within 30 days. If the Secretary of State is minded to revoke or modify the authorisation, he must give the local authority at least six months' notice unless immediate action is necessary in the interests of public health (2018, regulations 22–25).

CONTROL OVER PRIVATE WATER SUPPLIES

References

Water Industry Act 1991, sections 80–83
Water Supply (Water Quality) Regulations 2018
The Private Water Supplies Regulations 2016
Circular 24/91 DoE, Private Water Supplies
The Private Water Supplies (Scotland) Regulations 2006
The Private Water Supplies Regulations (Northern Ireland) 2017

Extent

These provisions apply to England and Wales. Similar provisions apply in Scotland and Northern Ireland.

Scope

Local authorities act as the regulators for private water supplies and have a number of statutory duties under the Private Water Supplies Regulations. These regulations place a duty on local authorities to conduct a risk assessment of each private water supply within their area and to undertake monitoring in order to determine compliance with drinking water standards.

Local authorities are required to provide certain information, including monitoring data, relating to private water supplies to the Secretary of State and Welsh Ministers (in practice the Drinking Water Inspectorate). This submission of data is due on 31 January of every year, for records relating to supplies in their area during the previous calendar year. Local Authorities must also within 12 months of the day it carried out a risk assessment, provide the Secretary of State with a summary of the results of that assessment.

A private water supply is a supply of water provided otherwise than by a water undertaker or water supply licensee and includes a supply provided for the purposes of the bottling of water. This definition is extended to cover water sources abstracted both outside and inside of premises, and water bottling in premises for use or consumption elsewhere (sections 93(1) and (2)). The Food Safety Act 1990 extended the definition to cover water supplied for food production (also see p. 182).

Local authorities are under the same general duty to monitor the wholesomeness and sufficiency of private supplies as they are in relation to public supplies (section 77(1)), but in this case they have direct and sole enforcement powers. The Private Water Supplies Regulations 2016 specify the scope of monitoring distributed by a person other than a water undertaker or licensed water supplies and for large supplies and supplies to commercial or public premises, the keeping of records and their notification to the Secretary of State (regulations 7 to 14).

Water Industry Act 1991: information

The local authority may serve a notice on any person requiring him to provide any information it may reasonably require in order to fulfil its responsibilities for public and private supplies (section 85(1)).

Improvement notices

Where a local authority is satisfied that a private water supply is not wholesome or sufficient for domestic or food production purposes, it may serve an improvement notice on the relevant person (section 80(1)).

Standards

Water from private supplies is to be judged as being wholesome for domestic purposes if it satisfies the requirements of the tables in schedule 1 of the Private Water Supplies Regulations 2016 (section 67 Water Industry Act 1991 and regulation 4).

Relevant person

The persons upon whom the notice should be served are the owners *and* occupiers of premises served by the supply, the owner *and* occupiers of premises where the source is situated *and* other persons who exercise powers of management or control of that source (section 80(7)).

Notice requirements

A notice must:

(a) identify the reasons why the supply is considered to be unwholesome and/or insufficient;
(b) specify the steps required to provide a wholesome and sufficient supply;
(c) allow a period of not less than 28 days for representation or objections to be received;
(d) state the effect of the notice;
(d) indicate the need for referral to the Secretary of State if representations or objections are made within the specified period (section 80(2)).

The notice may specify one or more of the following:

(a) the steps to be taken by the local authority themselves;
(b) the steps to be taken by the person specified in the notice within a specified period;
(c) the payments to be made to another relevant person or the local authority for expenses in meeting the local authority requirements;
(d) the payments to be made by the local authority to a person for the costs of meeting the notice requirements (section 80(3));

and may require:

(a) a supply of water to be provided by the water undertaker or another person; and
(b) steps to be taken to ensure that this is done (section 80(6)).

Operation of notices

Where no representation or objections are made to the local authority within the period specified in the notice, the notice becomes effective at the end of the period allowed for representation to be made (section 81(1)).

Where representations etc. are made and not withdrawn, then the local authority must refer the matter to the Secretary of State, who may then either confirm the notice with or without modification; or not confirm the notice.

Where the notice is confirmed, it becomes effective on a date as specified by the Secretary of State (sections 81(2), (3) and (4)).

Enforcement

The local authority's powers of enforcement where an improvement notice is not complied with are restricted to the undertaking of required works and the recovery of related costs (section 82).

Private Water Supplies Regulations 2016

These regulations replace 2009 regulations and implement European Council Directive 98/83/EC.

Regulations 7 to 14 of the Private Water Supplies Regulations 2016 also make it a duty of local authorities to monitor private water supplies, and take and analyse samples of certain private supplies. The regulations also lay down a sampling regime for private supplies (schedule 3).

Risk assessment

Every five years, the local authority must carry out a risk assessment of each private supply that supplies water to any premises (other than a supply to a single dwelling not used for any commercial activity) to establish whether there is a risk of supplying water that would constitute a potential danger to human health. It must also carry out a risk assessment of a private supply to a single dwelling not used for any commercial or public activity if requested to do so by the owner or occupier of that dwelling (regulation 6).

Provision of information

If a local authority considers that a private water supply in its area is a potential danger to human health, it must promptly take appropriate steps to ensure that people likely to consume water from it:

(a) are informed that the supply constitutes a potential danger to human health,
(b) where possible, are informed of the nature of the potential danger, and
(c) are given advice to allow them to minimise any potential danger (regulation 15).

Investigation

If a local authority considers a private water supply to be a potential danger to health, it must carry out an investigation as to why it is unwholesome or fails one of the parameters in the regulations (regulation 16).

If it is in the pipework of a single dwelling, the local authority must inform the people concerned and give necessary advice. If the local authority cannot solve the problem, it must grant an authorisation for different standards (in effect, a relaxation in standards (regulation 17) or serve a notice under regulation 18 or section 80 of the Water Industry Act 1991.

If an authorisation is granted, the person granted it must take action in accordance with the timetable in the authorisation. If the supply exceeds 1,000 m^3 a day or serves more than 5,000 people, it must send a copy of the authorisation to the Secretary of State (regulation 17(7)). The local authority must also inform people concerned of the authorisation (regulation 17(6)).

Notice

If any private supply of water intended for human consumption constitutes a potential danger to human health, a local authority must serve a notice on the relevant person.

The notice must:

(a) identify the private supply;
(b) state the grounds for serving the notice;
(c) prohibit or restrict the use of that supply;
(d) specify what other action is necessary to protect human health and to restore the quality of the water supply.

The local authority must promptly inform consumers of the issue of the notice and provide any necessary advice. The notice may be subject to conditions and may be amended by further notice at any time. The local authority must revoke the notice as soon as there is no longer a potential danger to human health. It is an offence to breach a notice served by the local authority (regulation 18). Any person aggrieved by such notice can appeal to the magistrates' court within 28 days of service of the notice. A notice remains in force unless suspended by the court. On an appeal, the court may either cancel the notice or confirm it, with or without modification (regulation 19). Penalties for breach of the notice are included in regulation 20.

The local authority must make and maintain records, and must send a copy of the records to the Secretary of State (regulation 14 and schedule 4).

FC144 Private water supply: notice procedure – Regulation 18 Private Water Supplies Regulations 2016

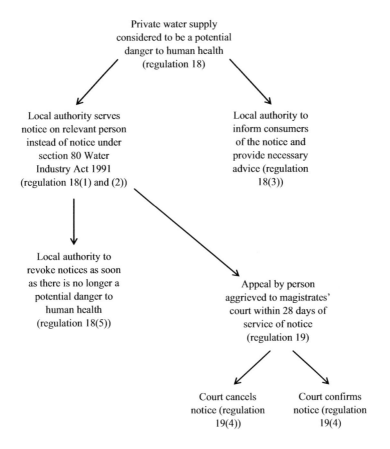

Notes
1. Offence to breach notice or fail to comply with it (regulation 18(6)).
2. Penalties – fine and/or imprisonment (regulation 20(1) and 20(2) for body corporate).
3. Local authority can charge fees for risk assessment, sampling, investigation, granting authorisation and analysing samples (schedule 5).

FC145 Private water supply: action in the event of failure – Part 3 Private Water Supplies Regulations 2016

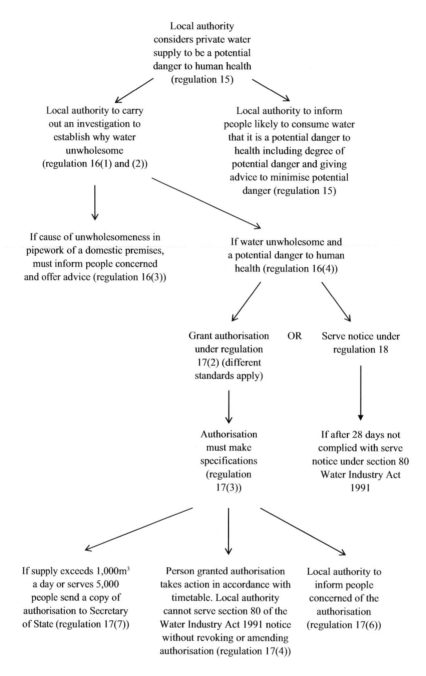

Local authority considers private water supply to be a potential danger to human health (regulation 15)

Local authority to carry out an investigation to establish why water unwholesome (regulation 16(1) and (2))

Local authority to inform people likely to consume water that it is a potential danger to health including degree of potential danger and giving advice to minimise potential danger (regulation 15)

If cause of unwholesomeness in pipework of a domestic premises, must inform people concerned and offer advice (regulation 16(3))

If water unwholesome and a potential danger to human health (regulation 16(4))

Grant authorisation under regulation 17(2) (different standards apply)   OR   Serve notice under regulation 18

Authorisation must make specifications (regulation 17(3))

If after 28 days not complied with serve notice under section 80 Water Industry Act 1991

If supply exceeds 1,000m³ a day or serves 5,000 people send a copy of authorisation to Secretary of State (regulation 17(7))

Person granted authorisation takes action in accordance with timetable. Local authority cannot serve section 80 of the Water Industry Act 1991 notice without revoking or amending authorisation (regulation 17(4))

Local authority to inform people concerned of the authorisation (regulation 17(6))

Notes
1. Local authority must keep progress of remedial action under review (regulation 17(8)).
2. Local authority can grant authorisation for up to three years with prior consent from Secretary of State and send copy to Secretary of State (regulation 17(9)).
3. Local authority may revoke or amend the authorisation at any time and in particular if the timetable for remedial action is not followed (regulation 17(10)).

WATER SUPPLY FOR NEW HOUSES

Reference

Building Act 1984, section 25

Extent

These provisions apply to England and Wales.

Scope

Before approving plans submitted for approval under the building regulations, the local authority must be satisfied that the water supply proposed will be sufficient and wholesome for domestic purposes and provided:

    (a) either by connecting to a piped supply provided by the water undertakers, or

    (b) if (a) is not reasonable, by otherwise taking water into the house, by pipe, e.g. from a well supply; or

    (c) if neither (a) nor (b) can be reasonably required, by providing a supply within a reasonable distance of the house (section 25(1)).

In deciding whether or not water is 'wholesome', the standards of section 67 of the Water Industry Act 1991 and of any regulations made under that section (i.e. the Water Supply (Water Quality) Regulations 2018 and the Private Water Supply Regulations 2016) apply here (section 25(7)).

Notices

A notice requiring prohibition of occupation in the event of the supply not being sufficient or wholesome, despite the passing of plans, can be served and must be in writing and served on the owner (section 25(3) and 92).

    A person who occupies the house or permits it to be occupied where the water supply is inadequate is liable on summary conviction to a fine not exceeding level 1 on the standard scale and to a further fine not exceeding £2 for each day on which the offence continues after he is convicted (section 25(6)).

CLOSURE OR RESTRICTION OF POLLUTED WATER SUPPLY

Reference

Public Health Act 1936, section 140

Extent

These provisions apply to England and Wales. Similar provisions apply in Scotland and Northern Ireland.

FC146 Water supply for new houses – section 25 Building Act 1984

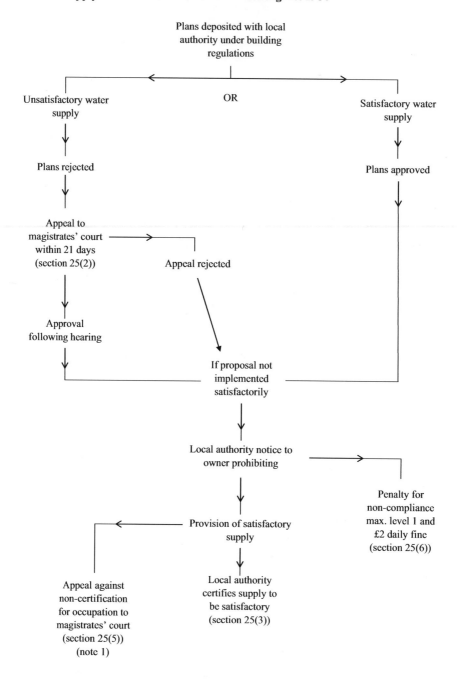

Notes

1. The court may either dismiss the appeal or authorise the occupation of the house, the latter having the same effect as certification by the local authority.
2. For closure/restriction of polluted supplies, see FC147.

FC147 Closure or restriction of polluted water supply – section 140 Public Health Act 1936

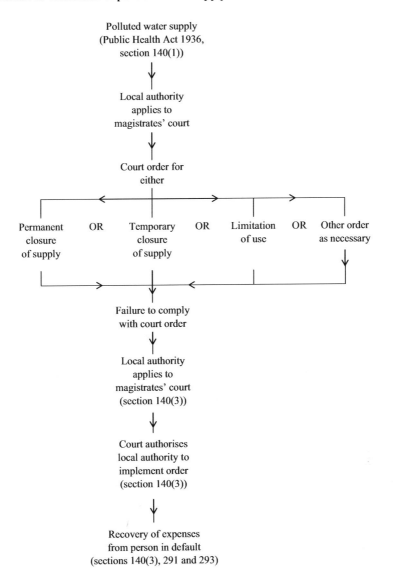

Scope

The section applies to wells, tanks or any other source of water supply not vested in the local authority which are used, or likely to be used, for domestic purposes or the preparation of food or drink for human consumption and which are so polluted as to be prejudicial to health (section 140(1)).

It would seem that this procedure is available to a local authority and some port health authorities[1] to deal with water supplied by statutory water undertakings operating under the Water Industry Act 1991 and therefore provides the only direct statutory remedy available to local authorities in respect of such public water supplies (see p. 639).

Persons responsible

The court order following applications by the local authority may be addressed to owners or occupiers of the premises to which the source of supply belongs or to any other person having control over it (section 140(1) and (2)).

Court orders

These may order any of the following:

(a) permanent closure or cutting off of the source; or
(b) temporary closure or cutting off of the source; or
(c) restriction of use for certain purposes only; or
(d) any other measure to prevent injury or damage to health of persons using water or consuming food or drink prepared from it (section 140(2)).

Works in default

If the court order is not complied with, the local authority can apply to the court to authorise them to do what is necessary to remedy the problem and it may claim expenses reasonably incurred in default of the order (section 140(3)).

Note
1. Bristol Port Health Authority Order 2010; Cornwall Port Health Authority Order 2010; Cowes Port Health Authority Order 2010; Portsmouth Port Health Authority Order 2010; Southampton Port Health Authority Order 2010; Hull and Goole Port Health Authority Order 2011; River Tees Port Health Authority Order 2016; Weymouth Port Health Authority Order 2017.

PART 8

PUBLIC SAFETY

CONTROLS OVER TRADING ON SUNDAYS AT LARGE SHOPS

References

Sunday Trading Act 1994 as amended by the Regulatory Reform (Sunday Trading) Order 2004

The Shops (Sunday Trading &c.) (Northern Ireland) Order 1997

Extent

These procedures apply only in England and Wales (section 9(4)). Similar provisions apply in Northern Ireland. There are no trading time restrictions in Scotland.

Scope

This Act deregulates Sunday trading except in relation to large shops where generally trading is prohibited except in accordance with the 'six-hour trading' procedures in FC148.

A local authority may designate its area to be a loading control area in order to control loading and unloading before 9 am on Sundays in accordance with procedures in FC149.

Large shops

A 'large shop' is one which has a relevant floor area exceeding 280 square metres.

The 'relevant floor area' is defined as being the internal floor area of so much of the shop as consists of or comprises in a building excluding any part of the shop which throughout the week ending with the Sunday in question is used neither for the serving of customers in connection with the sale of goods nor for the display of goods.

Outlets which offer a service, e.g. hairdressers, shoe repair shops, restaurants, public houses, etc., are therefore excluded from Sunday trading restrictions (schedule 1, paragraph 1).

Exempted large shops

Some specified types of large shops are exempted from the six-hour restriction. These are:

(a) farm shops;
(b) motor and cycle supply shops;
(c) stands at exhibitions;
(d) shops at railway stations and airports;
(e) petrol filling stations;
(f) pharmacists for the sale of medicines and medicinal and surgical appliances;
(g) shops servicing ocean-going ships (schedule 1, paragraph 3).

Shops occupied by persons observing the Jewish Sabbath are also exempt from the six-hour limit on Sunday trading provided the procedure identified in FC148 has been followed and the shop is closed on the Jewish Sabbath.

The certificate required to be sent to the local authority with the notification is one to be signed by an authorised person (a minister or secretary of a synagogue or a person nominated for the purpose by the Board of Deputies of British Jews) and indicates that the person giving the notice is of the Jewish religion. Such exemption is also available to persons of any religious body who regularly observe the Jewish Sabbath but certification is required from the religious body concerned (schedule 2, paragraphs 8 and 9).

The six-hour provision

Unless exempt or occupied by Jews, large shops may trade for six continuous hours on Sundays between 10 am and 6 pm. The choice of the six-hour trading period is a matter for each trader (schedule 1, paragraph 4 as amended).

Shops using the six-hour trading provision must display conspicuous notices both inside and outside of the building specifying the permitted Sunday opening times (schedule 1, paragraph 6).

Controls over loading and unloading

FC149 shows the procedure through which local authorities may, at their discretion, operate a scheme of control over loading/unloading of goods from vehicles before 9 am on Sundays at large shops that are permitted to open. There are no such restrictions on small shops.

The purpose of the procedure is to allow local authorities to prevent local residents from being caused undue annoyance and requires the local authority, once a loading control area has been declared, to sanction loading and unloading at each premises before 9 am and specifying appropriate conditions. A scheme can only apply to the whole of the local authority area.

Before adopting or subsequently revoking a scheme, the local authority must consult people likely to be affected, residents and traders, and must publicise its decision to adopt or revoke (section 2 and schedule 3).

Enforcement

Local authorities have a duty to enforce these provisions and must appoint inspectors who have investigatory powers under Schedule 5 of the Consumer Rights Act 2015 (schedule 2).

Definitions

Retail sale means any sale other than a sale for use or resale in the course of a trade or business, and references to retail purchase shall be construed accordingly.
Sale of goods does not include:

    (a) the sale of meals, or refreshments or intoxicating liquor for consumption on the premises on which they are sold; or
    (b) the sale of meals or refreshments prepared to order for immediate consumption off those premises.

Shop means any premises where there is carried on a trade or business consisting wholly or mainly of the sale of goods.
Stand in relation to an exhibition means any platform, structure, space or other area provided for exhibition purposes (schedule 1, paragraph 1).

FC148 Controls over trading on Sundays at large shops – schedules 1 and 2 Sunday Trading Act 1994

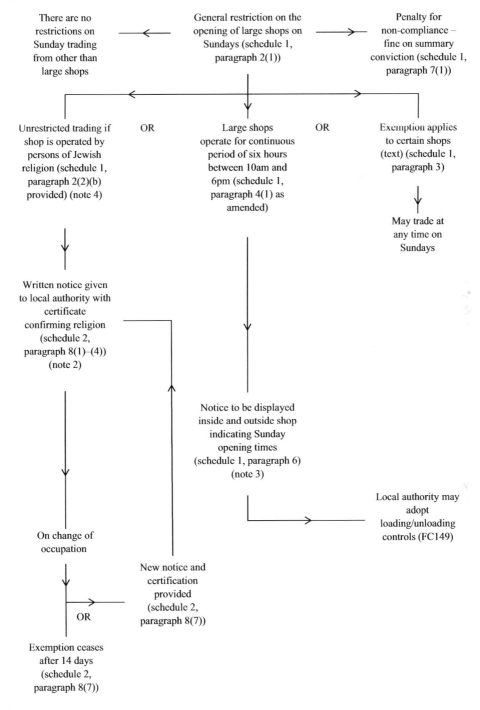

Notes

1. Penalty not exceeding level 5 for making false statements (schedule 2, paragraph 8(10)) and the local authority may cancel notice (schedule 2, paragraph 8(11)).
2. A fresh certificate is not required if previously provided for that occupier – this is at the local authority's discretion (schedule 2, paragraph 8(8)).
3. Fine for non-compliance (schedule 1, paragraph 7(2)).
4. These shops must be closed on the Jewish Sabbath – sundown on Friday to sundown on Saturday.

FC149 Controls over loading and unloading at large shops on Sundays – section 2 Sunday Trading Act 1994

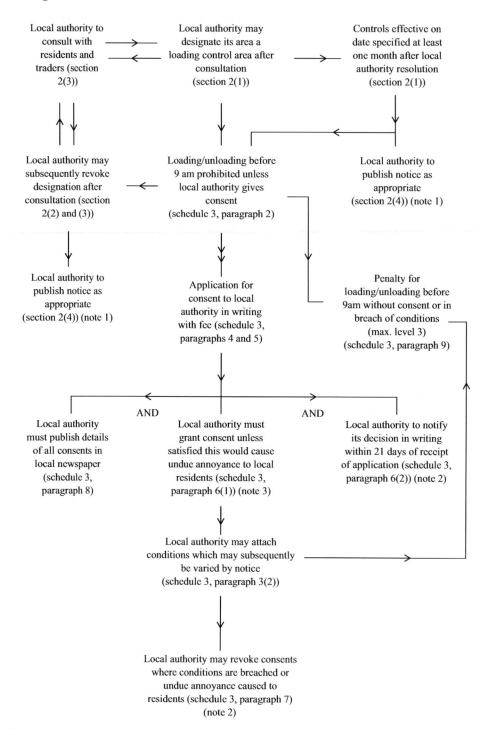

Local authority to consult with residents and traders (section 2(3))

Local authority may designate its area a loading control area after consultation (section 2(1))

Controls effective on date specified at least one month after local authority resolution (section 2(1))

Local authority may subsequently revoke designation after consultation (section 2(2) and (3))

Loading/unloading before 9 am prohibited unless local authority gives consent (schedule 3, paragraph 2)

Local authority to publish notice as appropriate (section 2(4)) (note 1)

Local authority to publish notice as appropriate (section 2(4)) (note 1)

Application for consent to local authority in writing with fee (schedule 3, paragraphs 4 and 5)

Penalty for loading/unloading before 9am without consent or in breach of conditions (max. level 3) (schedule 3, paragraph 9)

AND

AND

Local authority must publish details of all consents in local newspaper (schedule 3, paragraph 8)

Local authority must grant consent unless satisfied this would cause undue annoyance to local residents (schedule 3, paragraph 6(1)) (note 3)

Local authority to notify its decision in writing within 21 days of receipt of application (schedule 3, paragraph 6(2)) (note 2)

Local authority may attach conditions which may subsequently be varied by notice (schedule 3, paragraph 3(2))

Local authority may revoke consents where conditions are breached or undue annoyance caused to residents (schedule 3, paragraph 7) (note 2)

Notes
1. No particular means of publication is specified; this is for the local authority to decide.
2. There is no provision for appeal against a refusal of consent, conditions attached or revocation.
3. There is no requirement on a local authority to keep a register of consents.

STREET TRADING

References

Local Government (Miscellaneous Provisions) Act 1982, section 3 and schedule 4
Civic Government (Scotland) Act 1982
Street Trading Act (Northern Ireland) 2001

Extent

This procedure applies in England and Wales but does not apply in Greater London (for which see the London Local Authorities Act 1990). Similar provisions apply in Scotland and Northern Ireland.

Scope

Street trading is defined as being the selling or exposing or offering for sale of any article (or living thing) in a street, but the following are not considered to be street trading:

(a) trading by a person acting as a pedlar under the authority of a pedlar's certificate granted under the Pedlars Act 1871;

(b) anything done in a market or fair, the right to hold which was acquired by virtue of a grant (including a presumed grant) or acquired or established by virtue of an enactment or order;

(c) trading in a trunk road picnic area provided by the Secretary of State under section 112 of the Highways Act 1980;

(d) trading as a news vendor;

(e) trading which:

    (i) is carried on at premises used as a petrol filling station; or

    (ii) is carried on at premises used as a shop or in a street adjoining premises so used and as part of the business of the shop;

(f) selling things, or offering, or exposing them for sale, as a roundsman;

(g) the use for trading under Part VIIA of the Highways Act 1980 of an object or structure placed on, in or over a highway;

(h) the operation of facilities for recreation or refreshment under Part VIIA of the Highways Act 1980;

(i) the doing of anything authorised by regulations made under section 5 of the Police, Factories, etc. (Miscellaneous Provisions) Act 1916.

The reference to trading as a news vendor in (d) above is a reference to trading where:

(a) the only articles sold or exposed or offered for sale are newspapers or periodicals; and

(b) they are sold or exposed or offered for sale without a stall or receptacle for them or with a stall or receptacle for them which does not:

    (i) exceed 1 m in length or width or 2 m in height;

    (ii) occupy a ground area exceeding 0.25 m^2; or

    (iii) stand on the carriageway of a street (schedule 4, paragraph 1(2) and (3)).

Adoption

The procedures only operate through a specific resolution of adoption by the local authority (section 3).

Designation

Following adoption, the local authority may designate streets or parts of streets to be either:

(a) prohibited streets in which street trading is not allowed;
(b) licence streets in which licences to trade are required; and
(c) consent streets where the consent requirements operate.

These designations may be rescinded or changed from one type to another at any time using the full procedure (schedule 4, paragraph 2).

Notice of intention to designate is required with copies to the chief police officer and to the highways authority. If the designation is to be for licence streets, the consent of the highways authority is required.

The notice must contain a draft of the resolution and say that written representations may be made during a specified period not less than 28 days from publication of the notice (schedule 4, paragraph 2(6)).

Licences

Applications for street trading licences must be in writing and must give:

(a) full name and address of applicant;
(b) street, day and times of proposed trading;
(c) description of articles, stalls or containers;
(d) any other particulars required by the council, which can include two photographs of the applicant (schedule 4, paragraph 3(2) and (3)).

Unless one or more of the following grounds of refusal are applicable, the local authority must grant the licence and may even grant it if the grounds of refusal are available:

(a) that there is not enough space in the street for the applicant to engage in the trading in which he desires to engage without causing undue interference or inconvenience to persons using the street;
(b) that there are already enough traders trading in the street from shops or otherwise in the goods in which the applicant desires to trade;
(c) that the applicant desires to trade on fewer days than the minimum number specified in a resolution under schedule 4, paragraph 2(11);
(d) that the applicant is unsuitable to hold the licence by reason of having been convicted of an offence or for any other reason;
(e) that the applicant has at any time been granted a street trading licence by the council and has persistently refused or neglected to pay fees due to them for it or charges due to them under schedule 4, paragraph 9(6), for services rendered by them to him in his capacity as licence holder;

(f) that the applicant has at any time been granted a street trading consent by the council and has persistently refused or neglected to pay fees due to them for it;

(g) that the applicant has without reasonable excuse failed to avail himself to a reasonable extent of a previous street trading licence (schedule 4, paragraph 3(6)).

Also licences must be refused if the applicant is under 17 years of age or the location is covered by a control order (roadside sales) under section 7 of the Local Government (Miscellaneous Provisions) Act 1976.

The licence issued must state:

(a) the street in which and days and times between which the holder is able to trade; and

(b) the description of articles in which he may trade and may also state a particular location for trading (schedule 4, paragraph 4(1)).

These are known as the principal terms of the licence and a breach may result in prosecution (schedule 4, paragraph 4(1)–(3)).

In addition, subsidiary terms may be applied at the discretion of the local authority and these may include:

(a) specifying the size and type of any stall or container which the licence holder may use for trading;

(b) requiring that any stall or container so used shall carry the name of the licence holder or the number of his licence, or both; and

(c) prohibiting the leaving of refuse by the licence holder or restricting the amount of refuse which he may leave or the places in which he may leave it (schedule 4, paragraph 4(5)).

Licences may be revoked if:

(a) owing to circumstances which have arisen since the grant or renewal of the licence, there is not enough space in the street for the licence holder to engage in the trading permitted by the licence without causing undue interference or inconvenience to persons using the street;

(b) the licence holder is unsuitable to hold the licence by reason of having been convicted of an offence or for any other reason;

(c) since the grant or renewal of the licence, the licence holder has persistently refused or neglected to pay fees due to the council for it or charges due to them under schedule 4, paragraph 9(6), for services rendered by them to him in his capacity as licence holder; or

(d) since the grant or renewal of the licence, the licence holder has without reasonable excuse failed to avail himself of the licence to a reasonable extent (schedule 4, paragraph 5(1)),

and the local authority may also vary the principal terms by altering the days or times of trading or restricting the type of goods sold, subject to notice and appeal (schedule 4, paragraph 5(2)).

The licence street provisions are most appropriate for the formalised street market situations and imply a positive will to promote trading on behalf of the local authority to the extent that refusal powers are limited.

Consents

There are no specified particulars for street trading consent applications, although they must be in written form (schedule 4, paragraph 7(1)). Unless the applicant is under 17 years of age or the location is covered by a control order under section 7 of the Local Government (Miscellaneous Provisions) Act 1976 (roadside sales) in which situations refusal is mandatory, consents are entirely at the discretion of the local authority as it sees fit. If consent is given, the local authority may attach conditions at its discretion, including those to prevent obstruction and nuisance or annoyance.

Specific consent to trade from a vehicle or portable stall is required. Consent procedures are most applicable where trading is to be itinerant or infrequent and there is no appeal against refusals or conditions to be applied. There is no requirement in these provisions to hear applicants but, in the case of R v Bristol City Council, ex parte Pearce and Another (1984), it was indicated that the local authority should tell applicants of the contents of any objections and give them an opportunity to comment (schedule 4, paragraph 7).

Fees

The local authority may charge such fees as they consider reasonable for both licences and consents with different fee levels being possible for different types of licence/consents, periods, streets and articles. Fees may be paid in instalments and the initial application need only be accompanied by a deposit at the discretion of the local authority. Fees are returnable on surrender, revocation and refusal.

Separate and additional charges may be levied on licence holders for the collection of refuse, street sweeping and any other services rendered by the local authority (schedule 4, paragraph 9).

Definition

Street includes:

(a) any road, footway, beach or other area to which the public have access without payment; and
(b) a service area as defined in section 329 of the Highways Act 1980; and also includes any part of a street (schedule 4, paragraph 1(1)).

STREET COLLECTIONS

References

Police, Factories, & c. (Miscellaneous Provisions) Act 1916 (as amended)
Home Office model street collection regulations (articles below)
The Public Charitable Collections (Scotland) Regulations 1984
The Charitable Collections (Transitional Provisions) Order 1974 – (Model Regulations)

FC150 Street trading: designation of – schedule 4 of Local Government (Miscellaneous Provisions) Act 1982 (LG(MP)A)

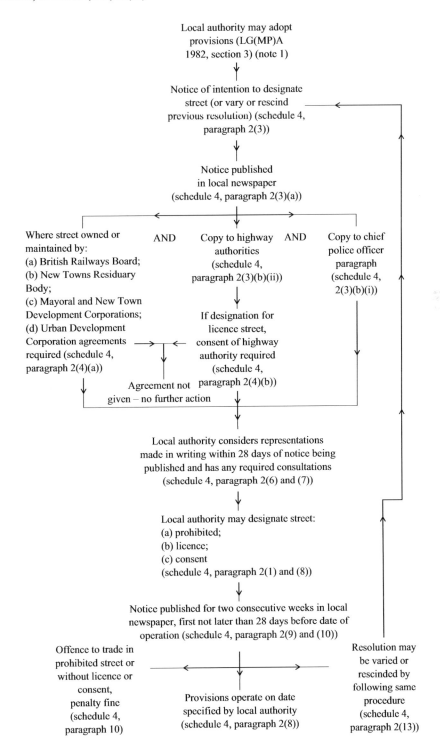

Note

1. No advertisement, publicity, etc. required at this stage.

FC151 Street trading: licences – schedule 4 of Local Government (Miscellaneous Provisions) Act 1982 (LG(MP)A)

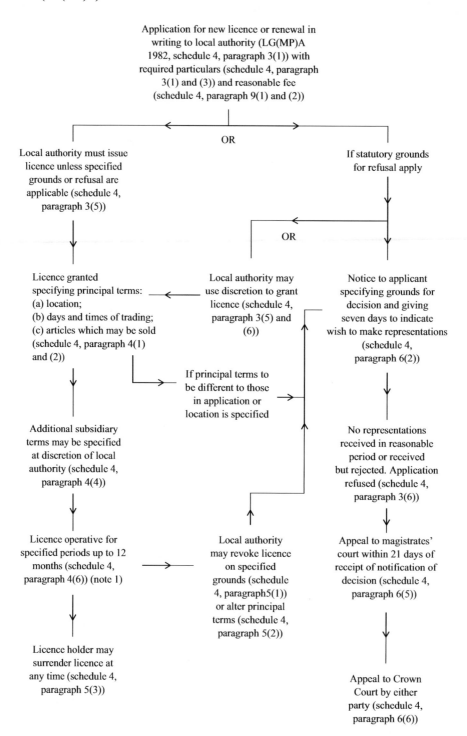

Application for new licence or renewal in writing to local authority (LG(MP)A 1982, schedule 4, paragraph 3(1)) with required particulars (schedule 4, paragraph 3(1) and (3)) and reasonable fee (schedule 4, paragraph 9(1) and (2))

OR

Local authority must issue licence unless specified grounds or refusal are applicable (schedule 4, paragraph 3(5))

If statutory grounds for refusal apply

OR

Licence granted specifying principal terms:
(a) location;
(b) days and times of trading;
(c) articles which may be sold (schedule 4, paragraph 4(1) and (2))

Local authority may use discretion to grant licence (schedule 4, paragraph 3(5) and (6))

Notice to applicant specifying grounds for decision and giving seven days to indicate wish to make representations (schedule 4, paragraph 6(2))

If principal terms to be different to those in application or location is specified

Additional subsidiary terms may be specified at discretion of local authority (schedule 4, paragraph 4(4))

No representations received in reasonable period or received but rejected. Application refused (schedule 4, paragraph 3(6))

Licence operative for specified periods up to 12 months (schedule 4, paragraph 4(6)) (note 1)

Local authority may revoke licence on specified grounds (schedule 4, paragraph5(1)) or alter principal terms (schedule 4, paragraph 5(2))

Appeal to magistrates' court within 21 days of receipt of notification of decision (schedule 4, paragraph 6(5))

Licence holder may surrender licence at any time (schedule 4, paragraph 5(3))

Appeal to Crown Court by either party (schedule 4, paragraph 6(6))

Note

1. The maximum penalty for breaching the principal terms of the licence is level 3. There is no offence committed in breaching the subsidiary terms but this could be taken into account in any consideration of renewal or revocation (schedule 4, paragraph 10(4)).

FC152 Street trading: consents – schedule 4 of Local Government (Miscellaneous Provisions) Act 1982 (LG(MP)A)

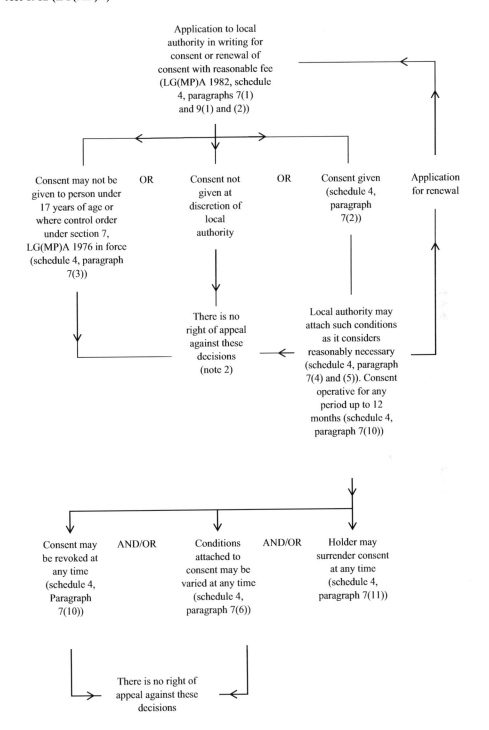

Notes

1. The maximum penalty for contravening a condition of consent is level 3 (schedule 4, paragraph 10(4)).
2. Although no appeal procedure exists, in the case of R v Bristol City Council, ex parte Pearce and Another (1984), the judge commented that the local authority should tell the applicant of the content of their objections and give him an opportunity to comment.

Extent

These provisions apply to England and Wales. Similar provisions apply in Scotland and Northern Ireland.

Scope

A licensing authority may, by a resolution, adopt the Home Office model street collection regulations. The resolution must send a copy to the Secretary of State (there is no need for confirmation) and insert an advertisement in two newspapers circulating within the area with a copy of the resolution and stating that a copy of the regulations will be provided free of charge to any person on application to the authority (section 5).

Model street collection regulations

A promoter must apply in writing at least one month before the date for a proposed collection to obtain a permit to collect in a street and public place (other than a collection taken at a meeting in the open air) (articles 2 and 3). Collections must be made on the day and between the hours stated in the permit (article 4). The collection may be limited to streets or public places as the licensing authority thinks fit (article 5).

The promoter must provide written authority to a collector before collection is carried out (article 6).

The licensing authority may waive the requirement for a collector to remain stationary if in connection with a procession (article 10).

Payment for services connected with the collection may be approved by the licensing authority.

Within one month after the date of any collection, the person to whom a permit has been granted must provide the licensing authority with:

(a) a statement showing the amount received and the expenses and payments in connection with the collection certified by that person and a qualified accountant;
(b) a list of the collectors;
(c) a list of the amounts contained in each collecting box; and
(d) a description of the proper application of the proceeds of the collection.

They must also publish in a newspaper or newspapers a statement showing the name of the person to whom the permit has been granted, the area to which the permit relates, the name of the charity or fund to benefit, the date of the collection, the amount collected and the amount of the expenses and payments in connection with the collection.

The licensing authority may extend the period of one month (article 16).

It is an offence to contravene any of the regulations and any person guilty of an offence will be liable on summary conviction to a fine not exceeding level 1 on the standard scale (article 18).

Definitions

Collecting box means a box or other receptacle for the reception of money from contributors.
Collection means a collection of money or a sale of articles for the benefit of charitable or other purposes.
Promoter means a person who causes others to act as collectors (article 1).

SEX ESTABLISHMENT LICENCES

References

Local Government (Miscellaneous Provisions) Act 1982, section 2 and schedule 3
Civic Government (Scotland) Act 1982
The Local Government (Miscellaneous Provisions) (Northern Ireland) Order 1985

Extent

These provisions apply to England and Wales. Similar provisions apply in Scotland and Northern Ireland.

Scope

A local authority may resolve that the provisions of the control of sex establishments in schedule 3 of the Act may apply to their area.

Once the resolution has been passed, the provisions come into force one month after the resolution is passed.

The local authority must publish notice that they have passed a resolution in two consecutive weeks in a local newspaper circulating in their area at least 28 days before the day of the provisions coming into force (section 2).

The licensing of lap dancing premises and similar venues is in line with other sex establishments (sex shops and sex cinemas).

The provisions allow the licensing authority to prescribe a wider range of conditions on the licences of sexual entertainment venues than those available under the Licensing Act 2003 and allow local people to oppose an application for a sex establishment licence if they have legitimate concerns that a sexual entertainment venue would be inappropriate, given the character of an area.

Licences must be renewed at least annually and local people have the opportunity to raise objections (if any) at that time.

Licensing provisions

Sex establishments must only operate under a licence granted by a licensing authority (paragraph 6). An application to operate a sex establishment must be made to the appropriate authority. An application must be made in writing and must include:

(a) the full name of the applicant or body corporate;
(b) his permanent address or address of registered or principal office, the full names and private addresses of the directors or other persons responsible for its management;
(c) his age;
(d) the full address of the premises;
(e) where an application relates to a vehicle, vessel or stall, where it is to be used as a sex establishment (paragraph 10).

The authority may grant to any applicant, and from time to time renew, a licence (with terms, conditions and restrictions) to use any premises, vehicle, vessel or stall specified in it for a sex establishment (paragraph 8).

The licence remains in place for one year and the authority may, where they think fit, transfer the licence to any other person on application (paragraph 9).

The applicant must give public notice of the application by publishing an advertisement in a local newspaper circulating in the appropriate authority's area not later than seven days after the date of the application. For premises, notice must be displayed for 21 days beginning with the date of the application on or near the premises and in a place where the notice can conveniently be read by the public. The notice must be prescribed by the authority.

A copy of an application for the grant, renewal or transfer of a licence must be sent to the chief officer of police within seven days.

Any person objecting to an application for the grant, renewal or transfer of a licence must give notice in writing of his objection to the authority, stating the grounds of the objection within 28 days of the date of the application. The authority then must give notice in writing of the general terms of the objection to the applicant. The authority must not, without the consent of the person making the objection, reveal the objector's name or address to the applicant.

The authority must have regard to any observations made by the chief officer of police and any objections of which notice has been sent to them before determining the application.

The authority must give an opportunity of being heard by a committee of the authority:

(a) before refusing to grant a licence, to the applicant;
(b) before refusing to renew a licence, to the holder;
(c) before refusing to transfer a licence, to the holder and the person to be transferred to.

Where the appropriate authority refuses to grant, renew or transfer a licence, they must give the applicant a statement in writing of the reasons for their decision (paragraph 10).

Licences remain in force where an application has been made for its renewal or transfer until the withdrawal of the application or its determination (paragraph 11).

Waiver

The authority may waive the requirement of a licence in any case where they consider that to require a licence would be unreasonable or inappropriate. Where the authority grants an application for a waiver, they must give the applicant for the waiver notice that they have granted his application. The authority may at any time terminate the waiver notice by giving 28 days' notice to the person in receipt of the waiver notice (paragraph 7).

Refusal/cancellation

Licences cannot be granted to:

(a) a person under the age of 18; or
(b) a person who is for the time being disqualified due to revocation of a licence; or
(c) a person, other than a body corporate, who is not resident in a European Economic Area state or was not so resident throughout the period of six months immediately preceding the date when the application was made; or
(d) a body corporate which is not incorporated in a European Economic Area state; or

(e) a person who has, within a period of 12 months immediately preceding the date when the application was made, been refused the grant or renewal of a licence for the premises, vehicle, vessel or stall in respect of which the application is made, unless the refusal has been reversed on appeal (paragraph 12(1)).

The authority may refuse an application for the grant or renewal of a licence if:

(a) the applicant is unsuitable to hold the licence by reason of having been convicted of an offence or for any other reason;

(b) on granting, renewing or transferring the licence, the business to which it relates would be managed by or carried on for the benefit of a person, other than the applicant, who would be refused the grant, renewal or transfer of such a licence if he made the application himself;

(c) the number of sex establishments, or sex establishments of a particular kind, in the relevant locality at the time the application is determined is equal to or exceeds the number which the authority considers is appropriate for that locality (which may be nil);

(d) the grant or renewal of the licence would be inappropriate, having regard:

(i) to the character of the relevant locality; or

(ii) to the use to which any premises in the vicinity are put; or

(iii) to the layout, character or condition of the premises, vehicle, vessel or stall in respect of which the application is made (paragraph 12(3)).

Where the holder of a licence granted dies, the licence shall be deemed to have been granted to his personal representatives and will remain in force for three months and then expire. However, the authority may, on the application of the representatives, extend or further extend the period of three months if the authority is satisfied that the extension is necessary for the purpose of winding up the deceased's estate and that no other circumstances make it undesirable (paragraph 15).

The authority may, at the written request of the holder of a licence, cancel the licence (paragraph 16).

Revocation

The authority may, after giving the holder of a licence an opportunity of appearing before and being heard by them, revoke the licence. Where a licence is revoked, the authority must, if required to do so by the person who held it, give him a statement in writing of the reasons for their decision within seven days. Where a licence is revoked, its holder shall be disqualified from holding or obtaining a licence in the area of the appropriate authority for a period of 12 months beginning with the date of revocation (paragraph 17).

Variation

The licence holder may apply to the authority to vary the terms, conditions or restrictions on application. The authority may vary or refuse the application (paragraph 18).

Charging

The authority may charge a reasonable fee for determining the licence (paragraph 19).

Offences

It is an offence for:

(a) a person to knowingly use, or knowingly cause or permit the use of, any premises, vehicle, vessel or stall without a licence;

(b) a licence holder of a sex establishment to employ in the business of the establishment any person known to him to be disqualified from holding such a licence;

(c) a licence holder or servant or agent of the holder without reasonable excuse to knowingly contravene, or without reasonable excuse, knowingly permit the contravention of, a term, condition or restriction specified in the licence (paragraph 20);

(d) any person, in connection with an application for the grant, renewal or transfer of a licence, to make a false statement which he knows to be false in any material respect or which he does not believe to be true (paragraph 21);

(e) a licence holder without reasonable excuse to knowingly permit a person under 18 years of age to enter the establishment or to employ a person known to him to be under 18 years of age in the business of the establishment (paragraph 23).

A person guilty of an offence is liable on summary conviction to a fine not exceeding £20,000 or level 3 on the standard scale (paragraph 22).

Powers of entry

An authorised officer of a local authority may, at any reasonable time, enter and inspect any sex establishment in respect of which a licence is for the time being in force, with a view to seeing:

(i) whether the terms, conditions or restrictions on or subject to which the licence is held are complied with;

(ii) whether any person employed in the business of the establishment is disqualified from holding a licence;

(iii) whether any person under 18 years of age is in the establishment; and

(iv) whether any person under that age is employed in the business of the establishment.

An authorised officer may enter and inspect a sex establishment if he has reason to suspect that an offence has been, is being or is about to be committed, if he has a warrant granted by a Justice of the Peace. The authorised officer must produce his authority if required to do so by the occupier of the premises or the person in charge of the vehicle, vessel or stall in relation to which the power is exercised.

Any person who, without reasonable excuse, refuses to permit an authorised officer of a local authority to exercise any such power shall be guilty of an offence and shall for every such refusal be liable on summary conviction to a fine not exceeding level 5 on the standard scale (paragraph 25).

Seizure

A person acting under the authority of a warrant may seize and remove anything found on the premises concerned that the person reasonably believes should be forfeited. The person who, immediately before the seizure, had custody or control of anything seized may request any authorised officer who seized it to provide a record of what was seized. The authorised officer must provide the record within a reasonable time of the request being made (paragraph 25A).

Appeals

Where an application or request for variation is refused or a licence holder is aggrieved by any term, condition or restriction on the licence or the licence is revoked, the applicant or licence holder can appeal to a magistrates' court within 21 days of the relevant date.

An appeal against the decision of a magistrates' court may be brought to the Crown Court. The decision of the Crown Court is final and either court can make such order as it thinks fit. The authority must carry out the order of the court.

Where a licence is revoked or an application for the renewal of a licence is refused, the licence remains in force until the time for bringing an appeal has expired or until the determination or abandonment of the appeal. Where the licence holder makes an application to vary the licence or the authority imposes a term, condition or restriction other than one specified in the application, the licence is free of it until the appeal is determined (paragraph 27).

Definitions

Relevant entertainment means any live performance or any live display of nudity which is of such a nature that, ignoring financial gain, it must reasonably be assumed to be provided solely or principally for the purpose of sexually stimulating any member of the audience (whether by verbal or other means). An audience includes an audience of one (paragraph 2A).

Sex cinema is any premises, vehicle, vessel or stall used to a significant degree for the exhibition of moving pictures, by whatever means produced, which:

- are concerned primarily with the portrayal of, or primarily deal with, relate to or are intended to stimulate or encourage, sexual activity; or acts of force or restraint which are associated with sexual activity; or
- are concerned primarily with the portrayal of, or primarily deal with or relate to, genital organs or urinary or excretory functions.

A sex cinema does not include a dwelling-house to which the public is not admitted.

A premises shall not be treated as a sex cinema only if they are licensed under section 1 of the Cinemas Act 1985, of their use for a purpose for which a licence is required; or of their use for an exhibition to which section 6 of that Act (certain non-commercial exhibitions) applies given by an exempted organisation within the meaning of section 6(6) of that Act (paragraph 3).

Sex establishment means a sex cinema, sex shop or sexual entertainment venue (paragraph 2).

Sex shop means any premises, vehicle, vessel or stall used for a business which consists to a significant degree of selling, hiring, exchanging, lending, displaying or demonstrating:

- sex articles; or
- other things intended for use in connection with, or for the purpose of stimulating or encouraging, sexual activity; or acts of force or restraint which are associated with sexual activity.

No premises shall be treated as a sex shop by reason only of their use for the exhibition of moving pictures by whatever means produced (paragraph 4).

Sexual entertainment venue means any premises at which relevant entertainment is provided before a live audience for financial gain of an organiser. For the provisions to apply, it is not necessary for the organiser to receive financial gain directly or indirectly from the performance or display of nudity (paragraph 2A).

Sexual entertainment venues: exempt premises – the following are not defined as sexual entertainment venues:

(a)   sex cinemas and sex shops;
(b)   premises at which the provision of relevant entertainment is such that:

(i)   there have not been more than 11 occasions on which relevant entertainment has been so provided which fall (wholly or partly) within the period of 12 months;
(ii)   no occasion has lasted for more than 24 hours; and
(iii)   no occasion has begun within the period of one month beginning with the end of any previous occasion on which relevant entertainment has been so provided.

PLEASURE BOAT LICENSING

Reference

Public Health Acts Amendment Act 1907, section 94

Extent

These provisions apply in England, Scotland, Northern Ireland and Wales.

Scope

A local authority may grant licences for pleasure boats and pleasure vessels (with terms and conditions) to be let for hire or to be used for carrying passengers for hire. Licences also apply to persons in charge of or navigating such boats and vessels. The local authority may charge an annual fee for each type of licence.

The period of the licence is determined by the local authority and the licence can be suspended or revoked where they think it is necessary or desirable in the interests of the public.

It is an offence to let for hire pleasure boats not licensed under these provisions and any person guilty of an offence is liable to a penalty not exceeding level 3 on the standard scale. However, it is a defence where the failure to comply is due to a reasonable excuse for the failure.

Any person aggrieved by the withholding, suspension or revocation of any licence may appeal to a court, held after the expiration of two clear days after such withholding, suspension or revocation, but the aggrieved person must give 24 hours' written notice of an appeal with details of the grounds of appeal. The court has the power to make an order as they see fit and to award costs. Costs are recoverable summarily as a civil debt. These provisions do not apply to pleasure boats etc. on any inland waterway owned or managed by the Canal & River Trust or licensed under other relevant regulations.

Definition

Let for hire means let for hire to the public (section 94).

FC153 Sex establishment licences – section 2 and schedule 3 Local Government (Miscellaneous Provisions) Act 1982

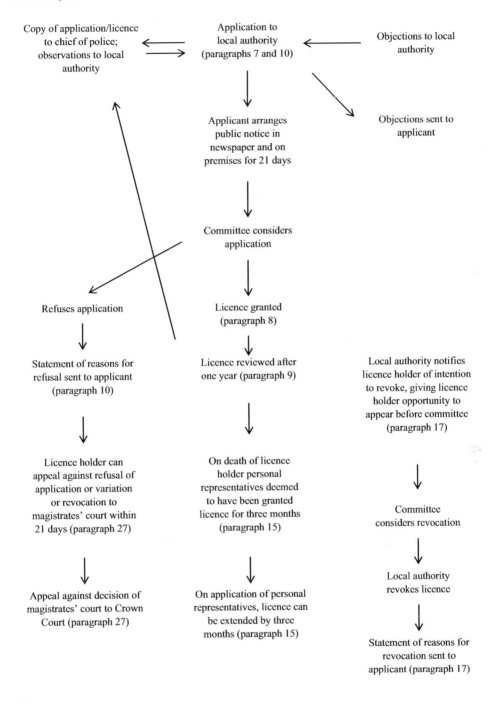

Copy of application/licence to chief of police; observations to local authority

Application to local authority (paragraphs 7 and 10)

Objections to local authority

Objections sent to applicant

Applicant arranges public notice in newspaper and on premises for 21 days

Committee considers application

Refuses application

Licence granted (paragraph 8)

Statement of reasons for refusal sent to applicant (paragraph 10)

Licence reviewed after one year (paragraph 9)

Local authority notifies licence holder of intention to revoke, giving licence holder opportunity to appear before committee (paragraph 17)

Licence holder can appeal against refusal of application or variation or revocation to magistrates' court within 21 days (paragraph 27)

On death of licence holder personal representatives deemed to have been granted licence for three months (paragraph 15)

Committee considers revocation

Appeal against decision of magistrates' court to Crown Court (paragraph 27)

On application of personal representatives, licence can be extended by three months (paragraph 15)

Local authority revokes licence

Statement of reasons for revocation sent to applicant (paragraph 17)

INDEXES

Flow Charts

Subject index of procedures

Taylor & Francis eBooks

www.taylorfrancis.com

A single destination for eBooks from Taylor & Francis
with increased functionality and an improved user
experience to meet the needs of our customers.

90,000+ eBooks of award-winning academic content in
Humanities, Social Science, Science, Technology, Engineering,
and Medical written by a global network of editors and authors.

TAYLOR & FRANCIS EBOOKS OFFERS:

A streamlined
experience for
our library
customers

A single point
of discovery
for all of our
eBook content

Improved
search and
discovery of
content at both
book and
chapter level

REQUEST A FREE TRIAL
support@taylorfrancis.com

 Routledge
Taylor & Francis Group

 CRC Press
Taylor & Francis Group